6th International Conference on Applications and Design in Mechanical Engineering (ICADME 2019)

IOP Conference Series: Materials Science and Engineering
Volume 670

Penang Island, Malaysia
26 - 27 August 2019

Editors:

Mohd Shukry Abdul Majid
Mohd Ridzuan Mohd Jamir
Tan Wee Choon
Anas Abdul Rahman

ISBN: 978-1-7138-0923-4
ISSN: 1757-8981

Printed from e-media with permission by:

Curran Associates, Inc.
57 Morehouse Lane
Red Hook, NY 12571

Some format issues inherent in the e-media version may also appear in this print version.

Additional copies of this publication are available from:

Curran Associates, Inc.
57 Morehouse Lane
Red Hook, NY 12571 USA
Phone: 845-758-0400
Fax: 845-758-2633
Email: curran@proceedings.com
Web: www.proceedings.com

TABLE OF CONTENTS

IOP Conf. Series: Materials Science and Engineering **670** (2019) 011001 doi:10.1088/1757-899X/670/1/011001

Preface

The International Conference on Applications and Design in Mechanical Engineering 2019 (ICADME 2019) was held on 26-27 August 2019 in Penang, one of the most popular tourist destinations in Malaysia. ICADME 2019 was organized by the Mechanical Engineering Programme, under the purview of School of Mechatronic Engineering, Universiti Malaysia Perlis.

The ICADME 2019 aims to provide a platform for participants from diverse field of Mechanical Engineering backgrounds to exchange ideas and findings to promote cooperation, networking as well as knowledge sharing in both engineering practice and research in higher education. The conference includes the fields of Applied Mechanics, Themofluids and Energy as well as Design and Manufacturing. The conference received more than 100 papers from Malaysia and other countries over the world. After a rigorous peer-review process, 81 papers were accepted. The accepted papers that focused on theory, applications, design and experimental studies relevant to mechanical engineering fields were invited to present in this conference.

We want to express our gratitude to all reviewers, both local and international, for their time and effort in reviewing the papers. Their feedbacks to the authors have been excellent and useful to ensure the high quality of papers for the conference. Our sincere thanks also go to all sponsors for their generosity. Last but not least, we would like to thank the organizing committee members for their hard work, cooperation and insights in ensuring the success of this conference.

Editors

Associate Prof. Ir. Dr. Mohd Shukry Abdul Majid
Ir. Dr. Mohd Ridzuan Mohd Jamir
Dr. Tan Wee Choon
Dr. Anas Abdul Rahman

IOP Conf. Series: Materials Science and Engineering **670** (2019) 011002 doi:10.1088/1757-899X/670/1/011002

Peer review statement

All papers published in this volume of *IOP Conference Series: Materials Science and Engineering* have been peer reviewed through processes administered by the proceedings Editors. Reviews were conducted by expert referees to the professional and scientific standards expected of a proceedings journal published by IOP Publishing.

IOP Conf. Series: Materials Science and Engineering **670** (2019) 012001 doi:10.1088/1757-899X/670/1/012001

Application of ductility exhaustion based damage model to predict creep rupture time of grade 92 steel

N A Alang[1] and N Ab Razak[1]

[1]Structural Materials and Degradation Focus Group, Faculty of Mechanical and Manufacturing Engineering, Universiti Malaysia Pahang, 26600, Pekan, Pahang, Malaysia.

E-mail: azuan@ump.edu.my

Abstract. To accurately predict the creep rupture time of notched bar becomes a challenge to academia and structural engineer due to complex stress-strain distribution around the notch throat. This paper presents a Finite Element (FE) simulation employing ductility exhaustion based damage model to predict creep rupture time of multiaxial notched bar Grade 92 steel. Three different notch acuity, $\eta = 2.5$, 5.0 and 12.0 were simulated and the FE predicted rupture time was compared to the available experimental rupture data. The reduction of creep ductility due to geometrical constraint is considered during the simulation by employing the void growth model. Further reduction in ductility in long term period arises from internal microstructural changes or damage is also accounted. Furthermore, empirical-type exponential prediction model coupled with skeletal stresses is applied to provide upper/lower bounds for short and long term rupture data. It is found that the FE prediction agreed well with the experimental data. At short-term, notched bar ruptured is controlled by the von-Mises stress while at long-term the rupture is controlled by the maximum principal stress.

1. Introduction

Grade 92 steel is widely used for a high-temperature application such as header and superheated pipes. After a certain period in service, damage in term of isolated creep voids have been observed mostly along the prior austenite grain boundaries (PAGB) and sub-grain [1]. Over the time, these microvoids joining up and then becoming oriented voids or microcracks. Regular inspection is necessary to monitor the progress of the damage. The remaining life of the damaged component can be estimated using two main predictive models namely time-based (life fracture rule [2]) and strain-based models. Mathematically, the models can be expressed as follows:

$$\omega = \sum \frac{t_n}{t_{r_n}} \quad (1)$$

$$\omega = \int_0^t \frac{\dot{\varepsilon}_c}{\varepsilon_f^*} dt \quad (2)$$

where t_n and t_{r_n} is the in-service duration and time to rupture, respectively. The time to rupture can be easily calculated for a uniaxial sample since the stress distribution along the gauge length is uniform and approximately equal to the remote stress. However, choosing the right stress to perform the prediction is paramount important for notched bar sample as the stress and creep strain distribution at the notched area is non-uniform and rather complex. The application of Eqn. (1) to the case of complex stress state such as in notched bar leads to underestimate result. The $\dot{\varepsilon}_c$ and ε_f^* in Eqn. (2) are the equivalent creep strain and multiaxial ductility, respectively. Damage predicted using the strain-based model in Eqn. (2) is based on the ductility exhaustion concept whereby the failure is assumed to occur when the accumulated creep strain reaches the critical value. For the case of notched bar, the local damage, ω is usually calculated numerically by incorporating Eqn. (2) into Finite Element (FE) analysis. Applying the model in Eqn. (2), however, requires careful determination of material ductility or failure strain because the overall prediction result is greatly depending on this value.

This study attempts to predict the creep rupture time of notched bar of Grade 92 steel using finite element analysis employing ductility exhaustion based damage model. The effect of geometrical constraint due to notch and local constraint on ductility is accounted. Furthermore, the upper/lower bounds of the data which covers both short and long term rupture data are estimated.

2. Methodology

2.1. Finite Element Analysis

The FE modelling is carried out using commercial FE Abaqus software v6.13, employing the elastic-plastic-creep material properties. The elastic deformation is modelled according to the Hooke's law while the plastic deformation is assumed to obey the power law relation with isotropic strain hardening behaviour. The total strain is given as:

$$\varepsilon_T = \varepsilon_e + \varepsilon_p + \varepsilon_c \tag{3}$$

The creep strain rate, $\dot{\varepsilon}_c$ is modelled to follow the Norton power law and can be expressed as:

$$\dot{\varepsilon}_c = A\sigma^n \tag{4}$$

Notched bar specimens having an acuity level (defined as net diameter/notch radius) of 2.5, 5.0 and 12.0 are modelled. The notched bar diameter, D and the net diameter, d is 5 mm and 3 mm, respectively. The modelling accounts the geometric and loading symmetric conditions, therefore, only a quarter of the specimen gauge length is analyzed as an axis-symmetric model. An appropriate boundary condition is applied including the prescribed pressure, P on the top surface as shown in Figure 1.

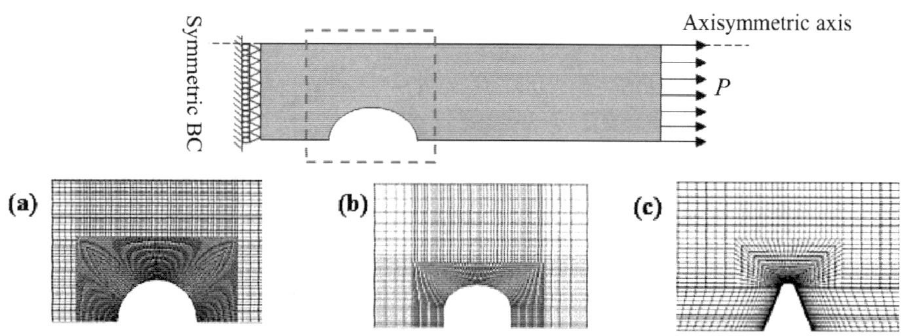

Figure 1. Details FE model with mesh: (a) acuity = 2.5, (b) acuity = 5.0 and (c) acuity = 12.

The model is meshed with continuum axis-symmetric quadrilateral elements with reduced integration scheme. The model consists of 11255-12472 nodes and 11025-12200 elements depending on the notch geometry. The details of FE mesh design at the notched throat area is shown in Figure 1. The failure strain value of Grade 92 steel is observed dependent on strain rate [3]. The best fit line of the available creep failure strain data is drawn according to Fermi' function [4] and is adopted in Abaqus through CREEP user subroutine. It should be noted that the lower plateau limit of failure strain is assumed to be 10% of its maximum value since the long-term data up to this point is currently not available. Further reduction of failure strain due to the notch constraint is estimated using Eqn. (5).

$$\frac{\varepsilon_f^*}{\varepsilon_f}=\sinh\left[\frac{2}{3}\left(\frac{n-0.5}{n+0.5}\right)\right]/\sinh\left[2\left(\frac{n-0.5}{n+0.5}\right)\frac{\sigma_m}{\sigma_e}\right] \tag{5}$$

Damage of the material is simulated using the ductility exhaustion based damage approach as described in Eqn. (2). The element is said to failure when the damage parameter, ω approaches unity at the gauss point. At failure point, the load-carrying capacity of the element is reduced to the value near zero. It is done by reducing the elastic modulus, E to 1.0% of its undamaged value. The progressive creep damage technique described above is implemented in Abaqus using the USDFLD subroutine. Table 1 shows the statics and creep properties of Grade 92 steel used in FE analysis.

Table 1. Statics and creep properties of Grade 92 steel at 600°C.

E (GPa)	*K* (MPa)	*N*	A_s (MPa^{-n}/h)	A_L (MPa^{-n}/h)	n_s	n_L
134	500.8	0.1	2.9×10^{-41}	8.7×10^{-26}	16.4	9.28

3. Estimating Upper/Lower Bounds of Notched Bar Rupture Data

The time to rupture against applied stress for Grade 92 steel under uniaxial stress state can be expressed as exponential relation:

$$t_r=F\exp^{-G\sigma} \tag{6}$$

where F and G are temperature-dependent material constant. The equation will be used as a basis in developing the model for predicting the rupture life of notched bar. The representative stress concept has been proposed [5] which is expressed in term of von-Mises, σ_{vm} and Maximum Principal Stress (MPS), σ_1 to estimate the rupture life of notched bar sample.

$$\sigma_{rep}=\alpha\sigma_1+(1-\alpha)\sigma_{vm} \tag{7}$$

The rupture life of notched bar specimen, t_r can be rationalised by replacing the uniaxial stress, σ in Eqn. (6) by the representative stress, σ_{rep} given in the Eqn. (7).

$$t_r=F\exp^{-G[\alpha\sigma_1+(1-\alpha)\sigma_{vm}]} \tag{8}$$

For the case where the rupture time is fully controlled by von-Mises stress (α=0), Eqn. (8) can be reduced to:

$$t_r=F\exp^{-G[\sigma_{vm}]} \tag{9}$$

While, when the MPS takes over (α=1), Eqn. (8) can be written as follows:

$$t_r=F\exp^{-G[\sigma_1]} \tag{10}$$

where, α is the rupture-controlled parameter. To apply the rupture life relation in Eqn. (8) to (10), the skeletal von-Mises and MPS is employed. The skeletal stress of various notch acuities can be calculated using the normalized stress factor recommended in the code of practice document [5]. Furthermore, the rupture data of Grade 92 steel at 600°C (approximately 38 data points) was collected from various sources and the best fit line was drawn to determine the constant F and G. Noted that the value of F and G in Eqn. (9) - (10) are 4.94×10^8 and 0.074, respectively.

IOP Conf. Series: Materials Science and Engineering **670** (2019) 012001 doi:10.1088/1757-899X/670/1/012001

4. Results and Discussion

Figure 2 compares the experimental data [1] with the simulation result for all cases. The over- and underestimate life curve with a factor of +/- 2 is also drawn to observe the variation of predicted result. Clearly, all the predicted rupture time falls within the factor of 2. It suggests the appropriateness of the employed techniques in predicting the rupture time.

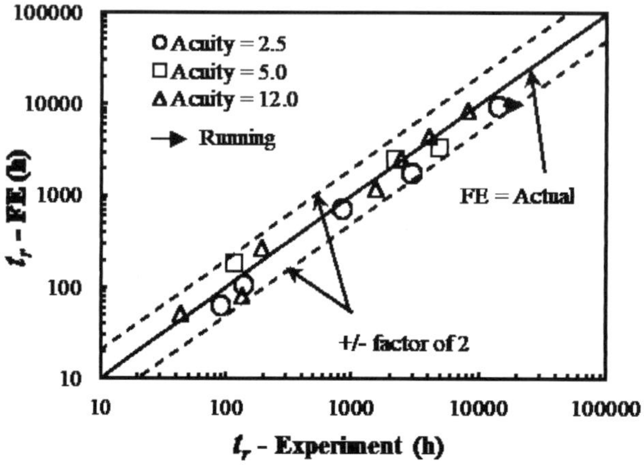

Figure 2. Comparison of rupture life between FE prediction and experimental data.

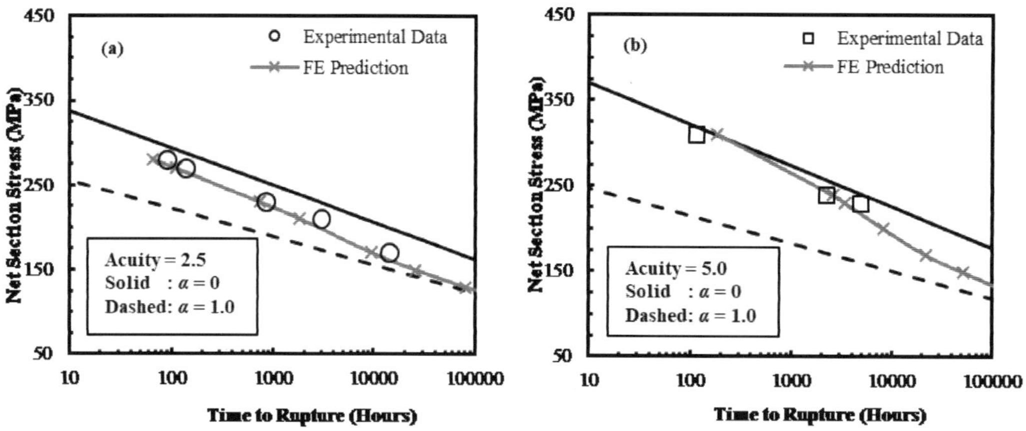

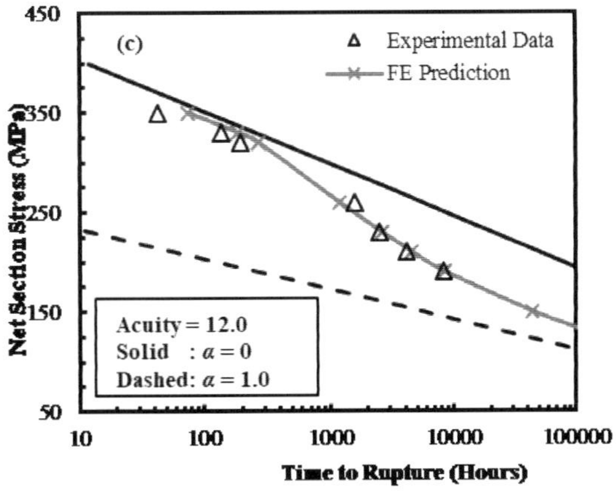

Figure 3. Creep rupture behaviour at short- to long-term time: (a) acuity = 2.5, (b) acuity = 5.0 and (c) acuity = 12.

Figure 3 compares the numerical prediction with experimental data [1] for notch acuity of 2.5, 5.0 and 12.0. Excellent prediction is shown using the employed technique for the acuity of 2.5 and 5.0. An exception for the acuity of 12.0, where the prediction is slightly overestimates the life at the high stress regime. However, it is clear that the approach employed is capable to capture a transition from von-Mises controlled at high stress to MPS-controlled at lower stress as shown in Figure 3(c). In fact, the prediction seems to agree with experimental data at a longer time. Also shown in Figure 3 is the upper/lower life curve calculated using Eqn. (9) and (10) with α = 0 and α = 1.0. Clearly, all the experimental data and FE prediction fall within these two extreme lines indicating the rupture is influenced by both von-Mises and MPS. It should be noted that few data for the acuity of 12.0 initially fall outside of the predicted upper/lower bounds. As all the data theoretically should be within the two lines, the F and G values used to construct the bound lines for acuity 12.0 is determined based on the upperbound of the uniaxial rupture data.

At higher stress, the creep rupture is controlled by a fixed bias between von-Mises and MPS ($0< \alpha < 1$). As time increases, the FE results shift continuously from von-Mises to MPS controlled. The transition from von-Mises to MPS controlled rupture is only can be seen at 10 000, 3 000 and 1 000 hours for acuity 2.5, 5.0 and 12.0, respectively. Shorter transition time is observed as the notched acuity increases. It justifies that the presence of a notch promotes faster drop in creep strength and consequently shorten the time to rupture [5]. It is clearly shown that at high stress, the specimen with a sharp notch has a longer life compared to the specimen with blunt notch. However, as the stress decreases, the predicted time to rupture of blunt notch appears to be higher than other notches. The phenomenon indicates that Grade 92 steel exhibits considerable notch strengthening effect at high stress but at low stress the notch weakening effect takes place.

5. Conclusion

Numerical simulation based FE has been performed to predict the time to rupture of the notched bar with different acuity ratio. The ductility exhaustion based damage model is employed. The reduction in ductility due to the notch constraint and internal local damage is accounted during the simulation. A good agreement between simulation result and experimental data is obtained which indicates the technique and the model employed is robust and capable to predict the rupture life for different constraint levels. It is predicted that at high stress, the rupture life is controlled by the von-Mises stress while at low stress, a transition from von-Mises to MPS-controlled rupture life is clearly observed. Even though the prediction result shows excellent agreement with the experimental data, however, the study is limited to only three different notch acuity. The limitation is due to the unavailability of rupture data at long-term time of other notch geometries or acuity. When the data becomes available, further validation of the simulation technique particularly on its capability to predict long-term behaviour should be made. In fact, the data which covers a wider range of acuity from blunt to the extreme crack-like notch must be analyzed.

Acknowledgement
The authors would like to thank the Universiti Malaysia Pahang for computational facilities and financial assistance under the Internal Research Grant (RDU181104 and RDU1803135).

6. References

[1] Panait C G 2010 Metallurgical evolution and creep strength of 9-12%Cr heat resistant steels at 600C and 650C *PhD Theses* L'Ecole Nationale Superieure Des Mines de Paris.
[2] Robinson E L 1952 Effect of temperature variation on the long-time rupture strength of steels *Trans. ASME* **74** p 777-780.
[3] Alang N A and Nikbin K 2018 An analytical and numerical approach to multiscale ductility constraint based model to predict uniaxial/multiaxial creep rupture and cracking rates *International Journal of Mechanical Sciences* **135** p 342-352.
[4] Payten W M, Dean D W and Snowden K U 2009 Creep-fatigue prediction of low alloy ferritic steels using a strain energy based methodology *Proceedings of the ASME, Pressure Vessels and Piping*, July 26-30, Prague.
[5] Webster G A, Holdsworth S R, Loveday M S, Nikbin K, Perrin I J, Purper H, Skelton R P and Spindler M W 2004 A code of practice for conducting notched bar creep tests and for interpreting the data *Fatigue Fracture Engineering Materials Structures* **27** p 319-342.

IOP Conf. Series: Materials Science and Engineering **670** (2019) 012002 doi:10.1088/1757-899X/670/1/012002

Preparation and characterization of chromium metal organic framework with trimesic or terephthalic acid in carbon dioxide adsorption

K C Chong *[1], P L Foo, [1], S O Lai,[1] H S Thiam[1], S S Lee[1], W J Lau[2]

[1]Lee Kong Chian Faculty of Engineering and Science, Universiti Tunku Abdul Rahman, Jalan Sungai Long, Bandar Sungai Long, 43300 Kajang, Malaysia.
[2]Advanced Membrane Technology Research Centre (AMTEC), Universiti Teknologi Malaysia, 81310 Skudai, Johor, Malaysia

E-mail: *chongkc@utar.edu.my

Abstract. Greenhouse gases (GHGs) have greatly changed the global climate in the recent year. Metal organic frameworks (MOFs) is one of the solutions which can be utilized as gas adsorbents to capture carbon dioxide (CO2), the main contributor of greenhouse effect. MOF possesses good CO2 adsorption ability attributed to its high surface area to volume ratio and pore size. Nevertheless, the success of the MOF is greatly depend on the fabrication parameter such as fabrication methods, type of metal oxide, type of organic linkers used and the molar ratio of the reactant. In this study, chromium (VI) oxide, trimesic acid and terephthalic acid were used to fabricate MOFs by solvothermal method to investigate the performance of CO2 adsorption. The characterization tests by scanning electron microscopy (SEM) and energy dispersive X-ray spectroscopy (EDX) shows that Cr-TMA-1:5 amongst the rest of the fabricated MOF possess the shape of small irregular diagonal with highest surface area to volume ratio which was ideal in CO2 adsorption. Furthers, the result CO2 adsorption test carried by thermogravimetric analyser (TGA) suggested the MOF Cr-TMA-1:5 possesses the highest the CO2 adsorption capacity recorded at 0.099 mmol/g. This had revealed that MOFs composed of chromium (VI) oxide and trimesic acid possess the potential to be applied in CO2 adsorption.

1. Introduction

MOF is a crystalline porous solid that made by organic acids and inorganic metals and it consists of multiple metal-ligand bonds [1]. These bonds enable the formation of three-dimensional organic-inorganic hybrid networks. There are numerous types of metal-ligand combinations that can be casted by varying the use of metals and acids as well as method of formation [2,3]. These provide endless combinations of MOFs to be formed [4]. MOFs are well-known for their characteristics of adjustable pore size, tuneable internal surface properties, enormous specific surface areas which can extend beyond 6000m2/g and ultrahigh porosity that can achieve up to 90% free volume [5]. Therefore, optimizing the combination of metal complexes and organic linkers can change the affinity of pores towards the desired gases to be adsorbed. Also, MOFs have been investigated as existing potential CO2 adsorbents mainly due to its physical structural characteristics. Generally, MOFs have high porosities, small pore sizes (ranging from scale of ultra-micro to meso), high specific surface area (>6000 m2/g), low densities (0.2-1 g/cm3), excellent thermal and mechanical stabilities. Besides,

MOFs are better than many other mesopores gas adsorbents as MOFs will not show hysteresis effect when the gas is removed from the micropores MOFs for regeneration [6]. Numerous MOFs with different metal ions and organic linkers have been synthesized and investigated in lab scale for CO2 adsorption by other researchers. According to a study, different starting materials which are cobalt ions, nickel ions, zinc ions and magnesium ions could yield various capacities for CO2 adsorption [7]. The metal ions in MOFs will be having interactions with molecules of CO2. Hence, it is suggested that metal with coordinative unsaturated sites in the framework shall be selected to fabricate an effective CO2 adsorbent. Besides, the property of the metal can also affect the stability of MOFs under humid conditions. The polarity of metal sites and CO2 must be considered as it will affect the interactions of adsorbate and adsorbent [8]. On the other hand, the selection of organic linkers is very vital in designing MOFs that can have good performance in adsorbing CO2. Generally, molecular size of linkers plays an important role in determining pore sizes between coordinated metal centers which can greatly influence the pore volume and surface area of the materials. In other words, longer molecules give higher porosity for better diffusion of CO2. Also, MOFs with functionalized organic ligands can adsorb CO2 better than MOFs with unfunctionalized pores. However, the selectivity of CO2 would be reduced if the pore sizes are too large [8]. Stoichiometry of reactants can also greatly affect the topographies and structural patterns of MOFs which may lead to different performances when they are applied for certain tasks. Hence, the molar ratio of organic and inorganic reactants used are very crucial in synthesizing MOFs [9].

2. Methodology

2.1. Materials

Metal oxides and organic acids were used in this study to fabricate MOF. In this study, the chosen metal oxide will be chromium (VI) oxide (CrO3) whereas organic acids that will be utilized were trimesic acid (TMA) and terephthalic acid (TPA). All the chemicals were obtained from Sigma Aldrich and used without further purification.

2.2. Synthesize and characteristic method

In this study, solvothermal method was applied to produce MOFs as it can form MOFs with nanoscale morphologies in short period of time [10,11]. First and foremost, predefined mass of metal oxide was dissolved in 12ml of deionized water. Later, the solution was mixed with predefined volume of organic acid and 12ml of ethanol solvent [12]. The mixture was stirred at room temperature until homogenous. Subsequently, the mixture was poured into a Teflon lined autoclave and heated in an oven at temperature range of 80°C for 72 hours. The mixture was later cooled to room temperature and crystalline powders was formed. Lastly, the MOF was collected and shall be kept at room temperature and dry condition for further application. The composition of MOFs fabricated were tabulated in Table 1 where it were labelled according to the molar ratio of metal oxide to organic acid.

Table 1. The absorption of the MOF.

Metal Organic Frameworks	Chromium oxide, (mmol)	Trimesic acid, (mmol)	Terephthalic acid, (mmol)
Cr-TMA-1:1	1	1	Nil
Cr-TMA-1:2	1	2	Nil
Cr-TMA-1:3	1	3	Nil
Cr-TMA-1:4	1	4	Nil
Cr-TMA-1:5	1	5	Nil
Cr-TPA-1:1	1	Nil	1
Cr-TPA-1:2	1	Nil	2

Cr-TPA-1:3	1	Nil	3
Cr-TPA-1:4	1	Nil	4
Cr-TPA-1:5	1	Nil	5

2.2.1. Scanning Electron Microscope (SEM)

SEM (Hitachi S3400N) was used and it will be carried out under accelerated voltage of 12kV. SEM was used to observe the morphology of MOFs. Before the analysis was carried, the specimen was placed on sample holder and fixed on the holder with carbon tape. Later, a layer of gold will be sputtered on the surface of sample by sputter coated machine. The gold metal layer can increase the conductivity of sample as well as enhance the focused beam of electrons to penetrate through the sample and hence clearer image can be obtained.

2.2.2. Energy Dispersive X-ray (EDX)

The distinct atomic structures of MOFs will be formed by using different materials and mole fraction of metal and organic linkers. EDX was used to determine the elemental composition of MOFs. In this analysis, five different points will be chosen randomly for characterization and the average results was obtained based on the unique set of peaks obtained.

2.3. CO2 Gas Adsorption Study

The CO2 gas adsorption study was carried by thermogravimetric analyser (Simultaneous thermal analyser (STA) 8000) to study the CO2 adsorption performance of the MOFs (Figure 1). Initially, the initial mass of MOF sample was weighted and placed in the sample pan of TGA. Subsequently, the moisture content of the sample needs to be removed by increasing the temperature of TGA to 150°C and nitrogen gas was flowed at the rate of 0.02 L/min. Before purging the gas of CO2 to measure the CO2 adsorption capacity of MOF, the sample of MOF was being cooled down from 150° C to room temperature. Lastly, the final mass of MOF sample was weighted and the difference of initial and final mass of MOF sample was the CO2 adsorption capacity of the MOF.

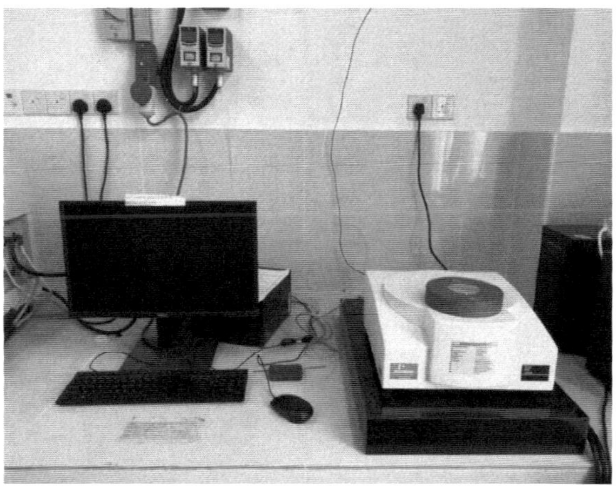

Figure 1. CO2 gas adsorption study by thermogravimetric analyser (TGA).

3. Results and Discussions

3.1. Morphologies of Metal Organic Framework

The morphologies of fabricated Cr-MOFs from TMA or TPA were captured from SEM analysis with magnification scale of 2,000 times as shown in Figure 2 and Figure 3. Both Cr-TMA and Cr-TPA MOF possess similar morphology which consists of randomly dispersed irregular diagonal shape of particles. Additionally, it was also noted the degree of crystallinity significantly enhanced and the sizes of crystal structure obtained was relatively smaller as the molar ratio of metal oxide to organic acid decreased. The comparison of the morphologies obtained between Cr-MOFs from TMA or TPA, the crystal structure of Cr-TMA series was relatively more oriented than Cr-TPA series. These may lead them to have higher surface area to volume ratio and hence have better gas adsorption.

(a) (b) (c) (d) (e)

Figure 2. SEM Morphologies of Cr-TMA MOF, (a) Cr-TMA-1:1, (b) Cr-TMA-1:2, (c) Cr-TMA-1:3, (d) Cr-TMA-1:4, and (e) Cr-TMA-1:5.

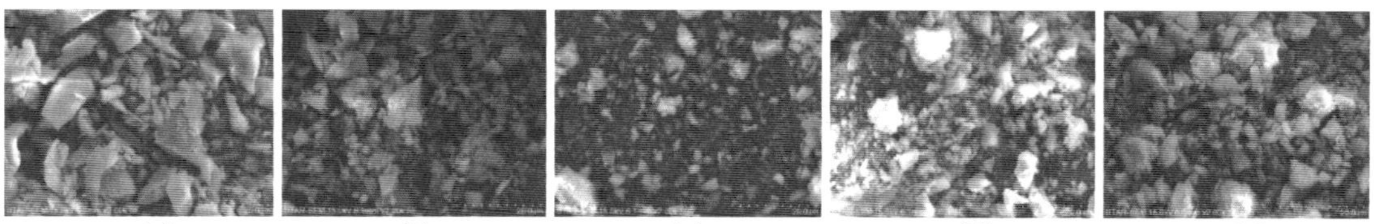

(a) (b) (c) (d) (e)

Figure 3. SEM Morphologies of Cr-TPA MOF, (a) Cr-TPA-1:1, (b) Cr-TPA-1:2, (c) Cr-TPA-1:3, (d) Cr-TPA-1:4, and (e) Cr-TPA-1:5.

3.2. EDX Analysis

The elemental content of each MOF sample was measured in term of weight percentage by employing EDX and the results was depicted in Table 2. There were only three elements being detected which were chromium, carbon and oxygen. These detected elements were coherent to the materials used to fabricate MOFs that were chromium (IV) oxide, terephthalic acid or trimesic acid. Due to the inhomogeneity of the samples, three readings were obtained from different spots for each sample to yield an average reading. The weight percentage of the chromium was observed decreases as the molar ratio of metal oxide to organic acid decreases (Table 2).

Table 2. EDX analysis.

MOFs	Carbon	Oxygen	Chromium	Carbon
Cr-TMA-1:1	16.29	40.84	42.87	16.29
Cr-TMA-1:2	46.13	42.26	11.61	46.13
Cr-TMA-1:3	46.60	43.11	10.30	46.60
Cr-TMA-1:4	50.60	39.21	10.19	50.60
Cr-TMA-1:5	48.18	45.00	6.82	48.18
Cr-TPA-1:1	48.92	39.59	11.49	48.92
Cr-TPA-1:2	41.08	48.45	10.47	41.08

Cr-TPA-1:3	38.83	50.89	10.27	38.83
Cr-TPA-1:4	44.95	45.05	9.99	44.95
Cr-TPA-1:5	44.94	48.93	6.12	44.94

3.3 *Gas Adsorption Study*

The gas adsorption study result reveal that the overall adsorption performance of Cr-TMA MOFs is better than Cr-TPA MOFs. It is interesting to note that both Cr-TMA and Cr-TPA MOFs illustreated the similar trend where the performance CO2 adsorption also increases as the molar ratio of metal oxide to organic acid decreases. Amongst all the MOFs fabricated in this study, Cr-TMA-1:5 has the highest adsorption capacity recorded at 0.095 mmol/g. This may be attributed by the morphologies of the MOFs. As the surface area to volume ratio of MOF, the CO2 adsorption increases. Hence, this has revealed that there was a great relationship between the molar ratio of metal to organic ligands, the types of organic ligands used during fabrication and the performance of CO2 adsorption.

Table 3. Gas adsorption study.

MOFs	Adsorption (mmol/g)	MOFs	Adsorption (mmol/g)
Cr-TMA-1:1	0.025 ± 0.0012	Cr-TPA-1:1	0.007 ± 0.0004
Cr-TMA-1:2	0.037 ± 0.0035	Cr-TPA-1:2	0.010 ± 0.0007
Cr-TMA-1:3	0.051 ± 0.0027	Cr-TPA-1:3	0.013 ± 0.0005
Cr-TMA-1:4	0.089 ± 0.0039	Cr-TPA-1:4	0.018 ± 0.0003
Cr-TMA-1:5	0.095 ± 0.0032	Cr-TPA-1:5	0.021 ± 0.0006

4. Conclusions

Various molar ratio of metal to organic ligands were used to fabricate ten different MOFs that were chromium oxide to organic acid of terephthalic acid or trimesic acid at ratio of 1:1, 1:2, 1:3, 1:4 and 1:5 respectively. The morphologies of Cr-TMA-1:5 from SEM has exhibited the most oriented and highest surface area to volume ratio will have the highest CO2 adsorption at 0.099 mmol/g among all the MOFs as observed in the gas adsorption study. This can also be proven by the EDX composition analysis as the materials used to synthesize MOFs were seemed to be fully incorporated to the MOFs and coherent with the elemental analysis of EDX. These results show that MOFs of Cr-TMA and Cr-TPA can be further developed to fight against global warming by adsorbing CO2.

Acknowledgments
The authors would like to thank Universiti Tunku Abdul Rahman for providing financial support in this work under the UTARRF (Project No.: IPSR/RMC/UTARRF/2017-C1/C04)

5. References

[1] Adams R, Carson C, Ward J, Tannenbaum R and Koros W 2010 *Micropor. Mesopor. Mat.* **131** 13
[2] Choi K M, Jeong H M, Park J H, Zhang Y-B, Kang J K and Yaghi O M 2014, *ACS Nano,* **8** 7451
[3] Furukawa H, Cordova K E, Keeffe M O and Yaghi, O M 2013. *Science* **341** 974
[4] Sneddon G, Greenaway A and Yiu H H P 2014 *Adv. Energy Mater.,* **4** 1301873
[5] Zhou H C, Long J R and Yaghi O M 2012 *Chem. Rev.,* **112** 673
[6] Adatoz E, Ahmet K A and Seda K 2015 *Sep. Purif. Technol.* **152** 207
[7] Caskey S R, Wong F A G and Matzger A J 2008 *J. Am. Chem, Soc.,* **130** 10870

[8] Andirova D, Cogswell C F, Lei Y and Choi S 2016 *Micropor. Mesopor. Mat.*, **219** 276
[9] Seetharaj R, Vandana P V, Arya P and Mathew S 2016 *Arab. J. Chem.*, **12** 295
[10] Augustus E N, Nimibofa A, Kesiye I A and Wankasi D 2017 *J. Environ. Prot.*, **5** 61
[11] Stock N and Biswas S 2012 *Chem. Rev.*, **112** 933
[12] Dey C, Kundu T, Biswal B P, Mallick A and Banerjee R 2013 *Acta Cryst. B*, **70** 3

IOP Conf. Series: Materials Science and Engineering **670** (2019) 012003 doi:10.1088/1757-899X/670/1/012003

Theory and derivation of unsymmetrical rotor equation of the foil-air bearing rotor system and its validation with experiment data

M F Hassan[1]

[1] Advanced Structural Integrity and Vibration Research (ASIVR)
Faculty of Mechanical and Manufacturing Engineering, University Malaysia Pahang, Malaysia

Email: firdaus@ump.edu.my

Abstract. This manuscript presents the theory and derivation of rotor equation using rigid body modelling used to validate the modal transformation approach for solving foil-air bearing rotor system. Numerical coefficient of the rotor equation for both methods only different less than 3%. The reasonable correlation in term of orbit's orientation with experimental data, has proved that the rotor was correctly modelled and adapted with simultaneous solution technique and modal model of foil structure.

1. Introduction

The recently developed simultaneous solution for solving nonlinear FAB-rotor system [1] has pave the way for other methods to be integrated into. This is proven when the author managed to develop a modal-based method for modelling the bump foil structure of FAB in the simulations of FAB-rotor problems [2] and neural network identification to replace the full numerical FAB model [3]. The latest method [2] requires the bump foil structure of the FAB to be modelled in FE software to obtain its modal properties wherein the bump interaction and inertia are considered. In this study, the capability of the method is adapted into a basic symmetrical FAB-rotor system which widely used in most of FAB's studies.

However, for validation purpose with experimental data, the rotor must be modelled according to the configuration of the test rig. The rotor equation has been modelled using modal transformation in which needs the *eigenvalues* and *eigenvector* of the rotor, obtained by finite element method and verified by impact testing. Alternatively, basic rigid body modelling approach is also applicable if the facilities to perform FE works are unavailable. This method needs accurate information of the rotor's dimensions, mechanical properties as well as the reaction and external forces act on the system before the center of gravity and the mass moment of inertia of the rotor are calculated. Since in the point of interest is at the FAB location, the actual mass of the rotor has to be shifted to equivalent mass at the FAB, denoted as $m_{eq,J}$.

2. Theory and derivation of the rotor equation using basic rigid body modelling

As can be seen in figure 1, the unsymmetrical test rotor is supported by a precision ball bearing at the left end and a FAB at the other end with the rotor disc overhanging. The self-aligning bearing at O (figure 1 (top)) is considered as a pivot and the bending stiffness of the coupling (between the left end of the shaft and the air motor) is neglected. Considering the operating speed range (up to around 30 krpm), the polar moment of inertia of the rotor and the gyroscopic effect is assumed to be negligible. The validation of the rotor's model was is done by impact tests [4]. The rotor equations derived from both methods are combined with the air film and foil equations of the FAB using the new foil modal model approach as in [2].

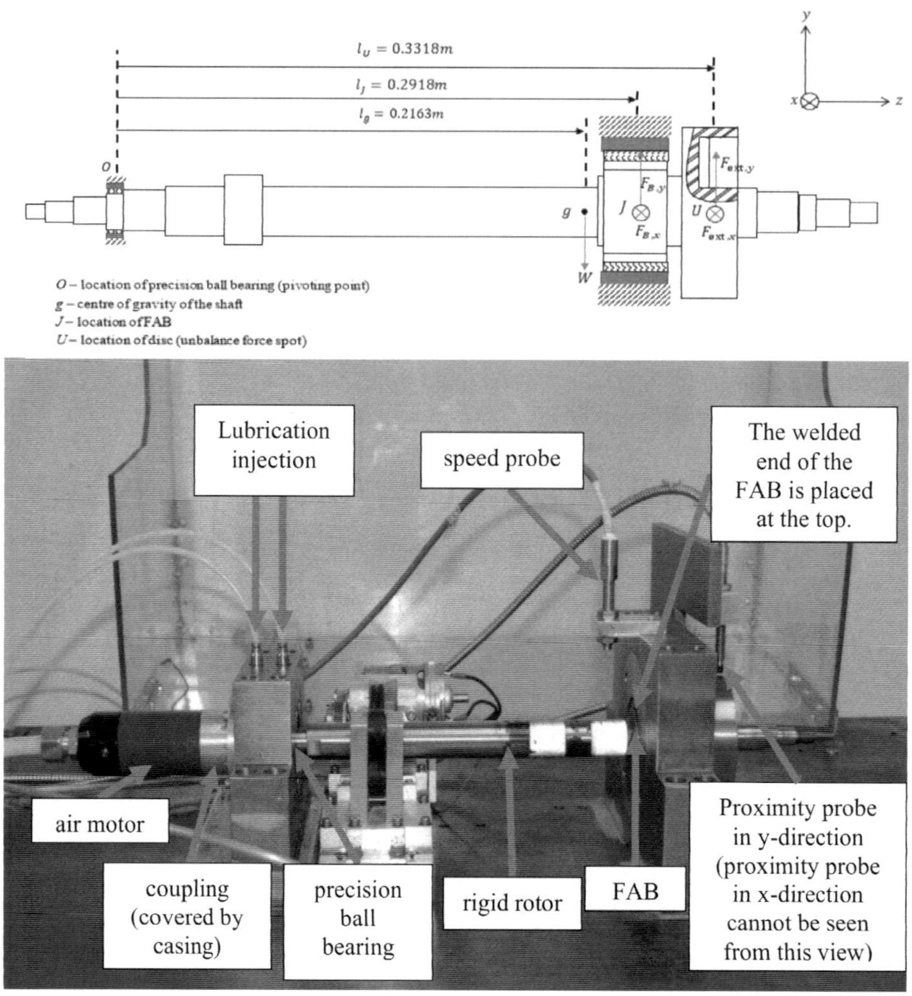

Figure 1. Test rig axial cut through (top) and FAB-rotor test rig configuration (bottom).

The equation of the test rotor as shown in figure 1 in *x*- and *y*- direction is given by:

$$F_{J,x}l_J + F_{U,x}l_u = I_O\ddot{\theta}_{x,z} \tag{1a}$$
$$F_{J,y}l_J + F_{U,y}l_u - Wl_g = I_O\ddot{\theta}_{y,z} \tag{1b}$$

where $F_{J,y}$, $F_{J,x}$ are hydrodynamic forces of the FAB, $F_{U,y}$, $F_{U,x}$ are the rotor unbalance forces at the disc, W is static load of the shaft, l_J, l_U are the distances of the FAB and disc from the pivot O, I_O is the mass moment of inertia of the shaft about a vertical or horizontal axis through the pivot point O, and $\ddot{\theta}_{x,z}$, $\ddot{\theta}_{y,z}$ are the angular accelerations in planes *xz* and *yz* respectively.
Dividing the eqs. (1a,b) by l_J results in the rotor equations as:

$$F_{J,x} + F_{U,x}\frac{l_u}{l_J} = \frac{I_O}{l_J}\ddot{\theta}_{x,z} \tag{2a}$$

$$F_{J,y} + F_{U,y}\frac{l_u}{l_J} - W\frac{l_g}{l_J} = \frac{I_O}{l_J}\ddot{\theta}_{y,z} \tag{2b}$$

It is known that $\ddot{y}_J = l_J \sin\ddot{\theta}_{y,z} \approx l_J\ddot{\theta}_{y,z}$ and $\ddot{x}_J = l_J \sin\ddot{\theta}_{x,z} \approx l_J\ddot{\theta}_{x,z}$ and note that $\frac{I_O}{{l_J}^2}$ term is defined as equivalent mass at point J, $m_{eq,J}$. Thus, the rotor motion can be expressed as:

$$m_{eq,J}\ddot{x}_J = F_{J,x} + F_{U,x}\frac{l_u}{l_J} \tag{3a}$$

$$m_{eq,J}\ddot{y}_J = F_{J,y} + F_{U,y}\frac{l_u}{l_J} - W\frac{l_g}{l_J} \tag{3b}$$

Converting eqs. (3a,b) to non-dimensional time and displacement with $\tau = \Omega t/2$, $(\ )'$ denoting derivative of $(\ )$ with respect to τ, and $\tilde{y}_J = \frac{y_J}{c}$, $\tilde{x}_J = \frac{x_J}{c}$ yields:

$$\tilde{x}_J''(\tau) = 4\left[F_{J,x} + F_{U,x}\frac{l_u}{l_J}\right]/m_{eq,J}c\Omega^2 \tag{4a}$$

$$\tilde{y}_J''(\tau) = 4\left[F_{J,y} + F_{U,y}\frac{l_u}{l_J} - W\frac{l_g}{l_J}\right]/m_{eq,J}c\Omega^2 \tag{4b}$$

where Ω is the rotational speed of the rotor and c is the radial clearance between the shaft and the top foil. The mass moment of inertia I_O of the rotor about the vertical or horizontal axis through the pivot O (figure 1) is calculated based on two different shapes of the shaft: (i) uniform solid cylinder of radius R, length L and centre of mass distant d_O from point O $(\frac{1}{4}MR^2 + \frac{1}{12}ML^2 + M{d_O}^2)$; (ii) thin circular hoop of radius $R_{average}$ with centre distant d_O from point O $(\frac{1}{2}M{R_{average}}^2 + M{d_O}^2)$. With reference to figure 1, the total mass of the shaft is 1.94405 kg, l_g is 0.2163 m, l_J is 0.2918 m, l_U is 0.3318 m. I_O was calculated to be 0.1131 kgm^2 using the procedure above. Hence, the equivalent mass at J, $m_{eq,J} = \frac{I_O}{{l_J}^2} = 1.3282$ kg. With reference to eq. (4b), the term $W\frac{l_g}{l_J}$ is equal to $W_{eq,J}$ where $W_{eq,J}$ is the equivalent static load at J, 14.1367 N. Hence, the final rotor equation derived from basic rigid body modelling can be written as:

$$\tilde{x}_J''(\tau) = 4\left[F_{J,x} + 1.1367F_{U,x}\right]/1.3282c\Omega^2 \tag{5a}$$

$$\tilde{y}_J''(\tau) = 4\left[F_{J,y} + 1.1367F_{U,y} - 14.1367\right]/1.3282c\Omega^2 \tag{5b}$$

It has therefore been proven that the rigid body rotor equations derived from rigid body modelling agrees with the rotor equation obtained using modal transformation approach [5], with the numerical coefficients differing by less than 3%. For validation with experimental data, since the displacement probes are located at the disc (figure 1), the computed journal displacements (eqs. (5a,b) were converted to the dimensional displacements at the disc using the following formulae

$$x_U(\tau) = c\frac{l_U}{l_J}\tilde{x}_J(\tau) \text{ and } y_U(\tau) = c\frac{l_U}{l_J}\tilde{y}_J(\tau) \tag{6a,b}$$

3. Coupling of rotor equation with air film and foil

The rigid rotor eqs. (5a,b) presented in section 2.1 are then coupled with the FAB modal model and air film equation as follows:

$$\boldsymbol{\psi}' = \mathbf{g}_{\mathrm{RE}}(\boldsymbol{\psi}, \mathbf{q}_{\mathrm{f}}, \boldsymbol{\varepsilon}) \tag{7a}$$

$$\begin{bmatrix} \mathbf{q}_{\mathrm{f}} \\ \mathbf{q}_{\mathrm{f}}' \end{bmatrix}' = \begin{bmatrix} \mathbf{q}_{\mathrm{f}}' \\ -\dfrac{4}{\Omega^2}\left[(\Omega/2)\mathbf{D}\mathbf{q}_{\mathrm{f}}' + \boldsymbol{\Delta}\mathbf{q}_{\mathrm{f}} + \mathbf{H}_{\mathbf{y}_{\mathrm{f}}}^{\mathbf{T}}\mathbf{f}_{\mathrm{p}}(\boldsymbol{\psi}, \mathbf{q}_{\mathrm{f}}, \boldsymbol{\varepsilon})\right] \end{bmatrix} \tag{7b}$$

$$\begin{bmatrix} \boldsymbol{\varepsilon} \\ \boldsymbol{\varepsilon}' \end{bmatrix}' = \begin{bmatrix} \boldsymbol{\varepsilon}' \\ \dfrac{4}{m_{eq,J} C \Omega^2}\begin{bmatrix} F_{J,x}(\boldsymbol{\psi}, \mathbf{q}_{\mathrm{f}}, \boldsymbol{\varepsilon}) + \epsilon m_{eq,J} u \Omega^2 \cos 2\tau \\ F_{J,y}(\boldsymbol{\psi}, \mathbf{q}_{\mathrm{f}}, \boldsymbol{\varepsilon}) + \epsilon m_{eq,J} u \Omega^2 \sin 2\tau - 14.1367 \end{bmatrix} \end{bmatrix} \tag{7c}$$

where $m_{eq,J}$ refers to the coefficient and equivalent mass at the FAB location. The time history of the response to a rotational unbalance $m_r u$ was obtained by solving state-space from equations of eqs. (7a,b,c) in Matlab using the implicit time domain integrator *ode23s*. This gave the full nonlinear response. Further understanding for solving this simultaneous equation of the rotor equation with air film and foil modal model can be referred to [5].

4. Results and discussion

The orbit response (y_U and x_U) at the disc of the FAB-rotor system computed from simulation using both rotor's modelling approaches and its comparison with experimental data are shown in figure 2. The predicted orbit responses were steady-state results obtained after integrating over 30 shaft revolutions at given shaft speed and the experimental orbit response is limited to one and two revolutions since it was taken from some specific speeds from *coasting down signal*. The orbit responses from both modeling approaches (rigid body modeling – *first column* & modal transformation – *second column*) are almost identical with little discrepancies. This remarkable similarity is strong enough to conclude that both modelling approaches are valid to be used.

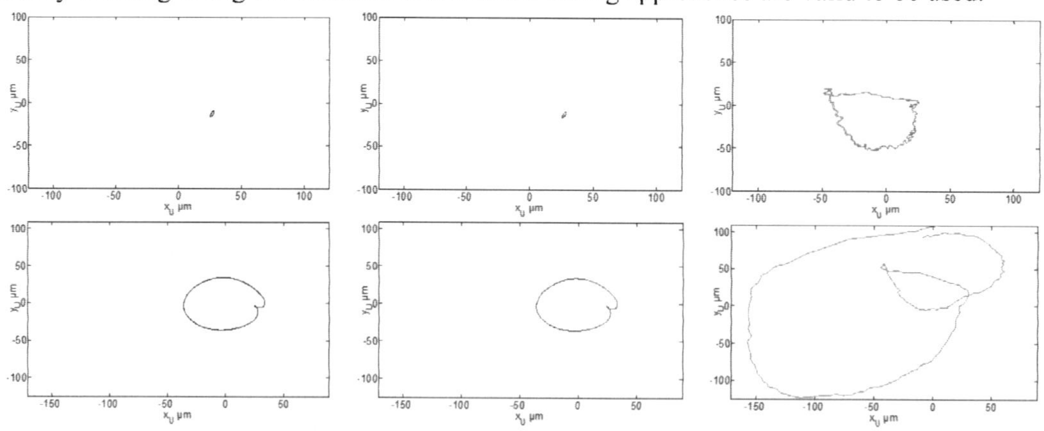

Figure 2. Steady-state orbit response of the FAB-rotor system obtained from simulation with rigid body modelling (first column), modal transformation modeling (second column) and experiment (third column) at low (4,015 rpm) and intermediate (15,031 rpm) speeds under 10 g mm unbalance.

In details, it is clearly seen that the predicted orbit response is an elliptical periodic orbit and operates within small amplitude and linear range. At 15,031 rpm the predicted orbit is not elliptical and a frequency analysis [5] shows that it exhibits period-two motion. The experimental results show that at those three speeds, the pattern of the orbit response evolves in a manner similar to the theoretical results, with a significant difference in amplitude, approximately 3 to 4 times bigger. The remarkable similarity in the pattern of the predicted and theoretical orbits at 15,031 rpm should be noted. It is suspected that the difference in amplitude between the theoretical and experimental results is due to the run-out of the rotor at the measurement position (the disc U). It is noted that, even though the rigid body modelling does not require FE platform as in modal transformation approach, it is not valid to be used if the rotor speed of the FAB-rotor system exceeds the first bending moment, as the rotor starts to behave as a flexible structure (no longer rigid body). This issue can be handled by using modal transformation approach e.g. using modal coordinate instead of physical coordinate for the rotor equation, as it has been successfully applied by the author in his publication [4].

5. Conclusion

This manuscript has revealed and shown that the rotor system running on foil-air bearing can be modelled by rigid body and modal transformation approaches. Both approaches lead to the same rotor equation with less than 3% different in their numerical coefficient. The validation with experimental results that shows a reasonable correlation in term of orbit's orientation, has proved that the rotor equation was accurately modelled and adapted with simultaneous solution technique and modal model of foil structure. However, the advantage of the modal transformation approach over rigid body approach is it can consider a number of modes when the rotor no longer behaves as a rigid body at high speed operation. On the contrary, rigid body modelling is easy to be developed since it does not require FE platform but only requires knowledge on structural dynamic modelling for rotating structure.

Acknowledgments
The author acknowledges the financial support from University of Malaysia Pahang for internal grant RDU1703192 and the Engineering and Physical Sciences Research Council for funding EP/I029184/1.

References

[1] Bonello P and Pham H 2014 The efficient computation of the nonlinear dynamic response of a foil–air bearing rotor system *Journal of Sound and Vibration* **333(15)** p. 3459-3478.
[2] Hassan M B and Bonello P 2017 A new modal-based approach for modelling the bump foil structure in the simultaneous solution of foil-air bearing rotor dynamic problems *Journal of Sound and Vibration* **396** p. 255-273.
[3] Hassan M F B and Bonello P 2017 A neural network identification technique for a foil-air bearing under variable speed conditions and its application to unbalance response analysis. *Journal of Tribology* **139(2)** p. 021501.
[4] Bonello P and Hassan M B 2018 An experimental and theoretical analysis of a foil-air bearing rotor system. *Journal of Sound and Vibration* **413** p. 395-420.
[5] Hassan M F B 2017 *Nonlinear Analysis of Rotating Machinery Running on Foil-Air Bearings* The University of Manchester (United Kingdom).

IOP Conf. Series: Materials Science and Engineering **670** (2019) 012004 doi:10.1088/1757-899X/670/1/012004

Fundamental study on the optimal path of centrifugal pendulum vibration absorbers in automatic transmissions

T Ryu[1,4], T Nakae[1,4], K Matsuzaki[2,4] and J Ozaki[3]

[1] Faculty of Science and Technology, Oita University, 700 Dannoharu, Oita-shi, Oita, Japan
[2] Graduate School of Science and Engineering, Kagoshima University, Korimoto 1-21-24, Kagoshima-shi, Kagoshima, Japan
[3] Graduate School of Engineering, Oita University, 700 Dannoharu, Oita-shi, Oita, Japan
[4] JSPS KAKENHI Grant Number 15K05866
E-mail: ryu&iota-u.ac.jp

Abstract. In recent years, high-power and low-cylinder engines have seen increasingly widespread use. These engines produce a strong torsional vibration due to engine combustion. Furthermore, automatic transmissions necessitate the use of a lock-up clutch system that directly links the engine and gear train when the engine rotational speed is low. Using a torsional spring called a lock-up damper in the torque converter is not sufficient to absorb the torsional vibration. Therefore, the centrifugal pendulum vibration absorber (CPVA) has been developed to suppress the torsional vibration. The natural frequency of a CPVA is proportional to the engine rotational speed, and it is expected to suppress the torsional vibration over a wide range of speeds. However, the natural frequency of the CPVA varies because of the nonlinearity when the vibration amplitude of the CPVA is large. In this study, the optimal path of the CPVA to suppress the torsional vibration of the engine was investigated.

1. Introduction

Automatic vehicle transmissions commonly use a torque converter to transmit the engine torque. When the difference between the rotational speeds at the input and output sides of the torque converter is small, a lock-up clutch begins to operate. With the operation of the lock-up clutch, the engine torque is transmitted directly to the gear train, which worsens the ride quality because of fluctuations in the torque due to engine combustion. To reduce the torsional vibration, a torsional damper called a lock-up damper can be installed behind the lock-up clutch. However, the torsional vibration cannot be sufficiently suppressed by this type of damper.

To address these problems, a centrifugal pendulum vibration absorber (CPVA) is often attached in the torque converter. The CPVA is a tuneable absorber that can tune its natural frequency according to the rotational speed. CPVAs are widely used for a long time. The natural frequency of a CPVA can be controlled by changing the path of the mass. Some researchers have examined the effect of the path on the vibration suppression performance [1]–[4]. However, the optimal path to suppress the torsional vibration has not yet been examined. In this study, the optimal path to suppress torsional vibration for automatic transmissions was designed using a genetic algorithm (GA).

IOP Conf. Series: Materials Science and Engineering **670** (2019) 012004 doi:10.1088/1757-899X/670/1/012004

2. Theoretical analysis

2.1. Analytical model

Figure 1 shows the analytical model of the automatic transmission power train, which is composed of an engine, a torque converter, an automatic transmission gear train, and a drive shaft. The power train is modelled as a four-degree-of-freedom system with inertias $J_1,\ldots,J_6$, torsional spring constants $K_1,\ldots,K_4$, and torsional damping constants $C_1,\ldots,C_4$. The angular displacements are given by $\theta_1,\ldots,\theta_4$. The fluctuation of the engine torque is described by $T\cos\omega t$ with amplitude T and frequency ω. The frequency is given by

$$\omega = \frac{N\Omega}{2}, \tag{1}$$

where N is the number of cylinders and Ω is the speed of rotation of the engine.

To suppress the torsional vibration, the CPVA is attached at point A, as shown in Figure 1. Modal analysis was applied to the vibration model in figure 1, and the vibration mode was defined as a one-degree-of-freedom system for the target to suppress the torsional vibration. Figure 2(a) shows the analytical model applied to consider the suppressive effect of the CPVA. J and K are the modal mass (inertia) and the modal stiffness, respectively. The damping coefficient C was defined to yield a modal damping ratio equal to that of the system in Figure 1. Figure 2(b) shows a CPVA with mass m and length r attached to the main system. The distance between the centre of rotation O and the position P at which the CPVA is attached is R.

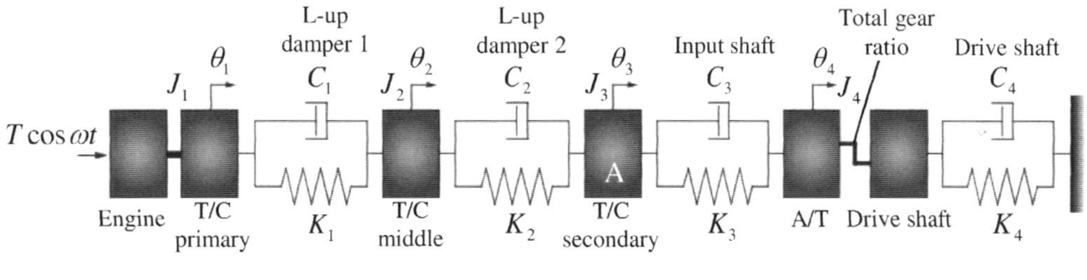

Figure 1. Analytical model.

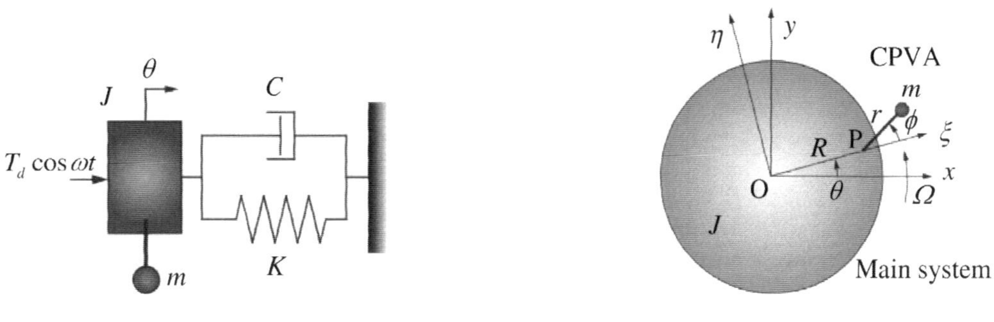

(a) (b)

Figure 2. Centrifugal pendulum vibration absorber.

2.2. Equations of motion

The equations of motion of the system shown in figure 2 are given by

$$\left.\begin{aligned}&\left\{J+m\left(R^2+r^2+2Rr\cos\phi\right)\right\}\ddot{\theta}+m\left(r^2+Rr\cos\phi\right)\ddot{\phi}\\&\quad-mRr\dot{\phi}\left(2\dot{\theta}+2\omega+\dot{\phi}\right)\sin\phi+C\dot{\theta}+K\theta=T_d\cos\omega t\\&m(r^2+Rr\cos\phi)\ddot{\theta}+mr^2\ddot{\phi}+mRr(\Omega+\dot{\theta})^2\sin\phi=0\end{aligned}\right\}, \tag{2}$$

where θ is angular displacement from the steady rotation; ϕ is the angular displacement of the CPVA in the rotational coordinate frame $\mathrm{O}-\xi\eta$, which is fixed to the main system; and T_d is the torque amplitude after the application of the modal analysis. The damping of the CPVA is neglected. In this case, the path of the CPVA is circular. Then the length of the pendulum r is constant of r_p.

The equation of motion when only the CPVA is considered is given by

$$mr^2\ddot{\phi}+mRr\Omega^2\sin\phi=0. \tag{3}$$

When the amplitude of ϕ is small, the linearized natural frequency is given by

$$\omega_p=\sqrt{\frac{R}{r}}\Omega=n\Omega. \tag{4}$$

Equation (4) shows that the natural frequency of the CPVA is proportional to the rotational speed Ω. If the order n is set to $N/2$, the CPVA can suppress the torsional vibration. However, when the amplitude ϕ is large, the natural frequency of the CPVA cannot maintain a constant value, worsening the suppressive effect.

To address this problem, the optimal path of the CPVA is examined. The path r is designed by the following polynomial equation of ϕ:

$$r(\phi)=r_p(1+r_2\phi^2+r_4\phi^4+r_6\phi^6+r_8\phi^8). \tag{5}$$

The equations of motion are given by

$$\left.\begin{aligned}&(J+mR^2+mr^2+2mRr\cos\phi)\ddot{\theta}+(mr^2+mrR\cos\phi+mRr_\phi\sin\phi)\ddot{\phi}\\&\quad+2mrr_\phi\dot{\phi}(\Omega+\dot{\theta}+\dot{\phi})+mR\dot{\phi}^2(r_{\phi\phi}\sin\phi+r_\phi\cos\phi)+mR\dot{\phi}(2\Omega+2\dot{\theta}+\dot{\phi})(r_\phi\cos\phi-r\sin\phi)\\&\quad+C\dot{\theta}+K\theta=T_d\cos\omega t\\&(r^2+Rr\cos\phi+Rr_\phi\sin\phi)\ddot{\theta}+(r^2+r_\phi{}^2)\ddot{\phi}+r_\phi r_{\phi\phi}\dot{\phi}^2\\&\quad+rr_\phi\left\{\dot{\phi}^2-(\Omega+\dot{\theta})^2\right\}+R(\Omega+\dot{\theta})^2(r\sin\phi-r_\phi\cos\phi)=0\end{aligned}\right\}, \tag{6}$$

where $r_\phi=\mathrm{d}r/\mathrm{d}\phi$, $r_{\phi\phi}=\mathrm{d}^2r/\mathrm{d}\phi^2$.

The equation of motion when only the CPVA is considered is given by

$$(r^2+r_\phi{}^2)\ddot{\phi}+r_\phi r_{\phi\phi}\dot{\phi}^2+rr_\phi(\dot{\phi}^2-\Omega^2)+R\Omega^2(r\sin\phi-r_\phi\cos\phi)=0, \tag{7}$$

and the linearized natural frequency is given by

$$\omega_p=\sqrt{\frac{R-2r_2(R+r_p)}{r_p}}\Omega. \tag{8}$$

Equations (6) can be normalized as

$$\left.\begin{array}{l}(1+\mu_R+\mu_r+2\mu_{Rr}\cos\phi)\theta''+(\mu_r+\mu_{Rr}\cos\phi+\mu_{Rr1}\sin\phi)\phi''+2\mu_{rr1}\phi'(z+\theta'+\phi')\\ \quad+\phi'^2(\mu_{Rr2}\sin\phi+\mu_{Rr1}\cos\phi)+\phi'(2z+2\theta'+\phi')(\mu_{Rr1}\cos\phi-\mu_{Rr}\sin\phi)+\theta+2\zeta\theta'=\delta\cos z\tau\\ (\mu_r+\mu_{Rr}\cos\phi+\mu_{Rr1}\sin\phi)\theta''+(\mu_r+\mu_{r1})\phi''+\mu_{rr2}\phi'^2\\ \quad+\mu_{rr2}\left\{\phi'^2-(z+\theta')^2\right\}+(z+\theta')^2(\mu_{Rr}\sin\phi-\mu_{Rr1}\cos\phi)=0\end{array}\right\}, \tag{9}$$

where

$$\begin{array}{l}\mu_R=\dfrac{mR^2}{J},\ \mu_r=\dfrac{mr^2}{J},\ \mu_{r1}=\dfrac{mr_\phi^{\ 2}}{J},\ \mu_{r2}=\dfrac{mr_{\phi\phi}^{\ 2}}{J},\ \mu_{Rr}=\dfrac{mRr}{J},\ \mu_{Rr1}=\dfrac{mRr_\phi}{J},\ \mu_{Rr2}=\dfrac{mRr_{\phi\phi}}{J},\\ \mu_{rr1}=\dfrac{mrr_\phi}{J},\ \mu_{rr2}=\dfrac{mr_\phi r_{\phi\phi}}{J},\ \delta=\dfrac{T_d}{K},\ \zeta=\dfrac{C}{2\sqrt{JK}},\ \omega_n^{\ 2}=\dfrac{K}{J},\ z=\dfrac{\omega}{\omega_n},\ \tau=\omega_n t\end{array}, \tag{10}$$

and the prime represents the derivative with respect to τ.

3. Numerical analysis

Table 1 gives the natural frequencies obtained from the vehicle model shown in figure 1. Table 1 also gives the resonance rotational speeds of the three-cylinder engine ($N=3$). The target vibration mode is the second mode. Table 2 lists the standard parameter values.

It is known that a CPVA with an epicycloidal path is tautochronic in a centrifugal field [5]. The suppressive effect of such a CPVA was also examined here. In the calculation of the CPVA with an epicycloidal path, the ratio between the base circles that form the epicycloidal path was defined such that $\omega_p=n\Omega$.

To design the CPVA with the optimal path, a GA was used. The initial population was set to 100 with 16 bits, and the fitness function was defined as the reciprocal of the mean amplitude between 900 and 1400 rpm. Roulette wheel selection and two-point crossover were used to produce new generations. Mutations were applied every five generations. The considered variables are r_4, r_6, and r_8, and their optimal values are given in table 3. r_2 was set to zero to keep the linearized natural frequency in Equation (8) constant.

The blue, green, and red lines in figure 3 show the characteristics of the free vibration of the CPVAs with the circular, epicycloidal, and optimal paths, respectively. The abscissa represents the ratio of the nonlinear natural frequency to ω_p and the ordinate represents the pendulum amplitude. The natural frequency of the CPVA with the epicycloidal path is independent of the amplitude of the pendulum. The natural frequency of the CPVA with the optimal path is also close to an order of 1.5 when the pendulum amplitude is small. The blue, green, and red lines in figure 4 show the frequency response curves when CPVAs with the circular, epicycloidal, and optimal paths are attached, respectively. The shooting method was used to calculate each frequency response curve. The frequency response curve when the CPVA is not attached is also shown as a black line. The abscissa shows the rotational speed of the engine, and the ordinate shows the peak-to-peak (pk-pk) amplitude of the angular displacement of the main system. The amplitude was reduced relative to that of the baseline model by attaching the CPVA with the circular path but was still large at low rotational speeds. The CPVA with the epicycloidal path yielded greater amplitude reductions than the CPVA with the circular path at high rotational speeds but very large amplitudes at low rotational speeds. Although the CPVA with the epicycloidal path is tautochronic in a centrifugal field, the amplitude did not become zero. The CPVA with the epicycloidal path is a linear system when the equation of motion is formulated as a function of the arc length along the path. In this system, the equations of motion are nonlinearly coupled with the angular displacements. This is the reason the amplitude of the main system does not become zero. When the CPVA with the optimal path is attached, the amplitude of the main system is suppressed.

IOP Conf. Series: Materials Science and Engineering **670** (2019) 012004 doi:10.1088/1757-899X/670/1/012004

Figure 5 shows the circular, epicycloidal, and optimal paths. The ordinate in this plot is shown on a finer scale than the abscissa. Although the three paths are very similar, the precision of the path is important to efficiently suppress the torsional vibration.

Table 1. Natural frequencies and resonance rotating speed.

Mode	1st	2nd	3rd	4th
Natural frequency [Hz]	10	33	82	240
Resonance rotating speed [rpm]	400	1320	3280	9600

Table 2. Standard parameters.

μ_R	0.0561
μ_r	0.0111
μ_{Rr}	0.0249
δ	0.0159
ζ	0.232
n	1.5

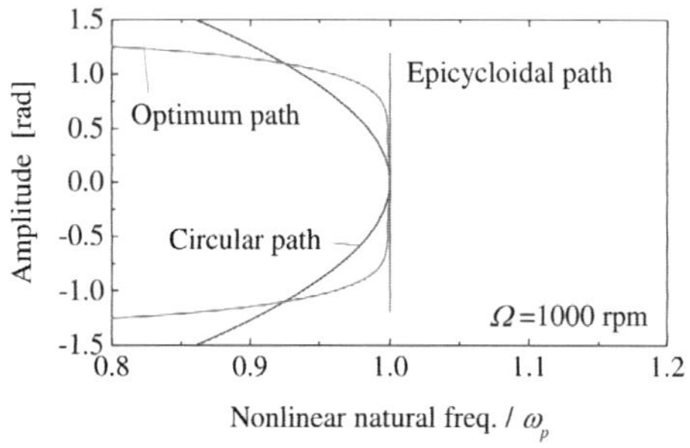

Figure 3. Characteristics of free vibration.

Table 3. Optimal values.

r_4	−0.02102
r_6	−0.02962
r_8	0.03138

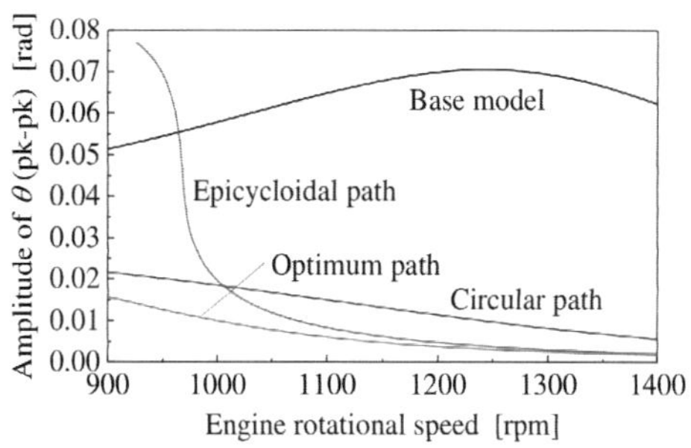

Figure 4. Frequency response curve.

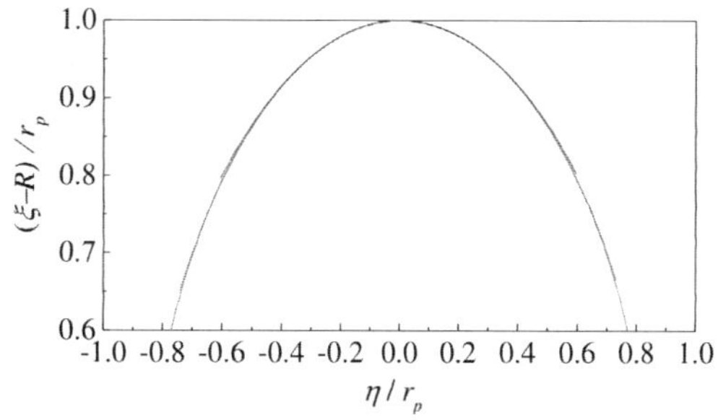

Figure 5. CPVA paths.

4. Conclusions

In this study, the optimal path of a CPVA to suppress the torsional vibration caused by engine combustion was investigated. The optimal path was calculated using a GA. The frequency response curves of CPVAs with circular, epicycloidal, and optimal paths were evaluated. It was found that the CPVA with the optimal path can more effectively reduce the torsional vibration than the CPVAs with the circular and epicycloidal paths.

5. References

[1] Denman H H 1992 *Journal of Sound and Vibration* **159**(2) 251.
[2] Alsuwaiyan A S and Shaw S W, 2002 *Journal of Sound and Vibration* **252**(5) 791.
[3] Haddow A G and Shaw S W, 2003 *Nonlinear Dynamics* **34** 293.
[4] Newland D E, 2017 *Proceedings of ICSV24* 1.
[5] Rana N C 2016 Classical Mechanics Mc Graw Hill 232.

IOP Conf. Series: Materials Science and Engineering **670** (2019) 012005 doi:10.1088/1757-899X/670/1/012005

Ultimate strength analysis of CFRP confined concrete using finite element method (FEM)

Z Z Mukhtar*[1], A H H AbWahid[2], A A Bakar[2], A Fitriadhy[2] , M S Abdul Majid[3], A M Saat[6]

[1]Universiti Kuala Lumpur Malaysian Institute of Marine Engineering Technology,Bandar Teknologi Maritim,Jalan Pantai Remis, 32200 Lumut, Perak, MALAYSIA.
[2]School of Ocean Engineering, Universiti Malaysia Terengganu, 21030 Kuala Nerus, Terengganu,MALAYSIA.
[3]School of Mechatronic, Universiti Malaysia Perlis, Pauh Campus, 02600, Perlis, MALAYSIA.

E-mail: *zaimi@unikl.edu.my

Abstract. This paper presents a comparison with experimental data of CFRP confined concrete column damage. The study investigates the behaviour of CFRP confined concrete beams against the experimental results. It also investigates the influence of different parameters on the behaviour of the FRP confined concrete column. The ABAQUS code was used to develop finite element models for simulation of the damage behaviour of the beams. The concrete was modelled using a plastic-damage model which is a perfect bond model and were evaluated for the concrete-FRP confinement interface. The results showed that when the thickness of carbon-fibre reinforced plastics (CFRP) proportional to the load capacity of the beam for both shear and compressive strength. The maximum load increases with an incremental of CFRP thickness. From the simulation results, the load- deflection relationships have been analysed and the FEM results agreed well with the experiments data. Finally, the study shows that the differences between the results of finite element analysis and experimental tests are in an acceptable range.

1. Introduction

Fibre reinforced plastics (FRP) is one of the most used strengthening techniques to reinforced concrete beams due to advantages, such as high strength to self-weight ratio, high tensile strength, large fatigue resistance capacity and high durability, [1-2]. Carbon and glass Fibre reinforced plastics (CFRP and GFRP) composites offer the solution to enhance shear and flexural capacities and ductility, as well as improve the mode of rupture, [3-4].

Study of the behaviour of CFRP reinforced concrete structures has become a very important. In terms of experimental application, several studies were performed to study the behaviour of retrofitted beams and how various parameters can influence its behaviour [5-6]. While experimental methods of investigation are very useful in obtaining collapse behaviour of FRP and reinforced concrete. The use of numerical models helps in developing a good understanding of the behaviour at lower costs, [7].

In this research, nonlinear finite element analysis models for strengthening of reinforced concrete beams with carbon fibre reinforced plastics (CFRP) have been presented. The collapse behaviour in shear and flexure of reinforced concrete beams strengthened with carbon Fibre reinforced plastics (CFRP) is highly affected by the way in which these composites are applied (bonded) to the beam, [3,4,6].

The influence of some important parameters on the overall response of the strengthened RC beams has been investigated in order to achieve the optimum utilization of such strengthening technique in terms of load capacity and possible deflection values. The research investigate nonlinear finite element models of reinforced concrete beams externally strengthened with carbon Fibre reinforced plastics (CFRP) using the finite element modelling software (ABAQUS). Finally, validate the finite element models by comparing with experimental tests available in the literature.

2. Characterizations Of Material

Experimental data was obtained from previous research[8] and the parameter setting for beam modelling have been set as follows:

Table 1. Main Particulars of RC Beam.

Particular	Parameter
High × Width × Length	300 mm × 150 mm × 1960 mm
Tension Reinforced	12 mm × 2
Compression Reinforced	10 mm × 2
Shear Reinforced	8 mm / 100 mm

2.1 *Mechanical Properties of Material*

There are three types of material such as concrete, steel and cfrp that had been modeled using Abaqus. Concrete is composed of coarse aggregate, fine aggregate, cement and water. CFRP materials generally have much higher strength than the yield strength of steel [9]. The CFRP materials are not homogeneous and their properties are dependent on many factors. Below are the mechanical properties that have been applied for finite element analysis :

Table 2. Mechanical Properties.

Particular	Properties		
	Concrete	Steel	CFRP
Mass Density, ρ	2400 kg/m^3	7800 kg/m^3	1600 kg/m^3
Young's Modulus, E	25084 MPa	205000 MPa	165000 MPa
Poisson's Ratio, ν	0.15	0.3	0.25

3. Methodology

ABAQUS software was employed to model and simulate damage collapse of the CFRP beam columns. Finite element analysis looking at load– deflection curves of retrofitted beams and control beams. The result from a control beam will be validated by comparing with experimental results. The analysis of linear finite element simulation output results will be compared with all cases and the control beam.

3.1 Finite Element Modeling (FEM)

Finite element failure analysis was performed to model the nonlinear behaviour of the beams. The FEM package Abaqus/standard was used for the analysis. There are also numbers of study have been done by many researchers using Finite Element Analysis (FEA) software to predict and validate material behaviour. Concrete is modelled using a solid eight-node element, [10]. The elastic response of concrete is considered linear by referring to Hooke's Law:

$$\{\sigma\} = [E]\{\check{\varepsilon}\} \quad (1)$$

Where ; $\{\sigma\}$is the stress tensor, [E] is the constitutive matrix of elasticity, ε is the strain tensor and superscript 'e' indicates elastic strain.

ABAQUS is allowing user to specify a method of analysis to be applied. Concrete Damaged Plasticity is one of the common method to be used in analysing elastic concrete behaviour. The stress-strain relations under uni-axial tension and compression are taken into account according to equations below [11] :

$$\sigma_t = (1 - d_t) \cdot E_0 \cdot (\varepsilon_t - \varepsilon_t^{\sim pl}) \quad (2)$$

$$\sigma_c = (1 - d_c) \cdot E_0 \cdot (\varepsilon_c - \varepsilon_c^{\sim pl})$$

Where ; σ_t = stress in tension, σ_c= stress in compression, ε_t = strain in tension, ε_c = strain in compression, E_o = Initial (Undamaged) elastic stiffness, $\varepsilon_t^{\sim pl}$ = plastic strain in tension, $\varepsilon_c^{\sim pl}$ = plastic strain in compression.

The 4-Node linear tetrahedral elements method were used for the reinforced concrete, steel reinforcement and also CFRP in this model. The element configuration is shown in Figure 1 and Figure 2. The Node 3-D cohesive elements were used to model the interface layer.To show the effect of the bond model and the behaviour of CFRP, five combinations of bond model and CFRP model were analyses by using the orthotropic material model. One quarter of the beam was modelled and a fine mesh with 3 mm size is set to obtain results of sufficient accuracy [12]. Sample of elements applied in the numerical analysis by using ABAQUS are as follows:

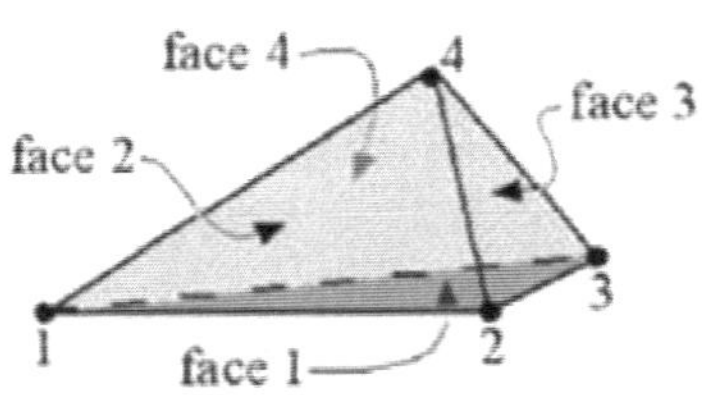

Figure 1. 4-node linear tetrahedral element.

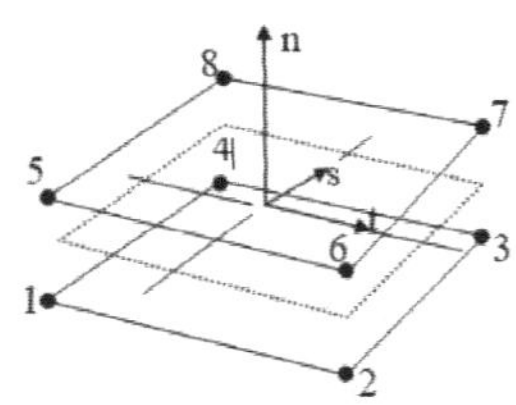

Figure 2. 8-node 3-D cohesive element.

A trust element was used to model steel reinforcement. CFRP composites can be modelled either using isotropic linear elastic model or orthotropic linear elastic model, [13]. In this study, the properties of the CFRP composites were nearly the same in any direction perpendicular to the fibres. Orthotropic linear elastic model for CFRP composites was used throughout this study.

Normally, the two parameters needed to describe the elastic behaviour of concrete are the Poisson's ratio and the modulus of elasticity. The Poisson's ratio of concrete under uniaxial tension and compressive stress ranges from 0.15 to 0.22. For this study, a value of 0.15 for concrete was adopted to perform the finite element analysis.

4. Result and Discussion

In this section, the load–deflection curves obtained for control beam and retrofitted beams from experiments and finite element analysis will be shown and discussed. For control beam, the comparison of FEA and experimental result for Load-Deflection Curve can be shown in Figure 4. From the result and graph below (figure 3), there is good agreement between FEM and experimental results for the control beam. The FEM analysis predicts the beam to be slightly stiffer and stronger, probably because of the assumed perfect bond between concrete and reinforcement. The good agreement indicates that the constitutive models used for concrete and reinforcement can capture the fracture behaviour well.

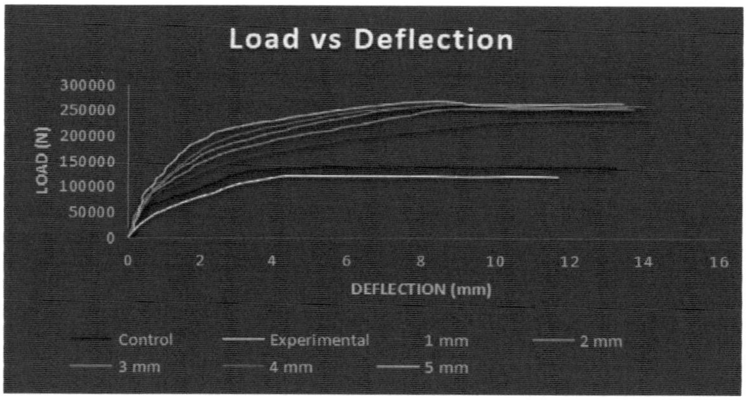

Figure 3. Load-Deflection Comparison.

When comparing all the results, it can be seen that the thickness of the CFRP significantly influences the behaviour of the beam. The thicker CFRP, the higher is the maximum load. The perfect bond models increasingly overestimate the stiffness of the beam. This is due to the fact that the perfect bond does not take the shear strain between the concrete and CFRP into consideration. This shear strain increases when cracks appear and causes the beam to become less stiff.

The maximum load for 1mm and 2mm thickness of CFRP before break are 228,143 N and 252,395 N. For 3mm and 4mm thicknesses, the maximum load before break are 260,470 N and 264,790 N. Lastly, the highest maximum load is 267,664 N and it is for 5mm of CFRP thickness.

5. Discussion and Conclusion

A finite element method was used to analyses the beams retrofitted with CFRP. Elastic orthotropic behaviours were used to represent the CFRP behaviour. A cohesive model was used to address the interfacial behaviour between CFRP and concrete. The following conclusions can be drawn from this study:

- The behaviour of the retrofitted beams is significantly influenced by the thickness of CFRP. This characteristic show in the experimental results as well as in simulation analysis. The ultimate load is found proportional with the thickness of the CFRP. It shows a good agreement compared [8] to the experimental of a full scale beam column.
- Finite element models using ABAQUS were successfully verified comparing with previously published experimental test results. Therefore, ABAQUS can be confidently used in analysis of RC Beams externally strengthened with Carbon Fiber Reinforced Plastics (CFRP) lamination. However, the finite element analysis show that the concrete beam slightly stiffer than the test data.

Finally, environmental factors that may affect the efficiency of RC beams strengthened with CFRP as seasonal temperature variation, creep, and shrinkage were not considered. The effect of these parameters during the beam life span can be studied.

References:

[1] Jayajothi P, Kumutha R and Vijai K 2013 Finite Element Analysis of FRP Strengthened RC Beams Using ANSYS *Asian Journal of Civil Engineering (BHRC)* **14** (4) 631-642.
[2] Camata G, Spacone E and Zarnic R 2007 Experimental and Nonlinear Finite Element Studies of RC beams Strengthened with FRP plates *Composites: Part B* **38** 277- 288.
[3] Vijayakumar A, Venkatesh B and Jayaprakash R 2012 Analytical Study on Various Types of FRP Beams by using ANSYS *Engineering Research and Applications* (*IJERA*) **2** (5) 593-598.
[4] Chansawat K, Potisuk T, Miller T H, Yim S C and Kachlakev D I 2009 FE Models of GFRP and CFRP Strengthening of Reinforced Concrete Beams *Advances in Civil Engineering* **9** 1-13.
[5] Basappa U and Rajagopal A 2013 Modeling of CFRP Strengthened RCC Beam Using the Nonlinear Finite Element Method *Journal of Structural Engineering* **40** (2) 169-184.
[6] Santhakumar R and Chandrasekaran E 2004 Analysis of Retrofitted Reinforced Concrete Shear Beams Using Carbon Fibre Composite. Electron *J Struct Eng* **4** 66–74.
[7] Elyasian I, Abdoli N and Ronagh H R 2006 Evaluation of Parameters Effective in FRP Shear Strengthening of RC Beams Using FE Method *Asian Journal of Civil Engineering* **7** (3) 249-257.
[8] Obaidat Y 2007 *Retrofitting of reinforced concrete beams using composite laminates* Jordan University of Science and Technology.
[9] Lu Y, Zhang H and Liu S 2008 Experimental study on tensile properties of steel plate bonded by CFRP *J. Wuhan Univ. Technol. Mater. Sci.* **23** 727–732
[10] Karabinis A I, Rousakis T C and Manolitsi G E 2008 3D Finite Element Analysis of Substandard RC Columns Strengthen by Fibre-Reinforced Polymer Sheets *Jurnal of Composite for Construction* **12** (5) 531 - 540
[11] ABAQUS/CAE User's Manual. Version 6.13 SIMULIA 2015
[12[Mukhtar Z Z, Bakar A A, Fitradhy A, Abdul Majid M S and Saat A M 2017 Study on Mesh Sensitivity Effect of FRP Confined Concrete Column Using Finite Element Method *IUP2017 Conference Proceeding*, University Malaysia Terengganu (UMT)
[13] Obaidat Y, Heyden S and Dahlblom O 2010 The Effect of CFRP and CFRP/ Concrete Interface Models when Modelling Retrofitted RC Beams with FEM. Accepted for publication in Composite Structures *Composites Science and Technology* **68** 925-932.

IOP Conf. Series: Materials Science and Engineering **670** (2019) 012006 doi:10.1088/1757-899X/670/1/012006

Corrosion behavior of hybrid coconut shell powder and silica carbide reinforced aluminium composite towards aggressive biofilm attack

M J Suriani[1,4], F Zulkifli[1], M A Khairul[2,5], M D Azaman[3,5] and M N K Jarkoni*[1,6]

[1]Faculty of Ocean Engineering Technology and Informatic, Universiti Malaysia Terengganu (UMT), 21030 Kuala Nerus, Terengganu, Malaysia.
[2]Mechanical Engineering Section, Universiti Kuala Lumpur, Malaysia France Institute, Section 14, Teras Jernang Road, 43650, Bandar Baru Bangi, Selangor, Malaysia.
[3]School of Manufacturing Engineering, Pauh Campus, Universiti Malaysia Perlis, 02600 Perlis, Malaysia.

E-mail: [4]surianimatjusoh@umt.edu.my, [5]khairulazhar@unikl.edu.my, [5]azaman@unimap.edu.my, *[6]khasbi@gmail.com.my

Abstract. This study enveloped innovative uses of a hybrid agro waste and silica carbide incorporated with aluminium as a new hybrid composite materials for marine application. In this study, the purpose of adding coconut shell powder (CSP) from the agro waste and silicon carbide (SiC) as reinforcement aluminium composites materials for combating corrosion problem. The CSP was incorporated into the molten aluminium alloys varies from 0%, 5%, 10%, 15% and 20% of weight percentages. The corrosion behaviour was determined by weight gain measurement (WGM), electrochemical impedance spectroscopy (EIS) and potentiodynamic polarization (PP). As a results, 5% CSP reinforcement in hybrid aluminum composites contribute to excellent corrosion resistance performance by shifting the polarization curve towards the positive region, increases the value of polarization resistance, R_p and reduces the I_{corr} value. It can be concluded that, by adding CSP into aluminium composite has significantly increase the corrosion resistance towards a severe biofilm attack.

1. Introduction

Corrosion has been recognized as the greatest issue worldwide as well as in Malaysia [1]. Nowadays, there are many types of hybrid metal composites reinforced with natural fiber and SiC used in the various industry [2-5]. Over the years, the great achievement has been witnessed in material engineering and material science through the development of hybrid aluminium composites from the natural green fiber. Furthermore, magnesium was added for improving wettability between the aluminium matrix and the reinforcement. Present discussion on hybrid composite ensures that the application and implementation of hybrid composites in various different fields seem to be practicable [6-7]. Previous study on the various aspects of hybrid composite application provides solid assumption regarding the influence of various different parameters on the hybrid composites performances.The method of fabricating for hybrid aluminium composites also add the credit. The development of composites production or fabrication includes the aspect of technique apply and degree of difficulty. Stir casting, powder metallurgy, spray atomization and co-deposition, plasma spraying, squeeze casting and compo-casting were the best developed methods for composite production [8-10].

2. Materials

The reinforcement used for aluminium matrix was pure SiC particles average range particles size within 30 μm and processed CSP about 425 μm consequential from controlled sieving of coconut shell. Magnesium powder was added for its role in cultivating wettability between the aluminium matrix and the reinforcement, increase strength through the solid solution and improve strain hardening ability. Hybrid materials composites were fabricated using stir casting method.

3. Methodology

The corrosion behavior was determine using Weight Gain Measurement (WGM), Electrochemical Impedance Spectroscopy (EIS) and Potentiodynamic polarization (PP. WGM was carried on followed the ASTM G31. The coupons were immersed in seawater solution accordingly for 35 days. Every 7 days interval, the coupons were withdrawn and dried at ambient temperature accordance with ASTM STP 1188, EIS test was conducted to calibrate the system impedance for a range of frequencies. The experiment was carried by Autolab PGSTAT302N controlled by NOVA software. The data from EIS test were collected and tabulated in Nyquist plot and the value of estimated polarization resistance (R_p) obtained. Meanwhile for PP, the potentials were scanned from -1.7 V to 0.01 V and a slow scan rate of 10 mV/s was used. The test was carried out at room temperature. The result from potentiodynamic polarization measurement was represented in a Tafel plot. From the experimental data obtained, the value of current density (I_{corr}) and the current potential (E_{corr}).

4. Results and discussion

It was observed that, there was an appearance of a milky white as depicted in Figure 1 for all samples as claimed by Bodunrin et al [11]. The inferior performance of single reinforced and CSP reinforced hybrid aluminium composites was clearly due to the electrochemical process during the immersion test which promoted pitting corrosion. The composites produced acting as an active anode. Corrosion of the composite is initiated by segregation of reinforcement interface, interfacial reaction, high dislocation density around reinforcement phase [15-16].

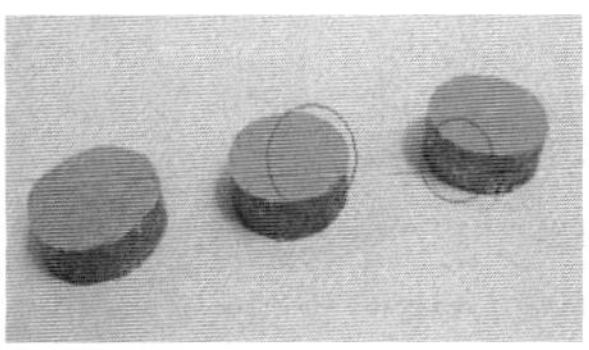

Figure 1. Milky white (corrosion deposit) appearance on the sample surfaces.

The average weight gain of the single reinforced hybrid aluminum composites for immersion week 1 is 0.0048 g followed by 0.0032 g, 0.0033 g, 0.0065 g and 0.0056 g for week 2 to week 5, respectively. The percentage of weight gain for all CSP reinforced hybrid aluminum composites is less than 0.2%. The weight gain for single reinforced hybrid aluminum composites reaches the highest at 0.34% in week 4. These results suggest that the trivial 1% weight gain of the composites is not significant. Once the weight gain is higher, it shows that the reaction on the surface is aggressive, leading to the reduction of metal thickness. Meanwhile, weight gain percentage less than 0.1%, this composite had the much longer lifespan to survive in the marine environment. Bodunrin et al. [11] revealed that the composites with calcined CSP and SiC less corrode over the time exposure. The increased weight in average range observed in CSP reinforced hybrid aluminum composites was due to the microbially influenced corrosion (MIC) on the sample surfaces [14]. As depicted on Figure 2 (i) and Figure 2(ii), the largest value of R_p on immersion week 2 is 23 944.1 Ω.cm^2 for 15% CSP-reinforced hybrid aluminum composites. In the week 5 of immersion, it is clear that 15% CSP reinforced hybrid composites recorded 22 057.4 Ω.cm^2, slightly decrease than the Week 2 reading. However, 5% CSP reinforced hybrid aluminum composites recorded the highest R_p value that is 52 551.8 Ω.cm^2 in the Week 5, increasing rapidly from 13 679.3 Ω.cm^2 in week 2. It is then followed by 15%, 10%, 20% and 0% CSP reinforced hybrid aluminum composites.

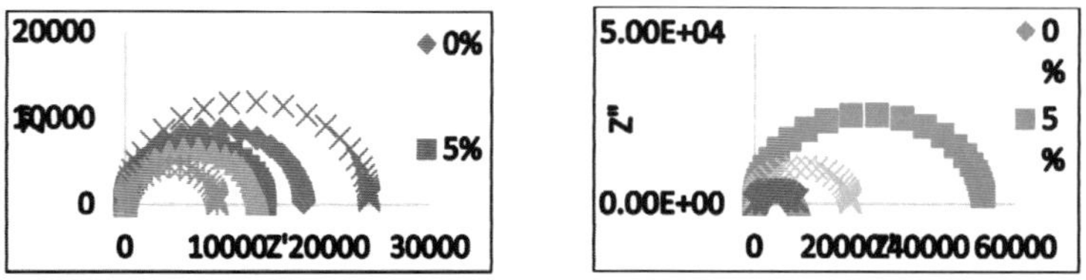

Figure 2. (i) and (ii) Nyquist plot of single reinforced and CSP reinforced hybrid aluminum composites for immersion week 2 and week 5, respectively.

The occurrence can be related to the adsorption of CSP molecules on the sample surface which aligns with the previous studies. The adsorption of CSP molecules on the sample surface influences the size of the surface coverage assigned. This happens due to the corrosion process that has been controlled by the charge transfer process [16]. Moreover, the impedance value of CSP reinforced hybrid aluminum composites is larger than the single reinforced hybrid aluminum composites due to the presence of CSP. From the EIS test, the diameter of semi-circle increases as the time for immersion test increases. In this study, the diameter of Nyquist plots starts to fluctuate, reduce and increase for CSP reinforced hybrid aluminum composites. Higher R_p value suggests better corrosion resistance as has been reported by previous researchers [17- 20]. This observation suggests that the presence of inhibitors hinder the adsorption to the metal matrix surfaces. It can be concluded that as the CSP weight percentage increase, the absolute quantity weakened. The 5% CSP reinforced hybrid aluminum composites demonstrate the highest R_p value in the week 5 immersion. As depicted in Figure 3, value for CSP reinforced hybrid aluminum composites shifted to the more electropositive region in comparison to the single reinforced sample. In particular, the Ecorr value for 0% CSP reinforced hybrid aluminum composites is -0.95764 V, followed by 15%, 20%, 10% and 5% CSP reinforced hybrid aluminum composites that shifted to a more electropositive region with reading -0.59459 V.

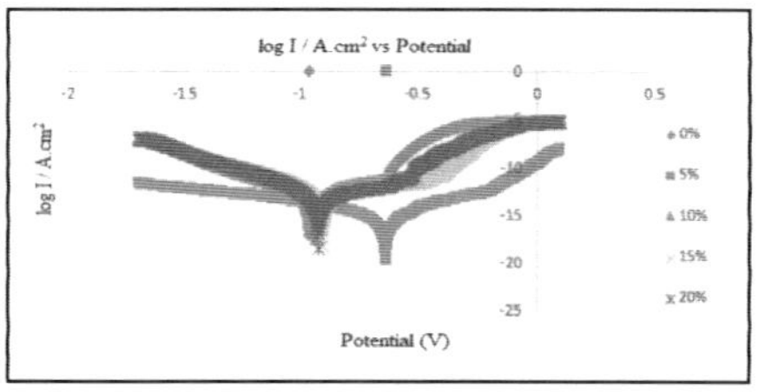

Figure 3. Tafel plot of single reinforced and CSP reinforced hybrid aluminum composites for immersion week 3.

Besides, 0% CSP reinforced hybrid aluminum composites has the least I_{corr} value of 4.28 μA/cm^2 followed by 15% (4.37 μA/cm^2), 20% (7.19 μA/cm^2), 10% (10.2 μA/cm^2) and 5% (34.7 μA/cm^2), respectively. The lowest I_{corr} value is shown by 15% CSP followed by 20%, 10% and 5% CSP respectively. There is no significant different of I_{corr} value for 0% and 15% whereas the prominent different can be seen from $\boldsymbol{E_{corr}}$. CSP reinforcement revealed that the inhibition takes place within the adsorption on the hybrid aluminum surfaces. This trend coincides with the research done by Vijayalakshmi et al [21]. As the 5% CSP reinforced hybrid aluminum composites show the lowest E_{corr} value due to the limitation amount of CSP molecules into the hybrid aluminum, it suggests that the presence of inhibitor molecules obstructs the absorption onto the hybrid aluminum surface. The work of Berenji et al [22] shows that the inhibition performance of natural fiber decreased at a specific point known as the limitation factor caused by the shifting of certain functional group among the hybrid aluminum and surroundings; the functional groups of natural inhibitor form protective films on the hybrid aluminum composites. The Tafel plot on week 3 immersion for potentiodynamic polarization mechanism is chosen because it shows the most electropositive value for E_{corr} which is reciprocal with the I_{corr} value compared to other immersion timelines. This trend aligns with Hajar et.al. [23].

5. Conclusion

It was found that through WGM, the single reinforced weighs more than CSP reinforced hybrid aluminum composites. The appearance of milky white corrosion deposit can clearly be seen due to localized and uniform corrosion that cause pitting corrosion on the composite surfaces. The inferior performance of single reinforced and CSP reinforced hybrid aluminum composites is because of the electrochemical process that happens during the immersion test that promotes localized corrosion and thus pitting corrosion. EIS study shows that 5% CSP reinforcement in hybrid aluminum composites contributes to excellent corrosion resistance performance. In EIS, 5% CSP reinforced recorded higher Rp value than others in week 5 that is about 52 551.8 Ω.cm^2, compared to week 2 with recorded value of 13 679.3 Ω.cm^2. The Rp value increases with longer immersion period. In PP measurement, 5% CSP reinforced hybrid aluminum composites shows the lowest E_{corr} values in week 1 and week 3. Both weeks show the most significant value with the lowest Ecorr values and the highest Icorr values. Overall, E_{corr} value of 5% CSP reinforced hybrid aluminum composites is shifted from electronegative region to a more electropositive region.

Acknowledgements
The authors would like to thank all the staffs of Maritime Technology Laboratory, Universiti Malaysia Terengganu (UMT) for the support throughout completing this study.

6. References

[1] Emilio B A 2018 Reliability of reinforced concrete structures subjected to corrosion-fatigue and climate changes *International Journal of Concrete Structures and Materials* **12** (10) 40-53.
[2] Alaneme K K, Bodunrin M O and Awe A A 2016 Microstructure, mechanical and fracture properties of groundnut shell ash and silicon carbide dispersion strengthened Aluminium matrix composites *Journal of King Saudi University* in Press.
[3] Alaneme K K, Ademilua B O, Bodunrin M O 2013 Mechanical properties and corrosion behavior of Aluminium hybrid composites reinforced with silicon carbide and bamboo leaf ash *Tribology in Industry* **35** (1) 25-35.
[4] Alaneme K K and Aluko A O 2012 Production and age-hardening behavior of borax pre-mixed SiC reinforced Al-Mg-Si alloy composites developed by double stir casting technique *The West Indian Journal of Engineering* **34** (1-2) 80-85.

[5] Alaneme K K and Bodunrin M O 2011 Corrosion behavior of alumina reinforced Aluminium (6063) metal matrix composites *Journal of Minerals and Materials Characterization and Engineering* **10** (12) 1153-1165.
[6] Bobic B, Mitrovic S, Babic M and Bobic I 2010 Corrosion of metal-matrix composites with aluminium alloy substrate *Tribology in Industry* **32** (1) 3-11.
[7] Boopathi M, Arulshri K and Iyandurai N 2013 Evaluation of mechanical properties of aluminium alloy 2024 reinforced with silicon carbide and fly ash hybrid metal matrix composites *American Journal of Applied Sciences* **10** (3) 219-229.
[8] Kim S W, Lee U J, Han S W, Kim D K and Ogi K 2003 Heat treatment and wear characteristics of Al/SiCp composites fabricated by duplex process *Composites Part B: Engineering* **34** 737-745.
[9] Daoud A, Abou-Elkhair M T and Rohatgi P 2004 Wear and friction behavior of near eutectic Al-Si+Zro_2 or WC particle composites *Composites Science and Technology* **64** 1029-1040.
[10] David Raja Selvam J, Robinson Smart D S and Dinaharan I 2013 Microstructure and some mechanical properties of fly ash particulate reinforced AA Aluminium alloy composites prepared by compocasting *Materials and Design* **49** 28-34.
[11] Bodunrin M O, Alaneme K K and Chown L H 2015 Aluminium matrix hybrid composites: a review of reinforcement philosophies; mechanical, corrosion and tribological characteristics *Journal of Materials Research and Technology* **4** (4) 434-445.
[12] Trowsdale A J, Noble B, Harris S J, Gibbins I S R, Thompson G E and Wood G C 1996 The influence of silicon carbide reinforcement on the pitting behavior of aluminum *Corrosion Science* **38** 177-191.
[13] Deuis R L, Green L, Subramaniam C and Yellup J M 1997 Corrosion behavior of aluminum composite coatings *Corrosion* **53** 880-890.
[14] Iwona B B, Bergel A, Mollica A, Flemming H C, Scotto V and Sand W 2000 *Simple methods for the investigation of the role of biofilm in corrosion* European Federation of Corrosion.
[15] Sekar N and Ramasamy R P 2013 Electrochemical impedance spectroscopy for microbial fuel cell characterization *Journal of Microbial and Biochemical Technology* **6** (4) 35-41.
[16] Zakaria K, Hamdy A, Abbas M and Elenien A O 2016 New organic compounds based on siloxane moiety as corrosion inhibitors for carbon steel in HCl solution: Weight loss, electrochemical and surface studies *Journal of Environmental Chemical Engineering* **1** 652-657.
[17] Sherif E M 2014 A comparative study on the electrochemical corrosion behavior of iron and X-65 steel in 4.0 wt% sodium chloride solution and after different exposure intervals *Molecules* **19** (7) 9962-9974.
[18] Alam M, Rahman M and Haque M 2007 Extraction of henna leaf dye and its dyeing effects on textile fibre *Bangladesh Journal of Scientific and Industrial Research* **42** (2) 217-222.
[19] Curioni M, Scenini F, Monetta T and Bellucci F 2015 Correlation between electrochemical impedance measurements and corrosion rate of magnesium investigated by real-time hydrogen measurement and optical imaging *Electrochimica Acta* **166** (1) 372-384.
[20] Hou J, Zhu G, Xu J and Liu H 2013 Anticorrosion performance of epoxy coatings containing small amount of inherently conducting PEDOT/PSS on hull steel in seawater *Journal of Material Science* **29** (7) 678-684.
[21] Vijayalakshmi P R, Rajalakhsmi R and Subhashini S 2010 Cocos nucifera shell as a potential inhibitor for mild steel corrosion in acidic medium *Asian Journal of Chemistry* **22** (6) 4537-4548.
[22] Berenji F, Rakhshandeh H and Ebrahimipour H 2010 In vitro study of the effects of henna extract (Lawsonia inermis) on Malassezia species *Jundishapur Journal of Microbiology* **3** (3) 125-128.
[23] Hajar H M 2017 *Performance of Henna extract as corrosion inhibitor for aluminum alloy in marine application* School of Ocean Engineering, Universiti Malaysia Terengganu.

New bio-based nanolubricants for turning of Inconel 718 towards improvement of tool wear resistance and specific cutting energy

Mohd Asyraf Mahboob Ali[1], Azwan Iskandar Azmi[2], Mohd Zahiruddin Mohd Zain[1], Muhamad Nasir Murad[1], Ahmad Nabil Mohd Khalil[1] and Mohd Shukor Abd Rahim[3]

[1]School of Manufacturing Engineering, Pauh Putra Campus, Universiti Malaysia Perlis (UniMAP), Pauh, 02000,Arau, Perlis, Malaysia.
[2]Faculty of Engineering Technology, UniCITI Campus, Universiti Malaysia Perlis, 02100,Pdg Besar, Perlis, Malaysia.
[3]Centre for Diploma Studies, UniCITI Campus, Universiti Malaysia Perlis, 02100, Pdg Besar, Perlis, Malaysia.

E-mail: azwaniskandar@unimap.edu.my

Abstract. Substitution of traditional metal cutting fluid with newly formulated bio-based nanolubricants for machining hard-to-cut material such as Inconel 718 can be advantageous in reducing the impact on environmental as well as on the overall operating or machining expenditure. Added to that, combination of nano-lubricants with Minimum Quantity Lubricant (MQL) system provides benefit towards controlling fluid consumption along with the aim of cooling and lubricating effectively at the cutting zone. Until now, research studies on the ability of bio-based nano-lubrications for cutting fluids application on hard metal alloys are still at infancy. Hence, in this work, newly formulated bio-based nano-lubrications were tested while machining Inconel 718 with uncoated carbide tool. Performances of different cutting fluids namely; Solcut oil, coconut oil, pure bio-based oil, bio-based oil + 0.2% Al_2O_3, bio-based oil + 0.5% Al_2O_3 and bio-based oil + 0.8% Al_2O_3 were evaluated with respect to tool wear and specific cutting energy. Overall, bio-based oil + 0.8% Al_2O_3 exhibited excellent wear resistant which prolong the cutting time or the tool life expectancy. Meanwhile, bio-based oil + 0.5% Al_2O_3 was superior in lowering the specific cutting energy for machining Inconel 718 as compared to other five types of cutting fluids.

1. Introduction

One of the recent research interests in improving cutting efficiency for metal cutting industries is the development of nano-lubricants. These new types of metal cutting fluids are aimed to enhance tribological characteristics during machining operation through reduction of internal and external heat generation within the cutting zone. Several studies [1–3] have been proved by the ability of nano-lubricants in many aspects of machining output. As explained by Talib and Rahim [3], internal heat generation was due to internal friction that occurs as result of material deformation process in the shear zone. Whereas, the friction due to metal-to-metal contact of chip surface and tool surface is another source of heat generation in metal cutting environment. Several research articles have highlighted that the addition of nanoparticles in a based oil led to reduction of wear and friction

IOP Conf. Series: Materials Science and Engineering **670** (2019) 012007 doi:10.1088/1757-899X/670/1/012007

mechanisms which were induced during cutting operation [1-3]. This is attributed to the formation of protective films of the smaller-size nanoparticles onto the other friction surface, rolling effect of the small spherical nanoparticles at the contact surfaces, and ability of nanoparticles to withstand compressive stress concentration due to high contact pressure [1, 4].

Many of industrial practice and previous research works combined the petroleum and synthetic based cutting lubricants with nano-particles for achieving efficiency in machining process. However, due to the adverse effects of petroleum and synthetic based cutting fluids on environment (such as toxicity and non-biodegradability) and human health, other potential alternatives are actively being explored by research communities [7]. Among many solutions, vegetable-based oil has been proposed and deemed suitable due to several important tribological characteristics. These characteristics are high viscosity index, greater molecular weight, lower volatility, higher flash point, higher boiling point, long and polar fatty acid carbon chains with rapid biodegradability [6, 7]. Furthermore, research on reformulation of additives and chemical modifications on the vegetable lubricants further improved their characteristics toward machining efficiencies [2,4,8]. Extending to this existing works, alterations on vegetable lubricants with addition of nanoparticles have also been explored by several researchers [1,2,9]. These lubricants are derived from edible and non-edible vegetable oils and possess high lubricity, viscosity index and flash point. Since they come from natural resources, bio-based oils can be class as renewable, low toxicity and biodegradability. However, these lubricants have poor cold flow properties and oxidation stability, which leads to polymerisation and degradation. The aforementioned drawbacks can be somehow overcome by chemical modifications of vegetable oils or by special formulation of fluids [10-11]. Thus, the present study aims to elucidate the effects of newly formulated bio-based nanolubrications on the growth of wear on cutting tool and specific cutting energy in the turning of Inconel 718 with uncoated cemented carbide inserts.

2. Experimental method

2.1. Preparation of bio-based lubricants

Three main ingredients, i.e. coconut oil, water and aluminum oxide (Al_2O_3) nanoparticles were used to prepare the bio-based lubricants. The amount of ingredient and method of mixing were decided based on preliminary experiments, which were repeatedly conducted until a stable cutting fluid was obtained. A batch of four bio-based lubricants consisting of 50% oil-emulsion and 50% of water were prepared. Virgin coconut oil was used as the based lubricant with two additives were added to it so as to produce the oil-emulsion. Coco-amido propylbetaine, CAPB was used as an emulsifying and bacterial control agent, while sodium dodecyl benzene sulfonate, SDBS as a thickener agent. The oil-emulsion was prepared at 2:1:0.5 ratio by volume fraction. CAPB and SDBS were first mixed using a magnetic stirrer at 350 rpm. Coconut oil was then slowly added into the solution and the mixing process continued for 10 minutes. Subsequently, distilled water was added into the oil-emulsion and the mixing speed was increased to 500 rpm. The pH value of lubricant was kept at 9.2 with addition of amino methane (Tris) as a pH buffer agent. Three out of the four lubricants were added with aluminium oxide (Al_2O_3) nanoparticles at 0.2%, 0.5% and 0.8% of concentration prior to sonication process of 3 hours at 15% amplitude.

2.2. Lubricant properties

Density, viscosity index (VI), kinematic viscosity and thermal conductivity for all samples were measured according to ASTM D4052, ASTM D2270, ASTM D445 and ASTM D5334, respectively. Density was determined by the ratio of weight over volume of oil using a pycnometer at 15°. Kinematic viscosity was measured by a viscometer machine at 40° and 100°. Viscosity index was calculated based on the results of kinematic viscosity at both tested temperatures. Thermal conductivity at three different temperature; 40°, 70° and 100° were measured using KD2 Pro thermal analyser. All measurements were repeated for three times before the average value were recorded in Table 1.

Table 1. Physicochemical properties and thermal conductivity

Coolant	pH	Density (g/cm^3)	Viscosity index (VI)	Kin. Vis @ 40°C	Kin. Vis @ 100°C	Thermal Conductivity (W/mK)		
						40°C	70°C	100°C
SolCut	8.17	0.909	71.5	14.96	3.25	0.233	0.215	0.204
Bio-based	6.1	0.913	130	24.9	5.12	0.187	0.163	0.154
Bio-based 0.2% Al_2O_3	9.2	0.985	1058.12	24.7	41.86	0.328	0.302	0.300
Bio-based 0.5% Al_2O_3	9.2	0.983	1109.91	22.32	40.05	0.333	0.355	0.352
Bio-based 0.8% Al_2O_3	9.2	0.986	1226.37	18.93	37.65	0.346	0.363	0.356
Coconut oil	9.2	0.989	1342.83	15.54	35.24	0.352	0.367	0.363

2.3. CNC machines, turning inserts and workpiece material
Machining experiments were conducted using Chevalier FCL-608 CNC turning machine with 2-axis in the x and z directions. The inserts used for the experiments were Sandvik CNMG 120408 H13A uncoated carbide inserts with nose radius of 0.794 mm. The toolholder was ECLNR-2020K12 type from CHAIN with (-7) rake angle. The turning process was performed on 50-mm-diameter and 100-mm-length Inconel 718 workpieces.

2.4. Lubricating condition and cutting parameters
All cutting processes were carried out under minimum quantity lubrication (MQL) condition. To ensure effective delivery of lubricant via MQL system, a 6-bar inlet air pressure was set. Based on this air pressure, the flow rate for the MQL was determined to be 80 ml/hr. The machining conditions were; 80 m/min cutting speed, 0.1 mm/rev feed rate and 0.1 mm depth of cut. During the cutting process, 0.2 mm diameter of material was removed in each cutting length of a 30 mm. The process was continued until 1 mm of the diameter of the workpiece has been removed.

2.5. Experimental measurement procedure
Tool wear and specific cutting energy are the machining outputs that have been selected. Specific cutting energy result were obtained through calculation based on following equations 1-3. For each cutting process, tool wear growth was consistently measured at a predetermined interval using Xoptron XST60 stereomicroscope at 35x optical magnifications.

$$MRR = 3.142 * D_{AVG} * d * f * N \quad (1)$$
$$Power = 3.142 * Velocity \quad (2)$$
$$Specific\ cutting\ energy = Power/MRR \quad (3)$$

Where; d: Depth of cut, f: Feed rate, N: Spindle speed, MRR: Material removal rate

3. Results and Discussion

3.1. Tool wear
The trend of tool wear growth (Figure 1) shows a typical tool wear curve for metal cutting process, in which tool wear increases with respect to cutting time. In accordance with ISO 3685:1993, the tool wear criterion of 0.3 mm was selected to determine the tool useful life. In specific, the tool useful life was decided when the maximum width of the flank wear land was equal or greater than a V_B of 0.3mm. Among the six types of cutting fluids, bio-based oil with 0.8% Al_2O_3 nanoparticle was effective and efficient for tool wear resistant as it recorded the longest cutting time of 43.3 min, Table

2. Whereas, the cutting time for bio-based oil + 0.8% Al_2O_3, bio-based oil + 0.5% Al_2O_3, bio-based oil + 0.2% Al_2O_3, bio-based oil, and coconut oil recorded an increase of 40.2%, 26.8%, 7.4%, 7.3% and 2.1%, respectively, as compared to that of the SolCut oil. The results showed that the benefits of using bio-based oil with 0.8% Al_2O_3 nanoparticles is with respect to a lengthened tool life for all test conditions. The cooling and lubrication characteristics of bio-based nanofluids improved the lubrication and wetting characteristics at the flank regions, and accordingly, it provided better heat and friction dissipation as discussed by previous studies [1,4,12,13]. As can be seen from Figure 2, friction and heat development generated from the friction between tool and workpiece, led to abrasion and material adhesion onto the cutting edge which causes the generation of BUE. Plus, the occurrence of notch wear was also restrained.

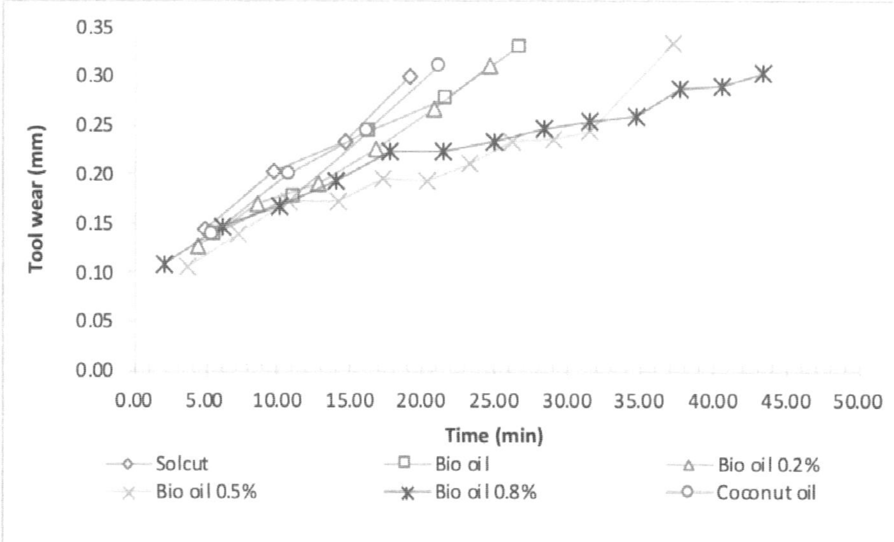

Figure 1. Tool wear progression against cutting time for difference cutting lubricants.

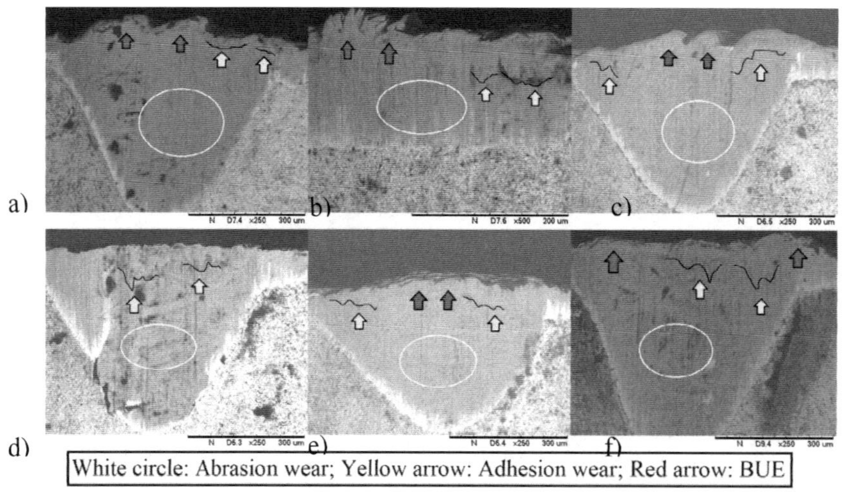

Figure 2. Flank surface of cutting tool; a) Solcut, b) Bio-based, c) Bio-based 0.2% Al_2O_3, d) Bio-based 0.5% Al_2O_3, e) Bio-based 0.8% Al_2O_3 and f) Coconut oil.

IOP Conf. Series: Materials Science and Engineering **670** (2019) 012007 doi:10.1088/1757-899X/670/1/012007

3.2. Specific cutting energy and coefficient of friction

The energy consumed in removing a unit volume of material is defined as the specific cutting energy, SCE (k). k is calculated by dividing the cutting power (W) with the volume of material removed, MRR (mm^3/s). Understanding value of k is an alternative approach to quantitatively measuring the efficiency of the metal cutting process or the machinability of a workpiece. Specific cutting energy is material dependent, as well as related to the cutting tool, machining parameters and type of machining; hence, it can represent very well the cutting phenomenon. Table 2 represents the value of k for all the lubricating condition. It was found that value of k was the highest in SolCut condition with 8.57 Ws/mm^3 and the lowest was for bio-based + 0.5% Al_2O_3 with 7.47 Ws/mm^3.

Table 2. Tool life and specific cutting energy for different cutting fluid

Coolant	Tool life (min)	Specific cutting energy (Ws/mm^3)
SolCut	19.20	8.57
Bio-based	23.56	8.50
Bio-based 0.2% Al_2O_3	23.62	8.38
Bio-based 0.5% Al_2O_3	35.28	7.47
Bio-based 0.8% Al_2O_3	43.30	7.54
Coconut oil	20.43	8.35

4. Conclusion

In this work, the cooling and lubrication capabilities of new formulated bio-based oil with addition of Al_2O_3 nanoparticles have been tested in machining of Inconel 718. Specifically, the tests have been carried out to study the effects on tool wear resistance and specific cutting energy for different formulation of bio-based oils. For tool wear; bio-based oil with 0.8% Al_2O_3 nanoparticles exhibited excellent wear resistant which prolong the tool life expectancy. Whereas, the inclusion of 0.5% Al_2O_3 nanoparticles in the bio-based oil was found superior in reducing the specific cutting energy. The effective cooling and lubrication properties of the new bio-based nanofluids were capable in lowering the friction and heat generation during machining hard to cut alloy.

Acknowledgements

The authors gratefully acknowledge the financial support of the Ministry of Higher Education Malaysia through Fundamental Research Grant Scheme (FRGS) no: FRGS/1/2016/TK03/UNIMAP/03/2.

5. References

[1] Talib N and Rahim E A 2018 *Tribol. Int.* **118** 89–104.
[2] Rapeti P, Krishna V, Mohana K and Gurram R 2018 *J. Clean. Prod.* **172** 2862–2875.
[3] Khalil A N M, Ali M A M and Azmi A I 2015 *Procedia Manuf.* **2** 130–134.
[4] Padmini R, Vamsi Krishna P and Krishna Mohana Rao G 2016 *Tribol. Int.* **94** 490–501.
[5] Sharma P, Sidhu B S and Sharma J 2015 *J. Clean. Prod.* **108** 72–79.
[6] Srikant R R and Ramana V S N V 2015 *J. Clean. Prod.* **108** 104–109.
[7] Kumar Gajrani K and Ravi Sankar M 2017 *Mater. Today Proc.* **4** 3786–3795.
[8] Tazehkandi A H, Shabgard M and Pilehvarian F 2015 *J. Clean. Prod.* **104** 422–435.
[9] Uysal A, Demiren F and Altan E 2015 *Procedia - Soc. Behav. Sci* **195** 2742–2747.
[10] Mobarak H M *et al.* 2014 *Renew. Sustain. Energy Rev.* **33** 34–43.
[11] Zainal N A, Zulkifli N W M, Gulzar M and Masjuki H H 2018 *Renew. Sustain. Energy Rev.* **82** 80–102.
[12] Sayuti M, Sarhan A A D and Salem F 2014 *J. Clean. Prod.* **67** 265–276.
[13] Ali M A M, Azmi A I, Khalil A N M and Leong K W 2017 *Int. J. Adv. Manuf. Technol.* **92** 117–127.

IOP Conf. Series: Materials Science and Engineering **670** (2019) 012008 doi:10.1088/1757-899X/670/1/012008

Effect of metal cutting fluid (MCF) conditions on tool wear and thrust force in drilling Ti-6Al-4V using coated carbide tools

R A Rashid[1], M N Murad[2], R Hamidon[2] and S Shariffuddin[2]

[1]Centre for Diploma Studies, Universiti Malaysia Perlis, Kampus UniCITI Alam, 02100 Padang Besar, Perlis, Malaysia.
[2]School of Manufacturing Engineering, Universiti Malaysia Perlis, Kampus Tetap Pauh Putra, 02600 Arau, Perlis, Malaysia.

Abstract. Titanium alloy (Ti-6Al-4V) is well known as difficult to cut material due to their low machinability rating. Machining Ti-6Al-4V generate high cutting temperature close to the cutting edge of the tool. The high cutting temperature, mostly during the deformation process and friction at the tool–chip and tool–workpiece interfaces. This circumstances will lead to the rapid tool wear. This study focuses on the growth of tool wear and development of the thrust force. Drilling tests were conducted using coated (TiAlN) carbide tool with diameter of 6 mm under two different condition of metal cutting fluid (MCF). The MCF used were minimum quantity lubrication (MQL) and minimum quantity cooling lubrication (MQCL). The outcomes of this study indicates MQL performed better coolant-lubricant effect during drilling Ti-6Al-4V as the results of thrust force development are consistent with the trends of tool wear growth.

1. Introduction

Ti-6Al-4V offer better strength than pure titanium and has been extensively used in aerospace industry due to their outstanding resistance to corrosion at high operating temperature. Also, their bio-compatibility, low elasticity and increased resistance to corrosion make this alloy desirable for bio-medical industry [1-3]. High cutting temperature were generated during machining Ti-6Al-4V close to the cutting edge of the tool. Low thermal conductivity of the material play a critical role toward the heat dissipation process, thus result to rapid tool wear and shorter tool life.

MCF are employed to reduce the friction or wear between tool-workpiece interfaces and carried away the heat generated in the cutting zone. MCF used also enhanced machining performance during drilling process. However, the application of MCF received a lot of question regarding to disposal, environmental and health issues, also lead to economical overhead [4-5]. As an alternative, MQL techniques and an extension of the MQL system, MQCL has been introduced to tackle issues associated with conventional flood cooling. Both technique used extremely low quantity of oil flow rates to reduce friction at the tool-chip and tool workpiece contacting regions. In addition, MQCL system used lubricant combined with compressed air to enhance the cooling action.

Other than environmental aspect, MQCL and MQL conditions improves machinability as well as prolong tool life, reduce cutting force, improves heat dissipation and surface roughness. MQCL condition has shown to be effective in reducing the temperatures and facilitate chip breaking. With the aid of cold air combine with MQL system, the condition lowered temperature rise during drilling process [6-7].

2. Experimental Details

A three axis CNC milling machine (Tong Tai EZ-5) were used to conduct drilling test under two MCF conditions as shown in Figure 1. Ti-6Al-4V size 94 x 70 x 14 mm was selected to perform the test. The plate was secure tightly on a special design jig to avoid chatter. Drilling tests were carried out using 6 mm diameter of coated (TiAlN) carbide tools with varying cutting speeds of 65, 75, 85 and 95 m/min. The feed rate were remain constant at 0.02 mm/rev. Table 1 shows the machining parameter in details.

Thrust force was measured using Kistler (Type 9129) dynamometer with charge amplifier. Xoptron stereomicroscope XST60 with IMT Mini 3 software was used to measure the growth of tool wear. The measurement of flank wear was made for every hole and stopped when any of the following criteria had reach.

i. Maximum flank wear, $V_{max} \geq 0.2$ mm
ii. Chipping ≥ 0.2

iii. Fracture or catastrophic failure

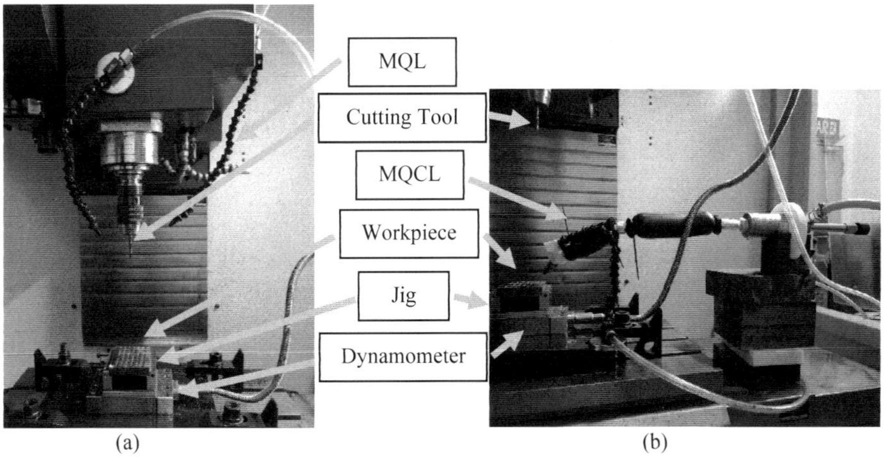

Figure 1. (a) Drilling with MQL; (b) Drilling with MQCL.

Table 1. Machining parameter

Items	**Descriptions**
Machine Tools/Equipment	3-axis CNC Milling Machine (Tong Tai – EZ 5)
Workpiece Material	Titanium alloys (Ti-6Al-4V)
Cutting Tool	WC-Co (TiAlN coated)
Point Angle (PA) & Helix Angle (HA)	140° & 30°
Cutting speed (m/min)	65 , 75, 85, 95 m/min
Feed rate (mm/rev)	0.02 mm/rev
Tool diameter	6 mm
Type of Cutting	Through hole (peck drilling)
MCF conditions	i. MQL** ii. MQCL** *** Both method using Karanja oils as medium lubrication*

3. Results and Discussion

3.1. Tool Wear

The effect of MQL and MQCL on the tool wear at four employed cutting speed is shown in Figure 2 and Figure 3 respectively. Tool wear, V_{bmax} was measured along the main cutting edge of the carbide tool after every drilled hole for both MQL and MQCL conditions. According to Asyraf [8], flank wear has direct influence on the machining cost and product quality. Once tool life has reach the criterion as stated earlier, the number of hole can be drilled at cutting speed of 65 m/min, 75 m/min, 85 m/min and 95 m/min under MQL condition are 9, 8, 7 and 4 holes while under MQCL are 10, 6, 5 and 4 holes respectively. The results show, cutting speed of 65 m/min achieved the highest number of holes compared to other cutting speed under both MQL and MQCL condition. The longest tool life is obtained under MQL at cutting speed 75 m/min, 85 m/min and 95 m/min as compared to MQCL. It can be said that MQL performed better than MQCL at cutting speed 75m/min and above. The result was in line with Raza [9] on tool wear pattern when machining Ti-6Al-4V, MQCL gave better performance in term of tool wear at lower cutting speed.

Figure 4, shows typical tool failure mode of cutting tool for both MQL and MQCL condition. The result show cutting tool experienced maximum flank wear, chipping, fracture and catastrophic failure during drilling experiment. Chipping was common phenomenon during drilling at higher cutting speed due to the immediate loss of sharp cutting edge. Also, adhesion wear has been observed to be the dominant mechanisms for tool failure at all cutting speeds due to the ability of the Ti-6Al-4V to retain its strength at elevated temperature.

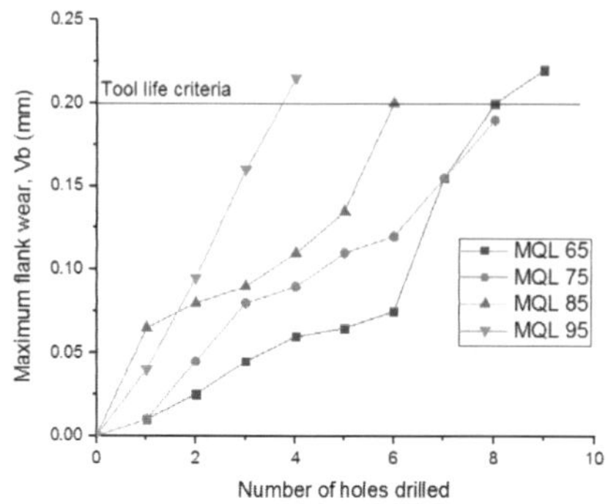

Figure 2. Maximum flank wear under MQL condition using coated (TiAlN) carbide drill

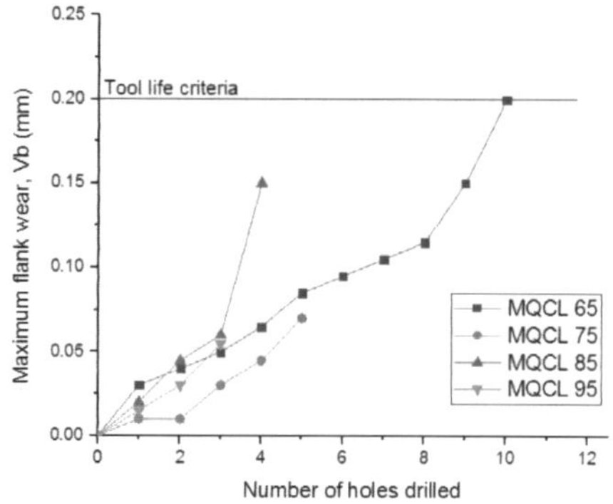

Figure 3. Maximum flank wear under MQCL condition using coated (TiAlN) carbide drill

Cutting Speed	MQL	MQCL
65 m/min	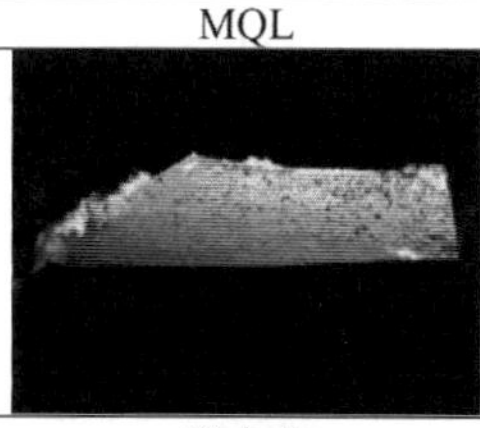 Hole 9	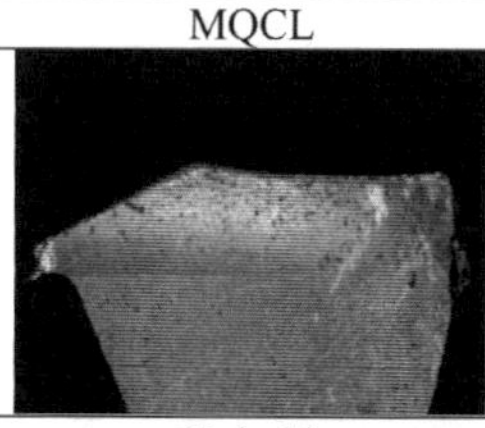 Hole 10
75 m/min	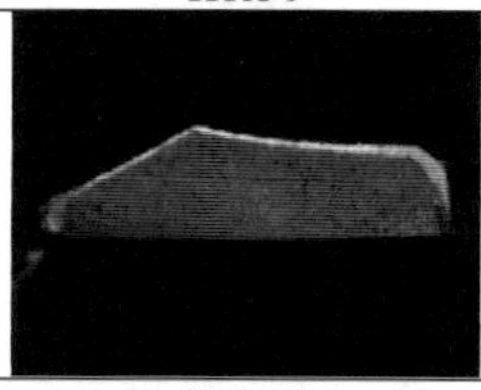 Hole 8	 Hole 6 (Fracture)
85 m/min	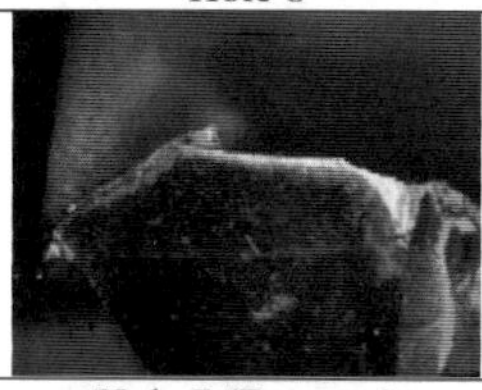 Hole 7 (Fracture)	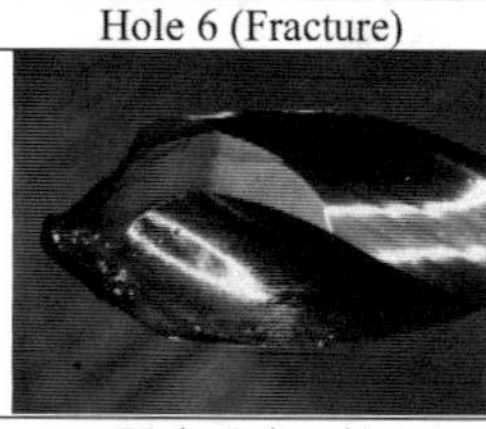 Hole 5 (break)
95 m/min	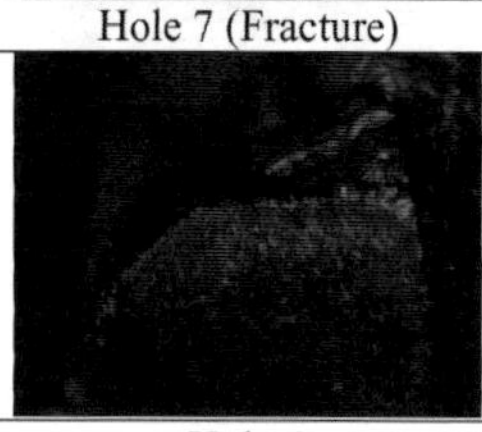 Hole 4	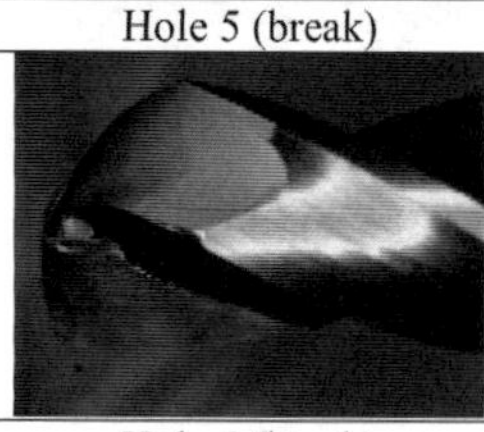 Hole 4 (break)

Figure 4. Typical pattern of tool wear under various cutting speed and feed 0.02 mm/rev under MQL and MQCL condition using coated (TiAlN) carbide drill

3.2 Thrust force

Figure 5 shows the progression of thrust force against number of drilled hole for both conditions, MQL and MQCL. It can be seen that thrust force increase with the increase of number of hole. This situation is due to the effect of flank wear. Thrust force increase as the tool wear increase as reported by Lin [10] on tool wear monitoring in drilling using force signals. The thrust force range from 100N to 450N. At first stage, under MQCL condition with cutting speed 95 m/min resulted lower thrust force of 104N compared to MQL condition 116N.

The result for thrust force versus cutting speed is shown in Figure 6. From the result, it indicates that cutting speed have significant effect on thrust force. The trend of thrust force is seem to be higher at lower cutting speed, and decrease as cutting speed increase. Thrust force increased as cutting speed decrease due to the difficulty to cut material, especially Ti-6Al-4V. According to Caliskan [11], at higher cutting speed, cutting temperature increase and soften the material, thus resulted to lower thrust force during machining process. Overall, during drilling process under MQL condition, temperature increase due to high cutting speed. Higher temperature facilitate material to be cut, thus lower the thrust force. , under MQCL condition, cold air lowered the temperature during drilling process.

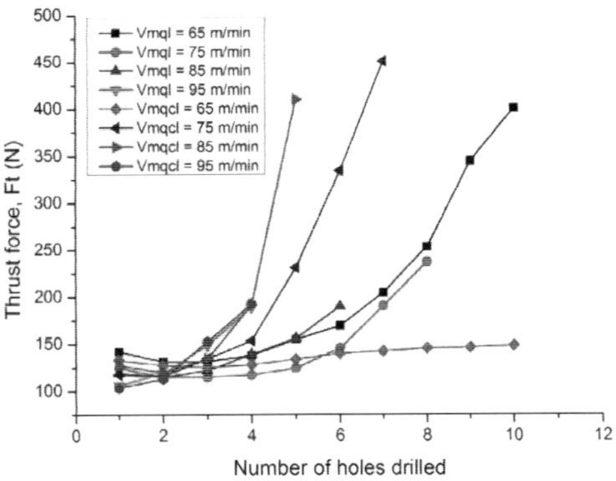

Figure 5. The progression of thrust force for MQL and MQCL conditions under various cutting speeds using coated (TiAlN) carbide drill.

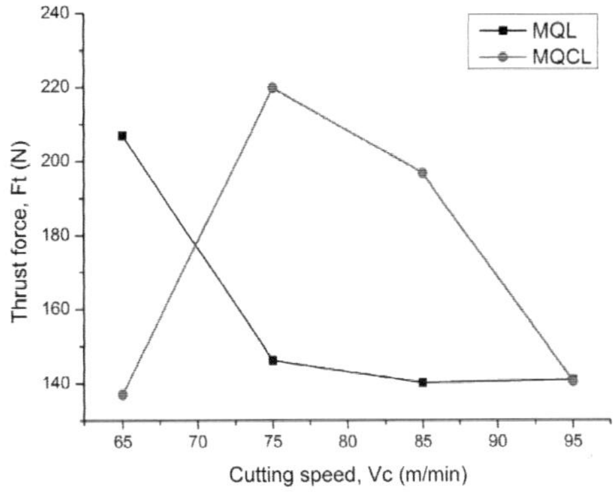

Figure 6. Thrust force comparison under MQL and MQCL conditions under various cutting speeds using coated (TiAlN) carbide drill.

4. Conclusion

MCF conditions of MQL and MQCL were tested in drilling Ti-6AL-4V. The results of this study indicated MQL and MQCL have significant influence on tool wear as well as the development of thrust force. At higher cutting speed, 75m/min and above, MQL provide longer tool life and tool wear compared to MQCL during drilling Ti-6AL-4V. As for thrust force, similar decreasing trend were display as cutting speed increase. The difference thrust force was observed for both condition. Lower thrust force was obtained under MQL condition.

References

[1] Guo Y B, Li W and Jawahir I S 2009 Surface integrity characterization and prediction in machining of hardened and difficult-to-machine alloys: A state-of-art research review and analysis *Machining Science and Technology* **13** (4) 437–470.

[2] Zhu Z, Sui S, Sun J, Li J and Li Y 2017 Investigation on performance characteristics in drilling of Ti6Al4V alloy *The International Journal of Advanced Manufacturing Technology* **93** (1–4) 651–660.

[3] Sharif S, Abd E and Sasahar H 2012 Machinability of Titanium Alloys in Drilling *Titanium Alloys-Towards Achieving Enhanced Properties for Diversified Applications* (Croatia : InTech)

[4] Ginting Y R, Boswell B, Biswas W K and Islam M N 2016 Environmental Generation of Cold Air for Machining *Procedia CIRP* **40** 648–652.

[5] Shokrani A, Dhokia V and Newman S T 2012 International Journal of Machine Tools & Manufacture Environmentally conscious machining of difficult-to-machine materials with regard to cutting fluids *International Journal of Machine Tools and Manufacture* **57** 83–101.

[6] Maruda R W, Legutko S, Krolczyk G M and Raos P 2015 Influence of Cooling Conditions on the Machining Process under MQCL and MQL Conditions *Tehnički vjesnik* **22** (4) 965–970.

[7] Pervaiz S, Rashid A, Deiab I and Nicolescu C M 2016 An experimental investigation on effect of minimum quantity cooling lubrication (MQCL) in machining titanium alloy (Ti6Al4V) *International Journal of Advanced Manufacturing Technology* **87** (5–8) 1371–1386.

[8] Asyraf M, Ali M, Azmi A I, Nabil A, Khalil M and Leong K W 2017 Experimental study on minimal nanolubrication with surfactant in the turning of titanium alloys *The International Journal of Advanced Manufacturing Technology* **92** (1–4) 117–127.

IOP Conf. Series: Materials Science and Engineering **670** (2019) 012008 doi:10.1088/1757-899X/670/1/012008

[9] Raza S W, Pervaiz S and Deiab I 2014 Tool Wear Patterns When Turning of Titanium Alloy Using Sustainable Lubrication Strategies *International Journal of Precision Engineering and Manufacturing* **15** (9) 1979–1985.

[10] Lin S C and Ting C J 1995 Tool wear monitoring in drilling using force signals *Wear* **180** (1–2) 53–60.

[11] Caliskan H and Altas E 2015 The Effect of Cutting Conditions on Cutting Forces In Milling of Ti6Al4V Alloy *International Journal of Engineering, Technology and Natural Sciences* **1** (1) 13-22

Evaluation of delamination damage and surface roughness in end milling flax fibre composites

A R Mustafa[1], A I Azmi*[1], M S Zakaria[1] and T C Lih[2]

[1]Faculty of Engineering Technology, UniCITI Campus, Universiti Malaysia Perlis, 02100, Pdg Besar, Perlis, Malaysia.
[2]School of Manufacturing Engineering, Pauh Putra Campus, Universiti Malaysia Perlis, Pauh, 02000, Arau, Perlis, Malaysia.

E-mail: *azwaniskandar@unimap.edu.my

Abstract. Flax fibre reinforced composites have been very attractive in numerous applications such as wall and automotive panels. With chemical treatments and modifications, their mechanical competent can be at par with the existing synthetic fibre composites. The flax fibre composites are normally fabricated to near-net shape, but very often these composites require a secondary cutting process such as milling to attain final geometrical and dimensional tolerances. Due to the continuous contact of the milling tool on the composite material during cutting process, several undesirable damage such as delamination, matrix cracking, fibre pull-out; existed which may lead to deterioration of their mechanical strength. In this study, parametric effects of milling parameters, namely; feed rate, spindle speed, and number of end mill flutes on delamination damage and surface roughness, were investigated. Taguchi L18 orthogonal array was used to design the experimental plan. Signal to Noise (S/N) ratio and response tables were implemented to analyse the experimental data. It was found that spindle speed and feed rate had equal effects on delamination damage and surface roughness, whereas the number of end mill flute had a marginal influence on the aforementioned machining outputs.

1. Introduction

Natural fibres are substances which are derived from plants and animals. These fibres can be spun into filament, thread or rope and further weaved into woven, knitted, matted or bound types. In recent years, natural fibres have been widely used as reinforcement materials in many types of polymer matrices [1]. Apart from having competitive mechanical properties, these fibres are 100% sustainable compared to their synthetic fibre counterparts. Hence, the application of these natural fibres for polymer composites have been increasing and becoming as an alternative or substitute for the conventional composite materials made from synthetic fibres such as E-glass and carbon fibres. Among many types of natural fibres, the use of flax fibres as reinforcement in polymer composites has gained popularity due to their high lignocellulosic contents [2]. This characteristic leads to decent mechanical properties that are comparable to that of the synthetic fibre reinforced composites.

It is widely known that the fabrication technique for the natural fibre composites follows the same approach as to that of synthetic fibre composites. Very often, these composites are fabricated near to the final shape, i.e. close to the finished product and as different assembly parts [3]. Due to this, machining operations are commonly employed to meet dimensional, design tolerance and assembly requirements. End milling is one of the machining practices highly employed as integral parts for composite manufacturing sequences. However, due to the continuous contact of the milling tool on the composite material during cutting process, several undesirable damage such as delamination, matrix cracking, fibre pull-out; existed [4]. Delamination is failure in a laminated material due to the separation of the layers of fibre reinforcement or fibre plies. Delamination failure that is due to machining process can be of several types, such as, fracture within the adhesive or resin, fracture within the reinforcement or debonding of the resin from the reinforcement. A number of factors govern the extent of end milling-induced damage in fibrous composite materials, which include cutting parameters, geometry and diameter of the tools, and also their material types [5]. Some studies in milling of natural fibre composites existed and recommended a low feed rate and a high cutting speed for reduction of delamination [6]. However, due to variations in the mechanical properties of natural fibre composites, inconsistent results were reported with respect to the milling induced delamination damage on the natural fibre composites in the prior experimental studies [6-7]. Therefore, in this paper, the main effects of end milling parameters such as cutting speed, feed rate and end mill flute on the delamination and surface roughness are studied through statistical analyses. Wide range of milling parameters were employed to distinguish the parametric effects of the aforementioned control factors.

2. Experimental method

2.1 Material details

In this study, woven type flax fibres and epoxy resins were used to fabricate the composite workpiece. Initially, the workpiece had a size of 300 x 300 mm at the fabrication stage. Later, machining samples of 94 mm × 70 mm as depicted were trimmed from that bigger plate. Each samples have a thickness of sample approximately 5 ± 0.5 mm. This thickness was achieved using five layers of woven flax fibres. Fabrication was carried out using vacuum assisted resin transfer moulding.. The flax fibre reinforced composites were tested for their following mechanical properties. It was found that the tensile strength and modulus were in the range of 60–80 MPa and 4.00–5.00 GPa, respectively.

Figure 1. The flax fibre reinforced epoxy composite.

2.2 Machining setup, cutting tool and experimental procedure

End mill slots were cut using a three axis CNC milling machine, by Tongtai EZ-5A model with FANUC controller. A high speed steel (HSS) end mill was used due to the fact that the effect of abrasive wear and friction from the flax fibres on the end mill can be very minimal. Two types of flutes were 2 and 4 number of flutes, in which these number of flutes are commonly used in research studies and also industries. Each sample of flax fibre reinforced epoxy composite was mounted on a vise prior to the machining experiments. End milling was performed under dry environment. After the milling experiments have completed, there were several distinguishable delamination damages that can be observed on top of the end milled slot. Xoptron XST60 stereomicroscope, was used to quantify these delamination damage zones (via delamination factor) around the milled slots. Meanwhile, the surface roughness of the end mill slots was determined using a Mitutoyo surface roughness measurer. Measurement was carried out on five positions along the slot and average value of surface roughness was calculated for subsequent experimental analyses.

IOP Conf. Series: Materials Science and Engineering **670** (2019) 012009 doi:10.1088/1757-899X/670/1/012009

2.3 Design of experiment

Taguchi method was implemented for execution of this experimental study. As mentioned in section B, there were 3 controlled factors such as number of flutes, feed rate and spindle speed, which were important for the parametric analysis. Table 1 shows these parameters and their levels.

Table 1. Parameters and their level the experiment for 2 and 4 fluted end mills.

Parameter	Level 1	Level 2	Level 3
(A) Number of flutes	2	4	–
(B) Spindle speed (rev/min)	1000	2000	3000
(C) Feed rate (mm/rev)	0.1	0.2	0.3

Based on parameter and level shown above, Taguchi L_{18} orthogonal array was used to apply the experiment layout. Taguchi L_{18} orthogonal array is a special type of array that allows the implementation of mixed level of experimental parameters. For the data analysis, signal-to-noise (*S/N*) ratios were calculated in order to identify the quality characteristics that were either nearing or deviating from desired value. The lower the better characteristic was chosen for the determination of *S/N* ratio. The formula is given by:

$$S/N_s = -10\log \frac{\sum_{i=1}^{n} y^2}{n}$$

Where, y = experimental value; n = no. of experiment.

Apart from the *S/N* ratio, statistical analyses were used to identify the significant factors that affect the experimental response. These analyses are presented in term of response tables.

3. Results and Discussion

3.1. Delamination factor and surface roughness

The outputs for both and delamination factor and surface roughness were obtained from experiment and tabulated in Table 2. The values for surface roughness were obtained through measurement of R_a using surface roughness measurer on the milled surface, whereas the delamination factors were obtained based on ratio between damage width to the required width, $F_d = W_{max}/W$, Figure 1. It is apparent that the highest surface roughness value, R_a for the experiment was 4.82 μm and the lowest value, R_a is 2.33 μm for experiment number 7 and number 16, respectively. As for the delamination factor, F_d, the highest value of delamination factor for top side is 1.08 mm whereas 1.01 mm is the lowest value of delamination factor.

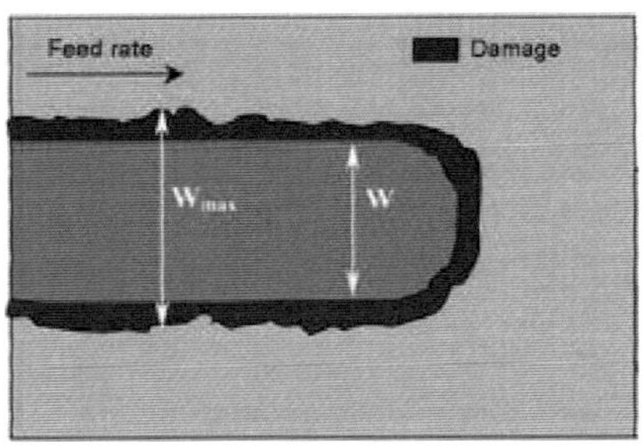

Figure 2. Schematic of delamination factor measurement.

Table 2. Delamination factor and surface roughness values from Taguchi L_{18} experiments

Exp	Number of Flute	Spindle Speed (RPM)	Feed rate (mm/rev)	Delamination Factor, F_d	S/N Ratio for F_d	Surface Roughness, R_a (μm)	S/N Ratio for Ra
1	2	1000	0.1	1.03	-0.29	3.51	-10.91
2	2	1000	0.2	1.06	-0.47	2.83	-9.03
3	2	1000	0.3	1.01	-0.10	2.65	-8.47
4	2	2000	0.1	1.06	-0.52	2.76	-8.82
5	2	2000	0.2	1.03	-0.23	3.23	-10.18
6	2	2000	0.3	1.06	-0.51	3.79	-11.58
7	2	3000	0.1	1.08	-0.67	4.82	-13.66
8	2	3000	0.2	1.04	-0.37	3.76	-11.49
9	2	3000	0.3	1.04	-0.32	2.94	-9.37
10	4	1000	0.1	1.04	-0.37	3.22	-10.17
11	4	1000	0.2	1.03	-0.22	3.62	-11.16
12	4	1000	0.3	1.03	-0.29	4.15	-12.35
13	4	2000	0.1	1.02	-0.19	4.30	-12.68
14	4	2000	0.2	1.01	-0.10	3.23	-10.19
15	4	2000	0.3	1.01	-0.10	2.87	-9.15
16	4	3000	0.1	1.05	-0.41	2.33	-7.35
17	4	3000	0.2	1.03	-0.25	2.43	-7.69
18	4	3000	0.3	1.05	-0.42	2.85	-9.09

3.2. Response table

Table 3 shows the response table for S/N ratio of surface roughness at each factor level. It can be seen that spindle speed and feed rate contributed equally on the surface roughness based on the calculated delta values (the difference between maximum and minimum value of each factor level). The lowest delta value in this table is 0.41 corresponding to the number of flute factor. This implies that this factor has the smallest influence on the surface roughness measurement as compared to the other two drilling factors. Consistent results were depicted for the delamination factor output as depicted in Table 4.

Table 3. Response table for S/N ratio of surface roughness.

Factors	Level 1	Level 2	Level 3	Delta	Rank
Number of flutes	-10.3909	-9.9809	-	0.4100	3
Spindle speed	-10.3491	-10.4323	-9.7763	0.6561	1
Feed Rate	-10.5974	-9.9580	-10.0023	0.6394	2

Table 4. Response table for S/N ratio of delamination factor.

Factor	Level 1	Level 2	Level 3	Delta	Rank
Number of flutes	-0.3864	-0.2616	-	0.1247	3
Spindle speed	-0.2897	-0.2760	-0.4062	0.1302	2
Feed Rate	-0.4072	-0.2728	-0.2919	0.1344	1

The aforementioned results appeared to be consistent with the reported study by Celik et. al. when the authors end milled jute fibre reinforced polymer composites [7]. According to the authors, at high feed rates, the end mill had insufficient time to cut most of the fibres and the uncut fibres remained intact due to softer nature of the flax fibres and also failure of the polymer metrics [7]. As a result, delamination damage occurred on each side of the cutting slot. At the same time, the increase in cutting speed also deteriorated the delamination on flax fibre composites. This was mainly attributed to the elevated cutting temperature due to friction at the contact surfaces of the tool and the flax fibre composite. Polymer matrix failed due to the increased cutting temperature, and hence the bond strength between the fibre and the matrix was reduced that lead to severe delamination [7]. Summaries of percentage contribution of R_a and F_d are shown in Figure 3.

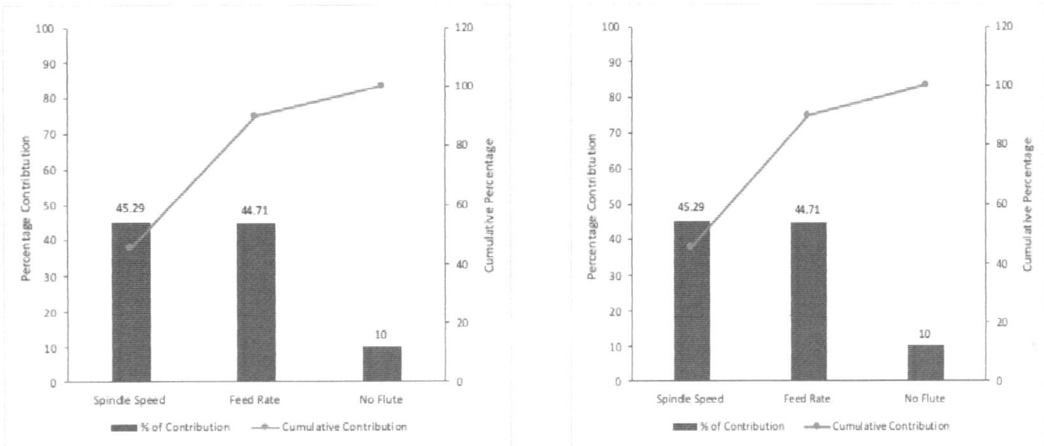

(a) (b)

Figure 3. Pareto ANOVA for pecentage contribution of experimental parameters on (a) R_a and (b) F_d.

4. Concluding Remarks

In this study, the effects of cutting parameters such as cutting speed and feed rate on delamination factor and surface roughness were evaluated in end milling process using two and four number of flutes end mill tools. The results can be summarised as following; delamination factor and surface roughness on the milled flax fibre composites were equally affected by the feed rate and spindle speed. The number of end mill flute had a marginal influence on the aforementioned machining outputs. The results were successfully analysed by using response table and Taguchi L18 orthogonal array.

Acknowledgements

The authors acknowledged the financial support of the Ministry of Higher Education Malaysia through Fundamental Research Grant Scheme (FRGS) no: FRGS/1/2016/TK03/UNIMAP/02/7 (UniMAP Project Code: 9003-00605).

References

[1] Pickering K L, Effendy M G A and Le T M 2016 *Compos A.* **83** 98-112.
[2] Yan L, Chouw N and Jayarama K 2018 *Compos B.* **56** 296-317.
[3] Lih T C, Azmi A I and Muhammad N 2016 *Mater Manuf Process.* **31** 1366-1376.
[4] Wang D, Onawumi P Y, Ismail S O, Popov H N, Silberschmidt V V and Roy A 2019 *Compos A.* **110** 188–195.

[5] Azmi A I, Lin R J T and Bhattacharyya D 2013 *Int J. Adv. Manuf Tech.* **64** 247–261.
[6] Harun A, Haron C H C, Ghani J A, Mohktar S and Sanuddin A 2016 *Proc. ASME*
[7] Celik Y H, Kilickap E and Kilickap A I 2019 *J Compos Mater.* 1-11.

IOP Conf. Series: Materials Science and Engineering **670** (2019) 012010 doi:10.1088/1757-899X/670/1/012010

Surface modification of nitinol by using electrical discharge coatings in deionized water

A F Mansor[1], R Jamaluddin[2], A I Azmi*[1], T C Lih[2] and M Z M Zain[2]

[1] Faculty of Engineering Technology, Universiti Malaysia Perlis, UniCITI Alam Campus, Sg Chuchuh, 02100 Padang Besar, Perlis, Malaysia
[2] School of Manufacturing Engineering, Pauh Putra Campus, 02600 Arau, Perlis, Malaysia

E-mail: *azwaniskandar@unimap.edu.my

Abstract. Nitinol, also known as nickel-titanium shape memory alloy, has a high potential in medical implantation due to its unique mechanical properties, shape memory effect, and low modulus of elasticity which close to bone's modulus. However, high composition of nickel element in this alloy is harmful whenever the element is released into human body that can induce inflammation and allergic reactions. The objective of this study is to develop a deposition layer of titanium oxide on nitinol surface through the study of electrical discharge coatings (EDC) parameters effects. In particular, deionized water and pure titanium rod were as the dielectric fluids and electrodes, respectively, during the EDC process. Variation settings of polarity, gap voltage and erosion depth were employed in this study. Experimental results examined the coating thickness and surface roughness through full factorial design with ANOVA analysis. It was determined that the polarity gave the highest significant effect on the deposition thickness. Meanwhile, the interaction between polarity and gap voltage was exhibited significant effect to the surface roughness. In additional, reverse polarity and low gap voltage were preferred in EDC process to provide good surface roughness and higher uniformity of the material deposition.

1. Introduction

Nitinol or also known as nickel titanium shape memory alloy is an intermetallic alloy that composed a nearly equal amount of nickel and titanium elements. This material has excellent mechanical properties, low modulus of elasticity and super-elasticity which make it fits as biomaterials. This alloy has a capability to return to its initial shape through heating process after undergone deformation through an external force. This occurs because of the transformation between two solid phases of this alloy which are martensitic (stable at low temperature) and austenitic (stable at high temperature). Due to this unique shape memory effect, this smart alloy has been extensively used in bone and dental implantations. The high recovery capacity of this alloy is preferable during implantation, i.e. for stress generation of teeth alignment in orthodontic process [1]. Despite of this advantage, the high content of nickel element in this alloy is harmful whenever excessive released into human body that can induce inflammation and allergic reactions. A study on the commercial orthodontic wire indicated that the Ni release is approximately in range 0.2 - 0.7$\mu g/cm^2$ [2]. However, it can be significantly increased with time [3] and then exceeded the maximum daily dietary intake of human body. With this respect, surface modification on the implant surface is essential to reduce or prevent the migration of Ni ions into the body. Since years ago, many researchers attempted to modify the Nitinol surface using high technological methods such as chemical vapor deposition (CVD) [4], physical vapor deposition (PVD) [5], ion implantation and deposition [6] and etc. Due to high operation cost, complicated procedures and machinery equipment [7], electrical discharge coatings (EDC) is seen as a viable alternative for surface modification.

EDC process is a revolutionary technique for surface modification which adapted from electrical discharge machining (EDM) process. Basically, EDM is used to erode workpiece material from spark generation in high frequency pulses. Most of the molten material from the erosion actions are flushed away from a sparking gap by high pressure dielectric fluids. However, the remaining molten metals tend to re-solidify on the workpiece surface and develop an alloying layer from the composition of the workpiece material, the electrode material and the breakdown molecules of the dielectric fluids. The alloying layer or also known as recast layer exhibits different properties than the workpiece material with undesired defects such as microcracks. The layer is extremely hard due to the rapid cooled/quenched phenomena in the dielectric fluids during the process. This can increase the corrosion resistance of the machined products. This feature is useful because it can increase the product lifetime from the coating's generation. In the previous study by M. Salmaliyan et al. [8], they reported that sparking energy significantly influenced the recast layer thickness, hardness and uniformity on H13 steel by adjusting the voltage and duty cycle during EDC process. However, they found that major cracks appeared at higher spark energy settings. Similar results were obtained by S. Algodi et al. [9] that revealed the hardness, voids and cracking formation occurred on 304 stainless steel as raising the input energy; i.e. current and pulse duration. J. Simao et al. [10] and K. Wu et al. [11] investigated the effect of polarity on the surface modification process for AISI H13 and SKD61, respectively. They concluded that the polarity had a significant influenced on the surface roughness and hardness due to different distribution of discharge energy at different polarity condition (negative and positive polarity of tool electrode). Meanwhile, A. Gill et al. [12] claimed that a higher uniformity of alloying layer can be achieved by using positive polarity of tool electrode. They also indicated that a large amount of material erosion from the electrode was not an effective approach to form an alloyed layer on the work material surface.

2. Materials and procedures

Medical grade Nitinol as specified by the ASTM F2035-12 was used as the workpiece material with dimensional of 70 mm x 70 mm x 5 mm. The chemical composition of the workpiece is shown in Table 1. Initially, the surface of the workpieces were prepared by the metallographic method and etched in Kroll reagent before undergo to the experimental trials. EDC experiments were performed using EA8 Mitsubishi die sinker machine utilising pure titanium as electrode. This electrode was cylindrical rod (Ø 20 mm x 10 mm) with purity of 99.9%. The dielectric fluids in this study was the deionized water.

Table 1. Chemical composition of as received nitinol.

Ni	Ti	C	O	N	H	Co	Cu	Cr	Nb	Fe
55.6	44.279	0.045	0.03	0.003	0.002	0.005	0.005	0.006	0.005	0.02

The experimental plan was based on 2 level full factorial design. Three parameters were selected in this study; polarity, erosion depth and gap voltage, as shown in Table 2. Figure 1 represents the schematic of the experimental design that consist of 8 factorial trials (point 1-8) and 6 replication of centre point trials (point 9). The effects of each parameter to the responses were identified from this analysis. Following the experimental trials, the samples were cut across the machined surface. Then, the cross-sectioned surfaces were ground with 800, 1000 and 2000 grit of SiC papers so as to obtain mirror finish on the sample. Then, the surfaces were polished by using monocrystalline diamond suspension. The recast layer thickness was observed and measured on the prepared surfaces under 1000x optical magnification using Xoptron X80 series optical microscopy. An average measurement was taken from 40 locations of each cross-sectioned sample. The recast layer thickness and surface roughness were analysed using statistical analysis of variance (ANOVA) so as to identify the parametric effects of the aforementioned parameters. Finally, the chemical composition of the recast layer was examined using Hitachi TM3000 table top scanning electron microscopy (SEM) machine on the top surfaces of the EDC samples.

Table 2. Experimental conditions.

No	Parameters	Low level	High level
1	Polarity	-	+
2	Erosion depth, mm	0.15	0.30
3	Gap voltage, volt	70	160
4	Pulse duration, ms	1.2	
5	Pulse interval, ms	1.2	
6	Peak current, amp	10	

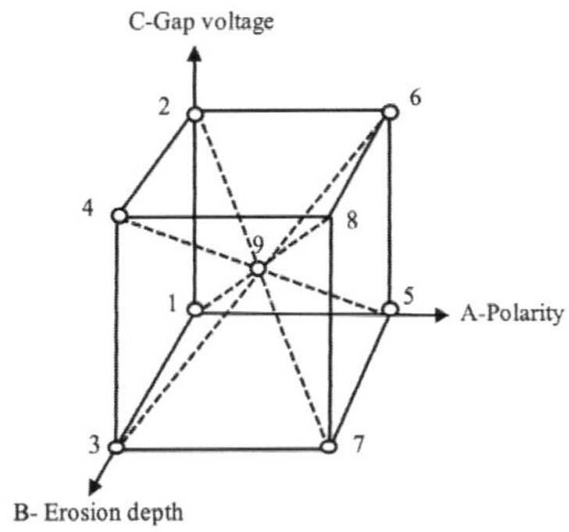

Figure 1. Schematic of the experimental design.

3. Results and discussion

Figure 2 represents the image of recast layer on the Nitinol surface after EDC operation under low settings condition (negative tool polarity, 0.15 mm erosion depth and 70 volt of gap voltage). It can be observed that the formation of the layer are uneven and showing large variation of the thickness. This phenomenon was also observed in all settings condition except the condition of reverse polarity. The inconsistency of the recast layer formation occurred due to the influence of high energy density to the material erosion of positive polarity of the workpiece. In fact, a uniform deposition was not affected by a higher amount of eroded material, but majority of the material deposition involved re-melting of the previous deposited material [13]. Thus, exceedingly high sparking energy can excessively deteriorate the previous deposited material and mostly discarded as small debris into the dielectric fluid. Although new material erosion and alloying deposition occurred during the new pulse cycle, however it was still unable to evenly deposit over the workpiece surface due the impact of the high energy density.

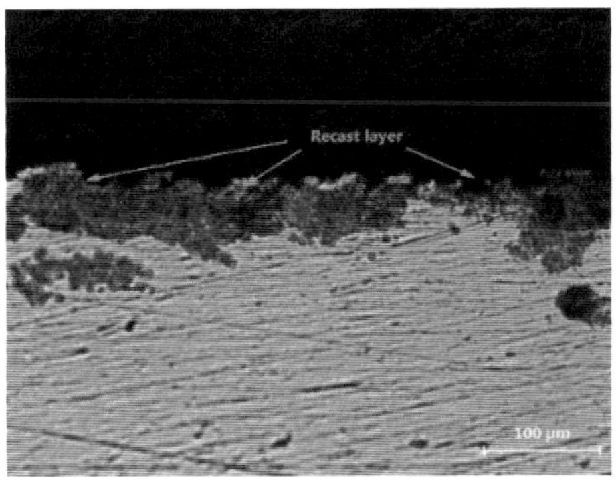

Figure 2. Image of recast layer under low settings condition.

Table 3 shows the effect of parameters towards the performance of recast layer thickness. The response was most significantly influenced by the gap voltage setting, following by the polarity at approximately 50 and 10 percent, respectively. None of interaction between parameters had any effect on the response. Figure 3 presents the effect of the significant parameters. It can be seen that the polarity has negative effect while shifting from the straight to the reverse polarity, see Figure 3(a). This was attributed to the high erosion of workpiece material and tendency of oxide formation from the dielectric breakdown. The negative oxidation state of oxygen from the water breakdown increased the accumulation of the ions to the positive workpiece and composing an oxide metal with the molten erosion material prior to deposition. However, the high energy density on the positive polarity of the workpiece can deteriorate the uniformity of the layer formation. In other hand, the energy also influenced by the gap voltage setting. Spark intensity increases with the larger gap voltage and resulted in higher thickness formation of the recast layer especially at straight polarity, as shown in Figure 3(c). Meanwhile, the erosion depth was only reflected almost 3 percent to the response. The reduction of the deposition thickness was resulted from the increasing of the erosion depth, as shown in Figure 3(b). In this study, erosion depth had raised the amount of debris which then suspended in the dielectric fluid. Due to no filtration of the particles, excessive accumulation of debris decreased the effectiveness of the plasma formation. Thus, the erosion was slightly affected, subsequently the migration of molten material was retarded, especially from the tool electrode.

Table 3. Effects lists of recast layer thickness.

Term	Standardized effects	Sum of squares	% contribution
A-Polarity	0.87	1.50	50.44
B-Erosion depth	-0.21	0.087	2.91
C-Gap voltage	0.39	0.30	10.22
AB	0.041	0.003	0.11
AC	-0.012	0.0003	0.01
BC	0.057	0.034	1.13

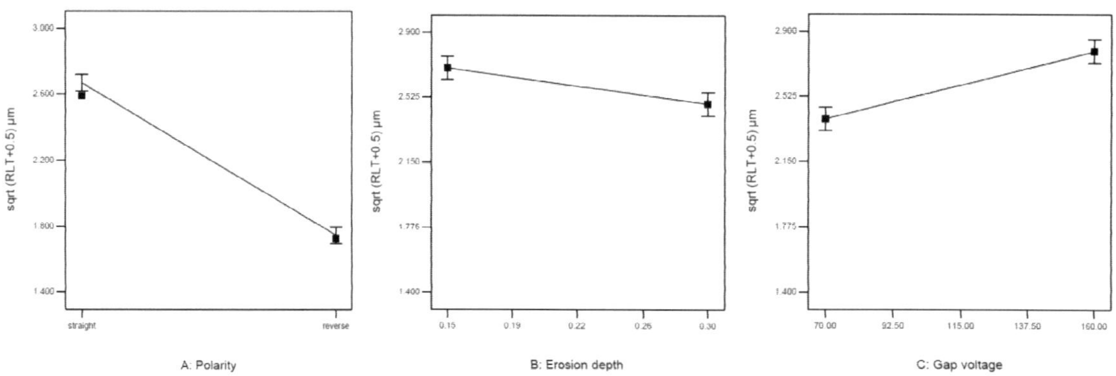

(a) (b) (c)

Figure 3. Effect graphs of parameters to the recast layer thickness.

Figure 4 represents the energy dispersive x-ray (EDX) mapping of chemical composition of the modified sample for the condition of low-level settings. The composition of all modified samples exhibited the increase of oxygen element than the unmodified sample. The oxidation might be occurred with the composition of titanium or nickel elements from the nitinol workpiece. However, a sacrifice pure titanium tool electrode had increased the amount of titanium and develop a higher formation of titanium oxide on the sample surface. Figure 5 depicts the evident of increment in the titanium and oxygen elements on the modified layer. Thus, this covered up the parent material which contain nickel element. However, the deposition composition was varied based on the process conditions.

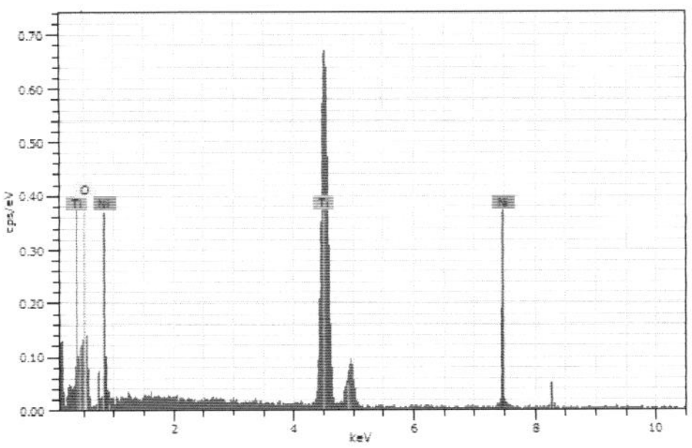

Figure 4. EDX mapping of a sample surface (straight polarity, 0.15mm of erosion depth and 70 volt of gap voltage)

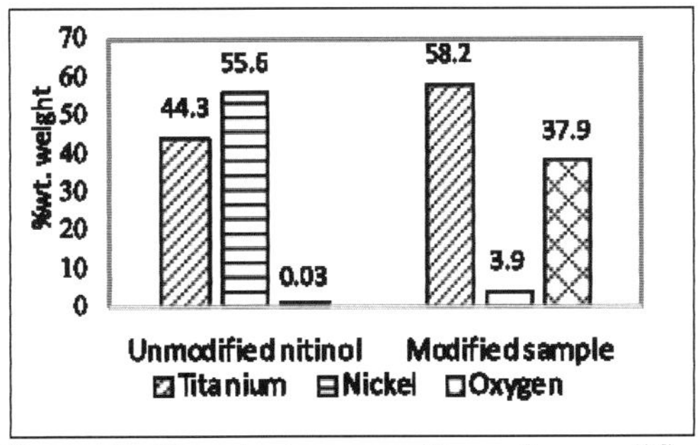

Figure 5. Chemical composition of unmodified nitinol and modified sample (under low level condition of all parameters)

4. Concluding remarks

In conclusion, oxide metal of titanium oxide can be formed on the nitinol surface by EDC process utilising deionized water. The variance analysis indicated the significant effects of polarity, gap voltage and erosion depth on the recast layer thickness. The main reasons of the recast layer formation were sparking energy and density. The reverse polarity setting encouraged more uniform formation of recast layer due to the less intensity of sparking energy on negative polarity of the workpiece.

Acknowledgments

The authors acknowledged the financial support from Fundamental Research Grant Scheme (FRGS/1/2017/TK03/UNIMAP/03/5 under UniMAP Project Code: 9003-00633 that was granted by Ministry of Education (Malaysia).

References

[1] Wadood A 2016 *Adv. Mater. Sci. Eng.*
[2] Arndt M, Brück A, Scully T, Jäger A and Bourauel C 2005 *J. Mater. Sci.* **40** 3659–3667.
[3] Shabalovskaya S, Anderegg J and Van Humbeeck J 2008 *Acta Biomater.* **4** 447–467.
[4] Zhang L *et al.* 2017 *NanoImpact.* **7**, 7–14.
[5] Guu Y H and Hocheng H 2001 *Mater. Sci. Eng. A* **318** 155–162.
[6] Liu X, Chu P K and Ding C 2004 *Mater. Sci. Eng. R Reports*, **47** 49–121.
[7] Chen Y F, Chow H M, Lin Y C and Lin C T 2008 *Int. J. Adv. Manuf. Technol.* **36** 490–500.
[8] Salmaliyan M, Malek Ghaeni F and Ebrahimnia M 2017 *Surf. Coatings Technol.* **321** 81–89.
[9] Algodi S J, Murray J W, Fay M W, Clare A T and Brown P D 2016 *Surf. Coatings Technol.* **307**, 639–649.
[10] Simao J, Lee H G, Aspinwall D K, Dewes R C and Aspinwall E M 2003 *Int. J. Mach. Tools Manuf.* **43** 121–128.
[11] Wu K L, Yan B H, Huang F Y and Chen S C 2005 *Int. J. Mach. Tools Manuf.* **45** 1195–1201.
[12] Gill A S and Kumar S 2015 *Mater. Today Proc.* **2** 1723–1730.
[13] Murray J W, Algodi S J, Fay M W, Brown P D and Clare A T 2017 *J. Mater. Process. Technol.* **243**, 143–151.

IOP Conf. Series: Materials Science and Engineering **670** (2019) 012011 doi:10.1088/1757-899X/670/1/012011

Thermal Comfort Study of a Classroom in Northern Malaysia: A CFD Approach

H Suhaila[1,2,a], I Suhaimi[1,2], M A S Mohamed[1], M T Mustaffa [1], A Kartini[1,2] and A Atikah[1]

[1]School of Manufacturing Engineering, Universiti Malaysia Perlis, Kampus Tetap Pauh Putra, 02600 Arau, Perlis, Malaysia.
[2]Clean Carbon Research Group, Universiti Malaysia Perlis, 01000 Kangar, Perlis, Malaysia.

Abstract. A conducive environment is essential in creating an ideal condition for the impartment of knowledge. One of the essential parts of such a conducive environment is to have a thermally comfortable condition. A study was conducted to assess the thermal comfort in one of the classrooms of a university located in the northern most part of Malaysia. The particular classroom was randomly chosen from a total of 4 classrooms of the same size and condition. Measurements of the dimensions of the room together with temperature and velocity were taken to make the computational model and as boundary conditions for the Computational Fluid Dynamics (CFD) simulation. The simulation was conducted to visualize and predict the temperature distribution and flow pattern inside the classroom. The results of the simulation showed that there exists a non-uniform circulation of cold air from the air-conditioning outlet to the rest of the classroom. The distribution of the cold air seems to be limited to only the location under the air-conditioning outlet making it thermally uncomfortable there. Hence, it is recommended to either reduce the airflow velocity coming out of the air-conditioning outlet or change the size or location of the air-conditioning outlets.

1. Introduction

Thermal comfort can be defined as the condition of mind which shows or exhibits a sense of contentment and satisfaction with the thermal environment [1]. It is one of the six key metrics of indoor air quality. One may experience thermal discomfort when the whole or one particular part of the body experiences an unwanted cooling or heating. Malaysia is a hot and humid country. Most of the time, air-conditioners are used indoors to provide a certain level of comfort from the hot environment outdoors. Some of the guidelines available for tropical countries like Malaysia are given in Table 1. In terms of indoor comfort, many studies through experiment and simulation had been conducted. de Dear et al [2] conducted a study on the thermal comfort in naturally ventilated apartment buildings in Singapore together with air conditioned ones. They found that 1/3rd of the occupants of the air-conditioned apartment felt cool thermal sensation while their counterparts in the naturally ventilated apartments felt warmer than the ISO thermal comfort standards. Chong et al [3] also investigated the airflow in a lecture room by using CFD. They compared their simulation using experimental data collected. Their study was to determine the effects of using two different turbulence models namely the RNG k-e and the Reynolds Stress Turbulence models on the results of the airflow namely air distribution, velocity and temperature. Ogoli [4] conducted a study on a natural ventilated education building in America. He used both surveys and laboratory experiments to study the effects of the environment, activities, clothing, age, gender as well as knowledge and experiences of the occupants, on determining the thermal comfort level. Yan and Chew [5] studied thermal comforts at four different hospitals in Malaysia. They used survey in their studies and found that 90% of the occupants feel at ease at the neutral temperature of 26.4°C. Haslinda et al [6] meanwhile, conducted a thermal comfort study in naturally ventilated residential terrace houses in Malaysia. She used CFD to simulate the air flow inside the houses. From the simulations, it was found that air did not exhibit a uniform circulation inside the house and that the air velocity was well above the acceptable level of thermal comfort set by ASHRAE. Rahman and Narahari [7] in their study found that in a lecture room with 198 occupants, there was a stagnant flow at the back of the room resulting in a temperature of up to 308K. This study was performed using CFD. Jain et al [8] also conducted a thermal comfort study of an office building in India. They used simulation method to determine the effect of air inlet locations in achieving thermal comfort in offices. Using computational fluid dynamics, they found the best combination of air-conditioner and window openings in achieving an optimal cooling temperature. This paper presents a thermal comfort study on a lecture room in Perlis, a state in Malaysia which has one of the highest temperatures in the country using computational fluid dynamics.

Table 1. Guidelines for temperature, wind speed and relative humidity applicable for Malaysia.

Guideline	Department of Standards Malaysia [9]	Department of Safety and Health Malaysia [10]	Chartered Institute of Building Services Engineers (CIBSE) Guide [11]
Temperature	24 - 26	23 – 26	25.5 - 27
Wind speed (m/s)	0.15 – 0.5	0.15 – 0.50	0.1 – 0.25
Relative humidity (%)	50 – 70	40 - 70	40 - 70

2. Measured Readings and Surveys

The following readings in Table 2 are the average of the measurements taken. It can be seen that it is slightly warmer in front of the room compared to the middle. All these values are within the temperatures, relative humidities and air velocities recommended by the guidelines in Table 1. However, when a survey was conducted among 84 students in the classroom (Table 3), it was found that most of them wish to have a warmer environment in the classroom. This is in contrast to the findings in Table 2.

Table 2. The average readings in the classroom.

Position inside room	Air Velocity	Temperature	Relative Humidity
Front	0.1 m/s	26 °C	52 %
Middle	0.12 m/s	25.5 °C	55 %
Back	0.1 m/s	26 °C	52 %

Table 3. Thermal preference results.

Thermal Preference	% of students
Warmer	60
No change	25
Cooler	15

3. Computational Fluid Dynamics (CFD) Modeling

The classroom chosen was measured to be 3 m in height, 7.62 m in width and 11 m in length. The readings of temperature, air speed and relative humidity for the room were taken at the time when the classroom was empty. The readings were taken one hour after the air-conditioning was turned on so as to stabilize the air flow inside the room. The readings for the velocity, temperature and relative humidity directly under the air-conditioner outlet are given in Table 4. These parameters were used as initial and boundary conditions in the simulations. The geometrical model and layout of the room is given in Figure 2. The air speed is the speed of the air entering the room from the air-conditioning system. There are 3 air-conditioners inside the classroom each located at the front, middle and back of the room. A Kestrel 2000 Pocket Weather Meter was used to measure the air speed, temperature and relative humidity. The room is considered empty as the reference data collected were taken when the room was unoccupied. The door and windows of the room are not included in the computational model as the room is assumed to be closed. The properties of the materials which are used in the simulation are given in Table 5.Only two boundary conditions were used in the model; wall and velocity inlet. Velocity inlet was used for the outlet vent of the air-conditioners while the ceiling, floor, and the surrounding walls of the classroom are classified as wall boundary condition. The velocity was set to a constant value of 2.6 m/s. The side walls of the classroom are made from brick while the side wall next to another classroom is made from gypsum. Constant heat flux was assumed for the walls with the air surrounding the classroom is set to be initially at 26 °C. The velocity of air and temperature is based on readings taken inside the classroom referring to Table 2.

4. Results and Discussions

To determine the distribution of the cold air inside the classroom, the air circulation at three different locations inside the classroom was obtained as shown by Planes 1, 2 and 3 in Figure 3. Plane 1 represents the location nearest to the gypsum wall; Plane 2 represents the location in the middle of the room while Plane 3 represents the location along the air-conditioning outlets. As expected the highest airflow velocity occurs at the air-conditioning outlets. The cold air only gets distributed once it reaches the floor of the classroom as evidenced by the contour plots in Figure 3 and 4. It can also be seen that the flow velocity decreases at the front and at back of the classroom resulting in low distribution of cold air at these locations. Rahman and Narahari [7] obtained about the same result where they found that the flow was stagnant at the back of the lecture room. Based on Figure 5 and 6, the warmest part of the room is walls of the classroom while the coldest is the location directly under the air-condition outlets. Based on the guidelines in Table 1, this coldest location is not within the thermal comfort guidelines outlined in Table 1. However, the rest of the room does fall within the thermal comfort temperature guideline. This would explain the findings of the survey and the measured temperature values in the previous section. The uneven temperature distribution maybe explained by the airflow velocity profile of Figure 3. Underneath the air-conditioning outlets, it can be seen that there is a rigorous circulation of the cold air, hence resulting in a concentration of cold air there. The cold air only gets distributed once it reaches the floor of the room. However, the amount of the cold air distributed and the temperature of the distributed air is higher than the air coming out of the air-conditioning outlet.

Figure 1. Inside of the classroom.

Table 4. Readings (averaged) after one hour of switching the air-conditioners. Readings taken directly under the air-conditioning outlet.

Parameter	Measured value
Wind speed	2.6 m/s
Air temperature	19.2 °C
Relative humidity	63%

Table 5. Material Properties [12].

Material	Density (kg/m³)	Viscosity (kg/ms)	Thermal conductivity (W/m.K)	Specific heat (J/kg.K)
Air	1.225	1.7894×10^{-5}	0.0242	1006.43
Brick	1600	-	0.69	840
Gypsum	1440	-	0.48	840
Carbon steel	7833		54	465

5. Conclusion

It was found from the simulation that there is a poor distribution of cold air inside the classroom which led to the space in the immediate vicinity of the air-conditioner to be colder than the recommended thermal comfort guidelines. A few recommendations can be made to improve the condition of the classroom based on the results obtained. Reducing the speed of the cold air coming from the air-conditioner outlet might be able to increase the temperature of the cold spots in the room. Another possibility would be to change the location and size of the air-conditioners that would make the cold air to be distributed more evenly in the room and less cold than the current temperature. Lastly, a less practical recommendation would be to make the ceiling higher so that the length that the cold air takes to reach the floor becomes longer hence enabling the cold air to have time to get distributed to the surroundings. These recommendations need to be studied further to determine their effectiveness in

IOP Conf. Series: Materials Science and Engineering **670** (2019) 012011 doi:10.1088/1757-899X/670/1/012011

promoting thermal comfort in the room. It is worth mentioning here that the simulation did not consider the affects of relative humidity as the aim was only to assess the flow profile inside the room. Therefore, for the next phase of the project, relative humidity conditions will be included in the simulation model so that an overall thermal condition for the room may be known.

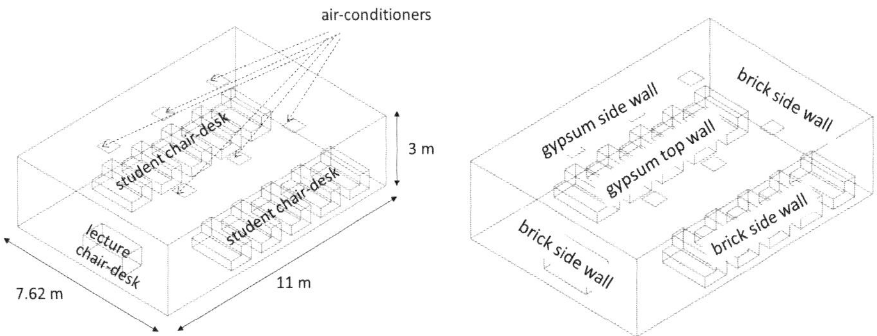

Figure 2. The geometrical model and layout of the classroom.

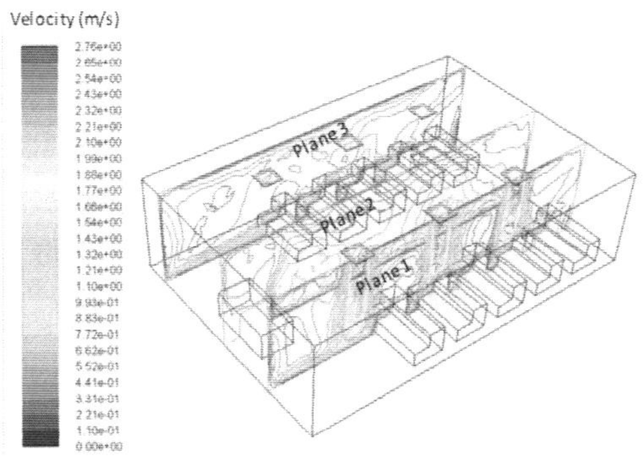

Figure 3. Velocity profile at selected locations in the classroom.

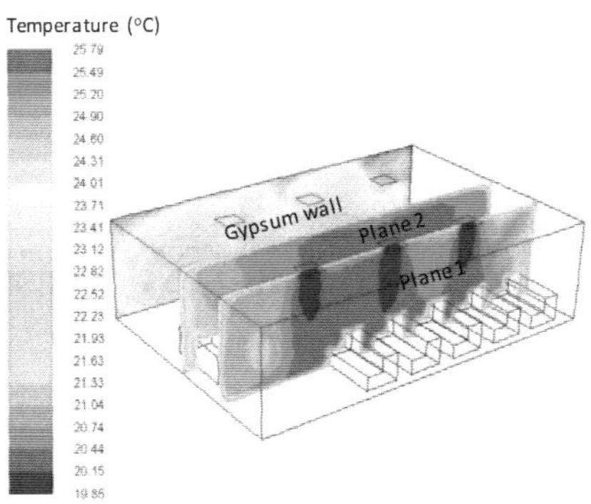

Figure 4. Temperature profile at selected locations in the classroom.

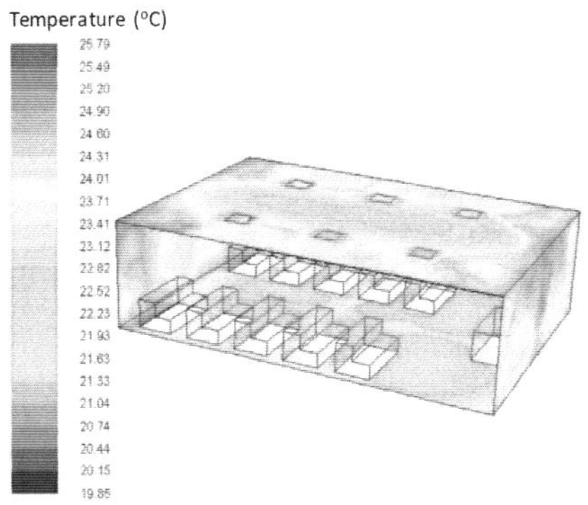

Figure 5. Temperature contour plots for the surrounding walls of the classroom.

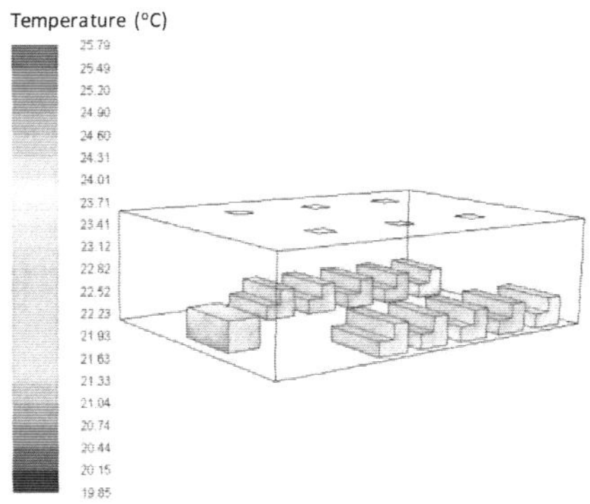

Figure 6. Temperature contour plots for the chair-desk in the classroom.

Acknowledgments
The author would like acknowledge the indirect support from the Fundamental Research Grant Scheme (FRGS) under the grant number of FRGS/1/2015/9003-00534 from the Ministry of Higher Education Malaysia.

References
[1] Olesen B W and Parsons K C 2002 Introduction to Thermal Comfort Standards and To the Proposed New Version of EN ISO 7730 *Energy and Buildings* **34** (6) 537-548
[2] de Dear R J, Leow K G and Foo S C 1991 Thermal Comfort in the Humid Tropics : Field Experiments in Air Conditioned and Naturally Ventilated Buildings in Singapore *Int J Biometeorology* **34** (4) 259-265.
[3] Chong J L S, Husain A and Tuan T B 2005 Simulation of Airflow in Lecture Rooms *Proceedings of the AEESAP International Conference* 7-8.
[4] Ogoli D 2007 Thermal Comfort in a Naturally-Ventilated Educational Building *ARCC Journal* **4** (2) 19-26.
[5] Yau Y H and Chew B T 2009 Thermal Comfort Study of Hospital Workers in Malaysia *Int. Journal of Indoor Environment and Health* **19** (6) 500-510.
[6] Mohamed Kamar H, Kamsah N, Md. Tap M and Mohd Salimin K A 2012 Assessment of Thermal Comfort in a Naturally Ventilated Residential Terrace House, *AIP Conference Proccedings* **1440** 247.
[7] Rahman M A and Narahari G A 2014 Analysis of Lecturing Room using Computational Fluid Dynamics Natural and Forced Ventilation *Int. Journal of Engineering Research & Technology* **3** (3) 653-659.
[8] Jain S K, Dadhich M, Sharma V, Sharma S K, Agarwal D and Gupta R 2015 Numerical Study on Thermal Comfort-A Case Study Using Taguchi Method *Int. Journal of Recent Advances in Mechanical Engineering* **4** (2) 121-136.
[9] Department of Standards Malaysia 2014 *Malaysia Standard MS 1525: 2014, Energy Efficiency and Use of Renewable Energy for Non-Residential Buildings – Code of Practice (Second Revisions).*
[10] Department of Occupational Safety and Health, Ministry of Human Resources Malaysia 1996 *JKKP:GP(1) 1/96 Guidelines on Occupational Safety and Health in the Office.*
[11] Chartered Institute of Building Services Engineers Guide (CIBSE) Guide 1986 *A Design Data* London.
[12] ASHRAE: Thermal Environmental Conditions for Human Occupancy 2004 *Standard 55-2004, American Society of Heating, Refrigerating, and Air-Conditioning Engineers* Atlanta, USA.

IOP Conf. Series: Materials Science and Engineering **670** (2019) 012012 doi:10.1088/1757-899X/670/1/012012

Investigation on the effect of drilling parameters on tensile loading of fibre-metal laminates

S Azwan[1], N I Sahira[1], M R Abdullah[1] and M Yazid Yahya[1]

[1] Centre for Advanced Composites (CfAC), School of Mechanical Engineering, Faculty of Engineering, Universiti Teknologi Malaysia, 81310 UTM Johor Bahru, Johor, Malaysia

Abstract. This study conducted to investigate the effect of different drilling parameters on the strength of vacuum infusion fibre reinforced epoxy resin-based fibre metal laminates (FML). The effect of several parameters such as the speed of drilling (1050, 2020, and 2750 rpm) and feed rate of (0.05, 0.1, and 0.15 mm/rev), and the thickness of FML on the strength of the composite materials were studied experimentally. The FML has been analysed under static loading. Drilling at lower spindle speed and lower feed rate results in better strength than higher spindle speed and feed rate. Then, the thicker FML gives more strength compare to the one with less thickness.

1. Introduction

Fibre Metal Laminates (FML) is an outstanding composite materials that have been used widely in aerospace owing to their excellent strength to weight ratio performance, low density and excellent corrosion resistance. However, the application FML in marine industry such as for small boat is still new and need a lot of research. Drilling is the most common machining process applied for joining these materials which not only reduces the structural integrity of the materials, but as well has the potential for a long-term performance deterioration [1] [2].

Various studies have been carried out on the effect of drilling process on FML, but most of the studies only focusing on the effect during the drilling process. Not many research purposing on the tensile behaviour of drilled specimen. Kılıçkap *et al.* [3] has studied on the effect of delamination factor on strength of glass fibre reinforced plastic (GFRP). The experiment was conducted at different drilling parameter such as cutting speeds and feed rate using tungsten carbide (WC) and Brad Spur cutting tools. It was determined that when the cutting speed and feed rate are increased, the delamination factor also increased which then lead to the reduction in strength of GFRP composites.

In addition, Pandit and Prasat [4] have studied on effect of drilling parameters on residual tensile strength of specimen and surface roughness of drilled hole using Taguchi method of optimization. The drilling parameters used were cutting speed, feed rate, drill point angle (DPA) and chisel edge width (CEW) for drilling of CFRP composites. As results, speed was the major factor contribute to changing in residual tensile strength followed by chisel edge width, drill point angle and feed rate. For surface roughness, drill point angle gave major contribution followed by speed, feed and chisel edge width.

Kumar *et al.* [5] have studied on the effect of thickness on tensile properties of a hybrid composite. The specimens of GFRP have been prepared at different thickness of 2mm and 3mm. It indicated that tensile strength mainly dependent on thickness of laminated polymer composites. If thickness increased by 1 mm, then 40-50% more load was required for fracture. So, increase in thickness of laminates resulted in increase of tensile strength.

In this present study, the effect of drilling and specimen parameter of FML, at different cutting conditions, was conducted. The drilling condition parameters were characterized by spindle speed and feed rate while the specimen parameter was categorized by the thickness of the specimen. After characterizing, specimens have been tested under stress load to investigate the tensile behaviour towards specimens.

2. Experimental materials and methods

2.1. Materials

FML is made up from hybrid combination of aluminium sheet and woven roving glass fibre. Aluminium sheet of 0.6 mm thickness and woven roving glass fibre of 450 gsm density were used in this research. Epoxy Epicote 2715 were used as a matrix and adhesive bonding. The fabrication process for the FML was vacuum infusion process which is one-time process, laminated the woven roving glass/epoxy at the same time bonded the laminate to the aluminium. FML 3/2 sample divided into two group which is 3A2F and 3A4F. 3A represented 3 layers of aluminium and (2F, 4F) represented 2 layers of woven roving glass/epoxy. As shown in **Table 1**, the thickness of woven roving glass/epoxy was maintained at ±0.37 mm regardless the number of fibre layers used. The sample were cut according to ASTM D3039 standard dimension for tensile test. Precise positioning and controlled FML movement during the cutting process must be taken to ensure defect free delamination. Schematic drawing of FML illustrated in **Figure 1**.

Table 1. Details of FML specimens.

Sample Designation	FML Thickness (mm)	Aluminium Thickness (0.6 X 3) (mm)	No. of Fibre Layer	Thickness of Single Fibre Layer (mm)
3A2F	3.28	1.8	4	0.37
3A4F	4.78	1.8	8	0.37

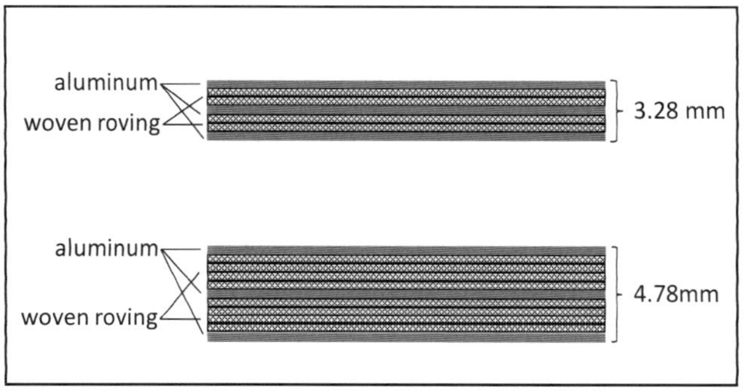

Figure 1. Schematic drawing of FML.

2.2. Drilling process

Universal Milling Machine DECKEL MAHO DMU 50, with the spindle rotational speeds (range from 20 to 18,000 rpm) and feeding rates maximum of 600 mm/min was used for drilling the specimens in this study. High Speed Steel (HSS) drill bit of 8 mm diameter was used during drilling process. The specimens were drilled under dry cutting conditions comprising of three spindle speeds (N= 1050,

2020, and 2750 rpm), and three feed rates (f = 0.05, 0.1, and 0.15 mm/rev) at the centre of specimen as shown in **Figure 2**. The levels defined for each of the process parameters are summarized in **Table 2**.

Table 2. Drilling parameters and their levels [6].

Drilling Parameter	Levels		
	1	2	3
Spindle Speed (N), (rpm)	1050	2020	2750
Feed Rate (f), (mm/rev)	0.05	0.1	0.15

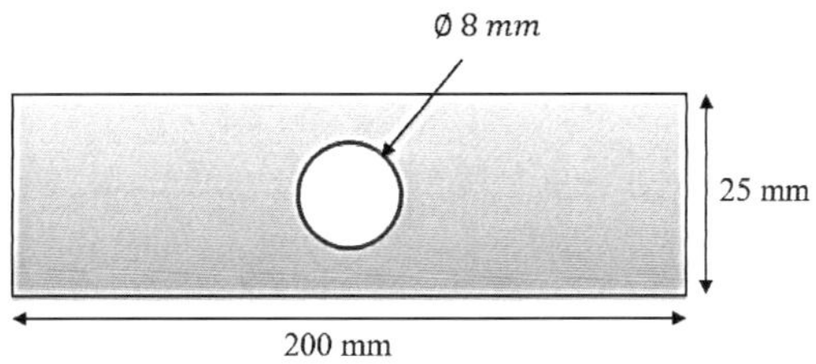

Figure 2. Illustration of drilled FML.

2.3. Tensile test

Tensile test experimental in which the specimen is mounted on 100kN Instron universal testing machine at a constant loading rate of 2mm/min complied with ASTM D3039 standard for tensile testing. The drilled hole was positioned in between of the extensometer pins. Each specimen was repeated three times to ensure the repeatability of test results and the average of obtained results were used for further analysis.

3. Result and discussion

In this study, the tensile behaviour of fibre reinforced epoxy resin-based fibre metal laminates subject to tensile loading were determined at different spindle speed of (1050, 2020 and 2750 rpm) and also at different feed rate of (0.05, 0.10 and 0.15 mm/rev) as shown in **Figure 3**.

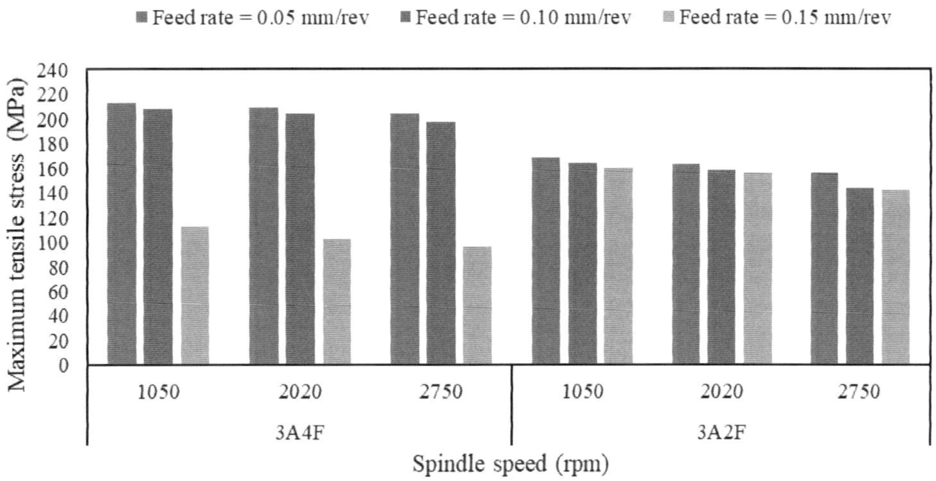

Figure 3. Maximum tensile stress obtained for 3A2F and 3A4F Fibre Metal Laminates (FML) at different spindle speed and feed rate.

From **Figure 3**, in sample 3A4F, at a constant speed of 1050 rpm, the maximum tensile stress value was reduced by 2.34% when the feed rate was increased from 0.05 mm/rev to 0.10 mm/rev. Meanwhile, by increasing the feed rate from 0.10 mm/rev to 0.15 mm/rev, drastic reduction of 46.23% for maximum tensile stress value of FML has occurred. This also resulted at spindle speed of 2020 rpm and 2750 rpm with 49.57% and 51.50% respectively. Moreover, at constant feed rate of 0.05 mm/rev, by increasing the spindle speed from 1050 rpm to 2020 rpm, the maximum tensile stress has been reduced by 2.12% only while when the spindle speed was increased by 2020 rpm to 2750 rpm, the maximum tensile stress value has been reduced by 3.53 %. Therefore, it can be discussed that in 3A2F FML, the increasing of spindle speed and feed rate only give a small reduction percentage to the maximum stress of the FML.

According to **Figure 3** also, for 3A2F, at a constant spindle speed of 1050 rpm, when the feed rate was increased from 0.05 mm/rev to 0.10 mm/rev, maximum tensile stress value was reduced by 3.28%. Then, by increasing the feed rate from 0.10 mm/rev to 0.15 mm/rev, the residual stress has decreased by 3.39%. The percentage of decrement for 1050 rpm at different feed rate were almost equal. Besides, at constant feed rate of 0.05 mm/rev, the maximum tensile stress has decreased by 4.08% when the spindle speed was increased from 1050 rpm to 2020 rpm. When increasing the spindle speed from 2020 rpm to 2750 rpm, the maximum tensile stress of FML is deducted by 4.26%.

Overall, the highest tensile stress value was obtained at the lowest range value which are 1050 rpm of spindle speed and 0.05 mm/rev of feed rate for both 3A2F and 3A4F. In comparison, the 3A4F maximum tensile stress value were bigger than in 3A2F. This is because, 3A4F is thicker than 3A2F. It contained much more fibre than in 3A2F. Hence, more load needed to break the sample. Plus, the load is distributed to bigger surface area. In additional, the reduction of maximum tensile stress due to increase in spindle speed and feed rate was because the stress cannot be transferred from outer plies to inner plies due to delamination. Delamination around the drilled hole fails to transfer stress from plies to plies causing major lost in overall stiffness [7]. Hence, the stress concentration is more severe in outer plies and lead to an early damage to the structure.

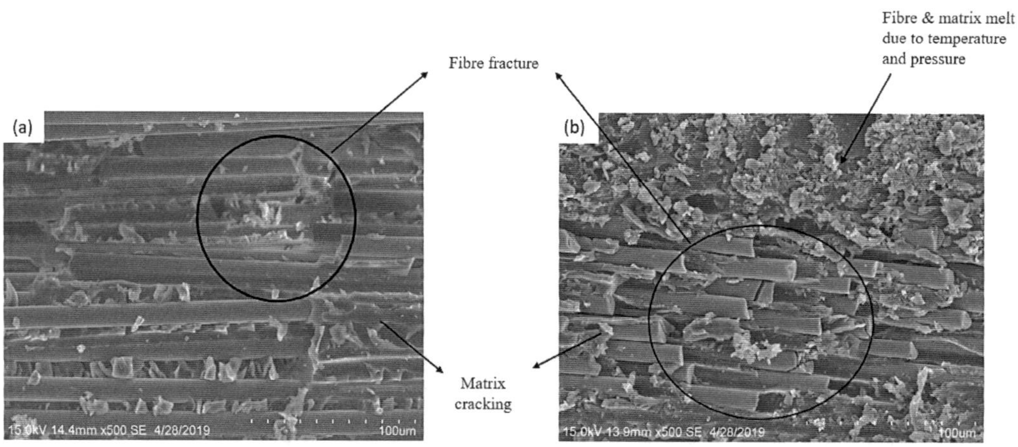

Figure 4. Microstructure of 3A4F: (a) sample S1F1 (b) sample S1F3

From both **Figure 4** (a) and (b), it has been noticed that the fibre breakages and matrix cracking are occurred more in S1F3 than the S1F1 of 3A4F specimen. This is due to the premature damage occurred during drilling and the application of the load applied during tensile test. From the defects happened in S1F3, it can be justified that the drastic reduction of residual strength obtained as spindle speed and feed rate were increased are due to the following defects. Plus, in drilling of composite materials, the temperature created while drilling the aluminium was higher at aluminium layer especially at high speed which has occurred in S1F3 where the speed was at maximum range.

The high temperature tool due to contact between aluminium and drill then made contact with GFRP panel made the fibre and matrix as a paste and spread it around [8].

4. Conclusion

In this study, fibre reinforced epoxy resin-based fibre metal laminates subjected to static tensile loading was studied experimentally. Several samples were prepared and drilled using HSS drill at different drilling parameter which are spindle speed and feed rate. Then, all the sample are analysed under static loading. It has found that the spindle speed and the feed rate are affecting the tensile behaviour of the FML. The obtained results for the maximum tensile stress show that the lower spindle speed and feed rate give greater strength to the FML structure. The thickness of the FML also plays a role in determining the strength of the drilled FML in which the thicker structure provide better maximum tensile stress to the structure.

References

[1] Khashaba U A 2004 Delamination in drilling GFR-thermoset composites *Composite Structures* **63** 313-327.

[2] Monivarman P and Michael Raj F 2015 Delamination analysis in drilling process of glass fiber reinforced plastic (GFRP) for marine composites *International Journal of Applied Engineering Research* **10** 2068-2073.

[3] Kilickap E, Yenigun B and Hisman Celik Y 2017 The effect of drilling parameters on strength of glass fibre-epoxy laminates by produced hand lay-up *e-Journal of New World Sciences Academy,* **12** 246-254.

[4] Pandit G D and Prasat K 2017 Experimental study of residual tensile of drilled composite *International Journal of Materials Science and Engineering* **5** (1) 35-46.

[5] Kumar M S, Krishna S G G and Rajanna S 2014 Study on effect of thickness and fibre orientation on a tensile and flexural properties of a hybrid composite *Journal of Engineering Research and Application* **4** (8) 56-66.

[6] Phadnis V A, Roy A and Silberschmidt V V 2012 Finite element analysis of drilling in carbon fiber reinforced polymer composites *Journal of Physics: Conference Series* **382** 1-7.

[7] Tao C, Qiu J, Yao W and Ji H 2016 The effect of drilling-induced delamination on tensile strength and prediction of residual strength of carbon fiber-reinforced polymer laminate *Journal of Composite Materials* **50** (24) 3373-3384.

[8] Abrao A M, Rubio J C C, Faria P E and Davim J P 2008 The effect of cutting tool geometry on thrust force and delamination when drilling glass fibre reinforced plastic composite *Materials and Design* **29** 508-513.

Acknowledgments

This research is fully supported by the Research University Grant from Universiti Teknologi Malaysia (Grant No. HIR - Q.J130000.2424.04G40). The authors fully acknowledged Ministry of Higher Education (MOHE) and Universiti Teknologi Malaysia for the approved fund which makes this important research viable and effective.

Evaporation lifetime and boiling curve on hemispherical stainless steel (304) surface

S Illias[1,a], S Hussain[1], N A Rosman[1], N S Abdullah[1], A I M Shaiful[1], M N B Omar[1], K A Ismail[1] and H Ani[2]

[1]School of Manufacturing Engineering, Universiti Malaysia Perlis, Pauh Putra Campus, 02600 Arau, Perlis, Malaysia.
[2]Department of Manufacturing and Materials, Kulliyyah of Engineering, International Islamic University Malaysia, PO Box 10, 50728 Kuala Lumpur, Malaysia.

Abstract. The purpose of this research is to study and investigate the evaporation lifetime and boiling curve on a hemispherical heated surface. The selected material was stainless steel (304). A nearly perfect and smooth hemispherical surface was developed by using EDM die sinker. For the test liquid, distilled water was used during the experimental work. The average droplet temperature was 31.36 °C corresponding to liquid subcooling ΔT_{sub} = 68.64K. Based on the theoretical calculation, the diameter of the water droplet was approximately 5.00 mm. Meanwhile, the impact height was approximately 65.0 mm corresponding to the theoretical impact velocity of 1.129 m/s. The material was heated using a digital hot plate which was able to give an accurate reading and stable temperature fluctuation during the heating process. The temperature ranged from a low temperature of T_w = 100 °C to a high temperature of T_w = 300 °C. As a result, the boiling curve showed a similar pattern of other experimental work that consists of two (2) important points which are the Critical Heat Flux (CHF) and Leidenfrost temperature.

1. Introduction

Boiling and evaporation are important elements in heat transfer studies. Apart from being used in our daily life, boiling is also important in heavy industrial area. For decades, complicated and highly advanced experimental work have been conducted by many researchers in order to study the boiling and evaporation process [1-5], to enhance the thermal protection system in nuclear reactor [6-9], to enhance the spray cooling technique [10-14] and many others. In heavy industrial area such as nuclear power plant, it is necessary to increase and upgrade the thermal protection and emergency cooling system. Therefore, scientists and engineers are working hard to increase the temperature limit in the boiling curve known as Leidenfrost temperature [15-16]. Recently, they used a structured and micro-structured surface in order to study this complicated phenomena. For instance, Kim et al. [17] studied the effects of wettability, height of surface roughness, and nano-porosity on Leidenfrost phenomena. As a result, it was concluded that nano-porosity (not solely high surface wettability) was an essential feature to enhance the LFP. Xiong et al. [18] investigated the evaporation of a small droplet impacting onto a hot stainless steel plate. The liquid included water, various pure and mixed hydrocarbon fuels with droplet sizes ranging from 0.07 to 1.8 mm. The wall temperatures varying from 63 to 605 °C covered the entire spectrum of heat transfer characteristics. They concluded that the maximum rate of heat transfer was independent of drop sizes for fuels and showed a slight decrease with increasing drop sizes in the case of water. At the Leidenfrost temperature, it showed that the heat transfer rate is maximum at D_o of approximately 0.5 mm. Kim et al. [19] observed the behavior of water droplets near the Leidenfrost point (LFP) on zirconium alloy surfaces with anodizing treatment and investigated the droplet cooling performance. As a result, the structured surface had delayed LFP due to the high wetting condition induced by strong capillary wicking forces on the structured surface. Other researchers also have contributed extensively to boiling and drop impact research [20-21].

In this article, the CHF and Leidenfrost point temperatures on a stainless steel hemispherical heated surface were investigated and found to have occurred at approximately T_{CHF} = 191.0-200.7 °C and T_L = 260-265.5 °C, respectively.

2. Experimental apparatus and procedure

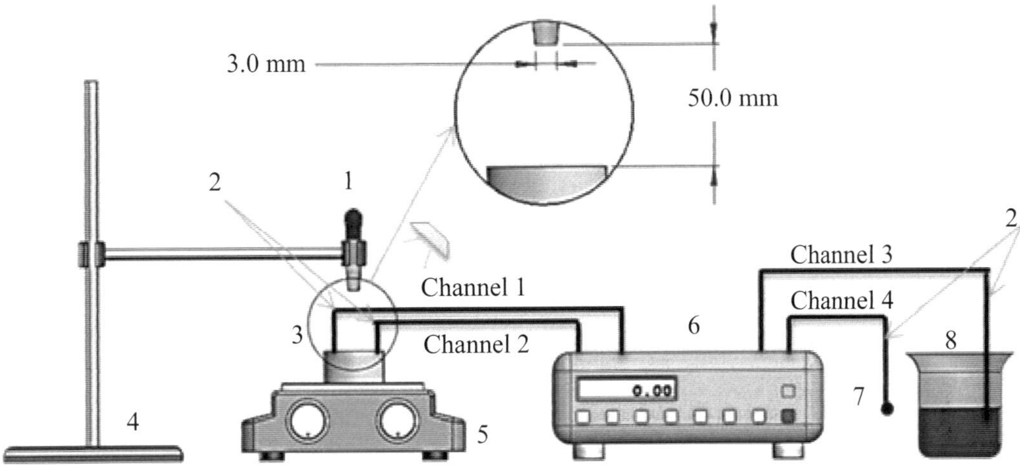

1. Manual droplet dispenser 2. Thermocouples 3. Hemispherical stainless steel 4. Retort stand 5. Hot plate 6. Temperature recording device 7. Distilled water 8. Beaker

Figure 1. The schematic diagram of the experimental apparatus.

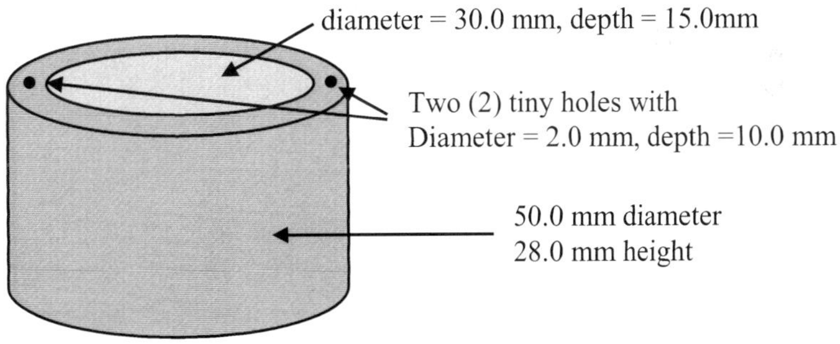

Figure 2. Technical drawing of the stainless steel hemispherical test surface.

Figure 1 shows a schematic diagram of the experimental apparatus while Figure 2 shows the dimensions used. The stainless steel hemispherical surface was approximately 28.0 mm in height and 50.0 mm in diameter. The surface temperature was measured by using two (2) type K thermocouples at separate locations on the test surface. There was a difference of less than 1 °C between the two (2) readings of the thermocouples. In the experimental work, distilled water was used as a test liquid. The radius and diameter of the droplet was experimentally and theoretically calculated using Eq. (1) and Eq. (2). The symbols V, r, σ, g, d_{needle} and ρ_{liq} represents the droplet volume, droplet radius, liquid surface tension, gravitational force, inner droplet dispenser diameter and liquid density, respectively. The result showed that the average data for a single water droplet was approximately 0.065 gram (65 mm^3) corresponding to the droplet radius and diameter of 2.494 mm and 5.00 mm, respectively. The experimental apparatus and procedure is similar with our previous report [21].

$$V = \frac{4}{3}\pi r^3 \quad (1)$$

$$d = \sqrt[3]{\frac{6\sigma d_{needle}}{\rho_{liq}\, g}} \quad (2)$$

3. Results and Discussions

Figure 3 shows the boiling curve plotted from the experimental work. The best three (3) readings were selected during the analysis. From the experimental data, the average reading of the evaporation lifetime was observed to be approximately 58.10 second at the low surface temperature of T_w = 100 °C. Then, the evaporation lifetime decreased as the surface temperature increased. At the surface temperature of T_w = 110 °C and 120 °C, the average evaporation lifetime measured was 30.10 sec and 11.31 sec, respectively. The average evaporation lifetime became shorter to about 5.04 sec and 5.19 sec, as the surface temperature increased to T_w = 130 °C and 140 °C, respectively. Finally, it was clearly observed that the average evaporation lifetime becomes close to zero reading when the surface temperature reached T_w = 191.0 up to 200.7 °C. At this surface temperature, the evaporation lifetime recorded was approximately 0.83 and 0.82 sec, respectively. This condition is known as CHF and normally occurred below 1 second. During CHF, the evaporation conditions and characteristics became very active compared to nucleation boiling regime. This active evaporation condition represents the steam generation in the nuclear power plant where the generated steam is used to spin the turbine. The evaporation lifetime increased a bit when the surface temperature rose to about 205 °C. Although the temperature different was only 5 °C, the boiling condition have shifted from nucleation boiling regime to transition boiling regime. At the surface temperature of 205 °C, the evaporation lifetime recorded was approximately 3.03 sec. Meanwhile, the evaporation lifetime at the surface temperature of T_w = 230 °C was approximately 18.66 sec which a bit higher compared to previous temperatures. Suddenly, the evaporation lifetime increase dramatically to a higher value of 117.6 sec at the surface temperature of T_w = 240 °C. This unique phenomenon represents the transition boiling regime in the boiling curve. Based on our observation, the highest data recorded during this experimental work was about 128.2 sec, which occur at the average surface temperature of T_w = 264.2 °C. This maximum condition of evaporation time is known as Leidenfrost temperature. At this surface temperature, the transition boiling condition had shifted to fully developed film boiling and a vapor film started to generate between liquid and hot surface soon after the droplet touched the hot surface. This vapor film hindered the heat transfer from the solid to the liquid. Interestingly, when the average surface temperature reached a much higher surface temperature of T_w = 301.98 °C, the evaporation lifetime recorded decreased to 121.3 sec. Similar experimental result which shows a reducing pattern after reaching LFP were also achieved by other researcher such as Kim et.al [19]. As a result, it is concluded that, the CHF and Leidenfrost point temperatures during the impact of a droplet on hemispherical stainless steel surface are approximately T_{CHF} = 191.0-200.7 °C and T_L = 260.0-265.5 °C, respectively.

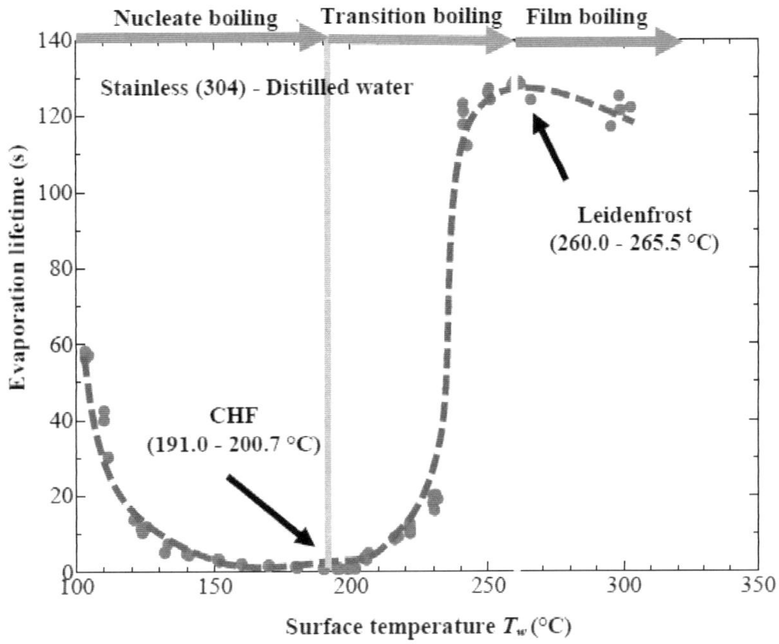

Figure 3. Surface temperature T_w (°C) vs Evaporation lifetime (s)

4. Conclusion

From the experimental work, it is confirmed that the CHF and Leidenfrost point (LFP) are about T_{CHF} = 191.0-200.7 °C and T_L = 260-265.5 °C, respectively. Therefore, we finally achieved our experimental objective.

5. Acknowledgments

The authors would like to thank the Ministry of Education Malaysia and the Research Management and Innovation Centre of Universiti Malaysia Perlis (project number: FRGS/1/2018/TK03/UNIMAP/02/11) for awarding a research grant to undertake this project.

6. References

1. Takata, Y., Hidaka, S., Cao, J. M., Nakamura, T., Yamamoto, H., Masuda, M., & Ito, T. (2005). Effect of surface wettability on boiling and evaporation. *Energy*, *30*(2-4), 209-220.
2. Tice, T. R., & Gilley, R. M. (1985). Preparation of injectable controlled-release microcapsules by a solvent-evaporation process. *Journal of Controlled Release*, *2*, 343-352.
3. Fathinia, F., Al-Abdeli, Y. M., & Khiadani, M. (2019). Evaporation rates and temperature distributions in fine droplet flash evaporation sprays. *International Journal of Thermal Sciences*, *145*, 106037.
4. Yang, Y., Cui, G., & Lan, C. Q. (2019). Developments in evaporative cooling and enhanced evaporative cooling-A review. *Renewable and Sustainable Energy Reviews*, *113*, 109230.
5. Patel, U., & Sahu, S. (2019). Effect of air turbulence and fuel composition on bi-component droplet evaporation. *International Journal of Heat and Mass Transfer*, *141*, 757-768.

6. Shama, A., Pouchon, M. A., & Clifford, I. (2019). Simulation of the microfluidic mixing and the droplet generation for 3D printing of nuclear fuels. *Additive Manufacturing*, *26*, 1-14.
7. Shanthanu, S., Raghuram, S., & Raghavan, V. (2013). Transient evaporation of moving water droplets in steam–hydrogen–air environment. *International Journal of Heat and Mass Transfer*, *64*, 536-546.
8. Trottier, A., Adams, F. P., Levinsky, A., & Roubtsov, D. (2018). Nuclear data sensitivity for reactor physics parameters in a lead-cooled reactor. *Annals of Nuclear Energy*, *120*, 333-347.
9. Lim, H. S., Tak, N. I., Lee, S. N., & Jo, C. K. (2018). Water-jacket reactor cavity cooling system concept to mitigate severe accident consequence of high temperature gas-cooled reactor. *Nuclear Engineering and Design*, *340*, 156-165.
10. Kim, J. (2007). Spray cooling heat transfer: the state of the art. *International Journal of Heat and Fluid Flow*, *28*(4), 753-767.
11. Liu, H., Cai, C., Yin, H., Luo, J., Jia, M., & Gao, J. (2018). Experimental investigation on heat transfer of spray cooling with the mixture of ethanol and water. *International Journal of Thermal Sciences*, *133*, 62-68.
12. Wang, J. X., Li, Y. Z., Li, J. X., Li, C., Xiong, K., & Ning, X. W. (2018). Enhanced heat transfer by an original immersed spray cooling system integrated with an ejector. *Energy*, *158*, 512-523.
13. Peng, C., Xu, X., & Liang, X. (2018). Experimental study on temperature variation patterns and deterioration of spray cooling with R21. *International Journal of Heat and Mass Transfer*, *121*, 1159-1167.
14. Lamini, O., Wu, R., Zhao, C. Y., & Xu, Z. G. (2018). Enhanced heat spray cooling with a moving nozzle. *Applied Thermal Engineering*, *141*, 921-927.
15. Bouillant, A., Mouterde, T., Bourrianne, P., Lagarde, A., Clanet, C., & Quéré, D. (2018). Leidenfrost wheels. *Nature Physics*, *14*(12), 1188.
16. Qiu, L., Dubey, S., Choo, F. H., & Duan, F. (2019). High jump of impinged droplets before Leidenfrost state. *Physical Review E*, *99*(3), 033106.
17. Kim, H., Truong, B., Buongiorno, J., & Hu, L. W. (2011). On the effect of surface roughness height, wettability, and nanoporosity on Leidenfrost phenomena. *Applied Physics Letters*, *98*(8), 083121.
18. Xiong, T. Y., & Yuen, M. C. (1991). Evaporation of a liquid droplet on a hot plate. *International Journal of Heat and Mass Transfer*, *34* (7), 1881-1894.
19. Kim, S. H., Ahn, H. S., Kim, J., Kim, M. H., & Park, H. S. (2014). Experimental study of water droplets on over-heated nano/microstructured zirconium surfaces. *Nuclear Engineering and Design*, *278*, 367-376.
20. Inada, S., Sumiya, H., Shinagawa, K., & Illias, S. (2006). MECHANISM ELUCIDATION FOR THE MINIATURIZATION-BOILIN PHENOMENON IN DROPLET COLLISION BOILING SYSTEM. In *International Heat Transfer Conference 13*. Begel House Inc..
21. Illias, S., Rosman, N. A., Abdullah, N. S., Hussain, S., Baharudin, M. E., Idris, M. A., & Ismail, K. A. (2019). Critical heat flux and Leidenfrost temperature on hemispherical stainless steel surface. *Case Studies in Thermal Engineering*, *14*, 100501.

IOP Conf. Series: Materials Science and Engineering **670** (2019) 012014 doi:10.1088/1757-899X/670/1/012014

Frontal impact on bus superstructure as per UNECE R29 and NCAP

M A A Afripin, A Z Zainudin, M A H F M Sahar and M Yusof*

School of Mechanical Engineering, Universiti Teknologi Malaysia, 81310 UTM Skudai, Johor, Malaysia.

Abstract: Frontal crash for the large vehicle is less severe compared to the lightweight vehicle due to the height position of the driver compartment in the vehicle which sits well above the impact area in any accident involving a passenger vehicle. Due to this reason, most of the coachbuilder does not include crumple zone in their design as what the car's manufacturer did. However, in the case of frontal collision between two large vehicles or with the rigid wall, the possibility of energy absorption of the structure is very low and the remaining energy will be transferred directly to the driver and occupants. In this respect, two prominent regulations for frontal impact, namely United Nation Economic Commission of Europe (UNECE) Regulation no. 29 and New Car Assessment Program (NCAP) is used to determine whether the structures are having enough strength to withstand the load produced by the impact. This paper deals with the finite element simulation of a frontal impact on bus superstructure by applying both of the regulations. Then, the results for both simulations are compared in terms of energy produced, structure deformations, maximum stress, and the corresponding plastic strains. It is found that the energy from the ECE R29 regulation is lower than NHTSA's NCAP with 55 kJ and 142 kJ respectively. However, the deformation observed from R29 simulation, the front structures is severely deformed. Meanwhile, the structures in the NCAP simulation are still intact and the steering wheel structures are still not in contact with any body parts of the driver.

1. Introduction

Road traffic accidents are always a huge issue in developing countries such as Malaysia. The high numbers of fatalities due to the road accident are recoded by a motorcyclist with compromised about 60 % of the total fatalities [1]. In addition, the number of accidents involving commercial vehicles such as bus also increases significantly. For bus, the cases of road traffic accidents increase significantly by 48% between the year 2006 to 2008 [2].

In the United States of America, an average of 55,000 buses crashed each year result in an average of 250 occupant fatalities and 14,000 injuries. The common types of bus-involved accidents are frontal collision, side impact, rear impact and rollover. The most severe accidents recorded by the Fatality Analysis, Reporting System (FARS) are rollover which contributed 44 pct. of fatal events involving road vehicle. However, the study of rollover is commonly done by previous researchers and following UNECE Regulation 66. However, the frontal impact on the bus is still at infancy [3]. The existing regulation for frontal impact is UNECE R29, which adapts from regulation specifically designed for heavy truck [4].

Studies on crashworthiness and safety of vehicles have gained concerns for the past years [5]. Frontal accident fatalities compared to all fatalities in bus accidents shows the highest in Netherland, 83.3% [6]. In this study, we are focusing on frontal crash of the bus. Accidents due to the frontal crash causing damage to the frontal bus structure and this will lead to the serious injury towards bus driver and in some cases, death on the spot. Providing the safety of the driver is crucial, since bus driver is a key person for keeping the control of the bus within the event of accidents, safety of the passengers is going to be ensured [5]. The safety of the bus driver is important as the driver that control the bus movement and keep their passenger in safe and comfortable journey. Thus, it is important to investigate the performance of the safety zone of a bus driver.

Most common methods used by researchers are by doing an analysis of dynamic simulation using various kinds of software [4, 7, 8]. Finite Element Method (FEM) is vital to get the data analysis.

To carry out the research, certain regulation need to be considered to meet the international standards. In this study, United Nation Regulation No.29 and New Car Assessment Program (NCAP) had been used. Finally, to provide assurance of protection in potentially serious injury crashes, the test procedures must be severe enough to represent a crash in which occupants could be seriously injured or killed [9].

2. Methodology

Frontal impact simulations have been executed following UNECE R29 and NCAP regulations.

2.1 Frontal Impact Assessment follows ECE-R29 and NCAP

The assessment technique used in the frontal impact study is ECE R29 and NCAP, which covered on frontal crash of the vehicle. The main difference between these two regulations are the type of loading which for ECE R29, the pendulum is used with particular angular velocity while for NCAP, the vehicle are moving towards a rigid wall with assigned velocity.

According to ECE Regulation 29, the vehicles with a gross mass of 7500 kg or more, the impact energy produce from the pendulum must be 55 kJ. There should be no contact between the structure and the driver to ensure the structure is meeting the requirement in the regulation. In view of this, the manikin should be used which is described in R29.

In the New Car Assessment Program (NCAP) frontal impact test under National Highway Traffic Safety Administration (NHTSA), a test vehicle is moving towards a rigid wall with a speed of 56 km/h perpendicular to the wall surface. In this study, similar size of manikin used in the R29 will be used in the NCAP frontal impact virtual simulation.

2.2 Material Properties

A series of tensile test are being performed prior to the simulation to obtain accurate input material properties for bus superstructure. The structure is assumed to behave in an elastic-plastic manner. Since the interested results are the plastic deformation of the structure, the input used are up to inelastic properties. The extensiometer is used as the measurement device to capture the elastic and plastic strain within the gauge length. Four specimens had been tested to ensure the best repeatability measurement produce from the test. The yield stress is 327 MPa and the elastic modulus measured based on the gradient of elastic regime is 210 GPa. Properties of mild steel with material density of 7800 kg/m^3 were used in this study. The strain rate effects of the structures are model using Johnson-Cook constitutive model [10] as shown in Equation 1.

$$\sigma = [327 + (534)\varepsilon^{(0.423)}][1 + (0.0756)\ln \dot{\varepsilon}] \tag{1}$$

The input strain rates dependent stress-plastic strain curves in this study are covered for rates ranging from 0.01 to 1600 s^{-1}. The resulting strain rate-dependent stress-true strain curves for different rates are displayed in Figure 1.

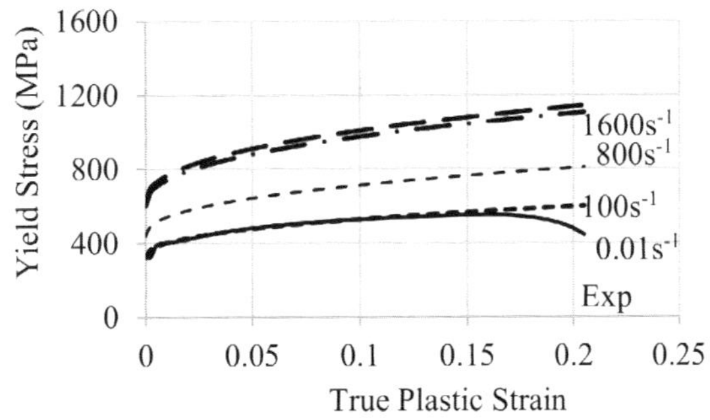

Figure 1. Stress-strain curves for different strain rates.

2.3 Loading and Boundary Conditions

The parameters used in this paper is partially based on previous researchers [5, 11, 12]. The pendulum has a striking surface of 2500 mm x 800 mm and is made of steel with evenly distributed mass of at least 1500 kg. The pendulum is suspended by two rigid beams of 1000 mm apart and 3500 mm long from the axis of suspension to the geometric centre of the impactor. Its striking surface is in contact with the foremost part of the vehicle and the vertical position of the pendulum's centre of gravity (H-point) is 50 + 5/-0 mm below the R-point of the driver's seat as shown in Figure 2.

The front parts of the bus body are meshed with beam element B31 (A 2-node linear beam in space) elements. The mesh size is chosen to be 50 mm. The boundary conditions are according to the actual test specified in the regulation. The pendulum mass moment of inertia is 2453 kg.m^2.

The configurations for frontal impact based on R29 and NCAP are shown in Figure 2 and . Figure *3* respectively.

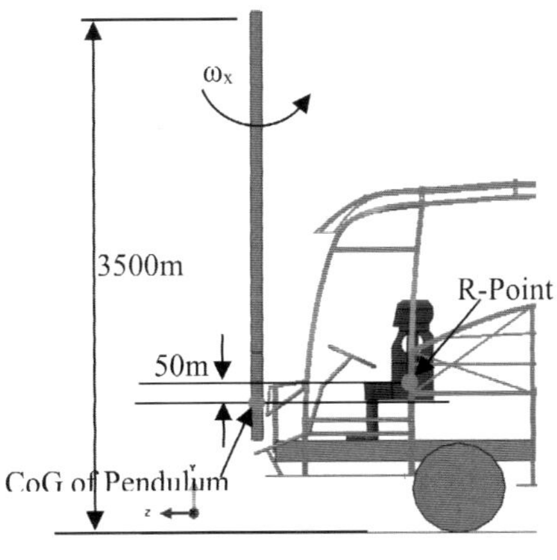

Figure 2. Simulation setup for UNECE R29.

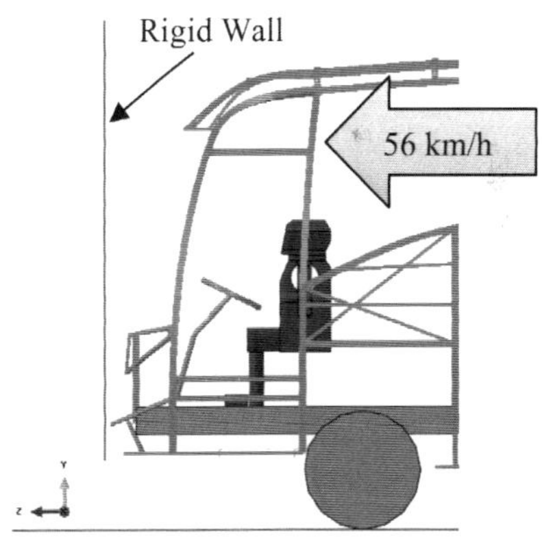

Figure 3. Simulation setup for NCAP.

3. Results and discussion

3.1 Deformation Response of Bus Superstructure

3.1.1 R29 Simulation. The evolution of the deform shape of the bus superstructure during the frontal impact simulation following UNECE R29 is shown in Figure 4. Figure 4 (a) and (b) shows the structure during initial condition and first impact respectively. It can be observed that in Figure 4 (c), the steering wheel is approaching manikin body. However, the maximum deformation is still in elastic region and the steering wheel structure is bouncing back up to the plastic deformation and a clearance can be seen in Figure 4 (d). The maximum von-Mises stress record is 551 MPa which exceeding the yield stress of quasi-static condition. The displacement of critical nodes of the steering wheel along the z - direction is calculated and the maximum displacement of 140 mm is recorded at 125 ms.

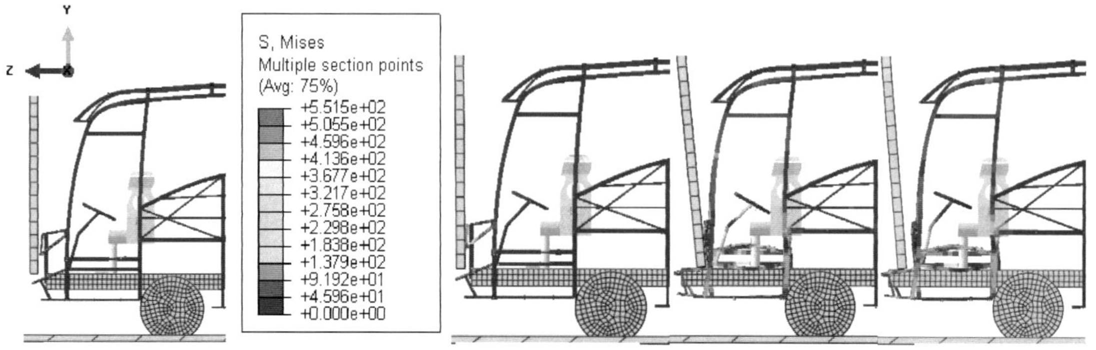

(d) t=0 ms (c) t=4 ms (b) t=125 ms (a) t=200 ms

Figure 4. The evolution of structural deformation during frontal impact following R29.

3.1.2 NCAP Simulation. The evolution of deformation shapes following NCAP requirements is shown in Figure 5. Figure 5 (a) showing the initial condition of the bus superstructure. At 3ms, the front structure is fully crush and start to crumple inside the driver compartment as shown in Figure 5 (c). Final deformation at t=198.3ms is shown in Figure 5 (d). During the deformation, the steering wheel doesn't infringe into the driver manikin indicating that the structure is passing the requirement by following the NCAP.

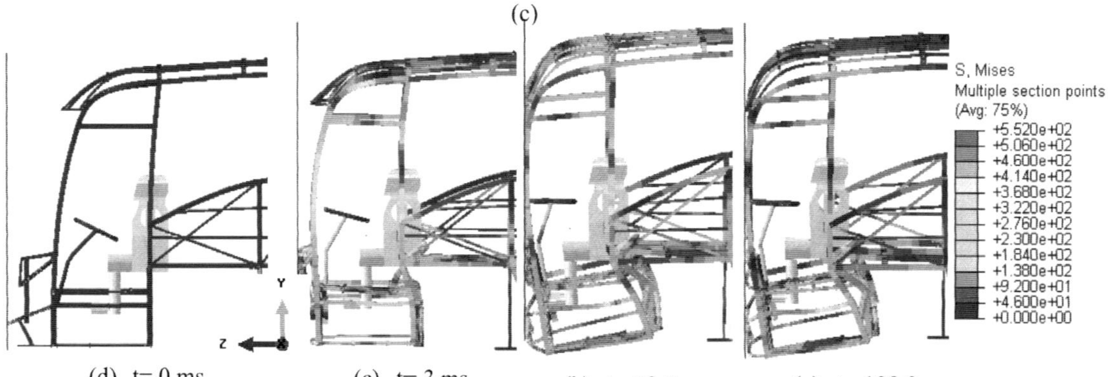

(d) t= 0 ms (c) t= 3 ms (b) t= 78.5 ms (a) t= 198.3 ms

Figure 5. The evolution of structure deformation during frontal impact following NCAP.

4. Conclusion

The frontal impact simulation of the bus superstructure by following UNECE R29 and NCAP have been demonstrated. Results show that:

- Energy produce from the frontal impact following NCAP requirement is much higher than UNECE R29 by the difference of 61 %.
- In R29 simulation, the energy from the impact are fully absorbed by the superstructure simulation while for NCAP, the energy is also borne by the chassis of the bus.

Acknowledgment
Authors wishing to acknowledge the MOE and tier 1 grant under UTM grant, vot 16H33 for funding this project.

References

[1] RMP 2009 *Statistical Report of Road Accidents Malaysia 2008*
[2] Solah M S, Ariffin A H, Isa M and Wong S V 2013 In-depth crash investigation on bus accidents in Malaysia *Journal of Society for Transportation and Traffic Studies* **3** (1):22-31.
[3] Jongpradist P 2015 Study on Frontal Impact Crashworthiness for Passenger Bus Manufactured in Thailand
[4] Mirzaamiri R, Esfahanian M and Ziaei-Rad S 2012 Crash test simulation and structure improvement of IKCO 2624 truck according to ECE-R29 regulation *International Journal of Automotive Engineering* **2** (3) 180-92.
[5] Ladammanavar D M, SN A and Prabha S 2017 Explicit Dynamic Analysis Of Automotive Bus Body Structure During Catastrophic Frontal Crash With Expulsion System *International Journal of Advances in Scientific Research and Engineering* **3** (5) 109-117.
[6] Matolcsy M 2007 The severity of bus rollover accidents *20th Conference ESV Proceedings 1-10*
[7] De Castro I, Jokuszies M, Altamore P and Lee W 2001 Simulation of occupant response in the ECE R29 safety test *SAE Technical Paper*
[8] Gendar D 2007 Numerical simulations for testing of commercial vehicle as per ECE-R29 regulations *SAE Technical Paper*
[9] Gabler HC and Hollowell WT 1998 *NHTSA's vehicle aggressivity and compatibility research program* NHTSA
[10] Vedantam K, Bajaj D, Brar N S and Hill S Johnson 2006 Cook Strength Models for Mild and DP 590 Steels *AIP Conference Proceedings* **845** (1) 775-8.
[11] Cerit M E, Guler M A, Bayram B and Yolum U 2010 Improvement of the energy absorption capacity of an intercity coach for frontal crash accidents *Proc 11th Int. LS-DYNA Users Conf.* 15-24
[12] Dwivedi P, Kulkarni A, Chalipat S and Pardeshi M 2011 Protection devices to improve frontal pendulum impact performance of heavy commercial vehicles *SAE Technical Paper*

Experimental validation of finite element modelling on tibia with osteogenesis imperfecta

H Y Tan[1], K S Basaruddin[1*], S F Khan[1], M H Mat Som[1], R Daud[1] and A R Sulaiman[2]

[1]Biomechanics and Biomaterials Research Group, School of Mechatronic Engineering, Universiti Malaysia Perlis, 02600 Pauh Putra, Perlis, Malaysia
[2]Department of Orthopaedics, School of Medical Science, Universiti Sains Malaysia, 16150 Kubang Kerian, Kelantan, Malaysia.

Abstract. The paper aimed to determine the validity of finite element analysis of tibia bone affected by Osteogenesis Imperfecta (OI) under various angle of bowing of the bone. Finite element (FE) model of human tibia bones were developed with various degree of bowing to simulate different severity of OI, and then the geometrical models of these FE models were used to develop bone specimen using fused deposition method (FDM) rapid prototyping technique. Compression tests were conducted to the bone specimen and FE analysis were performed by simulating conditions of the compression test, and then results from the compression test and FE simulation were compared. It have been found that both of the results corresponded to each other, therefore it is concluded that FE analysis is able to simulate the mechanical response of tibia bone with OI.

1. Introduction

Osteogenesis Imperfecta (OI), also known as brittle bone disease, is a genetic bone fragility disorder characterised by skeletal deformities. It is one of the most commonly found bone fragility disorder among children, with a statistical incidence of 1 in 10,000 to 1 in 20,000 newborns [1]. OI is a heterogeneous disorder and is categorised into four main types, based on radiographic, genetic and clinical criteria [2]. There are type I, type II, type III and type IV. OI type II is one of the most lethal form of OI and is considered as one of the most common severe skeletal dysplasias [3].

There are several patient-specific models which evaluate the fracture risk in tibia under various loadings [4]–[6]. Caouette at al. [7] employed finite element analysis (FEA) to investigate the fracture risk with respect to tibia deformity in children affected by OI under lateral and torsion impacts loads.. The results suggested that fracture risk for lateral and torsion impact is correlated to the level of tibia bowing. In order to investigate the fracture risk of OI-affected bones using FE analysis, the accuracy of FE models involving bones with OI needs to be validated to verify its accuracy in simulating the mechanical behaviour of bone specimens with a certain bowing angle [8].

Validation of FE model is conducted by comparing the results of FE analysis to another experiment which is not performed numerically, usually a physical experimental test. The geometry of the FE model should reflect the geometry of the physical experimental specimen, and then the results of the numerical simulation and physical experiment is to be compared, and if the results agree to each other, then the FE model is said to be validated with accurate result that reflects real life situation. For the physical experiment, it can be either by using the real bone such as what is performed by Dahan et al [9]. As an alternative, specimen fabricated using rapid prototyping technology to create a specimen that takes the shape of the bone meant to be studied as demonstrated by Johnson & Young and Hambli [10][11]. Johnson & Young performed experiment on skull model fabricated using rapid prototyping, and found that the deformation closely matches to FE analysis result. On the other hand Hambli did FE analysis on trabecular bone and compared the results with experimental results using real bone specimen, and managed to validate the accuracy of FE simulation as well. Both of these researches has shown that FE analysis is able to accurately predict the mechanical responses of complex geometries, this is very valuable because it would help to assist in pre-surgery evaluation on bones. Furthermore, research involving OI such as fracture risk prediction on femur has been done as well by Wanna et al [12]–[14] through the means of FE analysis.

Surgery involving bone of OI requires the prediction of when will the bent bone break to determine when to carry out orthopaedic surgery, such prediction is usually done by a doctor's evaluation based on their experience, which might not be reliable, and therefore the ability of FE analysis to predict the point failure of bone will be highly valuable to determine the best time to carry out surgery for OI treatment. As of now, there is still a clutter with regards to the validation of OI bone FE models, there is a research gap in this field with regards to tibia affected by OI, and therefore there is an interest in the validation of OI tibia FE models to investigate the accuracy of such FE model in predicting the mechanical behaviours of OI tibia. Hence, this research aim to develop FE models based on tibia affected by OI and validate of the accuracy of the FE model on predicting the occurrence of failure for the model.

2. Materials and methods

For the validation of FE simulation results by comparing it to experimental outcome using physical model, 3 sets of tibia geometrical models with different bowing angles to simulate different conditions of OI was developed. The bowing angles imposed onto the bones are 10°, 14°, and 20°. This allows us to have a glimpse on the effect of large and small bowing angles in response to applied forces, the bowing angle 14° was selected to reflect the bowing angle of a patient (male, 18 years old).

2.1. Development of 3D models of tibia

A 3D geometrical model of a standard tibia (Turbosquid, U.S.) is modified in ANSYS Workbench to artificially impose a certain amount of bowing onto the bone. The boundary condition was set up by having the distal end of tibiotalar joint as a fixed support, and a compressive force applied on the distal end of tibiofemoral joint. Further boundary condition is added to restrict the displacement of bone so that it only moves along the vertical axis to help in forming a smooth bowing angle. The geometrical models created are then used in FE simulation and experiment.

2.2. Fabrication of physical model

For the fabrication of physical tibia model, the same 3D geometrical model used in FE simulation is used to create identical physical models through rapid prototyping approach. The rapid prototyping method used was fused deposition modelling (FDM). The material selected for the fabrication of test model is polylactic acid (PLA), the model was scaled to the same size with the size of model used in FE simulation.

2.3. Compression test of physical model

Each of the rapid prototyped models were used to carry out compression tests to determine the responses, and the deformation for each of the models were recorded. In order to cater to the irregular shape of bone specimen, custom jigs were created to provide support for the bone specimens during compression test. Figure 1 shows the setup of the compression test.

IOP Conf. Series: Materials Science and Engineering **670** (2019) 012015 doi:10.1088/1757-899X/670/1/012015

Figure 1. Set up of compression test.

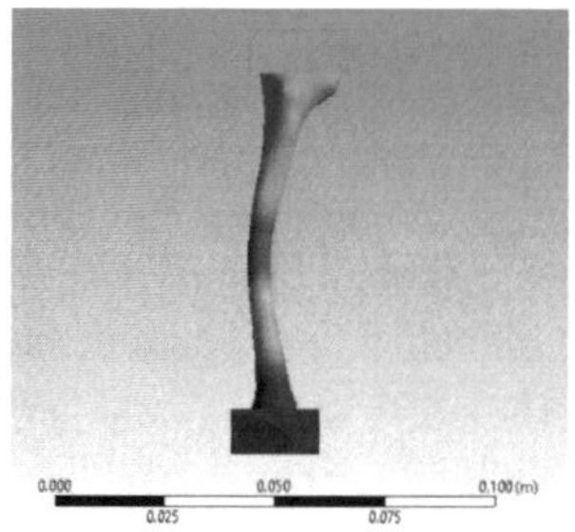

Figure 2. FE analysis in ANSYS.

2.4. FE model validation

The FE analysis was set up to simulate the testing conditions of the compression test machine, the jigs which were used in the compression test was included into the FE analysis as well to further simulate the actual conditions of the compression test. The jig at the lower part of the tibia model was set as fixed support while force was applied from on the jig at the top part of the tibia model. The material properties of the FE model is set as isotropic linear elasticity, Young's modulus and Poisson's ratio of the PLA material which are 1.73GPa and 0.36 respectively was applied to the FE model set up.

For the compression test, only the results up to the point before yielding occurs is taken into consideration as the FE analysis is performed under the assumption that the material is linear elastic, therefore the compression testing results after the material undergone plastic deformation would no longer be reflected in the FE analysis. The forces recorded during the experiment at certain interval points of deformation was applied into the FE analysis, and then the responses of both FE simulation and compression are compared for validation, at least 5 points of interval was recorded for each of the experiments.

3. Results and discussion

The deformation due to compressive force is represented in the form of total displacement, the results both compression test and FE simulation of each tibia model samples were plotted against each other to evaluate how closely does the FE simulation predicts the outcome of experiment. Figure 3 to 5 show the plot for tibia with bowing angle of 10, 14 and 20 degrees respectively.

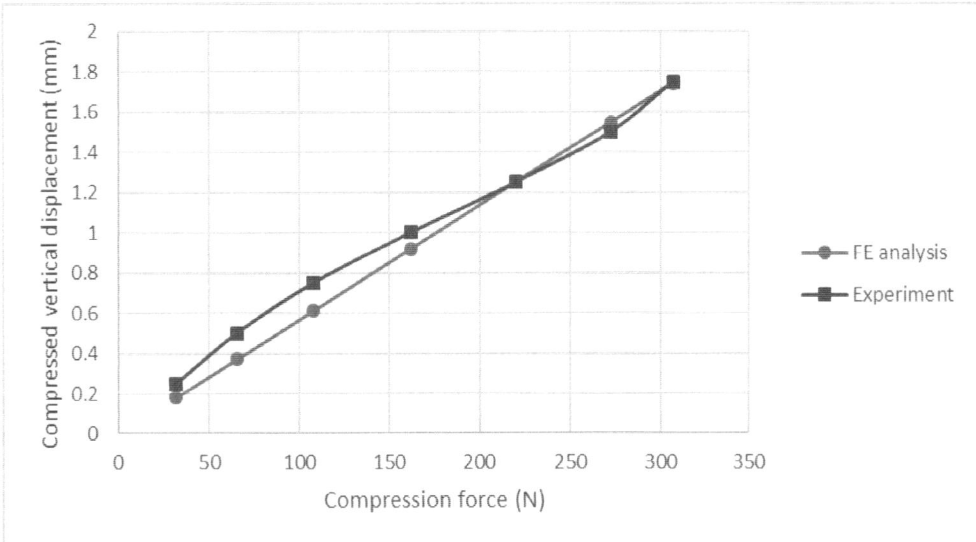

Figure 3. Results of deformation for simulation and experiment of 10° bowing angle.

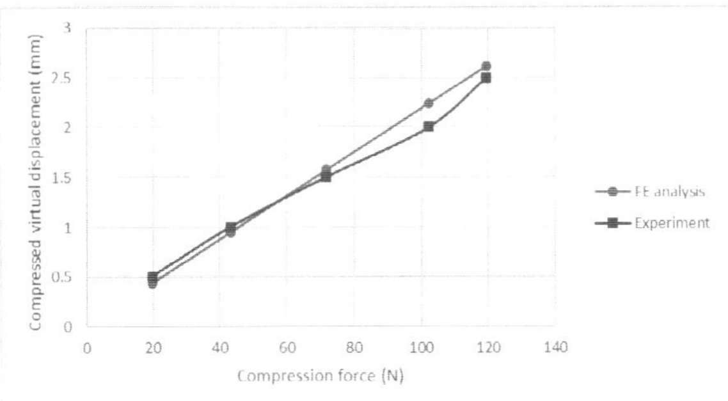

Figure 4. Results of deformation for simulation and experiment of 14° bowing angle.

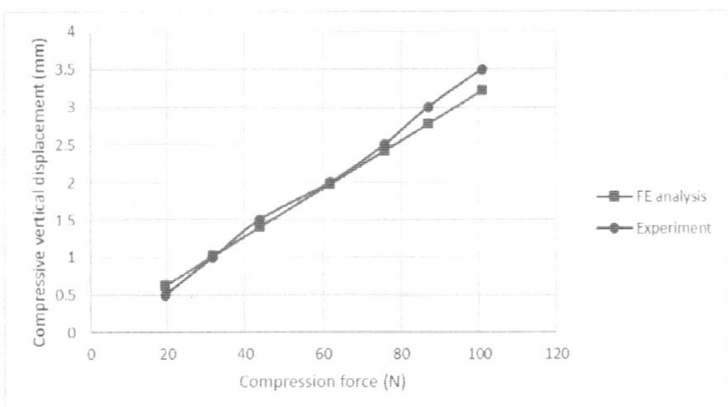

Figure 5. Results of deformation for simulation and experiment of 20° bowing angle.

FE simulations gave perfectly linear predictions of total displacement as the force applied increases. On the other hand, the normalised data from the experiments appears to be linear with slight deviations at each point. The plotted lines for all three sets of experiments crossed each other and appears to be overlapping at certain points, which suggests that the difference between simulation predictions and experimental results are very small at certain points.

For tibia model with bowing angle of 10°, the percentage difference of experimental displacement over FE simulation displacement result ranged from as small as 0.12% to as large as 37.65%. On the other hand, for tibia model with bowing angle of 14°, the smallest percentage difference is 4.45% while the largest percentage difference maxed out at 15.08%. Lastly for tibia model with bowing angle of 20%, the smallest percentage difference here is the smallest among all three sets of data which is 1.18%, while the largest percentage difference for this set of experiment is 20.08%.

It is worth mentioning that for all three sets of experiments, the largest percentage differences occurs at the first point of the plot where the deformation is the smallest, and once the deformation goes over 1mm and above the percentage differences for each of the points significantly drop to below 10%. The reason for higher errors at the lower end of the spectrum suggests that there are errors which occurred at the beginning of the experiment, as FE simulation is unaffected by such error, therefore the error likely lies in the earlier stages of the compression test. This is likely due to the pre-compressed force during the compression test setup that prevents the specimen from slipping away from the machine might induce a slight deformation during the initial stage, which is unavoidable, and therefore the initial readings up until 1mm contain larger errors.

Besides, due to the nature of FDM which has specific direction of printing, the specimen will be anisotropic instead of fully isotropic, resulting in slight difference in the behaviour of materials and cause some errors. Taking account of the slight errors due to FDM, and also the excess force during the setup for compression tests, percentage difference of below 10% between FE prediction and experimental results suggests that are highly correlated, and the errors are still tolerable. This finding is consistent with researches from other researches who validates FE simulations with experiment, which the authors obtained percentage differences of 2% to 7% [15], [16], which considered that such results validate the ability of FE simulation to predict mechanical behaviours of complex 3D geometry with good accuracy. It is also worth noting that in Su et al's finding, sub optimal set ups of FE simulation produced measurement difference of up to 49% for curved surfaces [15], suggesting that it is more complicated to predict complex shapes.

4. Conclusion

Even for geometry as complex human bone, FE analysis was able to predict the mechanical behaviour under compressive loading and produce results which matches closely with results from experimental tests. In order to achieve FE prediction with high accuracy, the simulation environment must depict the real environment as close as possible, by having appropriate boundary conditions. While the tolerance were high in the experiments of this paper, it was still able to produce satisfactory results, in the continuation of this work there are areas of improvement which might be able to further validate the usefulness of FE analysis, such as using rapid prototyping with smaller tolerance, minimizing parallax error in experimental set ups, and also fine tune the FE model by choosing more suitable type of mesh for different surfaces to further validate FE modelling results with higher confidence.

Acknowledgement

This work was financially supported by the Ministry of Education Malaysia under Fundamental Research Grant Scheme, FRGS/1/2016/TK03/UNIMAP/02/6.

References

[1] Hambli R and Allaoui S 2013 A robust 3D finite element simulation of human proximal femur progressive fracture under stance load with experimental validation *Ann. Biomed. Eng.* **41** (12) 2515–2527.

[2] Sillence S O, Senn A and Danks D M 1979 *J. Med. Genet.* **16** (2) 101–16.

[3] Alhousseini A, Mahaseth M, Zeineddin S, Jaiman S, Berman S, Bryant D, Tan S and Hernandez-Andrade E 2018 A Non-Lethal Osteogenesis Imperfecta Type II Mutation Gynecol *Obstet. Invest.* **48201** 1–5.

[4] Chevalier Y, Pahr D, Allmer H, Charlebois M, and Zysset P 2007 Validation of a voxel-based FE method for prediction of the uniaxial apparent modulus of human trabecular bone using macroscopic mechanical tests and nanoindentation *J. Biomech.* **40** (15) 3333–40.

[5] Schileo E, Taddei F, Cristofolini L and Viceconti M 2008 *J. Biomech.* **41** (2) 356–367.

[6] Mansor, Beh Y M, Mat Som M H, Basaruddin K S, Mustafa N, Yazid H and Salleh A F 2018

Reconstruction of cortical and cancellous bone in tibia with osteogenesis imperfecta *Journal of Telecommunication, Electronic and Computer Engineering* **10** (1-16) 115-119.
[6] Caouette C, Rauch F, Villemure I, Arnoux P J, Gdalevitch M, Veilleux L N, Heng J L and Aubin C E 2014. *J. Musculoskelet. Neuronal Interact* **14** (2) 205–212.
[7] Anderson A E, Ellis B J and Weiss J A 2007 *Comput. Methods Biomech. Biomed. Engin* **10** (3) 171–184.
[8] Dahan G, Trabelsi N, Safran O and Yosibash Z 2016 Verified and validated finite element analyses of humeri *J. Biomech.* **49** (7) 1094–1102.
[9] Johnson E A C and Young P G 2005 *J. Biomech.* **38** (1) 39–45.
[10] Hambli R 2013 *Bone* **56** (2) 363–374.
[12] Wanna S B C, Basaruddin K S, Mat Som M H, Sulaiman A R, Shukrimi A, Khan S F, Abdul Majid M S and Ridzuan M J M 2018 Fracture risk prediction on children with Osteogenesis Imperfecta subjected to loads under activity of daily living *IOP Conf. Ser. Mater. Sci. Eng.* **429** (1)
[13] Wanna S B C, Basaruddin K S, Mat Som M H, Mohamad Hashim M S, Daud R, Abdul Majid M S and Sulaiman A R 2017 Prediction on fracture risk of femur with Osteogenesis Imperfecta using finite element models: Preliminary study, *Journal of Physics: Conference Series* **908** (1) Article number 012022.
[14] Wanna S B C, Basaruddin K S, Som M H M, Salleh A F and Sulaiman A R 2018 Effect of loading direction on fracture of bone with osteogenesis imperfecta (OI) during standing *AIP Conference Proceedings* **2030**, Article number 020094.
[15] Su R, Campbell G M and Boyd S K 2007 *Med. Eng. Phys.* **29** (4) 480–490.
[16] Sutradhar A, Park J, Carrau D and Miller M J 2014 *Comput. Biol. Med.* **52** 8–17.

Normalization and weighting methods for precise and standardized sustainability assessment: recent practices in manufacturing

S Ahmad[1,2] , K Y Wong[1] , I Zahid[3], Z Hussain[4], M S K[5,2] and M Sarfraz[6]

[1]Department of Manufacturing and Industrial Engineering, School of Mechanical Engineering, Universiti Teknologi Malaysia, Skudai, Malaysia
[2]School of Mechanical and Manufacturing Engineering, National University of Sciences and Technology, Islamabad, Pakistan
[3]Department of Mechanical Engineering Technology, Government College University, Faisalabad, Pakistan
[4]Faculty of Management Sciences, National Textile University, Faisalabad, Pakistan
[5]Department of Applied Mechanics and Design, School of Mechanical Engineering, Universiti Teknologi Malaysia, Skudai, Malaysia
[6]Azman Hashim International Business School, Universiti Teknologi Malaysia, Skudai, Malaysia

Abstract. To address the challenges of sustainable manufacturing, precise, transparent and standardized sustainability assessment is essential. In this regard, normalization and weighting approaches are key tools to improve the precision, transparency and robustness of sustainability assessment efforts. However, mostly, the available literature on sustainability assessment discusses less about normalization and weighting. Moreover, previously, most of the related sustainability assessment and review studies were based on normalization and weighting for life cycle assessment (environmental dimension) only. Thus, this paper aims to present the recent status quo of normalization and weighting practices for sustainability assessment in manufacturing. Unlike previous review studies, this paper includes all three dimensions of sustainability (environment, economy and society). In order to achieve this objective, recent sustainability assessment studies (published in last 10 years) were reviewed and analyzed, from normalization and weighting viewpoints. The results showed that a majority of the reviewed studies which considered all three dimensions of sustainability were based on internal normalization. In contrast, the environmental assessment based studies were mostly grounded on external normalization. In addition, most of the reviewed studies considering all three dimensions of sustainability were concerned with normalization of input and/or output indicators, whereas environment based studies were mostly normalized for impact indicators. For weighting purposes, the analytic hierarchy process method was used most commonly, whereas the Delphi method and others were less frequently employed. Overall, more future work is required to increase awareness and usage of normalization and weighting methods for sustainability assessment in manufacturing.

1. Introduction

Manufacturing activities and products affects all three dimensions of sustainability; environment, economy, and society throughout their entire life cycle [1]. In order to evaluate and improve the performance, sustainability assessment is undertaken to support decision-making [2]. However, less literature is focused on how to make sustainability assessment efforts more precise, transparent, and standardized (comparable). Along with other reasons, the use of different and inconsistent normalization and weighting methods lead to differences and uncertainties in sustainability assessment results [3-5]. Moreover, most of the available literature on sustainability assessment discusses less clearly about normalization and weighting [3, 6], which makes it difficult to compare the results of different studies [7].

Due to inconsistent units of measurement, the data gathered for different indicators are not summed up together directly [8]. Resultantly, various normalization approaches are used which convert physical measurements into dimensionless scores. In this way, it increases the comparability of sustainability data and results, by transforming data into a compatible or comparable unique scale [3, 9]. Normalization is usually followed by weighting when relative importance of indicators or impact categories is needed [10]. If one indicator is more "important" than another, the former is assigned a higher weight than the latter [11]. Though, both normalization and weighting processes are challenging and complex tasks, they are crucial for more precise and more standardized sustainability assessment.

Previously, environment was the only concern for sustainability assessment studies which were normally based on life cycle assessment (LCA), etc. Because of this fact, most of the related review articles were also founded on normalization and weighting for LCA [5, 9, 10, 12]. However, the triple bottom line (TBL) concept of sustainability [13, 14] requires to consider all three aspects of sustainability, comprehensively. Moreover, previous studies were based on general discussions, whereas no study has reported the significance and application of normalization and weighting methods from the manufacturing viewpoint. Thus, this paper presents the review of these methods for sustainability assessment in manufacturing, from the TBL viewpoint. The objective is to present the status quo of normalization and weighting practices along with highlighting the usefulness of these approaches for more precise and more standardized sustainability assessment. For this purpose, sustainability assessment studies for manufacturing (published from 2009 to 2019) were reviewed and analyzed.

2. Review of normalization and weighting methods

Before presenting the review of various normalization and weighting methods, various related tools used in these studies are briefly described. Life cycle assessment (LCA) is used to assess the environmental burdens associated with a product, process, or service, etc. [15, 16]. The Eco-indicator 99 (H) V2.05/Europe EI 99 H/A method belongs to damage oriented impact assessment or endpoint methodologies (damage caused directly to human health, ecosystem and resources) [15], whereas the ReCiPe is a method used for both mid-point and end-point impacts' assessments [17].

Analytic hierarchy process (AHP) is especially suitable for complex decisions which involve the comparison of decision criteria (indicators) that are difficult to quantify. This gives a weighting for each criterion within a cluster (or level of the hierarchy) [18]. The Delphi method is effective in achieving consensus when there is uncertain information [19, 20] and identifying, selecting and validating factors and indicators [21, 22]. The reviewed studies were based on various types of indicators, such as input inventory indicators (material used, water used, energy used, etc.), output inventory indicators (emissions to air, emissions to water, etc.), and impact categories (midpoint, endpoint, etc.). The review of various normalization and weighting methods for sustainability assessment studies is presented in Table 1.

Table 1: Normalization and weighting methods for sustainability assessment in manufacturing

No.	Description of study	Normalization and weighting approaches	Reference
1	This research was based on sustainability evaluation, while considering technical, economic and environmental aspects for automotive coating technologies. The assessment was grounded on various input and output inventory indicators, along with some impact categories. However, the social dimension of sustainability was missed.	External normalization was done only for the environmental impacts that were based on the Singapore's population data, while using Eco-indicator 99 as an impact assessment method. The domain experts provided the relative importance (weight) of each indicator (criterion) and the weights were presented on a scale from 0 to 10. However, no details were provided on how these weights were calculated.	[23]
2	This study was based on the life cycle assessment (LCA) of a food manufacturing industry while including all life cycle phases.	LCA based impact categories were externally normalized. Normalization results were based on CML 2 baseline 2000 and Eco-indicator 99 (European	[15]

	The highest environmental burdens turned out to be from food ingredients and solid waste. However, economic and social dimensions were ignored.	data) as the reference points. The normalization was undertaken for both midpoint and endpoint impact categories. However, weighting of indicators or impact categories was not discussed.	
3	A methodology was developed for establishing Product Sustainability Index (ProdSI) for manufactured products based on all three dimensions of sustainability. It was tested for a fictitious case study while using equal weights.	Internal normalization was applied to convert measured data into dimensionless scores on a scale from 0 to 10. The score of 10 represents the best case and 0 shows the worst scenario. Only the input inventory related indicators were normalized.	[8]
4	This study reported the indicators based sustainability assessment approach for discrete manufacturing processes (grinding) while including all three dimensions of sustainability. The analysis was conducted for a quick comparison between different process variants.	The input and output inventory indicators were internally normalized whereas the impact categories were not discussed. Indicators were ranked based on degree of fulfillment from 1 to 10. The weights of the indicators were assigned in terms of percentage of their relevance, where all weights of the indicators add up to 100%.	[6]
5	LCA of wheat gluten powder and derived packaging film was reported in this paper. Results showed that the impacts of the wheat cultivation and gluten drying phase are significant in the ReCiPe midpoint assessment method.	External normalization of LCA based impact categories was undertaken. ReCiPe method was used for this purpose which was based on European data as a reference point. Normalization was conducted based on the percentage contribution of impact categories.	[24]
6	The LCA of cheese manufacturing was conducted in the USA, which overlooked the economic and social dimensions of sustainability. The results showed that aquatic ecotoxicity has the largest relative impact.	Impact categories which were measured through LCA were normalized with external reference. The reference data were based on the Impact 2002+ US midpoint assessment framework.	[17]
7	The paper was based on sustainability assessment for small-scale manufacturing: caddisfly jewelry production. All three dimensions of sustainability were included in order to investigate the performance of two manufacturing strategies.	Internal normalization was undertaken in which indicators were normalized in the form of a percentage. Both input and output inventory based indicators were included. However, the impact categories were not included. No weights were assigned to the indicators.	[25]
8	Sustainability indicators were used for finishing operations (material removal category) in order to assess the performance. All three dimensions of sustainability were considered along with quality as a fourth aspect.	Both input and output inventory based indicators were normalized. The approach was based on internal comparison of indicators. Normalization helped to plot indicators with different units in the same curve. The performance was normalized on a scale between 1 and 10.	[26]
9	The study reported the assessment for a production work cell while conducting impact assessment of all three dimensions of sustainability. The approach was tested for a representative machining work cell producing stainless steel knives.	Social impacts were normalized by evaluating the difference in performance and local standard. The endpoint approach (ReCiPe) was used to normalize impacts. For weighting, both an objective statistical method and subjective pairwise comparisons (AHP method) were used. Experts ranked indicators' weights from 1 to 9 and normalized them from 0 to 1.	[27]
10	The sustainability index of manufacturing was calculated at organizational and operational levels for a plastic manufacturing firm by using fuzzy analytic hierarchy process (fuzzy AHP). All three dimensions of sustainability were included for input and output indicators.	Input and output indicators were internally normalized on a scale from 0 to 10. A high score means that the company has a relatively higher performance. For weighting, a group of experts rated each indicator on a scale of 1–10 in the context of the fuzzy AHP methodology.	[28]
11	This study reported the sustainability assessment of manufacturing processes of boiler, requiring less detailed data, time, and expert knowledge. Indicators of all three dimensions of sustainability along with technological aspects were included for comprehensive assessment.	Indicators were assigned performance scores by the experts based on a 5-point Likert scale, which were later normalized on a scale from 0 to 1. It was an internal normalization which was based on an internal comparison.Weights of indicators were assigned using the AHP method on a scale from 1 to 9.	[29]
12	This research developed an integrated sustainability assessment method that includes both stochastic and fuzzy uncertainties for the Malaysian food manufacturing industry. It considers both quantitative and qualitative indicators for all three dimensions of sustainability.	Industry data were collected for all input and output inventory indicators. These data were normalized internally on a scale from 0 to 1. For weighting purposes the Delphi method was used in which experts ranked the indicators based on a Likert scale. The weights were calculated on a scale from 0 to 1 for all indicators.	[30]

3. Analysis and Discussion

An analysis of the reviewed studies is presented in Table 2. Table 2 shows that comparatively the sustainability assessment studies were more focused on the environmental dimension of sustainability than the social or economic dimension. With respect to internal or external normalization, an interesting fact was observed. Except for [27] which was based on all three dimensions of sustainability and used external reference for normalization, all TBL based studies were grounded on internal normalization. For internal normalization, the performance data of an indicator were internally compared with the performance of other indicators. However, all the reviewed studies which were concerned with environmental assessment only, were based on external normalization.

The reason for this might be comparatively, external reference data were more readily available for environmental assessment, but they were normally not available for economic and social assessments. From the viewpoint of indicators' type, most of the TBL based reviewed studies were based on normalization of input and/or output indicators. However, the environment based assessment studies were generally normalized at the impact indicator level. In addition, with respect to normalization, comparatively the normalization scale from 0-1 and 0-10 were used more frequently than the other scales.

From the weighting viewpoint, several reviewed studies used the AHP method to assign weights to indicators. Only two reviewed studies employed the Delphi method and/or consultation approach in order to assign weights to indicators. Overall, not all the reviewed studies used weighted indicators. There were various reviewed studies which did not consider the weighting scheme (equal or not equal). Additionally, in the reviewed studies, four different scales (0-1, 1-9, 1-10 and percentage) were used for weighting of indicators. However, from the repeatability and consistency viewpoint, no scale was used more than twice.

Table 2. Analysis of reviewed studies for normalization and weighting methods

No.	Reference	Sustainability dimensions			Normalization								Weighting						
					Reference point		Type of indicators			Scale			Approach		Scheme		Scale			
		Environment	Economy	Society	Internal	External	Input	Output	Impact	0-1	0-10	Percentage	Consultation/ Delphi method	AHP based	Equal	Not equal	0-1	1-9	1-10	Percentage
01	[23]	√	√			√			√				√			√			√	
02	[15]	√				√			√											
03	[8]	√	√	√	√		√				√				√					
04	[6]	√	√	√	√		√	√								√				√
05	[24]	√				√			√			√								
06	[17]	√				√			√											
07	[25]	√	√	√	√		√	√												
08	[26]	√	√	√	√		√	√			√									
09	[27]	√	√	√		√	√	√	√	√				√		√		√		
10	[28]	√	√	√	√		√	√			√			√		√			√	
11	[29]	√	√	√	√		√	√		√				√		√		√		
12	[30]	√	√	√	√		√	√		√			√			√	√			
Total frequency		12	9	8	7	5	8	7	5	3	3	1	2	3	1	6	1	2	2	1

Overall, for sustainability assessment in manufacturing the normalization and weighting approaches are not clearly described and commonly used and their usage is quite inconsistent. This inconsistent and infrequent usage of normalization and weighting methods in manufacturing, in a way pinpointed various challenges to achieve precise, transparent and comparable sustainability assessment. There is still considerable discussion on how normalization and weighing should be done.

4. Conclusions

Normalization and weighting approaches can play a significant role to make sustainability assessment more precise and more standardized. However, for sustainability assessment based research in manufacturing, these approaches have not been given due attention. So, this study is aimed to highlight the importance of these approaches and present a review of various normalization and weighting methods for sustainability assessment in order to provide a comprehensive analysis and recent picture of normalization and weighting trends and their applications in manufacturing.

This study showed that from the normalization viewpoint, the majority of the reviewed studies were based on internal normalization, especially the studies considering all three dimensions of sustainability. However, the studies based on environmental assessment were mostly grounded on external normalization. This might be because most of the approaches for environmental assessment (LCA, etc.) were more mature and carried external reference data. From the indicators viewpoint, most of the TBL based studies were concerned with normalization of input and/or output indicators, whereas the environment based studies were generally normalized at the impact category level. For weighting, the AHP method was used more frequently than the Delphi or other methods. Overall, more future work is required in order to increase the awareness and usage of normalization and weighting methods to enable more precise and comparable sustainability assessment in manufacturing.

References

[1]. Tarne P, Traverso M and Finkbeiner M 2017 Review of Life Cycle Sustainability Assessment and Potential for Its Adoption at an Automotive Company *Sustainability* **9** (4) 670.

[2]. Sala S, Ciuffo B and Nijkamp P 2015 A systemic framework for sustainability assessment *Ecological Economics* **119** 314-25.

[3]. Olsthoorn X, Tyteca D, Wehrmeyer W and Wagner M 2001 Environmental indicators for business: a review of the literature and standardisation methods *Journal of Cleaner Production* **9** (5):453-63.

[4]. Saisana M, Saltelli A and Tarantola S 2005 Uncertainty and sensitivity analysis techniques as tools for the quality assessment of composite indicators *Journal of the Royal Statistical Society: Series A (Statistics in Society)* **168** (2) 307-23.

[5]. Pizzol M, Laurent A, Sala S, Weidema B, Verones F and Koffler C 2017 Normalisation and weighting in life cycle assessment: quo vadis? *The International Journal of Life Cycle Assessment* **22** (6) 853-66.

[6]. Linke B S, Corman G J, Dornfeld D A and Tönissen S 2013 Sustainability indicators for discrete manufacturing processes applied to grinding technology *Journal of Manufacturing Systems* **32** (4) 556-63.

[7]. Maxim A 2014 Sustainability assessment of electricity generation technologies using weighted multi-criteria decision analysis *Energy Policy* **65** 284-97.

[8]. Zhang X, Lu T, Shuaib M, Rotella G, Huang A, Feng S, et al. 2012 A metrics-based methodology for establishing product sustainability index (ProdSI) for manufactured products. In: Linke BS, editor. *Leveraging technology for a sustainable world.* Berlin Heidelberg: Springer 435-41.

[9]. Prado V, Wender B A and Seager T P 2017 Interpretation of comparative LCAs: external normalization and a method of mutual differences *The International Journal of Life Cycle Assessment* **22** (12) 2018-29.

[10]. Pennington D, Potting J, Finnveden G, Lindeijer E, Jolliet O, Rydberg T, et al. 2004 Life cycle assessment Part 2: Current impact assessment practice. *Environment International* **30** (5) 721-39.

[11]. Böhringer C and Jochem P E 2007 Measuring the immeasurable—A survey of sustainability indices *Ecological Economics* **63** (1) 1-8.

[12]. Reap J, Roman F, Duncan S and Bras B 2008 A survey of unresolved problems in life cycle assessment *The International Journal of Life Cycle Assessment* **13** (5) 374 - 88.

[13]. Hacking T and Guthrie P 2008 A framework for clarifying the meaning of Triple Bottom-Line, Integrated, and Sustainability Assessment *Environmental Impact Assessment Review* **28** (2) 73-89.

[14]. Hall T J 2011 The triple bottom line: what is it and how does it work? *Indiana Business Review* **86** (1) 4-8.

[15]. Calderón L A, Iglesias L, Laca A, Herrero M and Díaz M 2010 The utility of Life Cycle Assessment in the ready meal food industry *Resources, Conservation and Recycling* **54** (12) 1196-207.
[16]. Ahmad S and Wong K Y 2018 Sustainability assessment in the manufacturing industry: a review of recent studies *Benchmarking: An International Journal* **25** (8) 3162-79.
[17]. Kim D, Thoma G, Ulrich R, Nutter D and Milani F 2014 Life cycle assessment of cheese manufacturing in the United States *9th International Conference on Life Cycle Assessment in the Agri-Food Sector* 634-40.
[18]. Ahmad S, Wong K Y, Tseng M L and Wong W P 2018 Sustainable product design and development: A review of tools, applications and research prospects *Resources, Conservation and Recycling* **132** 49-61.
[19]. Delbecq A L, Van de Ven A H and Gustafson D H 1975 *Group techniques for program planning: A guide to nominal group and Delphi processes: Scott Foresman.*
[20]. Murphy M 1998 Consensus development methods and their use in clinical guideline development *Health Technology Assessment* **2** (3) 1-88.
[21]. Davies S, Romano P S, Schmidt E M, Schultz E, Geppert J J and McDonald K M 2011 Assessment of a novel hybrid Delphi and nominal groups technique to evaluate quality indicators *Health Services Research* **46** (6, pt.1) 2005-18.
[22]. Vatalis K I, Manoliadis O G and Mavridis D G 2012 Project performance indicators as an innovative tool for identifying sustainability perspectives in green public procurement *Procedia Economics and Finance* **1** 401-10.
[23]. Yang Q, Chua B, editors. Metrics-based sustainability evaluation of manufacturing technologies: surface coating case study *ASME 2009 International Design Engineering Technical Conferences and Computers and Information in Engineering Conference* San Diego, California.
[24]. Deng Y, Achten W M, Van Acker K and Duflou J R 2013 Life cycle assessment of wheat gluten powder and derived packaging film *Biofuels, Bioproducts and Biorefining* **7** (4) 429-58.
[25]. Armstrong J L, Garretson I C and Haapala K R 2014 Gate-to-Gate Sustainability Assessment for Small-Scale Manufacturing Businesses: Caddisfly Jewelry Production. *International Design Engineering Technical Conferences (IDETC)* Buffalo, USA V004T06A42.
[26]. Linke B, Das J, Lam M and Ly C 2014 Sustainability indicators for finishing operations based on process performance and part quality *Procedia CIRP* **14** 564-9.
[27]. Zhang H and Haapala K R. Integrating sustainable manufacturing assessment into decision making for a production work cell *Journal of Cleaner Production* **105** 52-63.
[28]. Ocampo L A, Clark E E and Promentilla M A B 2016 Computing sustainable manufacturing index with fuzzy analytic hierarchy process *International Journal of Sustainable Engineering* **9** (5) 305-14.
[29]. Kluczek A 2016 Application of multi-criteria approach for sustainability assessment of manufacturing processes *Management and Production Engineering Review* **7** (3) 62-78.
[30]. Ahmad S, Wong K Y and Zaman B 2019 A comprehensive and integrated stochastic-fuzzy method for sustainability assessment in the Malaysian food manufacturing industry *Sustainability* **11** (4) 948-71.

IOP Conf. Series: Materials Science and Engineering **670** (2019) 012017 doi:10.1088/1757-899X/670/1/012017

Evaluation of ΔJ-integral for a shallow crack in steel for pressure vessels under large scale yielding (LSY) condition

M S Muhamad Azmi[1*], Tomoyuki Fujii[2], Keiichiro Tohgo[2], M S M Hashim[1], A H Ismail[1] and Z M Razlan[1]

[1]School of Mechatronic Engineering, Universiti Malaysia Perlis (UniMAP), Pauh Putra Campus, 02600 Arau, Perlis, Malaysia.
[2]Department of Mechanical Engineering, Shizuoka University, 3-5-1, Johoku, Naka-ku, Hamamatsu 432-8561, Japan

E-mail: *safwanazmi@unimap.edu.my

Abstract. Fracture mechanics studies on fatigue crack growth under elastic-plastic condition is usually characterized by J integral range, ΔJ determined by an approximation from area under load-displacement hysteresis loop developed by Dowling and Begley. Despite being supported by some experimental data which reported that fatigue crack growth rates under elastic-plastic conditions have good correlations with ΔJ, the applicability of ΔJ on crack growth in a material subjected to cyclic bending remains inconclusive. In this study, finite element analysis of a four point-bending rectangular plate model with an edge crack length a/W between 0.1 to 0.5 was carried out under displacement amplitude δ_a between 0.1 to 1.5 mm to represent small to large scale yielding conditions. The elastic component was found to be the driven force of fatigue crack growth parameter in small scale yielding (SSY) condition while in large scale yielding (LSY) condition, the plastic component must be considered. It was also found that the simple method of ΔJ-integral calculation using load-displacement approach is overestimated for shallow crack under LSY condition.

1. Introduction

The J-Integral range, ΔJ calculated from a load-displacement approach which was introduced by Dowling and Begley has been used as a parameter to correlate with fatigue crack growth rate under elastic-plastic condition [1]. Several studies have been conducted to discussed ΔJ from theoretical point of view [2]-[4]. It was found that ΔJ values from finite element analysis of elastic-plastic crack growth under cyclic loading shows path independency using path-integral method [2]. The validity of simple estimation technique by Dowling and Begley on ΔJ calculation has been discussed by Azmi et al. [5]. They compare ΔJ value obtained from path-integral (ΔJ_{path}) and simple estimation method ($\Delta J_{p\text{-}\delta}$) on cyclic bending deformation and found that the stress distribution ahead of crack tip could be characterized using ΔJ calculated and elastic-plastic fatigue crack growth rate would be evaluated using ΔJ. In recent studies, fatigue crack growth tests have been conducted using CT specimen as well as finite element analysis to establish an expression of ΔJ [6][7]. This show that the evaluation of ΔJ on the fatigue crack growth property remains controversial. Furthermore, the applicability of the simple estimation technique to calculate ΔJ on shallow crack conditions on cyclic bending remains unclear. Thus, finite element analysis of a four-point bending rectangular plate specimen with various edge crack length was carried out in this study under small to large scale yielding conditions.

IOP Conf. Series: Materials Science and Engineering **670** (2019) 012017 doi:10.1088/1757-899X/670/1/012017

2. Methodology

2.1. Material

The material to be analyzed in this study was hot rolled SPV235 which is a steel for pressure vessels used for intermediate temperature services. The mechanical properties of the material are as follows; the Young's Modulus E = 215 GPa, Poisson's Ratio v = 0.32, yield stress σ_y = 247 MPa and work-hardening exponent n = 0.21. For the sake of simplicity, the monotonic stress-strain relation is used.

2.2. Analytical Procedure

Figure 1 shows finite element mesh of two-dimensional model for half region of rectangular plate specimen with an edge crack of a, having a width of W = 15 mm and length of L = 70 mm. An elastic-plastic finite element analysis was conducted on a four-point bending rectangular model using an outer and inner span of L_o = 60 mm and L_i = 30 mm, respectively. The edge crack length, a is modelled in range between 1.5 to 7.5 mm (a/W = 0.1 to 0.5) for the simulation processes. Due to symmetry, only half region of the model was analysed. The 8-noded isoparametric elements were used. The simulation was conducted under plain strain condition using ANSYS 15.0 Software [8]. In the symmetric plane, a rigid plane as a contactor was introduced to simulate crack open and closure behaviour as shown in Fig. 1(b). The coefficient of friction was set to be zero as the contact behaviour at the crack surface. The contact simulation was conducted using the augmented lagrangian method [9].

Fully reversed loading is used for the loading process. First, the displacement was applied up to $+\delta_a$, then the reversal displacement was applied up to $-\delta_a$. Finally, the displacement was again applied up to $+\delta_a$. The applied displacement amplitude δ_a was change in the range between 0.1 mm to 1.5 mm. Note that the direction of displacement δ to open the crack were set to be positive.

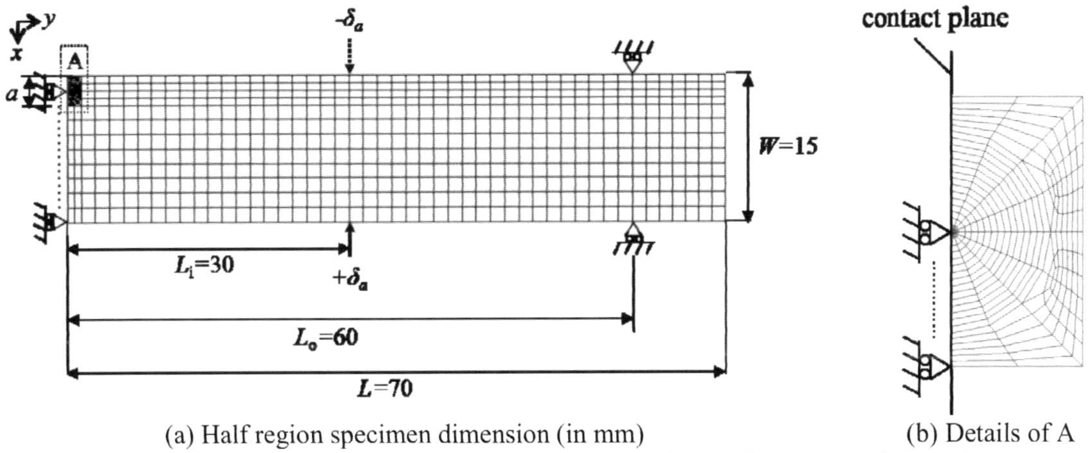

(a) Half region specimen dimension (in mm) (b) Details of A

Figure 1. Finite Element Mesh for half region of four-point bending rectangular specimen.

2.3. ΔJ Calculation Method

Using the load-displacement hysteresis loop from a simulation result, $\Delta J_{p\text{-}\delta}$ was calculated as in eq. (1) [10].

$$\Delta J_{p-\delta} = \Delta J_e + \Delta J_p \tag{1}$$

where, ΔJ_e and ΔJ_p represent an elastic and plastic component, respectively. The value of ΔJ_e was calculated by following equation.

$$\Delta J_e = \frac{(1-v^2)\Delta K_I}{E} \tag{2}$$

IOP Conf. Series: Materials Science and Engineering **670** (2019) 012017 doi:10.1088/1757-899X/670/1/012017

where, ΔK_I is stress intensity factor range, and the ΔK_I was calculated using the handbook [11]. Also, ΔJ_p was calculated by

$$\Delta J_p = \frac{2A_{pl}}{t(W-a)} \tag{3}$$

where, A_{pl} corresponding to plastic work obtained from load-displacement curve as defined by Azmi et al. [5] and t is the specimen thickness.
Following the original definition of J-integral for monotonic loading [12], the ΔJ_{path} was defined by

$$\Delta J_{path} = \int_\Gamma \left(\Delta W n_x - \Delta T_i \cdot \frac{\partial \Delta u_i}{\partial x}\right) ds \tag{4}$$

where, ΔT_i is the range of traction vector and Δu_i is the difference in displacement. The value of ΔW is the strain energy density given by following equation,

$$\Delta W = \int_0^{\varepsilon_{ij}} \Delta\sigma_{ij} d\Delta\varepsilon_{ij} \tag{5}$$

where, $\Delta\sigma_{ij}$ and $\Delta\varepsilon_{ij}$ are the range of stress and strain from the corresponding current values to the opening point as a reference point, respectively.

3. Results and Discussions

Figure 2(a) and (b) show typical load-displacement hysteresis loop for a/W = 0.1 with an applied displacement amplitude δ_a of 0.3 and 1.5 mm, respectively. In the first step, crack opening displacement was simulated followed by reversal displacement in second step with crack closure process. Then, crack opening deformation was analysed again in the last step. It was found that for small scale applied-displacement as in Fig. 2(a), the width of the hysteresis loop was very small and elastic deformation was dominant. On the other hand, for large scale applied-displacement shown in Fig. 2(b), the hysteresis loop was wide and plastic deformation was dominant. From Fig. 2, it is difficult to determine the crack opening and closure point based on load-displacement hysteresis loop in shallow crack condition regardless of applied-displacement condition. Therefore, a crack opening point (R-point) base on crack opening point were estimated on the third step of loading process as defined by Azmi et al. and the ΔJ-integral were calculated [5].

(a) δ_a = 0.3 mm (b) δ_a = 1.5 mm

Figure 2. Load-displacement hysteresis loop for a/W = 0.1.

IOP Conf. Series: Materials Science and Engineering **670** (2019) 012017 doi:10.1088/1757-899X/670/1/012017

Figure 3 shows the change in $\Delta J_{p\text{-}\delta}/\Delta J_{path}$ as a function of displacement amplitude by three different crack length i.e. $a/W = 0.1$, 0.3 and 0.5. It was found that in the case of shallow crack, the value of ΔJ calculated based on the simple method proposed by Dowling and Begley ($\Delta J_{p\text{-}\delta}$) did not agree with the ΔJ calculated based on path integral (ΔJ_{path}). However, the difference between these two approaches become small with increasing of crack length. In addition, regardless of crack length the difference also become smaller when the applied displacement δ_a is decrease.

Figure 3. Change in $\Delta J_{p\text{-}\delta}/\Delta J_{path}$ as a function of displacement amplitude δ_a.

Figure 4 shows the value of elastic and plastic component of ΔJ (ΔJ_e and ΔJ_p) versus applied displacement δ_a. It was found that the elastic component is dominant when the applied displacement δ_a is below 0.2 mm while plastic component is nearly zero in this condition. This shows that in this condition, the driving force of fatigue crack growth parameter is only affected by ΔJ_e while ΔJ_p can be neglected. In other words, this condition is under small scale yielding (SSY) condition. On the other hand, plastic component ΔJ_p has major contribution than elastic component ΔJ_e when the applied displacement δ_a is above 0.2 mm. This condition is also known as large scale yielding (LSY) condition.

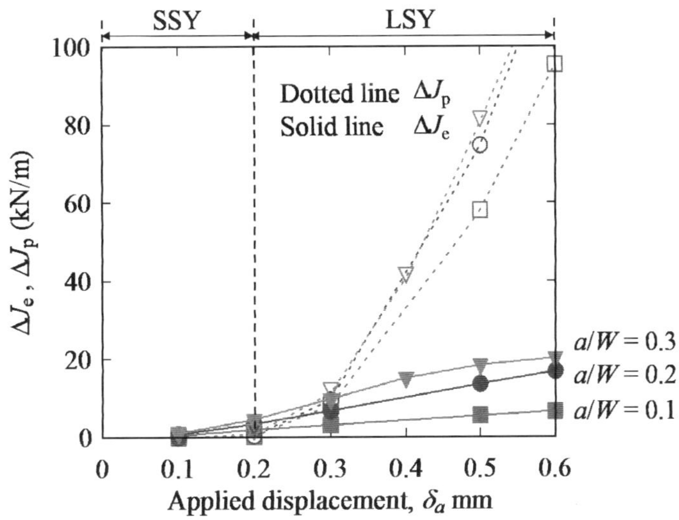

Figure 4. Elastic and plastic component ΔJ_e and ΔJ_p versus applied displacement δ_a.

IOP Conf. Series: Materials Science and Engineering **670** (2019) 012017 doi:10.1088/1757-899X/670/1/012017

4. Conclusions

In this study, finite element analysis of a four point-bending rectangular plate model with various edge crack length was carried out under fully reversed cycling loading from small to large scale yielding conditions. It was found that the load-displacement approach proposed by Dowling and Begley cannot be used for ΔJ calculation in shallow crack under LSY conditions. Moreover, plastic component ΔJ_p has major contribution towards ΔJ values under LSY conditions while in SSY condition, it can be considered that the fatigue crack growth parameter is driven by only elastic component ΔJ_e since the plastic component ΔJ_p is almost zero in this state.

Acknowledgements

The author would like to acknowledge the support from the Fundamental Research Grant Scheme (FRGS) under a grant number of FRGS/1/2018/TK03/UNIMAP/03/6 from the Ministry of Education Malaysia.

References

[1] Dowling N E and Begley J A 1976 *ASTM STP* **590** 82-103.
[2] Kubo S, Yasufo T, Nohara M, Ishimaru T and Ohji K 1989 Investigation on Path-Integral Expression of J-integral Range Using Numerical Simulations of Fatigue Crack Growth *JSME International Journal. Ser. I* **32** (2) 237-244.
[3] K Wütrich C 1982 The Extension of J-integral Applied to Fatigue Cracking *International Journal of Fracture* **20** (2) R35-R37.
[4] Tanaka K 1983 The Cyclic J-integral as a Criterion for Fatigue Crack Growth *International Journal of Fracture* **22** (2) 91-104.
[5] Muhamad Azmi M S, Fujii T, Tohgo K and Shimamura Y 2017 On the ΔJ to characterize elastic-plactic fatigue crack growth *Engineering Fracture Mechanics* **176** 300-307.
[6] Wang J, Jiang W, Li Y, Wang Q and Xu Q 2019 Numerical assessment of cyclic J-integral ΔJ for predicting fatigue crack growth rate *Engineering Fracture Mechanics* **205** 455-469.
[7] Li J, Su C Y, Lu L, Zhang P, Chang L, Miao X T, Zhou B B, He X H and Zhou C Y 2019 Investigation on fatigue crack growth behavior for commercial pure titanium at different crack tip plastic deformed levels *Theoretical and Applied Fracture Mechanics* **100** 1-13.
[8] Stolarski T, Nakasone Y and Yoshimoto S 2006 *Engineering Analysis with Ansys Software* Elsevier Butterworth-Heinemann.
[9] ANSYS mechanical APDL theory reference, Ansys Inc,2013, Version 15.0.
[10] Tohgo K 2004 *Zairyo-kyoudo kaiseki-gaku*, Uchida Roukakuho 129.
[11] Tada H, Paris P C and Irwin G R 1985 *The stress analysis of cracks handbook* Del Research 2-14.
[12] Rice J R 1968 *Journal of Applied Mechanics* **35** 379-386.

IOP Conf. Series: Materials Science and Engineering **670** (2019) 012018 doi:10.1088/1757-899X/670/1/012018

Chip Morphology of Stainless Steel in Oblique Cutting Process using Finite Element Modeling and Simulation

N C Maideen*, Salina Budin, Dzullijah Ibrahim, Lafishahzreen Maharam

Faculty of Mechanical Engineering, Universiti Teknologi MARA, Cawangan Pulau Pinang, 13500 Permatang Pauh, Penang, Malaysia.

*normariah@uitm.edu.my

Abstract. Finite Element Method (FEM) is a numerical modelling technique that can increase the understanding of characteristic and behavior of the machining process. In order to increase the machining efficiency, the suitable selection of cutting parameter is very important. Simulation can be used to compress a time frame and investigate quickly the change of parameter and their effects like in a real situation without producing a physical sample. In addition, simulation study also able to minimize material cost involved due to trial and error manufacturing techniques. Due to many advantages of simulation, this work conducts a study on oblique cutting process. 3D model of turning process of Stainless Steel 316L (workpiece) using tungsten carbide, WC (cutting tool) was developed by FEM software ABAQUS. Three different velocity of cutting tool which are 185,415 and 660 (m/min) was carried out as manipulated variables. Simulation of cutting process was conducted to examine the chip morphology and their stress distribution in the workpiece. Then, chip morphology was compared with previous study and the gap between these two techniques were discussed. From the results obtained, developed model was successfully produced similar chip morphology produced by experiments and able to mimic oblique cutting process.

1. Introduction

Machining is one of the material removal technique in which excess material is removed from a raw material to get desire shape or final geometry. Traditionally, in order to understand and optimize the parameters, it involved high cost and long lead time due to many experiments needs to be done. Turning process is one of the types of machining process that are not exceptional from receiving the bad effects due to understand the operation. Turning process is the metal cutting process that the material of the work piece will be removed by the cutting tool from the surface. In general, the running process can be divided into two types which is orthogonal cutting and oblique cutting. These two types of techniques most likely depending on the position of cutting tool to work piece during turning process. Oblique model is one of the ways to simulate the cutting process with the value of angle or (k) between edge of tool and the direction of cutting speed less than 90° [1].

In previous study, chip morphology is one of the famous parameters around the world to be study. The most common issue found includes industries and educational is about cost due to many repeating and testing to get desire condition in real situation. For instance, the observation of chip formation produced either favorable or unfavorable and was compared with the International Standard Organization (ISO) chip form classification [2]. Then, the study on chip formation in machining titanium alloy that influences by cutting speed was conducted and involved very high cost. Besides that, the influences of cutting speed to chip formation are the essential of information from metallurgical aspects and total damage [3]. Other research on effect of tool grades on Titanium and Stainless steel has produce more than 1400 chips roots [4]. This effort involved high cost to focus on that segmented formation. Moreover, the research about stainless steel on the independent variables for cutting speed and depth of cut was conducted in order to optimum the condition of machining [5]. Next, the cutting speed is more dominant factor contribute to the tool life compare to the feed rate [6]. In a negative side, the cost of this previous study definitely is high due to large quantity of raw materials, man power, and others. Time consume in this experiment is too long due to get optimizing result until achieved also are

considered and take into account.
Thus, Finite Element Modelling (FEM) is one of the reliable techniques to mimic the metal cutting simulation. FEM is an approximation technique of simulation that increase understanding machining process. It can perform complex geometries and wide variety output like chip formation and stress distribution. Definitely, the simulation technique is not completely mimic the real experimental due to reduce complexity and runtime. The most important is simulation can be used to compress time frame and investigate quickly the change of parameter and their effects without producing a physical sample. In this study, turning techniques operation been used between stainless steel 316L as work piece and tungsten carbide as cutting tool. A model of oblique cutting was built using Abaqus software to understand the machining process. Then, developed model will be compared with experiments conducted previously to look their gaps.

2. Finite Element Modeling and Simulation

2.1 Cutting tool model

The tool was developed as 3D deformable body and assigned material is tungsten carbide (WC). The rake angle, clearance angle and oblique angle are 45°, 0° and 20° respectively. The width of tool is 4mm. The size of mesh using approximate global size (0.0005). Figure 1 (a) shows a geometry of tool and Figure 1 (b) shows a developed took in mesh condition. Table 1 shows mechanical and thermal properties of WC.

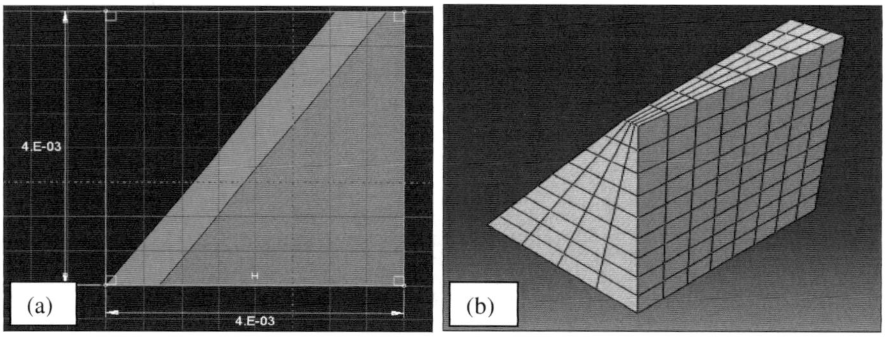

Figure 1. (a) Cutting tool geometry and (b) cutting tool mesh condition

Table 1. Mechanical and thermal properties of WC

Density, ρ (kg/m^3)	Young's Modulus, E (Pa)	Poisson's Ratio, ϑ	Thermal Conductivity, K (W/m.k)	Specific Heat, C (J/kg.k)	Expansion, α (k^-1)
15500	6.50E+11	0.21	50	203	5E-6 at 293K

2.2 Workpiece model

Work piece was developed as stainless steel as 3D deformable body. The measurement of work piece is (20x4x2) mm for (height, length and width). The model consists two partition lower and upper. Upper area for chip formation area while the lower area for the un-machined area (Figure 2 (a)). The depth of cut is 0.5mm and constant at all time. The type meshed used is C3D8R with combine element deletion to build chip formation. The mesh generated as shown in Figure 2(b). The properties materials are shown in Table 2.

IOP Conf. Series: Materials Science and Engineering **670** (2019) 012018 doi:10.1088/1757-899X/670/1/012018

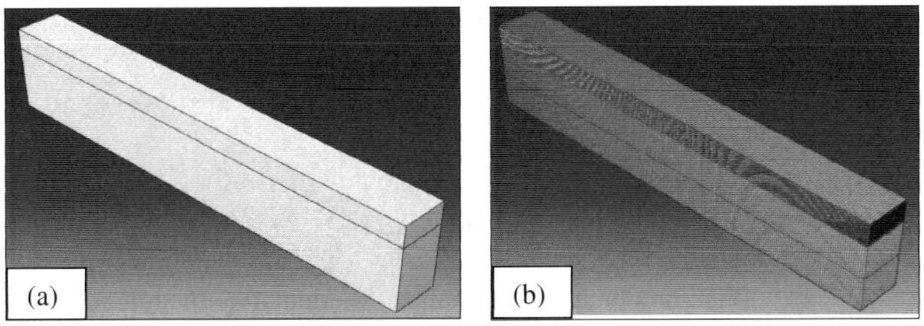

Figure 2. (a) 3D model of workpiece showing partition and (b) workpiece in mesh condition

Table 2. Mechanical and thermal properties of Stainless Steel 316L

Density, ρ (kg/m^3)	Young's Modulus, E (Pa)	Poisson's Ratio, ϑ	Thermal Conductivity, K (W/m.k)	Specific Heat, C (J/kg.k)	Expansion, α (k^-1)
8000	1.93E+11	0.27	16.3	500	~

Johnson cook (JC) model was applied in this simulation. This model was divided into two which are JC plasticity model and JC damage model/ All the reference strain rate at 1.00. The interaction between tool and work piece need to specify where the first contact occur. The zorev's stick slip contact friction was used with value 0.17 constant.

2.3 Boundary conditions
The type of interaction is general contact explicit. In this section, define the manipulated variable which are cutting speed. The tool set to move along the work piece with different speed in 3 simulation which are 185, 415, and 660, (m/min). Then, the work piece also was fixed by selecting the bottom of model. Full boundary condition is shown in Figure 3. Then, the job was created and submitted to run the simulation process.

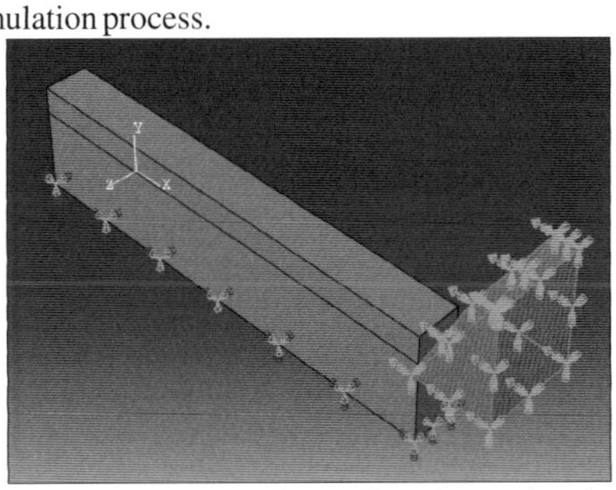

Figure 3. Boundary condition of cutting tool and workpiece model

3. Results and Discussions

3.1 Model and simulation of oblique cutting
Turning process was represented by modelling parts and interaction between cutting tool and work piece. The main objective is to model the moving cutting tool in axial direction towards work piece that is in static condition. Then, the progression of chip has shown agreeable with the position of cutting tool to work piece. Since the study using oblique cutting, the edge of the cutting tool that make contact with work piece are not immediately full contact. It starts contact from the right area and move along the edge to left side of cutting tool. Figure 4 (a) shows the machining process during simulation. The Figure 5 illustrates the progression of chip at three different state which are initial, middle and end.

IOP Conf. Series: Materials Science and Engineering **670** (2019) 012018 doi:10.1088/1757-899X/670/1/012018

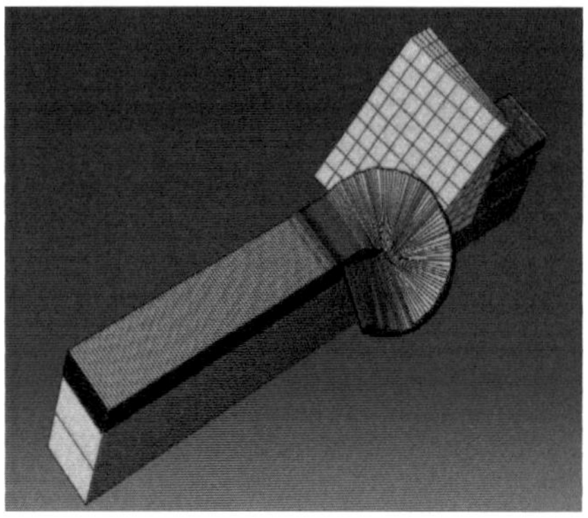

Figure 4. Simulation of oblique cutting

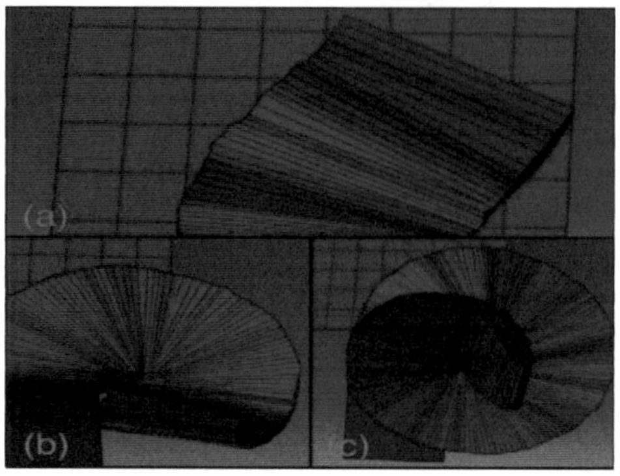

Figure 5. Progression of chip formation during oblique cutting (a) initial, (b) middle, and (c) end

3.2 Comparison between experiments and simulation (chip morphology)

For simulation validation, chip morphology from simulation was compared with chip morphology from experiments at three different speed (185 m/min, 415 m/min, and 660 m/min). Similarity gained from simulation and experiments is, continues chip was produced (Figure 6 and Figure 7). However, the result from the simulation is not showing the difference in terms of distance between curl for each velocity that show by the experimental. The distance between two curls is similar for every velocity. But, the width and thickness of chips are available to compare between both methods. The width and thickness of chip are recorded in this study. The simulation chip for width and thickness considered as ideal/constant value which are 2.0mm and 0.5mm respectively. Then, the width and thickness value for experimental obtained from direct measure on chip using Vernier caliper. The Figure 8 and Figure 9 shows the comparison for both techniques in term of chip width and chip thickness represents by percentage error.

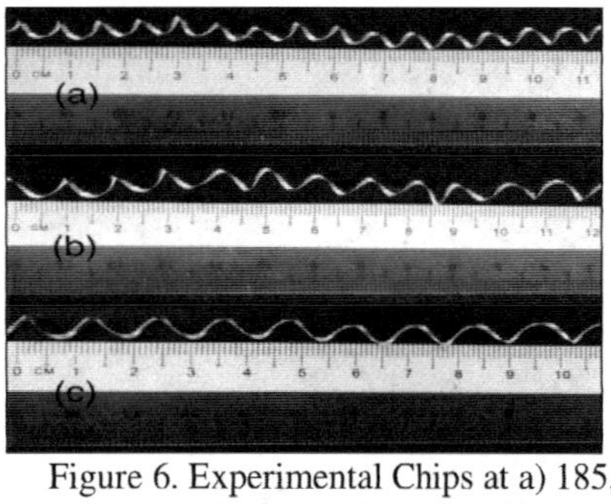

Figure 6. Experimental Chips at a) 185, b) 415 and c) 660, (m/min)

Figure 7. Simulation Chips at a)185, b) 415, and c) 660, (m/min)

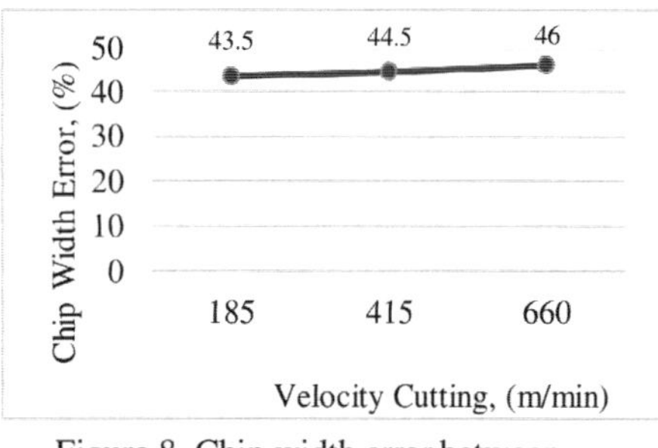

Figure 8. Chip width error between between two techniques

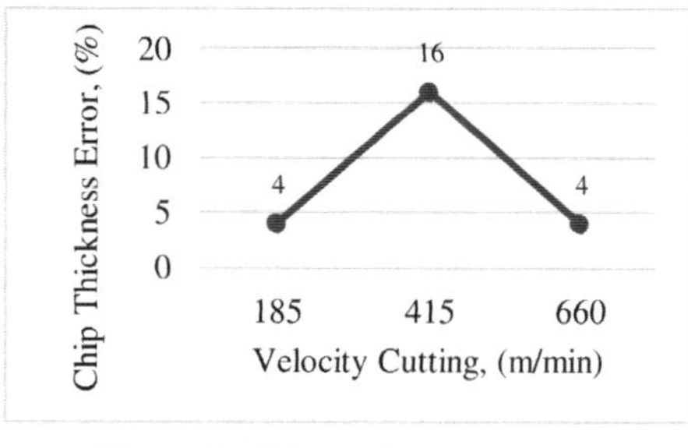

Figure 9. Chip thickness errors between two techniques

The difference of width around 43.5% to 46.0% error as shown in Figure 9. The percentage error is considered high since the normal acceptable percentage error less than 30%. So, it indicates that the size of work piece is important and effect the gap between FEM and experimental in term of width. In this study, the width was reduced half from the width of the cutting tool. Then, after modification of the workpiece width, the difference between simulations and experiments are quite close and less than 2.5% error. For the chip thickness error (Figure 9), the difference between experiments and simulations in the range 4% to 16%. It shows a good and closer similarity in thickness of chip between these two techniques. From the results, the range of chip width and thickness are acceptable and proved that developed model is reliable and able to mimic the actual process of oblique cutting.

4. Conclusion

In this study, a developed model of oblique cutting using FEM software Abaqus/Explicit was used to mimic the real machining process of turning between stainless steel (316L) workpiece and tungsten carbide cutting tool. The simulation results interested and concerned on chip morphology during the cutting. The developed model was indicated and proved that Abaqus/Explicit are able to mimic the oblique cutting process. The chip formation between simulation and experimental show the good correlation and agreeable through the type of chip and the way chip evolved.

References

[1] Pantale O, Rakotomalala R, Touratier M and Hakem N 1996 Three-dimensional numerical model of orthogonal and oblique metal cutting processes *Am. Soc. Mech. Eng. Pet. Div. PD* **75**.

[2] Perera G I P, Herath H M C M, Perera I M S J and Medagoda M G H M M P 2015 Optimising the machining parameters during straight turning of AISI 304 stainless steel and mild **7** (2).

[3] Ben Salem S and Bayraktar E 2012 Effect of cutting parameters on chip formation in orthogonal cutting *J. Achiev. Mater. Manuf. Eng.* **50** (1) 7–17.

[4] Awopetu O O and Ayodeji S P 2008 Effect of type of workpiece material on chip formation process *AU Journal of Tehnology* **11** (3) 181–186.

[5] Krolczyk G and Legutko S 2014 Investigations into surface integrity in the turning process of duplex stainless steel *Trans. FAMENA* **38** (2) 77–82.

[6] Talib N A 2010 Studying the effect of cutting speed and feed rate on tool life in the turning processes *Diyala Journal of Engineering Sciences* (1) 181–194.

Acknowledgement

The author would like to thank Universiti Teknologi MARA Cawangan Pulau Pinang for offering a good

Effect of synthetic fibres on tensile properties of Napier fibres reinforced epoxy composites

M J M Ridzuan[1*], M S Abdul Majid[1], A Khasri[2], E M Cheng[1], M H Sulaiman[1] and K S Basaruddin[1]

[1]School of Mechatronic Engineering, Universiti Malaysia Perlis, Pauh Putra Campus, 02600 Arau, Perlis, Malaysia.
[2]Faculty of Engineering Technology, Universiti Malaysia Perlis, Unicity Alam Campus, 02100 Padang Besar, Perlis, Malaysia

* denotes corresponding author
E-mail: *ridzuanjamir@unimap.edu.my, shukry@unimap.edu.my, azduwin@unimap.edu.my, emcheng@unimap.edu.my, hafissulaiman@unimap.edu.my, khsalleh@unimap.edu.my.

Abstract. This paper present the effect of synthetic fibres on the tensile properties of Napier fibres reinforced epoxy composites. The composites were fabricated using the vacuum infusion method, which the Napier grass fibre placed between the synthetic fibres, then mixed the epoxy resin with a ratio of fibres and epoxy was 70:30. There were four different types of composite samples were fabricated based on various volume fractions. The samples were manufactured using vacuum infusion process. The composites are tested using a universal testing machine according with ASTM standard D638. The tensile strength of the hybrid synthetic/Napier fibre reinforced epoxy composites increased as the content of the synthetic fibre content increased. The tensile properties of the hybrid carbon/Napier reinforced epoxy composites were observed higher than the hybrid glass/Napier reinforced epoxy composites.

1. Introduction

The rising concern towards environmental issues and global warming affects nowadays have been driven the researchers to examine the use of sustainable material as a substitute for regular polymer composites fabricated with engineered filaments which is synthetic fibres, such as aramid, glass and also carbon [1–4]. On the other hand, the natural fibres show extraordinary potential over synthetic fibres in view of their biodegradability, low cost, low density, and satisfactory specific properties [5,6]. Carbon fibre is most prominently used to strengthen composite materials, especially the class of materials known as carbon fibre and graphite reinforced polymers [7]. The diameter of the carbon fibre material is about 510 μm and generally made out from particles which is carbon atoms. This particles are bonded together along the parallel axis of the material fibre which gives higher strength depends on its size. This paper present the effect of synthetic fibres on the tensile properties of Napier fibres reinforced epoxy composites.

2. Materials and Method

2.1. Materials

The composites are the combining of Napier fibre and synthetic fibre which being prepared as a sandwich structure in dimensions of 320 x 300 x 0.1 mm per layer, thus fabricated using the vacuum infusion method. The Napier grass is a part of Napier species which required small amount of nutrients supplement to growth where harvested from a local plantation located at Bukit Kayu Hitam, Kedah, in Northern Peninsular of Malaysia [8]. The carbon and glass fibres was supplied in form of woven roll which offered high strength and stiffness. The matrix material, EpoxAmite 100 series resin was mixed with the slow hardener to a ratio of 3:1 as specified by the manufacturer.

2.2. Extraction of Napier Grass Fibre

The Napier grass fibres were extracted by using a conventional water retting process. Water retting is a microbial process that produces homogenous and high-quality fibre, where the stems are immersed in water tanks, and their progress being monitored daily. Prior to retting, the Napier leaves are removed from the internodes of each stem. Initially, the stems are cleaned and crushed with a mallet before immersed under tap water for approximately three to four weeks, for separation and decomposition process. The fibres are then manually extracted from the stem internodes up to the removing the excessive cellulose from the fibre.

2.3. Hybrid Composite Fabrication

The hybrid synthetic/Napier reinforced epoxy composite laminates are fabricated by a resin infusion technique powered by a vacuum pump which familiarly called as vacuum infusion process [9]. First step is by preparing a sandwich structure for this hybrid synthetic/Napier fibres reinforced epoxy composite, where the Napier grass fibre is placed as a core between the carbon fibre, which is layering on the upper and bottom faces of the sandwich structure. Then, the reinforcement plies is laminated over a glass mould. After settling the arrangement of materials over the composite as shown in the schematic of the vacuum infusion process, the resin then infused into the lamination plies.

2.4. Tensile Test

The composites are tested using an electronic extensometer to measure the tensile properties of the hybrid composite vary with various volume fraction. An INSTRON micro-tester with a 250 kN load cell is utilized, and the samples are tested at a crosshead speed of 0.5 mm/min. The specimens is prepared according to ASTM D638 standards where, each of the specimens are machined to dimensions of 165 x 19 x (3.2±0.4) mm^3 with a gauge length of 115 mm for accurate strain measurement [10]. Five identical test specimens were used for each sample in this experiment. The result for tensile strength and modulus of the hybrid composite were achieved and recorded for analysis.

3. Result

3.1. Tensile Strength

The tensile strength of hybrid carbon/Napier epoxy composites and hybrid glass/Napier epoxy composites is obtained as shown in Figure 1. The figure shown clearly indicates that the tensile strength of the hybrid composite is significantly improved with the incorporation of the carbon fibre compared with the incorporation with the glass fibre. For Sample So, the tensile strength was observed by 21.09 MPa which functioned as a reference for both hybrid composites. Thus, Figure 1 shows that the tensile strength of the hybrid carbon/Napier epoxy composites were obtained by 103, 162, 245 and 316 MPa for S1, S2, S3 and S4 respectively. The tensile strength of the hybrid carbon/Napier epoxy composites increased as the volume content of the carbon fibres increased. Therefore, this strength degradation trend was measured in percentages (%), thus represents strength increment by 142, 121, 124, and 143%, respectively for S1, S2, S3 and S4 results of hybrid glass/Napier epoxy composites compared with hybrid carbon/Napier epoxy composites. This shows that the hybridization with the carbon fibre improved the tensile strength of the hybrid composites compared with glass fibre.

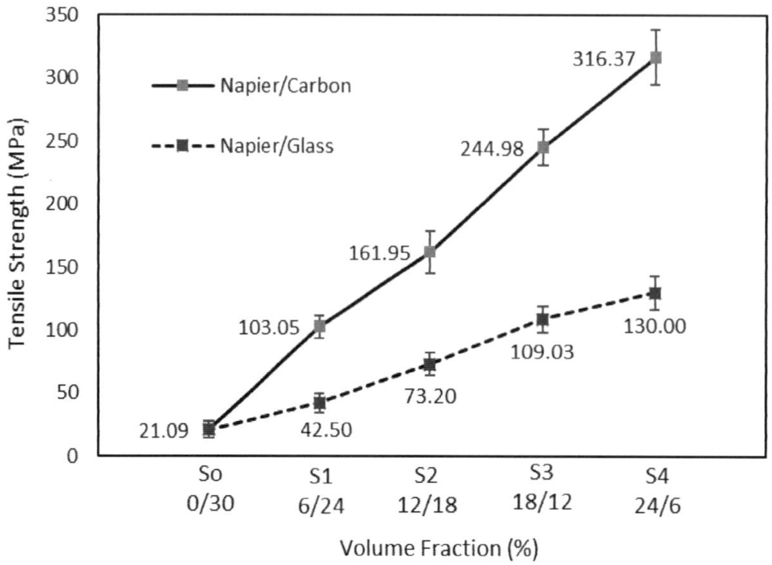

Figure 1. Tensile strength of hybrid synthetic/Napier reinforced epoxy composites

3.2. Tensile Modulus

The tensile modulus of hybrid carbon/Napier epoxy composites and hybrid glass/Napier epoxy composites is obtained as shown in Figure 2. There are slight difference with the tensile modulus between hybrid carbon/Napier epoxy composite and hybrid glass/Napier epoxy composite depending on the line graph. Whereas the tensile modulus of Napier fibres reinforced epoxy composites (So) increased from 1.72 to 6.18 GPa with a content of 6% volume fraction of carbon fibre (S1), representing an increase of 259% compared with the content of 6% of glass fibre (S1) by 88% of increment from 1.72 to 3.22 GPa. The tensile modulus of the hybrid carbon/Napier epoxy composites also increased by 8.4, 10.4 and 11.47 GPa for S2, S3, and S4 samples, respectively. This represent strength reduction of hybrid glass/Napier epoxy composites by 48, 37, 27, and 20% for S1, S2, S3 and S4 samples, respectively compared with the result of sample hybrid carbon/Napier epoxy composites. Therefore, the carbon fibre resulted higher stiffness in the reinforcement of hybrid composites compared to the glass fibre.

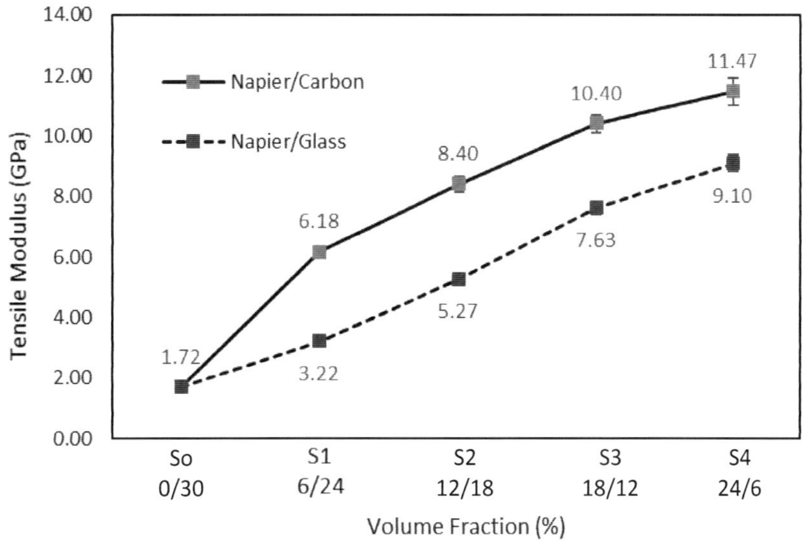

Figure 2. Tensile modulus of hybrid synthetic/Napier reinforced epoxy composites

4. Conclusion

The conclusion obtained based on the experimental results. The tensile strength of the hybrid synthetic/Napier fibre reinforced epoxy composites increased as the content of the synthetic fibre content increased, thus strengthening the interfacial bonding between the fibres and the epoxy of the composites. Therefore the larger load was required to break the interlocking structure of the hybrid composites. The tensile strength of the hybrid carbon/Napier epoxy composites were observed higher than hybrid glass/Napier epoxy composites. Obviously, the carbon fibre was point out to be well performance of the tensile strength compared to glass fibre.

Acknowledgement

This research was financed by the Ministry of Education, Malaysia, through the Fundamental Research Grant Scheme; (Ref: FRGS/1/2019/STG07/UNIMAP/02/2). The authors would also like to acknowledge the School of Mechatronic Engineering, Universiti Malaysia Perlis (UniMAP) for the equipment and technical assistance. The authors are much gratitude to the dedicated staff for the fruitful discussions and input to the project.

References

[1] Mohammed L, Ansari M N M, Pua G, Jawaid M and Islam M S 2015 A Review on Natural Fiber Reinforced Polymer Composite and Its Applications *Int J Polym Sci* 1–15. doi:10.1155/2015/243947.

[2] Thakur V K, Thakur M K and Gupta R K 2014 Review: Raw Natural Fiber–Based Polymer Composites *Int J Polym Anal Charact* **19** 256–71. doi:10.1080/1023666X.2014.880016.

[3] Azwa Z N, Yousif B F, Manalo A C and Karunasena W 2013 A review on the degradability of polymeric composites based on natural fibres *Mater Des* **47** 424–42. doi:10.1016/j.matdes.2012.11.025.

[4] Saba N, Jawaid M, Alothman O Y and Paridah M T 2016 A review on dynamic mechanical properties of natural fibre reinforced polymer composites *Constr Build Mater* **106** 149–59. doi:10.1016/j.conbuildmat.2015.12.075.

[5] Fartini M S, Abdul Majid M S, Ridzuan M J M, Amin N A M and Gibson A G 2016

Compressive properties of Napier (Pennisetum Purpureum) filled polyester composites *Plast Rubber Compos* **45** 136–46. doi:10.1080/14658011.2016.1149911.

[6] Ridzuan M J M, Majid M S A, Afendi M, Rahman M N and Azduwin Y K 2016 Thermal, composition and morphological properties of untreated and alkali-treated napier grass fibres *J Eng Appl Sci* **11** 2485–9. doi:10.3923/jeasci.2016.2485.2489.

[7] Safri S N A, Sultan M T H, Jawaid M and Jayakrishna K 2018 Impact behaviour of hybrid composites for structural applications: A review *Compos Part B Eng* **133** 112–21. doi:10.1016/j.compositesb.2017.09.008.

[8] Revati R, Majid M S A, Ridzuan M J M, Basaruddin K S, Rahman Y M N, Cheng E M, et al. 2017 In vitro degradation of a 3D porous Pennisetum purpureum/PLA biocomposite scaffold *J Mech Behav Biomed Mater* **74**. doi:10.1016/j.jmbbm.2017.06.035.

[9] Ridzuan M J M, Abdul Majid M S, Khasri A, Basaruddin K S and Gibson A G 2019 Effect of moisture exposure and elevated temperatures on impact response of Pennisetum purpureum/glass-reinforced epoxy (PGRE) hybrid composites *Compos Part B Eng* **160** 84–93. doi:10.1016/j.compositesb.2018.10.029.

[10] Ridzuan M J M, Majid M S A, Hafis S M and Azduwin K 2017 The effects of alkali treatment on the mechanical and morphological properties of Pennisetum purpureum/glass-reinforced epoxy hybrid composites *Plast Rubber Compos* **46** 421–30. doi:10.1080/14658011.2017.1368966.

IOP Conf. Series: Materials Science and Engineering **670** (2019) 012020 doi:10.1088/1757-899X/670/1/012020

Mechanical properties of carbon and glass fibre reinforced composites produced by additive manufacturing: A short review

N W Y Omar, N A Shuaib, M H J Ab Hadi and A I Azmi

Faculty of Engineering Technology, Universiti Malaysia Perlis, Uniciti Alam Campus, 02100 Padang Besar, Perlis, Malaysia.

Abstract. Fibre reinforced composites are widely used in various sectors such as aerospace, wind energy and automotive. Due to its versatility and low cost for rapid prototyping and production applications, additive manufacturing technology has grown exponentially over the past few years. In this paper, performances of glass fibre and carbon fibre reinforced composites in additive manufacturing are reviewed from the perspective of mechanical properties. From the review, the reinforcements generally improve mechanical properties, in particular for tensile modulus and tensile strength. The paper presents a benchmark of additive manufacturing technologies for composite material as well as the spotlights of further research in the usage of carbon and glass fibres in rapid prototyping processes.

1. Introduction

The usage of fibre reinforced composites is well known on account of their outstanding mechanical properties, light weight and long-life span. These engineered materials are progressively used in construction and transportation sector for considerably low energy demand and impact to the environment [1, 2]. In the field of composite materials, low cost manufacturing technology have been an important topic of research. A wide variety of composite processing methods, such as hand lay-up [3] , injection moulding [4] and compression moulding [5] have been developed. However, most of these methods has long processing time and high energy intensity. The high cost of processing significantly restricted the application of composite material. Clearly, it is highly desirable to use new and more effective manufacturing technology such as additive manufacturing for composites. Additive manufacturing (AM) is described as a method of incorporating materials to fabricate objects from CAD models in successive layers [6]. AM technology has grown exponentially and continues to grow over the past few years because of its versatility and low manufacturing cost. Besides, the technology has customisability in manufacturing complex, geometries with micrometre solution and monolithic structures [7]. Most studies in fibre reinforced thermoplastic composites field only focused on conventional process, characterising the products and determining potential applications. Only a few studies were conducted on application of fibre reinforced composites in additive manufacturing. A thorough analysis or review on the mechanical properties of the studies needs to be conducted, in order to understand the correlation between type of reinforcement and manufacturing technique with the mechanical performance of the products. The focus of this review paper is on mechanical properties of carbon and glass fibre composites fabricated through AM techniques. Findings will provide a comprehensive perspective on fibre reinforced composite product performance in previous additive manufacturing studies, which will provide a useful reference for future studies.

2. Three-dimensional printing

3D printing uses 3D haptic physical models layer by layer based on CAD models [8]. Fibre reinforcement in 3D printed parts can significantly enhance the polymer matrix properties. The main concerns of these composites are the fibre orientation and void content of composites. To manufacture polymer composites, several printing techniques have been employed. Some of these methods are well established, such as laminated object manufacturing (LOM), stereolithography (SL), selective laser sintering (SLS), extrusion and fused deposition modelling (FDM). In the production of composite products, each technique has its own benefits and limitations. The choice of manufacturing process depends on the materials, processing speed and resolution, the costs and quality aspects of final products [8].

3. Mechanical properties of fibre reinforced composites

In literature, the performance of materials is usually represented by mechanical properties. Usually, the properties reported are tensile and flexural properties. These characteristics are important for determining material capacity, particularly under mechanical demanding conditions that are linked to engineering performance. Studies on the fibre content of carbon fibre reinforced polymer composites were conducted via 3D printing [9–11]. Carbon fibre was placed between layers of 3D printed polymer to improve strength as well as fatigue life. A study, showed that increasing the number of carbon fibre layers led to larger areas of void, which negatively affected the tensile strength [11]. Poor bonding between PLA and carbon fibre can significantly affect mechanical characteristics, however surface adhesion, tensile and flexural strength can be improved by surface treatment of carbon fibre bundles with methylene dichloride and PLA pellets [9].

Tensile properties of continuous glass fibre reinforced with PLA composite was also investigated [12]. Based on rule of mixture, by adding 4% of glass fibre, tensile strength of the composite should be 83 MPa. The findings of the tensile test showed the effectiveness of the method to achieve about 71 MPa of tensile strength. Mechanical properties of different weight percentages of pure polypropylene (PP) reinforced with glass fibre and addition of maleic anhydride polyolefin (POEg-MA) were also investigated [13]. Despite lowering its flexibility, the addition of glass fibre improved the tensile modulus and tensile strength of the composite. However, upon addition of POEg-MA, the composite exhibited decrease in modulus and strength while increased flexibility. Ning et al. [14] studied acrylonitrile butadiene styrene (ABS) composites with different fibre loading of carbon fibre in an FDM-based printer. The study reported improvement of tensile strength and modulus of the carbon fibre reinforced product. However, when compared to pure plastic samples, the modified material showed reduced ductility and yield strength.

Yang et al. [15] manufactured composite specimens using 10 wt% fibre content of continuous carbon fibre (CCF) and ABS thermoplastic via 3D printing technique. These specimens have improved their flexural strength and tensile strength to 127 MPa and 147 MPa, respectively. The results are close to the CCF/ABS composites fabricated from injection moulding with the same fibre content. Zhong et al. [16] investigated process-ability of ABS matrix composites strengthened by glass fibre with different glass fibre volume contents. The researchers noted that the filament tensile strength and surface rigidity could considerably enhanced by addition of glass fibres. A study on glass fibre reinforced polypropylene (PP) composites fabricated via FDM process showed some improvement of tensile modulus and strength, compared to pure PP [17]. Fabrication of carbon fibre reinforced ABS composites using compression moulding and FDM was compared by Tekinalp et al. [18]. At nearly all fibre content, results indicated that the tensile strength of parts manufacturing via FDM have lower tensile strength than the compression moulding. Both products showed notably increases in tensile strength as well as Young's modulus upon adding the carbon fibre. Table 1 presents the overview of changes in tensile properties of fibre reinforced products fabricated using additive manufacturing techniques compared to the control sample in each study. The negative value of percentage represents reduction in the properties.

Table 1. Percentage of changes in tensile properties in fibre reinforced products.

Source	Matrix	Reinforcement	Additive manufacturing Techniques	Changes (%)	
				Tensile strength (MPa)	Tensile modulus (MPa)
Ning et al. [14]	ABS	Carbon fibre powder (100 μm, 150 μm)	FDM	23.53	25
Tekinalp et al. [18]	ABS	Short carbon fibre (3.2 mm, after mixing: 0.26 mm)	FDM	94.44	550
Li et al. [9]	PLA	Continuous carbon fibre	FDM	225	-
Ferreira et al. [19]	PLA	Short carbon fibre (60 μm)	FFF	-2.38	97.37
Chabaud et al. [20]	Polyamide	Continuous glass fibre	FFF	3400	-
Liao et.al [21]	Polyamide	Continuous carbon fibre (6-7 μm)	FDM	87.50	240
Sano et al. [22]	Resin	Glass fibre powder	SLA	150	426.32
Li et.al [23]	PF	Glass fibre powder	SLS	11.34	-
Kleijnen et al. [24]	PP	E-glass	SLS	-2.56	11.43
Kleijnen et al. [24]	PP	E-glass	SLS	-18.97	60
Matsuzaki et.al [25]	PLA	Carbon fibre powder	FDM	350	375
Blok et al. [26]	Nylon	Continuous carbon fibre	FFF	33.33	33.26
Blok et al. [26]	Nylon	Short carbon fibre	FFF	150.15	148.65
Shofner et al. [27]	ABS	Carbon nanofibres (100 μm)	FDM	39.03	-
Ning et al. [28]	ABS	Carbon fibre powders	FDM	23.53	31.58
Akhoundi et al. [12]	PLA	Continuous glass fibre	FDM	66.67	-
Yang et al. [15]	ABS	Continuous carbon fibre	FDM	390	99.95

FDM: fused deposition modelling, SLS: selective laser sintering, SLA: stereolithography, FFF: fused filament fabrication, PP: Polyproplyene, PF: Phenol formaldehyde

4. Discussion

Additive manufacturing (AM) of polymer fibre composites is a robust manufacturing paradigm where it is possible to produce customised components with considerably enhanced mechanical characteristics compared to unreinforced polymers. From the review, the most common technique for fabricating polymer composites is fused deposition modelling (FDM). In general, the thermoplastics used for FDM are currently limited to amorphous polymers or the ones with low crystallinity levels, such as acrylonitrile butadiene styrene (ABS) or polylactic acid (PLA) [2, 29]. It was reported that, laminated object manufacturing (LOM) is superior in terms of dimensional accuracy, speed and cost efficiency [30]. Stereolithography (SLA) offers superior dimensional precision and surface finish. In terms of material usage, selective laser melting (SLM) is only suitable for metals such as steel and aluminium. Various polymers, metals and allow powders can be fabricated using selective laser sintering (SLS). These method is commonly used for advanced applications in various industries, including tissue scaffolds, lattices, aerospace and electronics [30]. As depicted in Table 1, a review of the studies carried out on printing of reinforced composites shows the different matrix, the size of the reinforcement and the resulting tensile properties. While prior studies use various properties to characterise the product, tensile properties are the main focused in this paper, as the properties are commonly reported in the literature. This is done for a fair comparison. It is presented in Table 1 that most of the published works on application 3D printer process was using carbon fibre, in comparison with glass fibre. This is despite the fact that in the supply chain, the quantity glass fibre is considerably greater compared to carbon fibre [31]. The usage of carbon fibre can be due to its high strength to weight ratio and lightweight. Significant work in carbon fibre composites is motivated by the higher price and embodied energy of carbon fibre compared to glass fibre composites [32].

Several studies reported on the use of mixed thermoplastic polymer and very short carbon fibre (~0.1 mm). The matrix and the reinforcement are usually screw extruded to form the filament used in 3D printing processes. The use of short fibre can improve the strength of the printed product. However, the high shear mixing leads to fibre breakage therefore further reduces the fibre length in the filaments. As a result, the strength in the printed parts is weaker. The strength remains low in comparison with CFRP products fabricated by conventional composite manufacturing methods such as resin infusion and autoclave [33, 34]. From Table 1, it can be clearly seen that the addition of fibre mostly improved tensile properties. The improvement can be related to parameters such as manufacturing technique, type of reinforcement and fibre loading. The material and energy consumption of AM processes, particularly involving fibre reinforced composite are rarely reported in literature. These facets are important ensure resource efficient additive manufacturing processes.

5. Conclusions

From literature, it is clear that the addition of carbon and glass fibre in additive manufacturing of thermoplastics could improve mechanical properties of the product. Studies should be expanded on using recycled fibres, which have less embodied energy and cheaper. The use of recycled fibres could reduce the environmental and financial impact of virgin fibre production. Studies required to fully characterise physical and mechanical behaviour of fibre reinforced 3D printed structures from aspects other than tensile and flexural properties. Such studies will open a new market for application of the structures in various sectors. Besides, material and energy consumption associated with the products should be considered, ideally from life cycle assessment perspective.

Acknowledgement

The author would like to acknowledge the support from the Fundamental Research Grant Scheme (FRGS) under a grant number of **FRGS/1/2018/TK03/UNIMAP/02/15** from the **Ministry of Education Malaysia**

References

[1] Cousins D S, Suzuki Y, Murray R E, Samaniuk J R and Stebner A P 2019 Recycling glass fiber thermoplastic composites from wind turbine blades *J. Clean. Prod.* **209** 1252–63.

[2] Yang Y, Boom R, Irion B, Van Heerden D, Kuiper P and De Wit H 2012 Chemical engineering and processing : process intensification recycling of composite materials **51** 53–68.

[3] Hakim A, Halim N H A, Salleh Z and Taib Y M 2015 *Journal of Mechanical Engineering,* **12** 71-84.

[4] Stoeffler K, Andjelic S, Legros N, Roberge J, and Schougaard S B 2013 Polyphenylene sulfide (PPS) composites reinforced with recycled carbon fiber *Compos. Sci. Technol.* **84** 65–71.

[5] Saliu Hafsat R, Ishiaku U S, Yakubu M K, Kolawole E G, Adefila S S, Abu Bakar M B, Moh'd Ishak Z A 2015 The effect of epoxy concentration and fibre loading on the mechanical properties of abs/epoxy-coated kenaf fibre composites *Open J. Compos. Mater.* **5** 41–8.

[6] Huang S H, Liu P, Mokasdar A, Hou L 2013 Additive manufacturing and its societal impact: A literature review *Int. J. Adv. Manuf. Technol.* **67** 1191–1203.

[7] Parandoush P and Lin D 2017 A review on additive manufacturing of polymer-fiber composites *Compos. Struct.* **182** 36–53.

[8] Wang X, Jiang M, Zhou Z, Gou J and Hui D 2017 3D printing of polymer matrix composites: A review and prospective *Compos. Part B Eng.* **110** 442–58.

[9] Li N, Li Y and Liu S 2016 Rapid prototyping of continuous carbon fiber reinforced polylactic acid composites by 3D printing *J. Mater. Process. Tech.* **238** 218–25.

[10] Mori K, Maeno T and Nakagawa Y 2014 Dieless forming of carbon fibre reinforced plastic parts using 3d printer *Procedia Eng.* **81** 1595–1600.

[11] Van Der Klift F, Ueda M, Todoroki A, Hirano Y, Matsuzaki R and Koga Y 2016 3D printing of continuous carbon fibre reinforced thermo-plastic (CFRTP) tensile test specimens 2016 *Open J.*

Compos. Mater. **6** 18–27.

[12] Akhoundi B, Behravesh A H, Nabipour M and Saed A B 2017 Additive manufacturing of glass-fiber reinforced composites using fdm 3d printer *The 25th Annual International Conference on Mechanical Engineering, ISME2017.*

[13] Sodeifian G, Ghaseminejad S and Yousefi A A 2019 Preparation of polypropylene/short glass fiber composite as fused deposition modeling (FDM) filament *Results Phys.* **12** 205–22.

[14] Ning F, W Cong, J Qiu, Wei J and Wang S 2015 Additive manufacturing of carbon fiber reinforced thermoplastic composites using fused deposition modeling *Compos. Part B Eng.* **80** 369–78.

[15] Yang C, Tian X, Liu T, Cao Y and Li D 2017 3D printing for continuous fiber reinforced thermoplastic composites: Mechanism and performance *Rapid Prototyp. J.* **23** 209–15.

[16] Zhong W, Li F, Zhang Z, Song L and Li Z 2001 Short fiber reinforced composites for fused deposition modeling *Mater. Sci. Eng. A* **301** 125–30.

[17] Carneiro O S, Silva A F and Gomes R 2015 Fused deposition modeling with polypropylene *Mater. Des.* **83** 768–76.

[18] Tekinalp H L, Kunc V, Velez-Garcia G M, Duty C E, Love L J, Naskar A K, Blue C A and Ozcan S 2014 Highly oriented carbon fiber-polymer composites via additive manufacturing," *Compos. Sci. Technol.* **105** 144–50.

[19] Thiago R, Ferreira L, Cardoso I, Assis T and Bürger D 2017 Experimental characterization and micrography of 3D printed PLA and PLA reinforced with short carbon fibers *Compos. Part B* **124** 88–100.

[20] Chabaud G, Castro M, Denoual C and Le Duigou A 2019 Hygromechanical properties of 3D printed continuous carbon and glass fibre reinforced polyamide composite for outdoor structural applications *Addit. Manuf.* **26** 94–105.

[21] Liao G, Li Z, Cheng Y, Xu D, Zhu D, Jiang S, Guo J, Chen X, Xu G, and Zhu Y 2018 Properties of oriented carbon fiber/polyamide 12 composite parts fabricated by fused deposition modeling *Mater. Des.* **139** 283–92.

[22] Sano Y, Matsuzaki R, Ueda M, Todoroki A and Hirano Y 2018 3D printing of discontinuous and continuous fibre composites using stereolithography *Addit. Manuf.* **24** 521–27.

[23] Li Z, Zhou W, Yang L, Chen P, Yan C, Cai C and Li H 2019 Glass fiber-reinforced phenol formaldehyde resin-based electrical insulating composites fabricated by selective laser sintering *Polymers* **11.**

[24] Kleijnen R G, Sesseg J P W, Schmid M and Wegener K 2017 Insights into the development of a short-fiber reinforced polypropylene for laser sintering *AIP Conf. Proc.* **1914.**

[25] Matsuzaki R, Ueda M, Namiki M, Jeong T and Asahara H 2016 Three-dimensional printing of continuous-fiber composites by in-nozzle impregnation *Nat. Publ. Gr.*

[26] Blok L G, Longana M L, Yu H and Woods B K S 2018 An investigation into 3D printing of fibre reinforced thermoplastic composites *Addit. Manuf.* **22** 176–86.

[27] Shofner M L, Lozano K, Rodríguez-Macías F J and Barrera E V 2003 Nanofiber-reinforced polymers prepared by fused deposition modeling *J. Appl. Polym. Sci.* **89** 3081–90.

[28] Ning F , Cong W, Jia Z, Wang F and Zhang M 2016 Additive manufacturing of CFRP composites using fused deposition modeling: effects of process parameters **1989.**

[29] Chacón J M, Caminero M A, García-Plaza E and Núñez P J 2017 Additive manufacturing of PLA structures using fused deposition modelling: Effect of process parameters on mechanical properties and their optimal selection *Mater. Des.***124** 143–57.

[30] Ngo T D, Kashani A, Imbalzano G, Nguyen K T Q and Hui D 2018 Additive manufacturing (3D printing): A review of materials, methods, applications and challenges *Compos. Part B Eng.* **143** 172–96.

[31] Shuaib N A and Mativenga P T 2015 Energy Intensity and Quality of Recyclate in Composite Recycling *ASME Manufacturing Science and Engineering Conference MSEC2015.*

[32] Shuaib N A and Mativenga P T 2017 Carbon footprint analysis of fibre reinforced composite recycling processes. *Procedia Manuf.* **7** 183–90.

[33] Longana M L, Ong N, Yu H and Potter K D 2016 Multiple closed loop recycling of carbon fibre composites with the HiPerDiF (High Performance Discontinuous Fibre) method *Compos. Struct.* **153** 271–7.

[34] Fu S Y and Lauke B 1996 Effects of fiber length and fiber orientation distributions on the tensile strength of short-fiber-reinforced polymers *Compos. Sci. Technol.* **56** 1179–90.

IOP Conf. Series: Materials Science and Engineering **670** (2019) 012021 doi:10.1088/1757-899X/670/1/012021

Implementation of multi-component dusty-gas model for species transport in quasi-three-dimensional numerical analysis of solid oxide fuel cell. Part I: hydrogen fuel

W C Tan[1,2], H Iwai[2], M Kishimoto[2] and H Yoshida[2]

[1]School of Mechatronic Engineering, Universiti Malaysia Perlis, 02600 Pauh Putra, Perlis, Malaysia.
[2]Department of Aeronautics and Astronautics, Kyoto University, Nishikyo-ku, Kyoto, 615-8540, Japan.

Abstract. Quasi-three-dimensional numerical model of solid oxide fuel cell, which assumes constant physicochemical properties within the cell components in the thickness direction, typically employs a simple gas diffusion model for species transport in the porous electrodes, such as the Fick's model. In this study, a three-dimensional grid system is introduced in the anode layer and coupled with the quasi-three-dimensional solid oxide fuel cell model. The multi-component dusty-gas model is implemented to solve the conservation of species on this three-dimensional grid system. The results with the developed model are compared with experimental data obtained under hydrogen fuel. The obtained results show that the dusty-gas model can accurately predict the transport of gas species in the porous anode.

1. Introduction

Solid oxide fuel cell (SOFC) is one of the high-efficiency fuel cells that are operated at high temperatures. It can convert the chemical energy of fuel to electrical energy through electrochemical processes. The performance of an SOFC is strongly related with the transport of mass and species in the channels and porous electrodes, transport of charge carriers in the cell, and transport of thermal energy. A numerical analysis allows researchers to understand such transport phenomena in SOFCs.

A quasi-3D numerical model was developed in the authors' group to study the effect of cell aspect ratio on cell performance [1] and the effects of the air-flow configuration in a short stack [2]. The developed quasi-3D model can be used to analyze the transport of gas species in the streamwise and spanwise directions in the channels. It is also useful for analysis involving significant temperature gradient in the streamwise and spanwise direction and/or various flow configurations of a stack. The developed quasi-3D model applied Fick's model (FM) for the transport of gas in the porous anode. The species transport in a porous medium can also be expressed by the Stefan–Maxwell model (SMM), dusty-gas model (DGM), or binary-friction model (BFM). The SMM considers the interactions of gas species in a gas mixture [3]. The DGM extends the SMM by considering the Knudsen diffusion, which is important when the pore size of the porous medium is comparable to or smaller than the mean free path of the gas species. The BFM also extends the SMM, which takes into account diffusive slip as well as viscous slip phenomena [4,5] on the porous wall. Vural et al. [6] reported that the BFM requires the most computation time compared to the SMM and DGM. Suwanwarangkul et al. [7] compared the FM, SMM, and DGM in predicting concentration

overpotential in electrodes and recommended the DGM for use in a multi-component system consisting of H_2, H_2O, CO, and CO_2. The DGM can be simplified by assuming either constant pressure gradient [8,9] or linear pressure gradient [10] to reduce the computation cost. However, the simplified DGM with a constant pressure gradient is only applicable to thin porous electrodes. Pisani [11] proposed rewriting the DGM in terms of molar densities instead of molar fractions to express the non-uniform pressure within the porous medium. García-Camprubí and co-authors [12–14] adopted the suggestion by Pisani and rewrote the DGM equation to express the total molar flux of each species in the gas mixture. The advantages of the DGM include the ability to consider an unlimited number of gas species in the mixture, and to capture the non-uniform pressure gradient in the gas mixture.
In this work, we implement a gas diffusion model based on the DGM to the anode in a quasi-3D numerical model by newly introducing a 3D grid system to predict the species transport in the anode. The simulation results are compared to the experimental data from a hydrogen-fueled six-cell stack in reference [1].

2. Numerical modeling

A cell unit consists of top and bottom separators, fuel and air channels, and a positive-electrolyte-negative assembly (PEN) is considered in this study. Each component has only one mesh in the y-axis. Note that both fuel and air channels consist of solid- and fluid-phase meshes. Additionally, two layers of mesh located at the top and bottom of the cell unit are introduced as boundary layers. Hence, a cell unit with a dimension of $89.0 \times 3.40 \times 60.0$ mm^3 consists of $101 \times 9 \times 71$ meshes in the quasi-3D model. The unique feature of this study is the introduction of the anode mesh, which is an additional 3D grid system for the anode in the PEN. This additional grid system is solely used in the gas diffusion analysis in the anode. The anode mesh consists of 16 layers with 18 grid points including two grid points for the coupled boundary, which are located at the anode surface and the anode-electrolyte interface. All the anode mesh layers have an equal thickness along the y-axis. The size of the anode mesh in the xz-plane is the same as that of the meshes used in the channels. Values at the coupled boundaries are determined by considering the following data: molar fraction of species x_i and total pressure P from the fuel channel, and molar flux of species N_i due to the electrochemical reaction at the anode-electrolyte interface. The molar flux of species at the anode surface is evaluated by the implemented DGM model and converted to mass flow rate, and feedbacked to the fuel channel as a mass sink or source term S_{Y_i}. The electrochemical reaction is assumed to occur at the anode-electrolyte interface to provide the molar flux at the coupled boundary of the anode-electrolyte interface. Note that the FM is applied in the thin porous cathode to reduce the computation cost because no chemical reaction occurs in the cathode. Details of the quasi-3D numerical modeling have already been described in our previous reports [1,2], therefore only essential equations are summarized below.

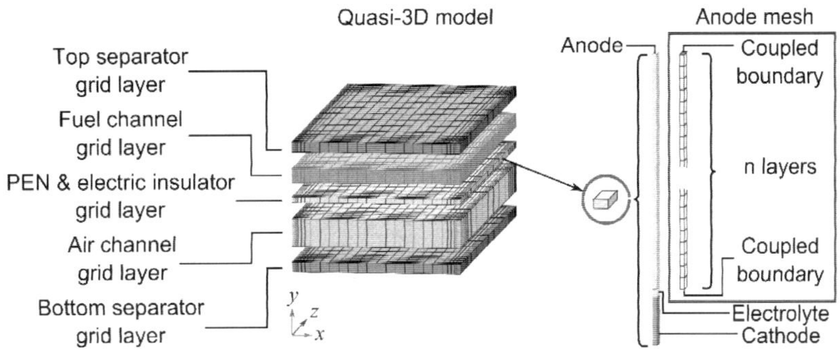

Figure 1. Schematic view mesh layers in the y-direction for the quasi-3D model with an anode mesh.

2.1. Governing equations

The numerical analysis was based on conservation equations of the mass, momentum, species, and energy.

Mass conservation:

$$\frac{\partial(\rho u)}{\partial x}+\frac{\partial(\rho w)}{\partial z}=\sum S_{Y_i} \tag{1}$$

Momentum conservation:

$$\left[\frac{\partial}{\partial x}\left(\frac{\rho uu}{\varepsilon}\right)+\frac{\partial}{\partial z}\left(\frac{\rho wu}{\varepsilon}\right)\right]=-\varepsilon\frac{\partial P}{\partial x}+\frac{\partial}{\partial x}\left(\mu\frac{\partial u}{\partial x}\right)+\frac{\partial}{\partial z}\left(\mu\frac{\partial u}{\partial z}\right)$$
$$+\frac{1}{3}\frac{\partial}{\partial x}\left(\mu\frac{\partial u}{\partial x}\right)-\frac{2}{3}\frac{\partial}{\partial x}\left(\mu\frac{\partial w}{\partial z}\right)+\frac{\partial}{\partial z}\left(\mu\frac{\partial w}{\partial x}\right)-\varepsilon\frac{\mu}{K}u-\varepsilon\frac{\rho f}{\sqrt{K}}|U|u \tag{2}$$

$$\left[\frac{\partial}{\partial x}\left(\frac{\rho uw}{\varepsilon}\right)+\frac{\partial}{\partial z}\left(\frac{\rho ww}{\varepsilon}\right)\right]=-\varepsilon\frac{\partial P}{\partial z}+\frac{\partial}{\partial x}\left(\mu\frac{\partial w}{\partial x}\right)+\frac{\partial}{\partial z}\left(\mu\frac{\partial w}{\partial z}\right)$$
$$+\frac{1}{3}\frac{\partial}{\partial z}\left(\mu\frac{\partial w}{\partial z}\right)-\frac{2}{3}\frac{\partial}{\partial z}\left(\mu\frac{\partial u}{\partial x}\right)+\frac{\partial}{\partial x}\left(\mu\frac{\partial u}{\partial z}\right)-\varepsilon\frac{\mu}{K}w-\varepsilon\frac{\rho f}{\sqrt{K}}|U|w \tag{3}$$

Energy conservation:

(fluid phase in the channels filled with metal foam)

$$\frac{\partial(\rho C_p uT_f)}{\partial x}+\frac{\partial(\rho C_p wT_f)}{\partial z}=\frac{\partial}{\partial x}\left(\lambda_f^{eff}\frac{\partial T_f}{\partial x}\right)+\frac{\partial}{\partial z}\left(\lambda_f^{eff}\frac{\partial T_f}{\partial z}\right)+h_{sf}a_{sf}(T_s-T_f) \tag{4}$$

(solid phase in the channels filled with metal foam)

$$0=\frac{\partial}{\partial x}\left(\lambda_s^{eff}\frac{\partial T_s}{\partial x}\right)+\frac{\partial}{\partial z}\left(\lambda_s^{eff}\frac{\partial T_s}{\partial z}\right)+h_{sf}a_{sf}(T_f-T_s) \tag{5}$$

(solid phases in the separator and the cell)

$$0=\frac{\partial}{\partial x}\left(\lambda_s\frac{\partial T_s}{\partial x}\right)+\frac{\partial}{\partial z}\left(\lambda_s\frac{\partial T_s}{\partial z}\right)+Q \tag{6}$$

Species conservation:

$$\frac{\partial}{\partial x}(\rho uY_i)+\frac{\partial}{\partial z}(\rho wY_i)=\frac{\partial}{\partial x}\left(\rho D_{i,m}^{eff}\frac{\partial Y_i}{\partial x}\right)+\frac{\partial}{\partial z}\left(\rho D_{i,m}^{eff}\frac{\partial Y_i}{\partial z}\right)+S_{Y_i} \tag{7}$$

Charge conservation

$$\frac{\partial}{\partial x}\left(\sigma_s^{eff}\frac{\partial\phi}{\partial x}\right)+\frac{\partial}{\partial y}\left(\sigma_s^{eff}\frac{\partial\phi}{\partial y}\right)+\frac{\partial}{\partial z}\left(\sigma_s^{eff}\frac{\partial\phi}{\partial z}\right)=S_\phi \tag{8}$$

Both fuel and gas channels are filled with the metal foam, which acts as a current collector in the equivalent circuit model [15]. The developed model adopts the geometric parameters representing the porous metal foam in the channels from Bhattacharya et al. [16] and Calmidi and Mahajan [17]. An in-house code is developed and used for the calculations. The grid independence in the quasi-3D model was confirmed by conducting preliminary calculations in our previous works [1,2].

2.2. Species transport in the anode mesh

The anode mesh as shown in Figure 1 is introduced to determine the transport of species in the anode of the SOFC. The multi-component DGM can be expressed in terms of the total flux of species *i* as a function of molar densities as follows [14]:

$$N_i=-\Gamma_i^{DGM}\nabla X_i+v_i^{p^{DGM}}X_i+v_i^{N^{DGM}}X_i, \tag{9}$$

$$\Gamma_i^{DGM}=1\Big/\left[\frac{RT}{P}\sum_{j\neq i}\left(\frac{X_j}{D_{i,j}^{eff}}\right)+\frac{1}{D_{K,i}^{eff}}\right], \tag{10}$$

$$v_i^{p^{\mathrm{DGM}}} = -\frac{K\nabla P}{D_{\mathrm{K},i}^{\mathrm{eff}}\mu} \Bigg/ \left[\frac{RT}{P}\sum_{j\neq i}\left(\frac{X_j}{D_{i,j}^{\mathrm{eff}}}\right)+\frac{1}{D_{\mathrm{K},i}^{\mathrm{eff}}}\right] = -\Gamma_i^{\mathrm{DGM}}\frac{1}{D_{\mathrm{K},i}^{\mathrm{eff}}}\left(\frac{K\nabla P}{\mu}\right), \tag{11}$$

$$v_i^{N^{\mathrm{DGM}}} = \frac{RT}{P}\sum_{j\neq i}\left(\frac{N_j}{D_{i,j}^{\mathrm{eff}}}\right) \Bigg/ \left[\frac{RT}{P}\sum_{j\neq i}\left(\frac{X_j}{D_{i,j}^{\mathrm{eff}}}\right)+\frac{1}{D_{\mathrm{K},i}^{\mathrm{eff}}}\right] = \Gamma_i^{\mathrm{DGM}}\frac{RT}{P}\sum_{j\neq i}\left(\frac{N_j}{D_{i,j}^{\mathrm{eff}}}\right). \tag{12}$$

where D_{ij}^{eff} and $D_{\mathrm{K},i}^{\mathrm{eff}}$ are the effective binary diffusion and the effective Knudsen diffusion coefficient of gas species. The molar flux of species i in the porous medium depends on the three contributors as described in Equation (9). These are, from the left to right, global molecular diffusion, flux due to the pressure gradient, and flux induced by the flux of other species. The species conservation within the anode mesh can be expressed as follows:

$$\nabla \cdot N_i = S_{x_i}. \tag{13}$$

Equation (13) is solved implicitly by using Equations. (9) to (12). First, the molar densities of species at the anode surface are calculated from the partial pressure of species in the fuel channel. At the same time, the molar flux of species at the anode-electrolyte interface is determined from the local current density. Then, the pressure gradient, molar densities gradient, and molar densities are used to compute Γ_i^{DGM}, $v_i^{p^{\mathrm{DGM}}}$, and $v_i^{N^{\mathrm{DGM}}}$. The species molar source/sink term S_{x_i} in Equation (13), within each mesh volume V is associated with the chemical reaction in the anode.

2.3. Calculation conditions

The boundary conditions are given in Table 1. Note that the electric potential difference between the top and bottom separators is the terminal voltage of the stack and is iteratively tuned to achieve a pre-determined average current density. Grid independence analysis was conducted for the anode mesh. The results suggest that the 16 layers with 18 grid points for the anode mesh are sufficient.

Table 1. Boundary conditions.

Main mesh	u	W	T_f	T_s	Y_i	ϕ
Inlet	$u = u_{in}$	$w = 0$	$T_f = T_{in}$	$\frac{\partial T_s}{\partial x} = 0$	$Y_i = Y_{i,in}$	$\frac{\partial \phi}{\partial x} = 0$
Outlet	$\frac{\partial u}{\partial x} = 0$	$\frac{\partial w}{\partial x} = 0$	$\frac{\partial T_f}{\partial x} = 0$	$\frac{\partial T_s}{\partial x} = 0$	$\frac{\partial Y_i}{\partial x} = 0$	$\frac{\partial \phi}{\partial x} = 0$
Wall	$u = 0$	$w = 0$	$\frac{\partial T_f}{\partial z} = 0$	$\frac{\partial T_s}{\partial z} = 0$	$\frac{\partial Y_i}{\partial z} = 0$	$\frac{\partial \phi}{\partial z} = 0$
Anode mesh	P		x_i		N_i	
Anode surface	$P = P_{\mathrm{channel}}$		$x_i = x_{i,\mathrm{channel}}$		$N_{i,y}$	
Electrolyte interface	$\frac{\partial P}{\partial y} = 0$		$\frac{\partial x_i}{\partial y} = 0, i \neq H_2, H_2O$		$N_i = 0, i \neq H_2, H_2O$; $N_{H_2} = -I/2F$; $N_{H_2O} = I/2F$	
Wall	$\frac{\partial P}{\partial x} = 0, \frac{\partial P}{\partial z} = 0$		$\frac{\partial x_i}{\partial x} = 0, \frac{\partial x_i}{\partial z} = 0$		$N_i = 0$	

3. Results and discussion

The numerical results from the developed quasi-3D model implemented with the DGM in this work are compared with the numerical results from our previous work using the FM [1]. Both numerical results are compared with a six-cell short stack experiment conducted under diluted hydrogen fuel at temperatures of 973, 1023, and 1073 K. The details of the short-stack experiment can be found elsewhere [18,19]. The developed model adopts the real microstructure properties of the electrodes quantified by Brus et al. [20,21]. Figure 2 shows the comparison of the current-voltage characteristics.

The prediction of present SOFC numerical model with the implementation of the DGM in the anode is shown by the solid lines while that of the FM is expressed by the dashed lines. No significant difference in the results between the DGM and FM is found. Therefore, the FM appears to be sufficient in the case of hydrogen fuel, as no chemical reaction occurs in the anode.

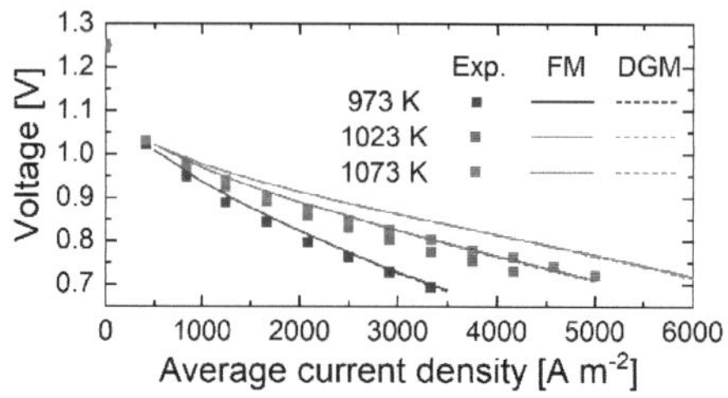

Figure 2. Comparison of current-voltage characteristics.

4. Conclusions

The multi-component dusty-gas model (DGM) was introduced in the quasi-three-dimensional (quasi-3D) numerical model of solid oxide fuel cell (SOFC) with the aids of anode mesh to predict species transport in the anode. The SOFC numerical model with the DGM was successfully validated with experimental results from a six-cell stack fueled with hydrogen. The developed model with the DGM was able to qualitatively capture the trends observed in the experiment.

Acknowledgments

This work was partially supported under "Fundamental study on rapid evaluation method of SOFC durability" by the New Energy and Industrial Technology Organization (NEDO, Japan), and by the Council for Science, Technology and Innovation (CSTI), Cross-ministerial Strategic Innovation Promotion Program (SIP), "Energy Carrier" (Funding agency: JST).

5. References

[1] Tan W C, Iwai H, Kishimoto M, Brus G, Szmyd J S and Yoshida H 2018 Numerical analysis on effect of aspect ratio of planar solid oxide fuel cell fueled with decomposed ammonia *J. Power Sources* **384** 367–378.
[2] Tan W C, Iwai H, Kishimoto M and Yoshida H 2018 Quasi-three-dimensional numerical simulation of a solid oxide fuel cell short stack: Effects of flow configurations including air-flow alternation *J. Power Sources* **400** 135–146.
[3] Tseronis K, Kookos I K and Theodoropoulos C 2008 Modelling mass transport in solid oxide fuel cell anodes: a case for a multidimensional dusty gas-based model *Chem. Eng. Sci.* **63** 5626–5638.
[4] Kerkhof P J A M 1996 A modified Maxwell-Stefan model for transport through inert membranes: the binary friction model *Chem. Eng. J. Biochem. Eng. J.* **64** 319–343.
[5] Kerkhof P J A M and Geboers M A M 2005 Analysis and extension of the theory of multicomponent fluid diffusion *Chem. Eng. Sci.* **60** 3129–3167.
[6] Vural Y, Ma L, Ingham D B and Pourkashanian M 2010 Comparison of the multicomponent mass transfer models for the prediction of the concentration overpotential for solid oxide fuel cell anodes *J. Power Sources* **195** 4893–4904.
[7] Suwanwarangkul R, Croiset E, Fowler M W, Douglas P L, Entchev E and Douglas M A 2003 Performance comparison of Fick's, dusty-gas and Stefan-Maxwell models to predict the concentration overpotential of a SOFC anode *J. Power Sources* **122** 9–18.
[8] Fu Y, Jiang Y, Poizeau S, Dutta A, Mohanram A, Pietras J D and Bazant M Z 2015 Multicomponent gas diffusion in porous electrodes *J. Electrochem. Soc.* **162** F613–F621.

[9] Yoon K J, Lee S, An H, Kim J, Son J -W, Lee J -H, Je H -J, Lee H -W and Kim B -K 2014 Gas transport in hydrogen electrode of solid oxide regenerative fuel cells for power generation and hydrogen production *Int. J. Hydrogen Energy* **39** 3868–3878.

[10] Marković J and Omorjan R 2007 Applicability of linearized dusty gas model for multicomponent diffusion of gas mixtures in porous solids *Acta Period. Technol.* 75–84.

[11] Pisani L 2008 Multi-component gas mixture diffusion through porous media: A 1D analytical solution *Int. J. Heat Mass Transf.* **51** 650–660.

[12] García-Camprubí M and Fueyo N 2010 Mass transfer in hydrogen-fed anode-supported SOFCs *Int. J. Hydrogen Energy* **35** 11551–11560.

[13] García-Camprubí M, Izquierdo S and Fueyo N 2014 Challenges in the electrochemical modelling of solid oxide fuel and electrolyser cells *Renew. Sustain. Energy Rev.* **33** 701–718.

[14] García-Camprubí M, Sánchez-Insa A and Fueyo N 2010 Multimodal mass transfer in solid-oxide fuel-cells *Chem. Eng. Sci.* **65** 1668–1677.

[15] Iwai H, Yamamoto Y, Saito M and Yoshida H 2011 Numerical simulation of intermediate-temperature direct-internal-reforming planar solid oxide fuel cell *Energy* **36** 2225–2234.

[16] Bhattacharya A, Calmidi V V and Mahajan R L 2002 Thermophysical properties of high porosity metal foams *Int. J. Heat Mass Transf.* **45** 1017–1031.

[17] Calmidi V V and Mahajan R L 2000 Forced convection in high porosity metal foams *J. Heat Transfer* **122** 557.

[18] Brus G, Iwai H, Otani Y, Saito M, Yoshida H and Szmyd J S 2015 Local evolution of triple phase boundary in solid oxide fuel cell stack after long-term operation *Fuel Cells* **15** 545–548.

[19] Komatsu Y, Brus G, Kimijima S and Szmyd J S 2014 The effect of overpotentials on the transient response of the 300W SOFC cell stack voltage *Appl. Energy* **115** 352–359.

[20] Brus G, Iwai H, Sciazko A, Saito M, Yoshida H and Szmyd J S 2015 Local evolution of anode microstructure morphology in a solid oxide fuel cell after long-term stack operation *J. Power Sources* **288** 199–205.

[21] Brus G, Iwai H, Mozdzierz M, Komatsu Y, Saito M, Yoshida H and Szmyd J S 2017 Combining structural, electrochemical, and numerical studies to investigate the relation between microstructure and the stack performance *J. Appl. Electrochem.* **47** 979–989.

IOP Conf. Series: Materials Science and Engineering **670** (2019) 012022 doi:10.1088/1757-899X/670/1/012022

Implementation of multi-component dusty-gas model for species transport in quasi-three-dimensional numerical analysis of solid oxide fuel cell. Part II: direct ammonia fuel

W C Tan[1,2], H Iwai[1], M Kishimoto[1] and H Yoshida[1]
[1]School of Mechatronic Engineering, Universiti Malaysia Perlis, 02600 Pauh Putra, Perlis, Malaysia.
[2]Department of Aeronautics and Astronautics, Kyoto University, Nishikyo-ku, Kyoto, 615-8540, Japan.

Abstract. The Fick's model is applied to analyze the species transport in the porous electrodes for an in-house developed quasi-three-dimensional numerical model of solid oxide fuel cell. It may not be reliable when significant concentration difference is formed in the gas mixture, owing to chemical reactions such as steam methane reforming and ammonia decomposition. In this study, the multi-component dusty-gas model is implemented to solve the conservation of species on a three-dimensional grid system in the anode layer involving direct ammonia decomposition. The results with the developed model are compared with the available two-dimensional numerical analysis. The obtained results show that the dusty-gas model can accurately predict the transport of gas species in the porous anode even when a chemical reaction occurs.

1. Introduction

Solid oxide fuel cell (SOFC) is one of the high-efficiency fuel cells that are operated at high temperatures. It can convert the chemical energy of fuel to electrical energy through electrochemical processes. The supplied fuel is not limited to hydrogen. Fuel like methane and ammonia also can be directly supplied to SOFC owing to its high operating temperature. Furthermore, nickel that found in the anode of SOFC is a suitable catalyst to produce hydrogen from the supplied fuel.

A quasi-3D numerical model was developed in the authors' group to study the effect of cell aspect ratio on cell performance [1] and the effects of the air-flow configuration in a short stack [2]. However, the developed quasi-3D model was limited to hydrogen-fueled SOFCs owing to the simplicity of the implemented Fick's model (FM) in the porous anode. In the case of direct supply of methane- or ammonia-based fuels, chemical reactions such as steam methane reforming and ammonia decomposition occur in the porous anode, which results in a significant change of concentration in the gas mixture and thus, limits the implementation of the simple FM. Hence, a more accurate transport model like the dusty-gas model (DGM) is required to predict multi-component gas transport in the porous anode involving chemical reactions. García-Camprubí and co-authors [3–5] adopted the suggestion by Pisani [6] and rewrote the DGM equation to express the total molar flux of each species in the gas mixture. Our recent work [7] had successfully validated the developed quasi-3D numerical model with the implementation of multi-component DGM to the anode with experimental results from a six-cell stack fueled with hydrogen.

In this short communication, we implement the same gas diffusion model based on the DGM to the anode in a quasi-3D numerical model to predict the species transport in the anode from an ammonia-fueled cell in reference [8].

2. Numerical modeling

The cell unit as shown in Figure 1 is considered in this study. Note that the 3D grid system of anode mesh for the anode in the PEN is solely used in the gas diffusion analysis in the anode.

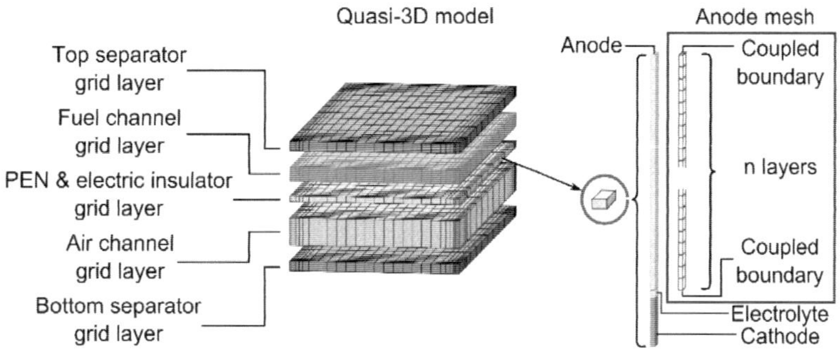

Figure 1. Schematic view of mesh layers in the *y*-direction for the quasi-3D model with an anode mesh.

Details of the mesh and the quasi-3D numerical modeling have already been described in our previous reports [1,2,7], therefore only essential equations are summarized below. The numerical analysis was based on conservation equations of the mass, momentum, species, and energy. The species conservation within the anode mesh based on the multi-component DGM [5] can be expressed as follows:

$$\nabla \cdot N_i = S_{x_i}. \tag{1}$$

The species molar source/sink term S_{x_i} in Equation (1), within each mesh volume V is associated with the chemical reaction in the anode. This study considers an SOFC single cell fueled with direct ammonia. S_{x_i} is associated with the ammonia decomposition reaction on the nickel-pore interfacial area in the anode, and the following reaction rate equation [9] is used in this study.

$$NH_3 \rightarrow \frac{1}{2}N_2 + \frac{3}{2}H_2 \tag{2}$$

$$R_{\mathrm{dec}} = S_{\mathrm{Ni-pore}} R_{\mathrm{dec}}^{\mathrm{Ni-pore}}, \tag{3}$$

$$R_{\mathrm{dec}}^{\mathrm{Ni-pore}} = 98.4 \exp\left(-\frac{1.20 \times 10^5}{RT_s}\right)\left(p_{\mathrm{NH_3}}\right)^{0.69}\left(p_{\mathrm{H_2}} + 750\right)^{-0.39}. \tag{4}$$

where $S_{\mathrm{Ni-pore}}$ is the volumetric density of the nickel-pore interfacial area.

2.1. Calculation conditions

The same boundary conditions that reported in part I [7] are considered. The source terms for molar of species in Equation (1) are given in Table 1. The mass generation between channel and anode was explained in our previous report [7], except for the species of ammonia is $N_{\mathrm{NH_3},y}W_{\mathrm{NH_3}}$.

Table 1. Source terms for molar of species.

	O_2	N_2	H_2	H_2O	NH_3

IOP Conf. Series: Materials Science and Engineering **670** (2019) 012022 doi:10.1088/1757-899X/670/1/012022

Molar generation in the cell from the electrochemical reaction [mol m^{-3} s^{-1}]	$-\frac{I}{4F}$	-	$-\frac{I}{2F}$	$\frac{I}{2F}$	-
Molar generation in the anode from the ammonia decomposition reaction [mol m^{-3} s^{-1}]	-	$\frac{1}{2}R_{dec}$	$\frac{3}{2}R_{dec}$	-	$-R_{dec}$

3. Results and discussion

The results of direct ammonia fuel obtained from the developed numerical model with the DGM are compared with the results obtained from a 2D numerical model by Kishimoto et al. [8] in Figure 2. This 2D simulation was conducted resolving the gas transport in the streamwise and cell thickness directions in the electrodes and fuel/air channels. Note that the quasi-3D analysis in this section is conducted without the electric insulators located at the upstream and downstream of the cell to apply the same calculation condition that was used by Kishimoto et al. [8]. Also, there is a slight difference between the quasi-3D simulation and the 2D simulation in the fuel and air channels; they are filled with porous metal form in the quasi-3D simulation, whereas they are void volume in the 2D simulation. Figure 2(a) shows the comparison of the two models in terms of the current-voltage curve. Although the characteristic of the current-voltage curves is different between these two models, the quasi-3D model with the DGM can reasonably reproduce the performance of the direct ammonia fueled cell. On the other hand, the distribution of the molar fraction of gas species in the fuel channel at a terminal voltage of 0.75 V shown in Figure 2(b) agrees with the 2D numerical results by Kishimoto et al. [8]. Also, the distribution of the ammonia decomposition rate in the anode is shown in Figure 2(c). It indicates that the ammonia decomposition reaction actively occurs in the entire thickness of the anode near the upstream edge of the anode but the reaction rate is significantly decreased towards the downstream.

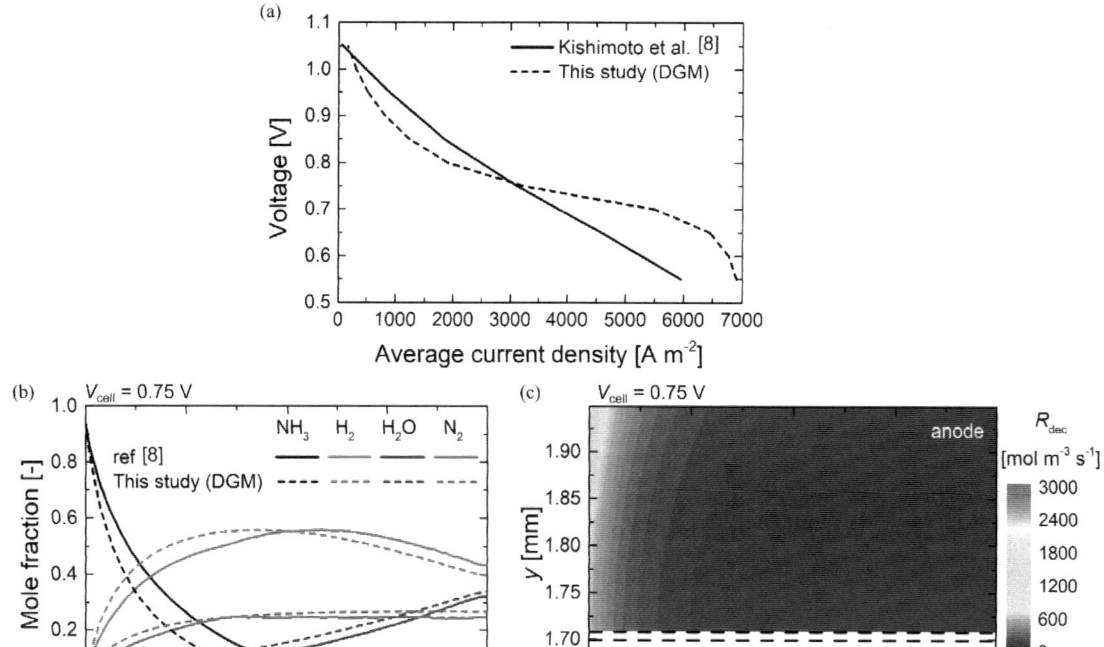

Figure 2. Comparison between numerical results with the DGM and literature in terms of (a) current-voltage characteristics, (b) distribution of molar fraction in the fuel channel, and (c) NH_3 decomposition reaction rate in the anode.

In principle, the FM is not applicable to the gas transport in porous electrodes when a significant concentration change occurs due to chemical reactions. However, such a situation can be numerically avoided if the chemical reaction is assumed to occur only near the anode surface; the thickness of the reaction area can be called the "active thickness". If the active thickness is substantially thin compared with the entire anode thickness, the FM can be used below the active reaction area. A similar situation is frequently assumed for an electrochemical reaction with a relatively thin active thickness in the vicinity of the anode-electrolyte interface. The electrochemical reaction in a Ni/YSZ anode was estimated with an active thickness of 4.22 μm [10]. Because the active thickness of the ammonia decomposition reaction $h_{\mathrm{dec,ave}}$ is not available in open literature, it is estimated in this study as the equivalent thickness from the anode surface that can decompose the same total amount of ammonia within the anode in the DGM as follows,

$$h_{\mathrm{dec,ave}} = \sum \left(R_{\mathrm{dec,DGM}} V_{\mathrm{local}}\right) / \sum \left(R_{\mathrm{dec,FM}} A_{\mathrm{local}}\right). \qquad (5)$$

$R_{\mathrm{dec,DGM}}$ and $R_{\mathrm{dec,FM}}$ are the ammonia decomposition rate under the DGM and FM, respectively. Note that the ammonia decomposition rate under the FM is calculated using the concentration of gas species at the anode surface. V_{local} and A_{local} are the local volume of the anode mesh and surface area of the cell, respectively. The $h_{\mathrm{dec,ave}}$ for the FM is iteratively calculated by Equation (5) under a terminal voltage of 0.75 V so that desirable amount of ammonia decomposition is achieved. This $h_{\mathrm{dec,ave}}$ is then applied for all the other terminal voltages under the analysis with the FM.

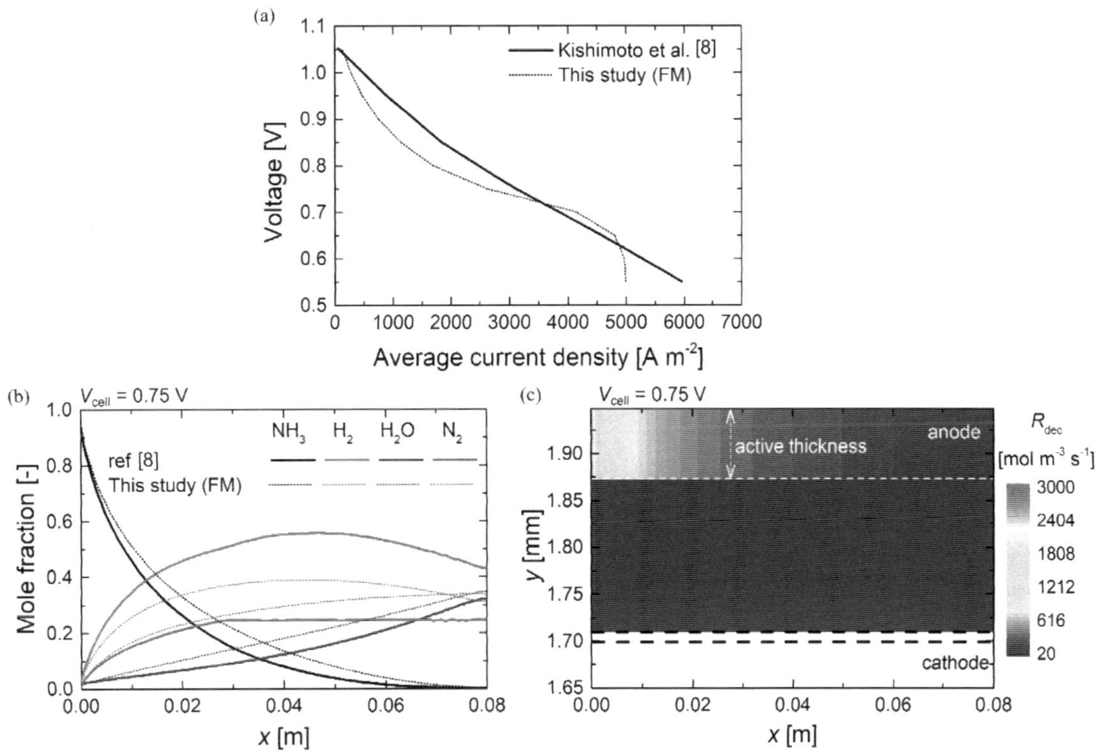

Figure 3. Comparison between numerical results with the FM and literature in terms of (a) current-voltage characteristics, (b) distribution of molar fraction in the fuel channel, and (c) NH3 decomposition reaction rate in the anode.

Figure 3 shows the comparison between the numerical results of the quasi-3D simulation with the FM employing $h_{\mathrm{dec,ave}}$ and the results obtained from a 2D numerical model. It is found that the characteristic of the current-voltage curve in Figure 3(a) from the FM are also apart from the 2D simulation. Figure 3(b) shows the distribution of the molar fractions of gas species in the fuel channel. It clearly shows that the differences between the two simulations in Figure 3(b) are larger than those shown in Figure 2(b). The average active thickness of the ammonia decomposition reaction in the quasi-3D simulation with the FM is estimated as 0.075 mm, which is approximately 31 % of the total anode thickness. It indicates that the assumption of the surface reaction is not realistic. The distribution of the local decomposition rate is shown in Figure 3(c). A significant difference can be observed when the figure is compared with Figure 2(c).

Noticeable differences between the DGM and FM model in the distribution of gas concentration in the fuel channel and the ammonia decomposition rate in the anode are found under direct ammonia fuel. The quasi-3D model with the DGM successfully captures the realistic non-uniform decomposition rate of ammonia in the anode. In conclusion, the implementation of the multi-component DGM improves the capability of the quasi-3D model, making it possible to analyze gas species transport in the anode with chemical reactions such as ammonia decomposition, which cannot be achieved using the FM.

4. Conclusions

The numerical results with the DGM and FM were compared with the 2D numerical results from literature under direct ammonia fuel. An active thickness of ammonia decomposition reaction was pre-determined from the numerical analysis with the DGM at a terminal voltage of 0.75 V, and used for the analysis with the conventional FM. It was found that the FM was not able to accurately reproduce the distribution of gas concentrations in the anode. Also, the pre-determined average active thickness was approximately 31 % of the anode thickness, which is beyond the applicable range of the assumption of surface reaction. On the other hand, DGM successfully reproduces the similar distribution of gas concentration in the fuel channel and captures the realistic non-uniform decomposition rate of ammonia in the anode. This study proved that the implementation of the DGM can accurately predict the transport of species involving reactions in the anode that significantly changes the concentration in the mixture, whereas the implementation of the FM is not suitable for analyzing direct ammonia fuel.

Acknowledgments

This work was partially supported under "Fundamental study on rapid evaluation method of SOFC durability" by the New Energy and Industrial Technology Organization (NEDO, Japan), and by the Council for Science, Technology and Innovation (CSTI), Cross-ministerial Strategic Innovation Promotion Program (SIP), "Energy Carrier" (Funding agency: JST).

5. References

[1] Tan W C, Iwai H, Kishimoto M, Brus G, Szmyd J S and Yoshida H 2018 Numerical analysis on effect of aspect ratio of planar solid oxide fuel cell fueled with decomposed ammonia *J. Power Sources* **384** 367–378.

[2] Tan W C, Iwai H, Kishimoto M and Yoshida H 2018 Quasi-three-dimensional numerical simulation of a solid oxide fuel cell short stack: Effects of flow configurations including air-flow alternation *J. Power Sources* **400** 135–146.

[3] García-Camprubí M and Fueyo N 2010 Mass transfer in hydrogen-fed anode-supported SOFCs *Int. J. Hydrogen Energy* **35** 11551–11560.

[4] García-Camprubí M, Izquierdo S and Fueyo N 2014 Challenges in the electrochemical modelling of solid oxide fuel and electrolyser cells *Renew. Sustain. Energy Rev.* **33** 701–718.

[5] García-Camprubí M, Sánchez-Insa A and Fueyo N 2010 Multimodal mass transfer in solid-oxide fuel-cells *Chem. Eng. Sci.* **65** 1668–1677.

[6] Pisani L 2008 Multi-component gas mixture diffusion through porous media: A 1D analytical solution *Int. J. Heat Mass Transf.* **51** 650–660.
[7] Tan W C, Iwai H, Kishimoto M, and Yoshida H 2019 Implementation of multi-component dusty-gas model for species transport in quasi-three-dimensional numerical analysis of solid oxide fuel cell. Part I: hydrogen fuel *Proc. 6th Int. Conf. on Applications & Design in Mechanical Engineering* (Penang) (submitted).
[8] Kishimoto M, Kume T, Iwai H and Yoshida H 2017 Numerical analysis of ammonia-fueled planar solid oxide fuel cells *ECS Trans.* **78** 2845–2853.
[9] Kishimoto M, Furukawa N, Kume T, Iwai H and Yoshida H 2017 Formulation of ammonia decomposition rate in Ni-YSZ anode of solid oxide fuel cells *Int. J. Hydrogen Energy* **42** 2370–2380.
[10] Nagasawa T and Hanamura K 2017 Prediction of overpotential and effective thickness of Ni/YSZ anode for solid oxide fuel cell by improved species territory adsorption model *J. Power Sources* **353** 115–122.

IOP Conf. Series: Materials Science and Engineering **670** (2019) 012023 doi:10.1088/1757-899X/670/1/012023

CFD simulations on a phase change thermal energy storage integrated with conducting fins

Muhammad Faris Abdul Razak[1], Nasrul Amri Mohd Amin[1,*], Mohd Shukry Abdul Majid[1], Nashrul Fazli Mohd Nasir[1], MD Tasyrif Abdul Rahman[1], and Anas Abdul Rahman[1]

School of Mechatronic Engineering, Universiti Malaysia Perlis, 02600 Pauh Putra Campus, Malaysia

Email: *nasrulamri.mohdamin@unimap.edu.my

Abstract. Global warming has become a major problem that leads to an abnormal climate change crisis. For most countries around the world, there be would be a rise in the average daily temperature. Therefore, many research works have been conducted in order to overcome this problem. One of the most viable solutions is to store thermal energy for later use. This can be done by integrating phase change material (PCM) into thermal energy storage (TES) system. To counter the increasing temperature problem in the residential house, PCM can be implanted in the wall of the house, for instance. This way, PCM can go through freezing process during the night and countering the heat through the wall by melting process during day time. However, due to the thermal resistance within these systems, the inadequate heat transfer problem had always become the problem of low thermal conductivity. Therefore, the integration of conducting fins is proposed to the system to increase the overall thermal conductivity.

1. Introduction

Phase change material (PCM) has a large latent heat-storage capacity that makes them the best storage medium for thermal energy system (TES). The utilization of PCM as a storage medium is one of the great choices especially for refrigeration system [1]. PCM works by absorbing or releasing thermal energy on the surrounding during the process of melting and solidifying. During melting process, PCM absorbs a large amount of heat from surrounding and vice versa during solidification process.

In a TES, the rate of the PCM volume per total system volume is characterized as the compactness ratio [2]. The obstacle in designing PCM encapsulation is to have a greater compactness ratio so that the greatest amount of latent energy can be utilized. Much of the time, the PCM should be encapsulated in order to maintain liquid phase PCM and also to prevent impurities enter from surrounding.

In any case, most PCM endure low thermal conductivity, which brings about a danger temperature for the electronic parts [3, 4]. To manage it, the fins are taken into the PCM to improve the thermal conduction [5, 6]. Some conducting related work has introduced composites for options [7, 8].

Packed bed of spheres has been used as thermal energy storage (TES) to enlighten the availability in space especially within the system incorporated with phase change material (PCM), nevertheless, due to the thermal resistance within these systems, the inadequate heat transfer problem had always become the problem of low thermal conductivity. Therefore, the integration of conducting fins is proposed to the system to increase the overall thermal conductivity. The main objective of this project is to study the effect of multiple physical design of the PCM in TES system towards the thermal performance of TES system.

2. Design Configuration

Three 3D models are developed using Solidwork software as shown in Figure 1.

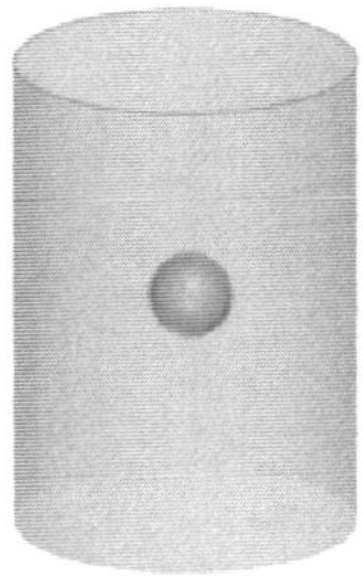

(a)

(b)

(c)

Figure 1. Assembly design of TES tank (a) for no fin [9, 10], integrated with (b) 2 fins and (c) 4 fins, tilted at different angles within a PCM sphere.

These models are then imported into Ansys CFX in .IGES format as preparation for simulation works. The different tilted angle positions are to be named as Orientation 1 (facing Inlet) and Orientation 2 (facing tank wall).

3. Simulation set up

The first step is to set boundary conditions that the wall of the tank is no slip type and adiabatic, sphere is thin Medium Density Poly Ethylene, heat transfer fluid is air at -20°C with velocity of $1ms^{-1}$, and the PCM is PCM0 (ice-water), and copper fin/s. There are four domains in total for the simulation which is the PCM and the HTF, encapsulation and the fin. Simulations are transient with total time of 5 minutes of freezing process, with optimised time steps to satisfy the convergence criteria of 10 x 10^{-4}, as has been found sufficient for most of the heat exchanger simulations [11].

4. Results and discussions

Result of the temperature distributions against the time steps (from CFX Solver) for phase change time (PCT) visualisation and the temperature distribution on the mid plane of PCM in sphere (from CFX Post) are shown in Figures 2 to 6.

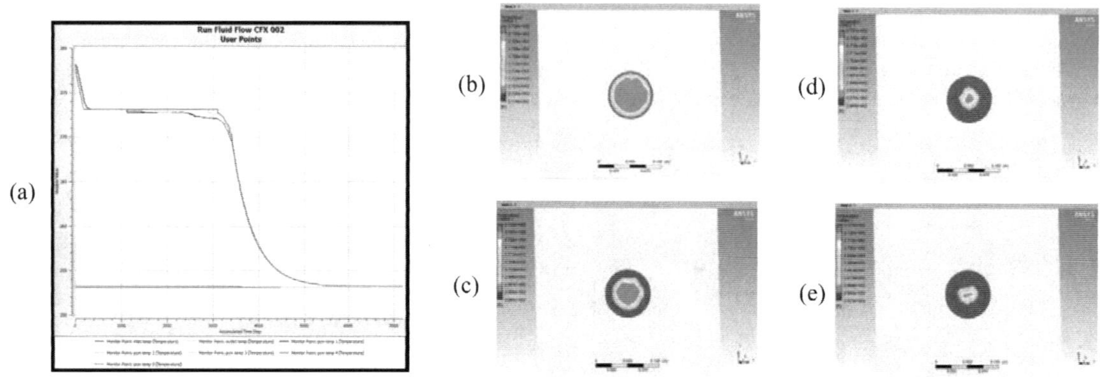

Figure 2. (a) PCT for PCM sphere with no fin, and temperatures contour for (b) 5%, (c) 25%, (d) 50%, & (e) 75% frozen mass fraction.

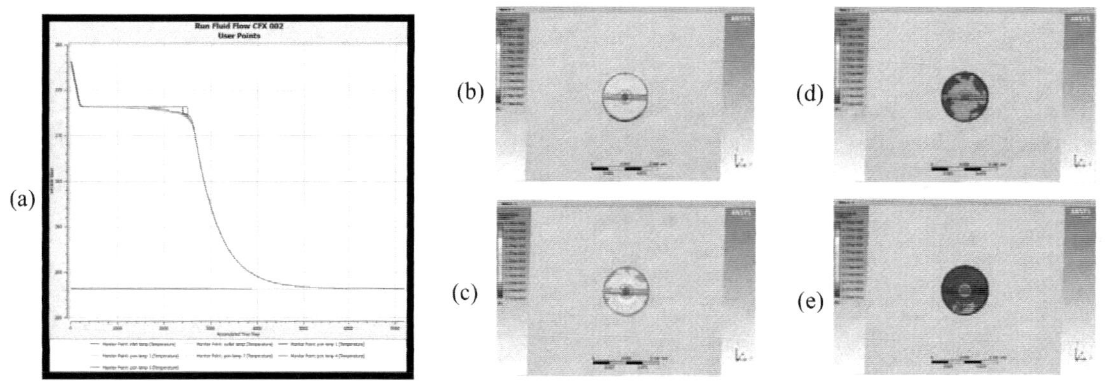

Figure 3. (a) PCT for PCM sphere with 2 fins at Orientation 1, and temperatures contour for (b) 5%, (c) 25%, (d) 50%, & (e) 75% frozen mass fraction.

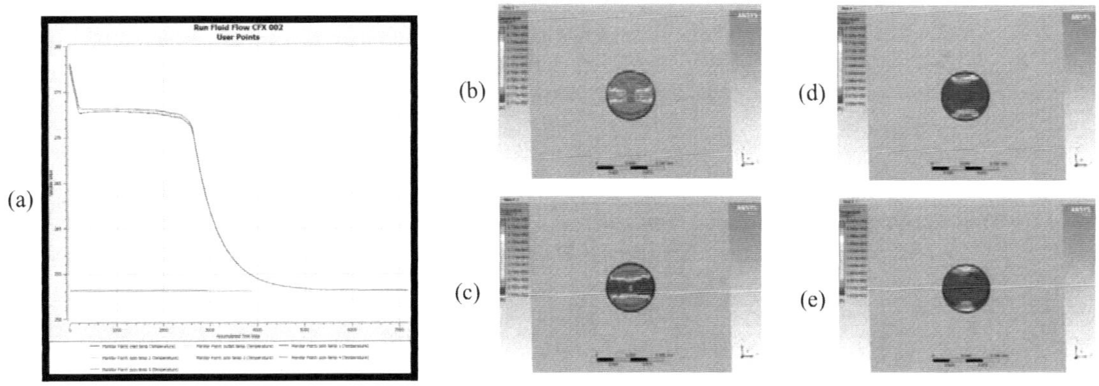

Figure 4. (a) PCT for PCM sphere with 2 fins at Orientation 2, and temperatures contour for (b) 5%, (c) 25%, (d) 50% & (e) 75% frozen mass fraction.

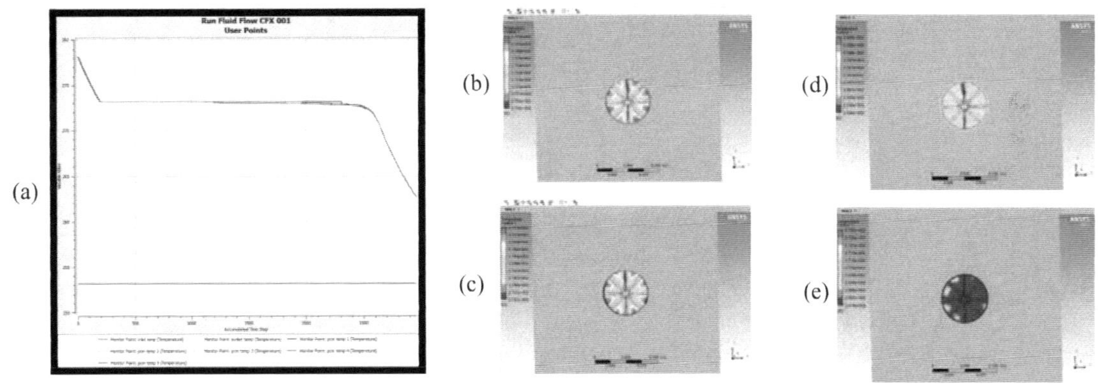

Figure 5. (a) PCT for PCM sphere with 4 fins at Orientation 1, and temperatures contour for (b) 5%, (c) 25%, (d) 50% & (e) 75% frozen mass fraction.

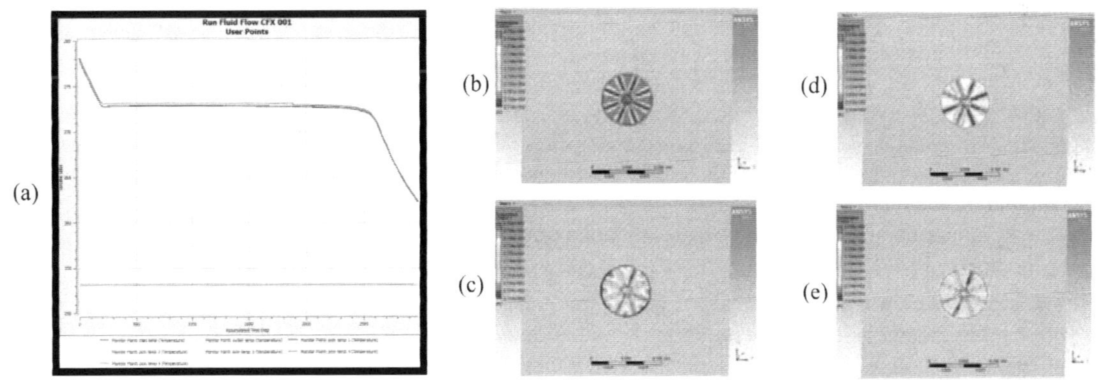

Figure 6. (a) PCT for PCM sphere with 4 fins at Orientation 2, and temperatures contour for (b) 5%, (c) 25%, (d) 50% & (e) 75% frozen mass fraction.

The PCT can be calculated based on the the time step interval as shown in Table 1. It is shown that the phase change time are different based on the number of fin used and the orientation of the fin inside the PCM sphere.

Table 1. Time step comparison

Design of PCM sphere	Intitial Time Steps	Final Time Steps	Time Step Interval
No fin	255	3414	3159
2 fins at Orientation 1	209	2510	2301
2 fins at Orientation 2	193	2376	2183
4 fins at Orientation 1	188	2519	2331
4 fins at Orientation 2	184	2492	2308

The shortest PCT is depicted from the PCM sphere with 2 fins at Orientation 1. While the longest time taken for phase change is by PCM sphere with no fin. Surprisingly, the PCT for 2 fins are shorter than the 4 fins in general. This is due to the large fin area perpendicular to the inlet of the tank. Comparing PCM sphere with fins at different orientations, it is found that the change in the HTF angle of attack (or the orientation) can result to as high as 5% different on the phase change time.

5. Conclusion

In summary, the best design configuration tested in this simulation project is the PCM sphere with 2 fins at Orientation 1. It is due to its shortest PCT or in other word, the highest thermal performance. This finding is contradicted with the hypothesis that the more high-conducting material (fins) used in the PCM sphere, the higher the thermal performance of the TES. The other consideration must be taken on the largest area exposed to the HTF angle of attack. The result also indicated that the correct orientation does affect thermal performance of the TES. Whereas in general, it is concluded that the use of high conducting material in PCM sphere shown a tremendous improvement in phase change time compared to no fin design.

Acknowledgements

The authors would like to acknowledge the School of Mechatronic Engineering, Universiti Malaysia Perlis, Perlis, Malaysia for the financial support.

References

[1] Liu M, Saman W and Bruno F 2011 Validation of a mathematical model for encapsulated phase change material flat slabs for cooling applications *Applied Thermal Engineering* **31** (14-15) 2340-2347.

[2] Amin N A M, Belusko M and Bruno F 2009 Optimisation of a phase change thermal storage system *World Academy of Science, Engineering and Technology* **56** 765–769.

[3] Zalba B, Marın J M, Cabeza L F and Mehling H 2003 Review on thermal energy storage with phase change: materials, heat transfer analysis and applications *Applied Thermal Engineering* **23** (3) 251–283.

[4] Dutil Y, Rousse D R, Ben Salah N, Lassue S and Zalewski L 2011 A review on phase-change materials: Mathematical modeling and simulations *Renewable and Sustainable Energy Reviews* vol. **15** (1) 112–130.

[5] Shatikian V, Ziskind G and Letan R 2008 Numerical investigation of a PCM-based heat sink with internal fins: Constant heat flux *International Journal of Heat and Mass Transfer* **51** (5-6) 1488–1493.
[6] Lamberg P and Sirén K 2003 Approximate analytical model for solidification in a finite PCM storage with internal fins *Applied Mathematical Modelling* **27** (7) 491–513.
[7] Majid M S A, Afendi M, Daud R, Amin N A M, Mohamad A, Cheng E M, Gibson A G and Hekman M 2014 General lifetime damage model for glass fibre reinforced epoxy (GRE) composite pipes under multiaxial loading *Key Engineering Materials* **594-595** 624-628.
[8] Fartini M S, Abdul Majid M S, Ridzuan M J M, Amin N A M and Gibson A G 2016 Compressive properties of Napier (Pennisetum Purpureum) filled polyester composites *Plastics, Rubber and Composites* **45** (3) 136-146.
[9] Aziz S, Amin N A M, Abdul Majid M S, Belusko M and Bruno F 2018 CFD simulation of a TES tank comprising a PCM encapsulated in sphere with heat transfer enhancement *Applied Thermal Engineering* **143** 1085-1092.
[10] Aziz S, Amin N A M, Abdul Majid M S, Bruno F and Belusko M 2018 Effectiveness-NTU correlation for a TES tank comprising a PCM encapsulated in a sphere with heat transfer enhancement *Applied Thermal Engineering* **143** 1003-1010.
[11] Shahril S M, Quadir G A, Amin N A M and Badruddin I A 2017 Numerical investigation on the thermohydraulic performance of a shell-and-double concentric tube heat exchanger using nanofluid under the turbulent flow regime *Numerical Heat Transfer; Part A: Applications* **7** (2) 215-231.

IOP Conf. Series: Materials Science and Engineering **670** (2019) 012024 doi:10.1088/1757-899X/670/1/012024

Orthodontics Anchorage in Mandibular Angle Fractures –A Brief Review

Siti Aisyah Abdul Halim[a], Ruslizam Daud[a], Ishak Ibrahim[a], Fauziah Mat[a], Nur Saifullah Kamaruddin[a], Rohaizar Ismail[b]

[a]Fracture and Damage Mechanics SIG, School of Mechatronic Engineering, Universiti Malaysia Perlis, 02600 Arau, Perlis, Malaysia
Email: sitiaisyahabdulhalim@rocketmail.com

[b]Klinik Pakar Bedah Mulut Dan Maksilofasial, Hospital Tuanku Fauziah, Jalan Tun Abdul Razak, 02600 Kangar, Perlis, Malaysia

Abstract. Maxillofacial fracture could be a quite common injury these days particularly in jaw angle. The healing process is physiologically complicated including each biological and mechanical aspects. Over the past decade, mandibular angle fracture simulation and modeling has been used to understand its details and mechanisms, to analyze specific clinical inquiry and to style healing approches. This paper aims to review the history and also the most recent work on biological and mechanical properties of miniplate . This paper provides a short review of mandibular angle fracture, miniplate evolution, biomaterials utilized in miniplates, structure of miniplates and finite element analysis of miniplate in three dimensional (3D) model.

1. Introduction

Modern life's increasing pace, the increasing frequency of violence, sports injuries, high-speed travel, traffic accidents, wars and industrial trauma have caused a maxillofacial fracture. A maxillofacial fracture refers to any fracture to the face or jaw caused by physical force. Research on orthodontic anchorage of miniplates materials, methods and models has risen in latest years and is anticipated to grow in the future owing to current development in the worldwide orthodontics anchorage industry and increasing demand for cosmetic dentistry.

The mandible forms the lower jaw and the only moveable bone of the facial skeleton. Hence, the mandible is vulnerable to fracture due to its mechanically weak components. Several factors influence the mandibular fractures such as direction, severity and impact of force, the presence of soft tissue bulk and occlusal loading pattern. Biomechanical characteristics such as anatomic structures, mass and bone density creating weak areas are responsible to fractured in mandibular (Mannan, Farooq, & Mukhtar, 2018). The occurrence of mandibular fracture can occur either in unilateral symphysis, body, angle, condylar or bilateral condylar (Zhou, Lv, Yang, Li, & Li, 2016)

The mandibular angle fracture accomplish more complication compare to other mandibular fractures due to complex anatomical and biomechanical aspects including abrupt change in curvature, attachment of the masticatory muscles, a thin cross-sectional area and the presence of the third molars (Ellis, 1996). Interesting trends for mandibular angle fractures were stated that the patients with impacted mandibular third molar are significantly to have mandibular angle fracture (Asif, Alam, Haque, Rashdan, & Mahmoud, 2018).

Mandibular angle fractures pose a unique challenge for surgeons and in medical industries because they have the highest reported post-operative complication rate than other mandibular fractures. The ideal treatment of mandibular angle fractures remains subject to considerable debate. Therefore, anchorage orthodontics of miniplates become expeditiously necessary management of mandibular fractures. Orthodontics anchorage can be grouped into two based on function; devices for symmetrical tooth movements in all planes of spaces and an alternative towards orthognathic surgery. Apart from this, orthodontics anchorage can be classified into two main categories based on its relation with bone. The first category consists of osseointegrated dental implants such as the orthodontic mini-implants, the retromolar implants and the palatal implants (Sugawara, 2014). The second category derives from the surgical mini-screws and miniplates (Sugawara, 2014). The differences between these both categorized is the size in diameter, smooth surface and stable differing from miniplates

Review the orthodontics anchorage evolution of miniplates for mandibular angle fractures, by discussing the existing miniplates modalities and defining the outcome measures in accordance with the recently published review on miniplates of mandibular angle fractures..

2. Miniplates Evolution

Michelet et. al introduced an original technique to reduce osteosynthesis, osteotomies and immobilization of fractures of the facial skeleton (Michelet, Deymes, & Dessus, 1973). Later on modified Michelet's technique of mandibular osteosynthesis by Champy consists of monocortical juxta-alveolar and sub-apical osteosynthesis without compression and without inter-maxillary fixation (Champy M, Lodde JP, 1978). Iii & Walker (1994) have used 2.0 mm self-threading screws placed double and 2.0 mm non-compression miniplates through a transoral incision with transbuccal trochar instrumentation. Ellis (1996) uses a single miniplate to demonstrate the use of either extraoral open reduction and internal fixation with reconstruction plate AO / ASIF or intraoral open reduction and internal fixation (Ellis, 1996), following treatment of mandibular angle fracture with single, malleable and thin 2.0 mm miniplate (J. & E., 1999).

Survey conducted among surgeons shown that a single miniplate on the superior border of the mandible has become the preferred method of treatment in Arbeitsgemeninschaft fur Osteosynthesefragen (AO) faculty (Gear, Apasova, Schmitz, & Schubert, 2005). One of the most important developments in miniplates began in 2005, when three dimensional (3D) miniplate for mandibular angle fracture fixation been suggested by some researchers (Al-Moraissi, El-Sharkawy, El-Ghareeb, & Chrcanovic, 2014; Al-Mozaiek, Darwich, & Darwich, 2016; Alkan, Çelebi, Özden, Baş, & Inal, 2007; Guimond, Johnson, & Marchena, 2005; Kalfarentzos, Deligianni, Mitros, & Tyllianakis, 2009; Khiabani, Keyhan, Razmdideh, Chaleh Chaleh, & Amirzade-Iranaq, 2018; Rastogi et al., 2017)

However, miniplate with Champy technique has gained the confidence of many surgeons. This miniplate was suitable for simple angle fracture which reduces surgical time and dissection, minimizes risk to the facial and inferior alveolar nerves and decreases complication rates (Saito & Murr, 2008).The development of modern plate osteosynthesis non-corrosive materials like vitallium or titanium, plate with screws as small as 0.8 mm in diameter and small self-locking systems had recommended by researcher (Sauerbier, Schön, Otten, Schmelzeisen, & Gutwald, 2008). Recent progress in the past studies has focused on materials and techniques to improve quality and anchorage; and other important development in miniplates have been focused in three dimensional model using finite element (FE) analysis and postoperative complications

3. Biomaterial used in miniplates

There are two ideal requirements for implant biomaterials like miniplate. First ideal requirement is biological properties. Biological properties included provide effective osseointegration, should not be harmful to hard and soft tissue, should not contain the toxic diffusible substance and should be free from potentially sensitizing agent that may cause an allergic reaction. Miniplate biomaterials also should have no carcinogenic potential and should be tasteless and ordorless. Second ideal requirements is physical properties which should be dimensionally stable and should possess adequate strength and resilience and resist biting or chewing (Bajaj et al., 2017).Miniplates are made of titanium or titanium alloys. Comercially pure titanium (C P Ti; ASTM F67) is widely been used as a material of choice for implants especially for miniplates. It is because this type of titanium usually suitable in mechanical properties and excellent biocompatibility. Mini screws were made of titanium alloy (Titanium-6Aluminum-4Vanadium; ASTM F-136) (Bajaj et al., 2017; Rangel Goulart, Takanori Kemmoku, Noritomi, & de Moraes, 2015). Along with this, there is some potential biomaterials for mandibular angle fractures fixation such as biodegradable miniplates. Biodegradable miniplates consists of self-reinforced poly(L-lactide-co-D,L-lactide) [P(L/DL)LA] copolymer BioSorb FX plates (Lee et al., 2010).

Hence, biomaterials with effective osteointegration, no carcinogenic potential, and resistance on biting or chewing should be chosen as material for miniplates.

4. Miniplates structure

Miniplates are manufactured with alloys of titanium or titanium and have different shapes and sizes. The miniplate structure consists of three parts, such as head, arm, and body. Within the mouth, the head part is exposed and physically situated outside the dental arches. The head exists in a diversity of forms, e.g. curved, bent and tubular.

Some head portion manipulated into the desired shape like a bendable stick. Meanwhile, The arm portion is transgingival and usually in rectangular or round shape (Sugawara, 2014).
There are four basic shapes for body portion which T, L, Y and I (straight). Even though modifications regarding maniplates heads were extensively explored, but there is a few research in modifications of the body portions (Sugawara, 2014). (Sugawara, 2014).

5. Finite Element Analysis of Miniplate Fixation

Finite Element Analysis (FEA) in dental started in 1973, finite element analysis were usud to investigate the state of stress for a homogeneous and non-homogeneous tooth (Thresher & Saito, 1973). Now, the popularity of FEA has drastically grown especially in miniplates and screws for mandibular angle fractures. At 2003, 8-noded hexahedral elements three dimensional (3D) mandible were generated in FEA by Cox et.al (Cox, Kohn, & Impelluso, 2003). Later on, in 2008 a fixation in mandibular angle fractures computational model developed by Arbag et.al using finite elements in order to calculate mechanical stress occurring in osteosynthesis plates (Arbag, Korkmaz, Ozturk, & Uyar, 2008).

In addition, FEA also capable to identify the optimize shape and fixation technique for a certain type of mandible fracture at corpus. In the future development of osteosynthesis materials and techniques, authors also agreed that FEA enables savings in term of time, material and animal experiments (Arbag et al., 2008).

Besides, FEA had been used to determine optimal modalities and plates configuration for intraoral and transoral approaches. Authors introduced four types of miniplate design such as 6-hole non compression miniplate, 6-hole single plate/Champy's technique, 3D strut plate and 2 parallel 4-hole non compression miniplates. This analysis included segmental displacement and Von Mises Stress evaluation of 3D reconstruction of the human mandible (Coskunses et al., 2015).In silico 3-dimensional FEA was introduced in year 2017. In silico FEA is an approach to identify an ideal design as it let on mechanical stress or strain distribution visualization and quantification of the marginal bone of dental implant fixtures and implant components (Ko, Rocha, & Larson, 2012). Yamaguchi et al. (2017) investigate the effect of fixation methods and different materials on the state of stress and displacement in reconstruction plate (Yamaguchi et al., 2017).

With topological optimization approach, Liu et.al created a three-dimensional (3D) FEA model virtual mandible using CT images with a mimic angle fracture and 1 mm gap between two bone segments in order to design a customized plate (Liu, Fan, Jiang, & Baur, 2017).

Ayali & Erkmen conducted 3D FEA for favourable mandibular angle fracture and unfavourable mandibular angle fractures (Ayali & Erkmen, 2017, 2018). Khiabani et al. also suggested 3D FEA to reach the best mastication power by determine the optimal method for mini-plating in fracture of the mandible angle. The usage of two miniplates in two different planes had been recommended by reason of the fracture of the lower angle of the jaw and reconstruction the optimal mastication force in patients (Khiabani et al., 2018).

FEA has a long history of in orthodontics , the development of 3D FEA has been an interesting research subject in the area. Throughout research in recent years, FEA technology has been constantly improving and providing a 3D FEA model of miniplates with an exceptional level of effectiveness, convenience and affordability. Several designs of 3D model of miniplates have been evaluated and also been tested. As such, more research on a better model in materials, surface treatment technologies, design parameters, and reliability analysis techniques is still needed to improve the outcomes.

6. Conclusions

The introduction of miniplates as orthodontic anchorage had a tremendous impact on the field of dentistry. The devices themselves are evolving that might support the best combination of ease of placement, least invasive procedure and best physical design properties to deliver optimum mechanical forces.

7. Acknowledgment

The author would like to acknowledge the support from the Fundamental Research Grant Scheme (FRGS) under a grant number of FRGS/1/2018/TK03/UNIMAP/02/8 from the Ministry of Higher Education Malaysia.

References

[1] Al-Moraissi E A, El-Sharkawy T M, El-Ghareeb T I and Chrcanovic B R 2014 Three-dimensional versus standard miniplate fixation in the management of mandibular angle fractures : a systematic review and meta-analysis *International Journal of Oral & Maxillofacial Surgery* **43** (6) 708–716.

https://doi.org/10.1016/j.ijom.2014.02.002

[2] Al-Mozaiek M Y A-S, Darwich K and Darwich M A 2016 3D Finite element analysis of miniplate fixation d techniques in mandibular angle fractures *Journal of the Indian Dental Associatio* **9** (12) 10–18.

[3] Alkan A, Çelebi N, Özden B, Baş B and Inal S 2007 Biomechanical comparison of different plating techniques in repair of mandibular angle fractures *Oral Surgery, Oral Medicine, Oral Pathology, Oral Radiology and Endodontology* **104** (6) 752–756. https://doi.org/10.1016/j.tripleo.2007.03.014

[4] Arbag H, Korkmaz H H, Ozturk K and Uyar Y 2008 Comparative evaluation of different miniplates for internal fixation of mandible fractures using finite element analysis *Journal of Oral and Maxillofacial Surgery* **66** (6) 1225–1232. https://doi.org/10.1016/j.joms.2005.11.092

[5] Asif J A, Alam M K, Haque S, Rashdan N H B M and Mahmoud M 2018 The correlation between mandibular third molar and mandibular angle fracture among patients in Hospital Universiti Sains Malaysia *Oral Surgery* **11** (3) 189–194. https://doi.org/10.1111/ors.12342

[6] Ayali A and Erkmen E 2017 Biomechanical evaluation of different plating methods used in mandibular angle fractures with 3-dimensional finite element analysis: favorable fractures *Journal of Oral and Maxillofacial Surgery* **75** (7) 1464–1474. https://doi.org/10.1016/j.joms.2017.02.028

[7] Ayali A and Erkmen E 2018 Three-Dimensional finite element analysis of different plating techniques for unfavorable mandibular angle fractures *Journal of Craniofacial Surgery* **29** (3) 1–5. https://doi.org/10.1097/SCS.0000000000004327

[8] Bajaj R, Shenoy U, Banerjee S, Hazare A, Karia H and Atulkar M 2017 Implants in orthodontics- a brief review *International Journal of Oral Health and Medical Research* **3** (5) 92–97.

[9] Champy M, Loddé J P, Schmitt R, Jaeger J H and Muster D 1978 Mandibular osteosynthesis by miniature screwed plates via a buccal approach *Journal of Maxillofacial Surgery* **6** (1) 14–21.

[10] Coskunses F M, Kocyigit I D, Atil F, Tekin U, Suer B T, Tuz H H et al. 2015 Finite-element analysis of a new designed miniplate which is used via intraoral approach to the mandible angle fracture: Comparison of the different fixation techniques *Journal of Craniofacial Surgery* **26** (5) e445–e448. https://doi.org/10.1097/SCS.0000000000001890

[11] Cox T, Kohn M W and Impelluso T 2003 Computerized analysis of resorbable polymer plates and screws for the rigid fixation of mandibular angle fractures *Journal of Oral and Maxillofacial Surgery* **61** (4) 481–487. https://doi.org/10.1053/joms.2003.50094

[12] Ellis E 3rd 1996 Treatment methods for fractures of the mandibular angle *The Journal of Cranio-Maxillofacial Trauma* **2** (1) 28–36. https://doi.org/10.1016/S0901-5027(99)80152-0

[13] Gear A J L, Apasova E, Schmitz J P and Schubert W 2005 Treatment modalities for mandibular angle fractures *Journal of Oral and Maxillofacial Surgery* **63** (5) 655–663. https://doi.org/10.1016/j.joms.2004.02.016

[14] Giovacchini F, Paradiso D, Bensi C, Belli S, Lomurno G and Tullio A 2018 Association between third molar and mandibular angle fracture: A systematic review and meta-analysis *Journal of Cranio-Maxillofacial Surgery* **46** (4) 558–565. https://doi.org/10.1016/j.jcms.2017.12.011

[15] Guimond C, Johnson J V and Marchena J M 2005 Fixation of mandibular angle fractures with a 2.0-mm 3-dimensional curved angle strut plate *Journal of Oral and Maxillofacial Surgery* **63** (2) 209–214. https://doi.org/10.1016/j.joms.2004.03.018

[16] Potter J and Ellis E 3rd 1999 Treatment of mandibular angle fractures with a malleable noncompression miniplate *Journal of Oral and Maxillofacial Surgery* **57** (3) 288–293.

[17] Kalfarentzos E F, Deligianni D, Mitros G and Tyllianakis M 2009 Biomechanical evaluation of plating techniques for fixing mandibular angle fractures: The introduction of a new 3D plate approach *Oral and Maxillofacial Surgery* **13** (3) 139–144. https://doi.org/10.1007/s10006-009-0163-7

[18] Khiabani K, Keyhan S O, Razmdideh R, Chaleh Chaleh Z and Amirzade-Iranaq M H 2018 Effect of different miniplate osteosynthesis in different mandibular angle fracture patterns on bite force: A 3D finite element analysis *Journal of Oral and Maxillofacial Surgery, Medicine, and Pathology* **30** (4) 324–329. https://doi.org/10.1016/j.ajoms.2018.02.014

[19] Ko C -C, Rocha E P and Larson M 2012 Past, present and future of finite element analysis in dentistry *Finite Element Analysis - From Biomedical Applications to Industrial Developments* (Lanczos 1962). https://doi.org/10.5772/38037

[20] Lee H B, Oh J S, Kim S G, Kim H K, Moon S Y, Kim Y K et al 2010 Comparison of titanium and biodegradable miniplates for fixation of mandibular fractures *Journal of Oral and Maxillofacial Surgery* **68** (9) 2065–2069. https://doi.org/10.1016/j.joms.2009.08.004

[21] Liu Y F, Fan Y Y, Jiang X F and Baur D A 2017 A customized fixation plate with novel structure designed

by topological optimization for mandibular angle fracture based on finite element analysis *BioMedical Engineering Online* **16** (1) 1–17. https://doi.org/10.1186/s12938-017-0422-z

[22] Mannan R, Farooq A and Mukhtar H 2018 Comparison of single vs double miniplates in the management of mandibular angle fractures *Pakistan Journal of Medical and Health Sciences* **12** (2) 662–664. https://doi.org/10.1016/j.arthro.2008.04.076

[23] Michelet F X, Deymes J and Dessus B 1973 Osteosynthesis with miniaturized screwed plates in maxillo-facial surgery *Journal of Maxillofacial Surgery* **1** (C) 79–84. https://doi.org/10.1016/S0301-0503(73)80017-7

[24] Rangel Goulart D, Takanori Kemmoku D, Noritomi P Y and de Moraes M 2015 Development of a titanium plate for mandibular angle fractures with a bone defect in the lower border: finite element analysis and mechanical test *Journal of Oral and Maxillofacial Research* **6** (3) 1–7. https://doi.org/10.5037/jomr.2015.6305

[25] Rastogi S, Paul S, Kukreja S, Aggarwal K, Choudhury R, Bhugra A et al. 2017 Treatment of mandibular angle fractures with single three-dimensional locking miniplates without maxillomandibular fixation: how much fixation is required? *Craniomaxillofacial Trauma & Reconstruction* **10** (03) 188–196. https://doi.org/10.1055/s-0037-1600904

[26] Saito D M and Murr A H 2008 Internal fixation of mandibular angle fractures with the Champy technique *Operative Techniques in Otolaryngology - Head and Neck Surgery* **19** (2) 123–127. https://doi.org/10.1016/j.otot.2008.04.006

[27] Sauerbier S, Schön R, Otten J E, Schmelzeisen R and Gutwald R 2008 The development of plate osteosynthesis for the treatment of fractures of the mandibular body - A literature review *Journal of Cranio-Maxillofacial Surgery* **36** (5) 251–259. https://doi.org/10.1016/j.jcms.2007.08.011

[28] Sugawara J 2014 Temporary skeletal anchorage devices: The case for miniplates *American Journal of Orthodontics and Dentofacial Orthopedics* **145** (5) 559–565. https://doi.org/10.1016/j.ajodo.2014.03.010

[29] Thresher R W and Saito G E 1973 The stress analysis of human teeth *Journal of Biomechanics* **6** (5) 443–449. https://doi.org/10.1016/0021-9290(73)90003-1

[30] Yamaguchi S, Anchieta R B, Guastaldi F P S, Tovar N, Tawara D, Imazato S and Coelho P G 2017 In silico analysis of the biomechanical stability of commercially pure ti and ti-15mo plates for the treatment of mandibular angle fracture *Journal of Oral and Maxillofacial Surgery* **75** (5) 1004.e1-1004.e9. https://doi.org/10.1016/j.joms.2016.12.043

[31] Zhou H, Lv K, Yang R, Li Z and Li Z 2016 Mechanics in the production of mandibular fractures : a clinical *Plos One* **11** (2) 1–11. https://doi.org/10.1371/journal.pone.0149553

Assessment of the Malaysian Tidal Stream Energy Resources

A Rahman, I Ibrahim, and M T A Rahman

School of Mechatronic Engineering, Universiti Malaysia Perlis, Pauh Putra Campus, 02600 Arau, Perlis, Malaysia.

Abstract. As a developing country, Malaysia has shown great interest and effort to commit and invest for green technology and green energy especially tidal energy resources. This research project is conducted to carry out an assessment study to analyse and compare several locations in Malaysia while using the optimal analytical approach in finding the best sites to harness tidal stream resource around Malaysia. Previous studies that only considered flow velocity at the deployment sites may not be accurate since geometry of the coastal region and the friction force exerted from the tidal flow with the seabed can also contribute to the suitability of the proposed tidal sites. Tidal Model Driver was used to extract tidal elevation data on specific harmonic constituents and tidal velocity attribute, and then compared to the data obtained from General Bathymetric Chart of the Oceans. The weighted decision matrix was used to analyse the potential of each location by considering essential criteria such as tidal velocity, depth of coastline, geometry of coastline and environmental impacts. Top locations with the highest score were chosen as potential location to invest in a tidal energy farm, which consists of Pulau Pangkor, Port Dickson, and Pulau Tioman among others. The locations that produce the highest yearly energy density are found to be Pulau Pangkor with 1.21 MWh/year followed by Pulau Pinang with 836.5 kWh/year. Notably, the inclusion of other factors besides tidal velocity such as the site characteristic and the amount of potential impact to the environment provided different outcome than published materials in this field.

1. Introduction

Tidal energy is extracted through a natural phenomenon called ocean tides, where the water in the seas and oceans rise and falls periodically [1]. Using the tidal stream approach, tidal stream turbine is planted on the oceans seabed where the current is usually strong to generate enough electricity. The purpose of this project is to carry out an assessment study on various potential locations for tidal energy farm in Malaysia's water since different location will exert different amount of velocity. Furthermore, previous study on Malaysia's tidal energy resources only consider the flow velocity as the main criteria in calculating the available power. Such method, however, is not comprehensive and require additional analytical assessment to determine the best sites to harness tidal energy resources. One critical factor to be considered when determining the potential locations for tidal energy farm in Malaysia is the geometry of the coastal region and the friction force exerted from the tidal flow with the seabed [2]. Planting a tidal turbine may disturb the natural conditions of the environment due to the changes of flow of water and the sediment transport [2]. Therefore, an assessment study needs to be conducted to determine suitable locations to invest in tidal energy farm without interrupting the environment. Besides, the chosen sites must also be geographically suitable to install arrays of turbine.

2. Methodology

This study follows the research and guidelines published by Claire Legrand, Black & Veatch Ltd [3]. Several other resource assessment study on tidal energy within Malaysia's open water had been done by Yun Seng Lim et. al. [4] and Saw Jun Sheng [5], which are used as the backbone of this project research.

2.1. Data collection

For the first part of this research, important parameters such as the value of harmonic constituents of the tide is obtained from the GEBCO (General Bathymetric Chart of the Oceans) database [6]. The harmonic constituents consist of M2, S2, N2, K2, K1, O1, P1 and Q1. The harmonic constituents from the TMD software are then compared with the value from GEBCO to determine the percentage in difference of both values.

2.2. Tidal Model Driver (TMD) simulation

The second part is mainly focused on the software simulation done by an ocean modelling software call TPXO or Tidal Model Driver (TMD). TMD is a database of global tides properties that uses equation such as the Laplace Tidal Equations and altimetry data to predict the properties of the oceans in the upcoming years.

2.3. Model used for simulation

In this project, the TMD's model used is the China Seas & Indonesia 2016 that was retrieved from OTIS Regional Tidal Solutions [7]. Nonetheless, the first model that was employed in the TMD software was the Pacific Ocean 2009, and a comparison of several characteristics between the two models are highlighted in **Table 1.**

Table 1. Comparison between models.

Characteristics \ Models	Pacific Ocean 2009	China Seas & Indonesian 2016
Bathymetry	GEBCO 1'	GEBCO
Resolution	1/12^{0}; 1585x2065 grid	1/30°; 1801x2161 grid
OBC	TPXO8	TPXO7.1
Constituents	M2, S2, N2, K2, K1, O1, P1, Q1, M4, MS4, MN4.	M2, S2, N2, K2, K1, O1, P1, Q1.

2.4. Site selection

An assessment needs to be done to find out suitable locations in Malaysia's open water for tidal array, in which several key factors need to be considered.[3].

2.4.1. Current velocity.

Malaysia's ocean can be classified as shallow water, where the average current velocity is approximately 1 m/s. **Table 2** shows that current velocity above 0.65 m/s are able to generate electricity, although the amount is very small. Therefore, locations that have velocities of more than 1 m/s are chosen as potential locations. The flow speed is chosen based on a simulation on the TPXO software for various locations throughout Malaysia's coastline.

2.4.2. Site characteristics.

According Adcock [2], there are 4 examples of sites where tidal turbine can be installed, which is illustrated in **Figure 1**. Before deciding a potential location to invest in tidal energy farm, the effect of tidal turbine on the environment needs to be considered. According to Legrand, a minimum of 5m on the top clearance is recommended to allow for recreational activities such as fishing activity, small boats and swimmers as well as avoiding floating garbage from hitting the turbine [3]. Likewise, a minimum of 5m is also needed for the bottom clearance as it is to allow potential damaging material to the tidal turbine to pass through along the seabed and to minimize turbulence [3].

Table 2. Mean electrical power generated per average tidal velocity [3].

Average bin velocity	Velocity occurrence likelihood	Electrical power per bin	Mean annual electrical power per bin
U_i	$f(U_i)$	$P(U_i)$	$P(U_i) \times f(U_i)$
m/s	%	kW	kW
0.05	0.3	0	0
0.15	0.9	0	0
0.25	1.5	0	0
0.35	2.2	0	0
0.45	3.0	0	0
0.55	4.0	11	0
0.65	5.2	19	1
0.75	6.4	30	2
0.85	7.4	44	3

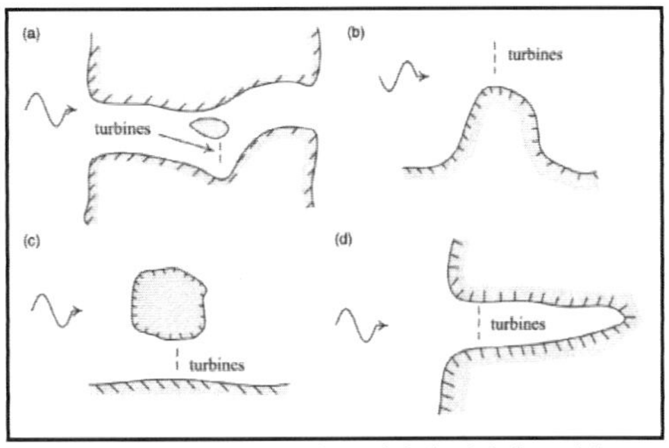

Figure 1. Different coastal geometries in which tidal turbine can be installed [2].

3. Results and Discussion

Weighted decision matrix was used to evaluate the potential locations from the selection of criteria based on ranking. The weighted factor and scoring degree will define whether the potential location meets the main criteria for the deployment of tidal energy farm. **Figure 2** illustrates the objective tree for tidal energy farm's key evaluation criteria.

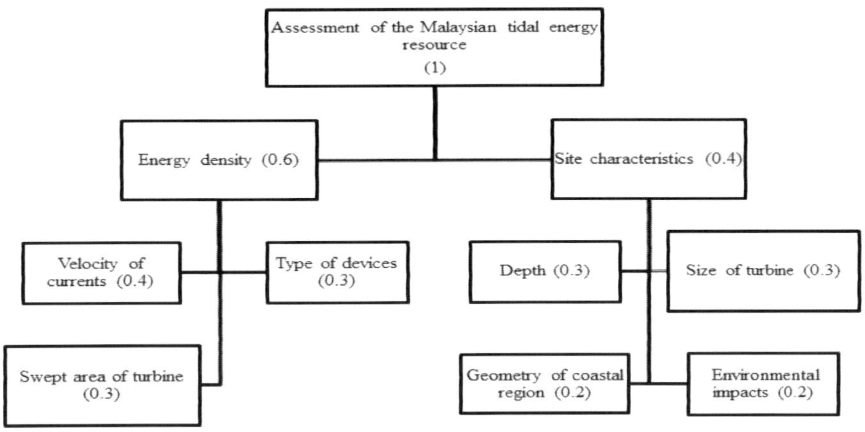

Figure 2. Objective tree for criteria on potential location for tidal farm in Malaysia.

Table 3. Weighted score for top selected locations.

Criteria	Weight Factor	Units	Pulau Pangkor		Port Dickson		Pulau Tioman		Pulau Redang		Pulau Langkawi	
			Score	Rating	Score	Rating	Score	Rating	Score	Rating	Score	Ratin g
Velocity of currents	0.24	m/s	7	1.68	5	1.20	2	0.48	1	0.24	3	0.72
Swept area of turbine	0.18	m^2	9	1.62	8	1.44	9	1.62	9	1.62	8	1.44
Type of devices	0.18	-	8	1.44	8	1.44	8	1.44	8	1.44	8	1.44
Depth	0.12	m	9	1.08	7	0.84	9	1.08	9	1.08	8	0.96
Size of turbine	0.12	m^2	9	1.08	8	0.96	9	1.08	9	1.08	8	0.96
Geometry of coastal region	0.08	-	8	0.64	5	0.40	6	0.48	7	0.56	7	0.56
Environmental impacts	0.08	-	3	0.24	3	0.24	4	0.32	5	0.40	4	0.32
Total rating			-	7.78	-	6.52	-	6.5	-	6.42	-	6.4

As shown in **Table 3**, the top selected locations are based on the locations that scored the highest rating on the weighted decision matrix. The top ten selected locations are Pulau Pangkor, Port Dickson, Pulau Tioman, Pulau Redang, Pulau Langkawi, Batu Pahat, Pulau Pinang, Melaka, Sibu and Pulau Banggi. After obtaining the top ten locations to invest in a tidal energy farm through the weighted decision matrix, the next step is to determine the energy density of each of the location monthly. The energy density is calculated using the following equations:

$$E_{av} = \sum_{h=1}^{8760} \left[\frac{1}{2} C_p A \rho V^3\right]_h \qquad (1)$$

where E_{av} = Yearly average energy (kWh), ρ = Water density (kg/m^3), C_p = power coefficient, A = swept area of tidal turbine (m^2) and V = Tidal velocity (m/s). The diameter of turbine is fixed at 5 meters and the swept area of turbine used the horizontal axis turbine swept area. The energy density of selected locations was calculated on a month to month basis as illustrated in **Figure 3.**

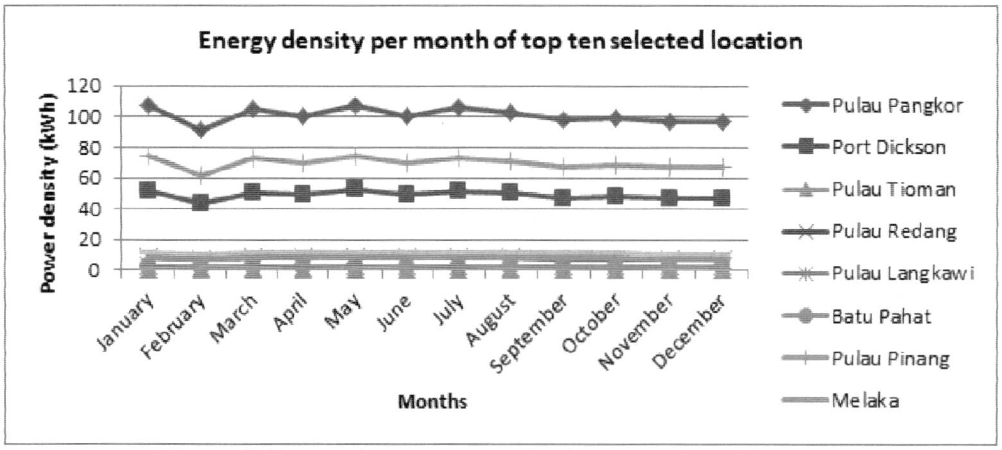

Figure 3: Energy density per month of top ten selected location.

4. Conclusion

This study has shown that there are several locations in Malaysia that are able to harness tidal energy resources from detailed analysis and comparison between the selected sites. In addition, an optimal analytical approach through the simulation of TMD software and surveying the entire locations/ coastline characteristic was done to all of the selected locations. The result of this research was found to slightly differ with other past researches. The top ten locations are Pulau Pangkor, Port Dickson, Pulau Tioman, Pulau Redang, Pulau Langkawi, Batu Pahat, Pulau Pinang, Melaka, Sibu and Pulau Banggi. This assessment study brings contribution to the community of those selected locations as it have been verified that those coastal region of selected locations have the potential to harness the tidal current energy resources. Furthermore, the potential stakeholders such as the investor, government and local community will be able to benefit from this renewable tidal current energy. Lastly, this assessment study also creates an opportunity for academic scholars to further proceed on detailed site study regarding turbine design and also array system at the proposed locations.

Acknowledgement

The first author would like to acknowledge the School of Mechatronic Engineering, UniMAP for the financial support. This project would never be possible without the help of my student, Izzudin Ibrahim, who is tireless in the pursuit of knowledge.

References

[1] Khan N, Kalair A, Abas N and Haider A 2017 Review of ocean tidal, wave and thermal energy technologies *Renew. Sustain. Energy Rev.* **72** (May 2016) 590–604.

[2] Adcock T A A, Draper S and Nishino T 2015 Tidal power generation - A review of hydrodynamic modelling *Proc. Inst. Mech. Eng. Part A J. Power Energy* **229** (7) 755–771.

[3] Legrand C and B & V Ltd 2009 *Assessment of Tidal Energy Resource* European Marine Energy Centre Ltd (EMEC).

[4] Lim Y S and Koh S L 2010 Analytical assessments on the potential of harnessing tidal currents for electricity generation in Malaysia *Renew. Energy* **35** (5) 1024–1032.

[5] Sheng S J 2018 *Assessment Study For The Modest Tidal Energy System in Malaysia* (June, 2018).

[6] Bonar P A J, Schnabl A M, Lee W K and Adcock T A A 2018 Assessment of the Malaysian tidal stream energy resource using an upper bound approach *J. Ocean Eng. Mar. Energy* **4** (2) 99–109.

[7] Egbert G D and Erofeeva S Y 2010 OTIS Regional Tidal Solutions web page *Oregon State University web pages* [Online]. Available: http://volkov.oce.orst.edu/tides/region.html. [Accessed: 19-Apr-2019].

IOP Conf. Series: Materials Science and Engineering **670** (2019) 012026 doi:10.1088/1757-899X/670/1/012026

A review of hip fracture analysis subjected to impact loading

A A M Zuki[1], F Mat[1], R Daud[1], N S Kamaruddin[1] and I Ibrahim[1]

[1] Fracture and Damage Mechanic Research Group, School of Mechatronic Engineering, Universiti Malaysia Perlis, Arau, Malaysia.

Abstact. A hip fracture not only affects the body but it also the economy of human. Before improvement and prevention have been developed, research need to be done in order to identify the factors that contribute to the hip fracture. The objective of this paper is to review on past studies that related to factors that influence the impact of force to the proximal femur and to evaluate the relation of the results between computational analysis and experimental method. Experimental and finite element (FE) method is the main approaches to predict the hip fracture of the femur bone. There are several parameters of impact force such as body mass index, the thickness of soft tissue, impact velocity and stiffness of human body that influence the effect of hip fracture during fall. Based on the previous study, force and stiffness result between FE analysis and experimental have shown good correlation. For that reason, the FE method can be applied to predict future result related to the hip fracture research.

1. Introduction

A hip fracture usually a result of short, low trauma fall, car crash, sport injuries. Fall in older adults particularly to the side contributes to the highest rate of fractures[1–3]. These fractures happen when the force exerted on the bone is stronger than the bone itself. The force is found to influence the extent of the fracture. The direction of force varies, as the falling is an unpredictable event [4]. Because of this, the pattern of the fracture are varies with different type of fall [5]. According to the Feldman and Robinovitch [6], the result from the experiment shows that 98% from fall that struck on pelvis affects the hip area. Fracture of the hip does not only effect on mobility and quality of life but also contribute to the increase risk of death ; the rate of mortality one year after hip fracture in the elderly patient is reported to be as high as 23% [7].

With the intention to provide a better understanding of hip fracture so that appropriate prevention and improvement can be made, researchers from previous studies have conducted an experiment and computational model of various conditions [3–5,7–12]. Most of these research related to the stress distribution [13,14], impact force, initiation of failure [3] and strain distribution [4]. In order to determine the accurate result, a constant value of factor that influences the magnitude of peak force that affected the proximal femur during falling were used. Prior to determining the effect of force on the proximal femur, some parameters need to be identified namely impact velocity, boundary condition (position of the femur), body condition (effective body mass, stiffness of pelvic, thickness of trochanter soft tissue), loading condition (falling height) and geometry of femur. These factors are important in order to better

IOP Conf. Series: Materials Science and Engineering **670** (2019) 012026 doi:10.1088/1757-899X/670/1/012026

understanding the mechanism of hip fracture because of its influence the compression rate of the proximal femur [15] as reported by previous studies [3,16–21].

Body mass and falling height of the body directly influence the peak force. However, these effects differ with the level of cushioning by soft tissue and clothing on the femur [9]. For Korean, the highest risk of hip fracture for both sexes was correlated with low BMI [22]. As for soft tissue, Stephen et al [23] stated that the peak force reduced linearly as the thickness of soft tissue increased. The objective of this paper is to review past studies related to factors that influence the impact of force to the proximal femur and to compare the results based on computational analysis and experimental method.

2. Method to predict the hip fracture

Computational analysis and experimental method are found to be the main approaches to predict the hip fracture of the femur bone. During the experiment, most of the tests employed drop test method where the body mass hit the top of pelvis spring. The pelvis mass bypasses the spring and directly impact the soft tissue and femur [1,3,15,18].

Most analysis that related to hip fracture from computational method is using finite element analysis (FEA) [19,24–28]. FEA has been widely used in the medical science field where FEA been applied to investigate the mechanical behavior of human body structures. With the objective to develop a finite elements model of the proximal femur, several steps and conditions need to be defined [20].The first step is creating the finite element model of the femur by identifying the geometry of the bone as well as the thickness of soft tissue. Most of the research nowadays used CT scan to determine the accurate geometry and material properties of the femur. Accuracy is important because elasticity and density of bone are one of the parameters on the load transfer mechanism. Otherwise, the data may lead to wrong conclusions as a result of the data inaccuracy. Secondly, the mesh of finite element model need to be created after the geometry of the bone is prepared. Then only the finite element analysis can be proceed further. To conduct analysis properly, mechanical properties, loading and boundary condition such as the position of femur, position and value of peak force need to be defined. These parameters may vary as they highly depend on the goal of the research.

3. Parameter of Impact Force

In order to investigate the effect of the impact force on the femur bone, some parameters that contribute to the degree of impact force need to be defined. By identifying and understanding these parameters, more accurate results can be obtained and a further improvement in the prevention of hip fractures can be done.

3.1. Body Mass Index (BMI)

BMI is dependent on the weight and height of the human. Every value of body mass index influences the energy and impact force delivered to the proximal femur during the fall. A person with a high body mass index is less likely to have hip fractures because of falling [29] where the result from past studies shows that mean BMI of participants with fracture and without fracture of the femur are 22.6 and 25.5 respectively [30]. People with a weight loss of 10% considerably exaggerated the chance of hip fractures among folks aged 65 years and over [21]. This is because energy absorption reduced due to less degree of padding of trochanteric soft tissue [9].

3.1.1 Effective body mass

Effective body mass can interpret as a half of body weight. The value of effective body mass that been used in the experiment from several researcher are in the range 32-35kg [3,8]. While the value of effective body mass in finite element analysis is from 20.9 kg until 35 kg [1,31]. Nevertheless, the past study of direct measurement in a young participant that related to the effective mass body shown the range of mass is from 24.1 kg to 50 kg [16,32].

3.2. Stiffness of Pelvis
In order to predict the impact force that hit the femur bone, the estimation of pelvis stiffness is required. This is because , during fall that involves direct impact, impact load is dependent on the effective stiffness of the body to receive the degree of force before it impacts the femur bone [12]. However, some of the previous studies show that during impact, pelvis stiffness only has minimum effect on peak force. From the previous study, a constant value of pelvis spring that act as pelvis compliance are 50kN/m [1,15]. This value also supported by international consensus statement for hip protector testing [15]

3.3. Thickness of trochanteric soft tissue
Skin, fascia lata, muscle, fat, and the trochanteric bursa act as soft tissues that cover the greater trochanter [23]. The thickness of trochanteric soft tissue is the overall sum of soft tissue that surrounds the greater trochanter. It influences the distribution and magnitude of the force as it acts as a protector of proximal femur during fall. Based on the result of previous research, the thickness of soft tissue over trochanteric was correlated to the peak force during sideways fall[3,12]. If the thickness of soft tissue is set as a big value, then the ability of energy absorption will increase and dissipate more energy [9,12]. As the thickness of trochanteric soft tissue increase, reduce the body's effective stiffness, consequently reducing the peak force applied to the hip amid an impact. In a variety of experiment, the chosen value of trochanteric soft tissue thickness was 19mm as it is in the range of standard in hip protector testing protocol [15].

3.4. Velocity of the Impact
Several factors influence the impact velocity of falling. One factor is the height and weight of the human body. The result of the research shows that the velocity of impact increases as the height and weight of person increase [9]. Another factor is the reaction of the body when people about to fall such as using their hand. This results in the time interval between pelvis impact and hand [6]. The increase in time interval reduces the impact velocity during falling. On the other hand, the impact force decreases during falling as the impact velocity decreases. The range of impact velocity in several previous study for experiment and finite element method were from 2 m/s until 3.17 m/s [1,6,8] and 1 m/s until 3m/s [3,13,14] respectively. Generally the range of impact velocity for computational and experiment were 3.35-4.34 m/s and 2.94-3.6 m/s respectively [15]. However, the value of impact velocity during falling is dependent on the boundary condition of the experiment and FEA.

4. Comparison between Experimental and Finite Element Method
The objective of this study was to compare the result between experimental and computational method related to the hip fractures cases. This comparison is to validate the capability of finite element analysis in predicting the result that related to the hip fracture. Both FEA and experiment results show no significant different in stiffness or yield load between augmented and control specimen[7]. In fact, the result of force and stiffness between experimental and FEA have shown good correlation[22,32] However, there are several result that show a moderate correlation between experimental and FEA result for bone stiffness and energy absorption[1,27].

The differences of correlation between FEA and experiment result, as mention here is due to the different boundary condition and different in finite element models such as material properties, geometry and velocities. Size of meshing element and geometry are also additional factors that influence the FEA results

IOP Conf. Series: Materials Science and Engineering **670** (2019) 012026 doi:10.1088/1757-899X/670/1/012026

obtained. In general, the FEA method is reliable and able to accurately predict the future result related to hip fracture research. This will help to reduce the cost, as the procedure and item of the experiment are very costly.

5. Conclusion

In this paper, information related to the method, parameter of impact force and the relationship between finite element analysis and experimental result have been gathered and discussed. These studies show that the combination of complex parameter of human body condition can influence the impact force during falling. However, the geometry of the proximal femur is complex. There are several different requirements in order to run the experiment and finite element analysis. However, the previous studies show that the results of both methods have good agreement with each other. This can be concluded that finite element analysis can be apply to predict hip fracture cases in future.

Acknowledgements

The authors would like to be obliged to Universiti Malaysia Perlis and Ministry of High Education (MOHE) of Malaysia for providing laboratory facilities and financial assistance under FRGS/1/2018/TK03/UNIMAP/02/22.

References

[1] Ariza O, Gilchrist S, Widmer R P, Guy P, Ferguson S J, Cripton P A and Helgason B 2015 Comparison of explicit finite element and mechanical simulation of the proximal femur during dynamic drop-tower testing *J. Biomech.* **48** 224–32

[2] Parkkari J, Kannus P, Palvanen M, Natri A, Vainio J, Aho H, Vuori I and Ja M 1999 <Parkkari 1999 Hip Fractures falls.pdf> 183–7

[3] Bahaloo H, Enns-Bray W S, Fleps I, Ariza O, Gilchrist S, Soyka R W, Guy P, Palsson H, Ferguson S J, Cripton P A and Helgason B 2018 On the Failure Initiation in the Proximal Human Femur Under Simulated Sideways Fall *Ann. Biomed. Eng.* **46** 270–83

[4] Zani L, Erani P, Grassi L, Taddei F and Cristofolini L 2015 Strain distribution in the proximal Human femur during in vitro simulated sideways fall *J. Biomech.* **48** 2130–43

[5] Yates J B and Fountain J R 2016 Fractures of the femur and tibial shaft *Surg. (United Kingdom)* **34** 444–52

[6] Feldman F and Robinovitch S N 2007 Reducing hip fracture risk during sideways falls: Evidence in young adults of the protective effects of impact to the hands and stepping *J. Biomech.* **40** 2612–8

[7] Basafa E, Armiger R S, Kutzer M D, Belkoff S M, Mears S C and Armand M 2013 Patient-specific finite element modeling for femoral bone augmentation *Med. Eng. Phys.* **35** 860–5

[8] Choi W J, Cripton P A and Robinovitch S N 2014 Effects of hip abductor muscle forces and knee boundary conditions on femoral neck stresses during simulated falls *Osteoporos. Int.* **26** 291–301

[9] Majumder S, Roychowdhury A and Pal S 2008 Effects of trochanteric soft tissue thickness and hip impact velocity on hip fracture in sideways fall through 3D finite element simulations *J. Biomech.* **41** 2834–42

[10] Kumar K C N, Tandon T, Silori P and Shaikh A 2015 Biomechanical Stress Analysis of a Human Femur Bone Using ANSYS *Mater. Today Proc.* **2** 2115–20

[11] Orwoll E S, Marshall L M, Nielson C M, Cummings S R, Lapidus J, Cauley J A, Ensrud K, Lane N, Hoffmann P R, Kopperdahl D L and Keaveny T M 2008 Finite Element Analysis of the Proximal Femur and Hip Fracture Risk in Older Men *J. Bone Miner. Res.* **24** 475–83

[12] Laing A C and Robinovitch S N 2010 Characterizing the effective stiffness of the pelvis during sideways falls on the hip *J. Biomech.* **43** 1898–904

[13] Sarai T, Inoue T, Fujiwara K and Kuramoto K 2012 Dynamic Finite Element Analysis of Impulsive Stress Waves Propagating from Distal End of Femur *Acta Med. Okayama* **66** 409–15

[14] Sarai T and Tokumoto A 2015 Dynamic Finite Element Analysis of Impulsive Stress Waves Propagating

from the Greater Trochanter of the Femur by a Sideways Fall. *Acta Med. Okayama* **69** 165–71

[15] Gilchrist S, Guy P and Cripton P A 2013 Development of an Inertia-Driven Model of Sideways Fall for Detailed Study of Femur Fracture Mechanics *J. Biomech. Eng.* **135** 121001

[16] Laing A C and Robinovitch S N 2008 The Force Attenuation Provided by Hip Protectors Depends on Impact Velocity, Pelvic Size, and Soft Tissue Stiffness *J. Biomech. Eng.* **130** 061005

[17] Marco M, Giner E, Larraínzar-Garijo R, Caeiro J R and Miguélez M H 2018 Modelling of femur fracture using finite element procedures *Eng. Fract. Mech.* **196**

[18] Gilchrist S, Nishiyama K K, de Bakker P, Guy P, Boyd S K, Oxland T and Cripton P A 2014 Proximal femur elastic behaviour is the same in impact and constant displacement rate fall simulation *J. Biomech.* **47** 3744–9

[19] Javir A V. and Kirkire M S 2017 Finite element analysis of Femur prosthesis *2017 Int. Conf. Nascent Technol. Eng. ICNTE 2017 - Proc.* 1–5

[20] Belaid D and Bouchoucha A 2015 Modeling of the mechanical behavior of the human femur: Stress analysis and strain *AIP Conf. Proc.* **1702** 1–5

[21] Alfaro-Acha A, Ostir G V., Markides K S and Ottenbacher K J 2006 Cognitive status, body mass index, and hip fracture in older hispanic adults *J. Am. Geriatr. Soc.* **54** 1251–5

[22] Kim S H, Yi S W, Yi J J, Kim Y M and Won Y J 2018 Association Between Body Mass Index and the Risk of Hip Fracture by Sex and Age: A Prospective Cohort Study *J. Bone Miner. Res.* **33** 1603–11

[23] S.N. R, T.A. M and W.C. H 1995 Force attenuation in trochanteric soft tissues during impact from a fall *J. Orthop. Res.* **13** 956–62

[24] Viceconti M, Qasim M, Bhattacharya P and Li X 2018 Correction to: Are CT-Based Finite Element Model Predictions of Femoral Bone Strength Clinically Useful? *Curr. Osteoporos. Rep.* 1

[25] Koivumäki J E M, Thevenot J, Pulkkinen P, Kuhn V, Link T M, Eckstein F and Jämsä T 2012 Ct-based finite element models can be used to estimate experimentally measured failure loads in the proximal femur *Bone* **50** 824–9

[26] Dall'Ara E, Luisier B, Schmidt R, Kainberger F, Zysset P and Pahr D 2013 A nonlinear QCT-based finite element model validation study for the human femur tested in two configurations in vitro *Bone* **52** 27–38

[27] Miura M, Nakamura J, Matsuura Y, Wako Y, Suzuki T, Hagiwara S, Orita S, Inage K, Kawarai Y, Sugano M, Nawata K and Ohtori S 2017 Prediction of fracture load and stiffness of the proximal femur by CT-based specimen specific finite element analysis: Cadaveric validation study *BMC Musculoskelet. Disord.* **18** 1–8

[28] Shen M, Jiang B, Kalra A, Shu Y, Zhou R and Jin X Finite Element Investigation of Injury Risks of Immature Pelvis and Femur in Pedestrian Impact 1–10

[29] Levine I C, Bhan S and Laing A C 2013 The effects of body mass index and sex on impact force and effective pelvic stiffness during simulated lateral falls *Clin. Biomech.* **28** 1026–33

[30] Pagani R C, Kunz R E, Girardi R and Guerra M 2014 Body mass index as a prognostic factor for fracturing of the proximal extremity of the femur: a case–control study *Rev. Bras. Ortop. (English Ed.* **49** 461–7

[31] Rajapakse C S, Hotca A, Newman B T, Ramme A, Vira S, Kobe E A, Miller R, Honig S and Chang G 2017 Patient-specific Hip Fracture **283**

[32] van den Kroonenberg A J, Hayes W C and McMahon T A 1995 Dynamic Models for Sideways Falls From Standing Height *J. Biomech. Eng.* **117** 309

[33] Van Den Kroonenberg A J, Hayes W C and McMahon T A 1996 Hip impact velocities and body configurations for voluntary falls from standing height *J. Biomech.* **29** 807–11

[34] Nishiyama K K, Gilchrist S, Guy P, Cripton P and Boyd S K 2013 Proximal femur bone strength estimated by a computationally fast finite element analysis in a sideways fall configuration *J. Biomech.* **46** 1231–6

IOP Conf. Series: Materials Science and Engineering **670** (2019) 012027 doi:10.1088/1757-899X/670/1/012027

A short review on the use of coconut shell powder as filler in cement concrete

S N I H A Nadzri[1], M T H Sultan[1,2,3 *], A U M Shah[1,2], and S N A Safri[2]

[1]Department of Aerospace Engineering, Faculty of Engineering, Universiti Putra Malaysia, 43400 UPM Serdang, Selangor Darul Ehsan, Malaysia
[2]Laboratory of Biocomposite Technology, Institute of Tropical Forestry and Forest Products (INTROP), Universiti Putra Malaysia, 43400 UPM Serdang, Selangor Darul Ehsan, Malaysia
[3]Aerospace Malaysia Innovation Centre (944751-A), Prime Minister's Department, MIGHT Partnership Hub, Jalan Impact, 63000 Cyberjaya, Selangor Darul Ehsan, Malaysia

*Corresponding author
E-mail: thariq@upm.edu.my

Abstract. Environmentally-friendly composites had been the interest of researchers nowadays. Therefore the use of natural sources to gradually replace the conventional materials had been rapidly developed. Besides the use of natural fibres as reinforcement in polymer matrix composites, the use of fillers from natural sources, such as rice husk and corn cobs, was implemented in ceramic matrix composites. Solid waste like coconut shell is one of the potential material to be used as filler in this composites. Coconut shell in powder forms is usually used together with cement to produce high strength, more durability and lightweight concrete for structural component in construction area. Based on this review paper, coconut shell concrete showed comparable mechanical properties to the conventional concrete, in terms of its compressive strength. These properties suggested coconut shell powder as potential material to replace course aggregate in concrete.

1. Introduction

In 21st century, many sectors receiving challenges to develop new innovation by using eco-friendly materials from agricultural waste products to be converted into biocomposites [1][2]. One of the sector is ceramic matrices composites (CMC) [3]. CMC is a compound ceramic substrate which are blended together with ceramic fibers. Producing an eco-friendly CMC, some natural fibers such as rice husk, groundnut husk and corn cobs is added to the fabrication [3].

Coconut tree, known as an economical plant, is one of the beneficial raw material to be embedded with CMC to develop an environmental friendly product. Every single part of this economical plant can be benefited which provides foods from its fruit and soft trunk, water and oil from its fruit, roof and broom from its leaves, and bridges from its trunk [3]. Furthermore, in construction sector, timber and wood can be replaced by coconut trunk as the main structure. Husk, leaves and shell of coconut can also be being used as low cost materials since they are known as agriculture waste products [3].

Compared to husk and leaves, coconut shell can be more effective to be used in composites, since it is a solid agro waste products [4][5]. Coconut shell can be used as filler material to produce a high strength to weight ratio CMC while reducing the use of conventional fillers [4]. Coconut shell has comparable mechanical properties to the mineral filler, which this plant sources also benefits in low cost, low density and renewability [6]. The use of coconut shell also can reduce the health hazard to a minimal level [6]. In addition, coconut shell has favourable properties in manufacturing fields, which it causes less abrasion to the machine leads to less possibility of machinery break down [6].

Since coconut shell is good in abrasive, which cause less damage to the machinery, it is known as one of the potential material for long lasting use. In terms of chemical composition, coconut shell properties are almost similar to hard wood [6][7][8]. Therefore, it was possible for the coconut shell powder to be used as a potential material in the development of hardwood industries, to improve the strength of building structure [6][7].

2. Coconut shell powder (CSP)

Coconut shells are generally found as waste materials in a semi sphere shape, which the water and inner soft part of the fruit had been removed. There are several processes in preparing coconut shell powder (CSP) from the waste coconut shell to be used as fillers in CMC. Figure 1 shows the preparation process of CSP cement.

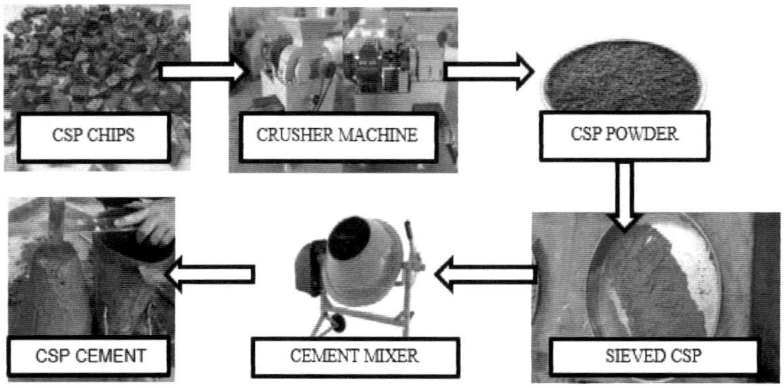

Figure 1. Procedures to prepare coconut shell powder (CSP) cement.

Hairy fibres surround the coconut shell was removed before the dry coconut shell was crushed into small pieces of coconut shell chips, [9][10][11]. A crusher machine was then used to produce CSP from the chips, continue with the sieving of CSP to obtain the desired size. The sieved CSP was oven dried at 105 °C to remove moisture [9][10][12]. The properties of prepared CSP were listed in Table 1. These properties suggested CSP as a potential material in developing a coconut shell concrete.

Table 1. Properties of coconut shell powder. [10]

Properties	Result
Specific gravity	1.33
Water absorption	23
Impact value	15.6%
Crushing value	2.58%

It was reported from the X-Ray Fluorescence (XRF) analysis that the chemical compound found in the CSP was also normally found in the concrete. Therefore, the use of CSP in concrete can be beneficial to the properties of concrete. Table 2 shows the chemical composition of CSP obtained from the XRF analysis [13].

Table 2. XRF analysis of coconut shell powder. [13]

COMPOUND	FULL NAME	CONCENTRATION
Coconut shell powder (g)		5
Wax (g)		5
C	Carbon	10.00 %
K_2O	Potassium Oxide	1.21%
SiO_2	Silicon Dioxide	0..98%
Cl	Chlorine	0.79%
Fe_2O_3	Iron (III) Oxide	0.35%
MgO	Magnesium Oxide	0.31%
Na_2O	Sodium Oxide	0.29%
CaO	Calcium Oxide	0.23%
MoO_3	Molybdenum (VI) Oxide	0.17%
S	Sulphur	0<LLD
Al	Aluminium	0<LLD
P	Phosphorus	0<LLD

3. Mechanical properties of coconut shell powder (CSP) composites

The IS 10262-1982 standard method was applied in blending CSP with cement in the laboratory to ensure the fabrication achieved the standard concrete grade M-20 [14]. A mixer was used to blend all together CSP, sand and water to produce the coconut shell (CS) concrete in various ratio of each material. Fabricated CS concrete can be used not only as construction material, but also in replacing the conventional pipelines. The properties evaluation of concrete need to be carried out on the 7th and 28th days in terms of compressive strength [15] [16] [17]. The coconut shell powder was normally used to replace the coarse aggregate in concrete. Table 3 shows the compressive strength of CS concrete reported by previous researchers.

Table 3. Compressive strength of coconut shell (CS) concrete.

Ratio CA:CS (%)	Compressive strength (N/mm^2)		Ref.
	7 days	28 days	
100:0 (C)	**24.6**	**28.3**	[18]
95:5	15.7	25.6	
80:20	9.2	22.2	
100:0 (C)	**11.11**	**22.33**	[19]
90:10	5.16	13.56	
80:20	7.82	9.33	
100:0 (C)	**20.53**	**27.58**	[20]
90:10	18.14	23.46	
50:50	11.05	15.48	
100:0 (C)	**21.25**	**28.54**	[21]
90:10	18.17	25.47	
50:50	10.15	14.63	
Control sample was not reported			[22]
90:10	18.87	21.25	
70:30	8.88	9.88	
Control sample was not reported			[23]
90:10	19.98	21.79	
80:20	13.56	15.25	

It was depicted from Table 3 that replacement of aggregate at 5 % and 10 % marked slightly lower compressive strength compared to the conventional 100 % course aggregate concrete. Except one study reported a distinct lower value to the conventional [23] [24], which is more than 40 % difference, other studies showed bright potential of replacing the course aggregate with CSP. Although the minimum percentage replacement used could not achieve the conventional strength, the small difference can be overcome in the next future [25] [26]. In all studies, several ratios were tested to analyse their effects to the compression properties. The table also listed the maximum percentage replacement of CSP in each study, to give the overview of the lowest compressive strength of concrete being produced with the respective amount of CSP [27].

4. Conclusion

The compressive strength of coconut shell concrete is comparable to the conventional, leads to a bright potential for its use as replacement to the course aggregate in concrete. The use of coconut shell powder will result in a more environmentally friendly composites, which its applications especially in pipeline under the water can reduce the harm to the living things in the sea. A lightweight composite can also be being produced, as well as increasing the market value of the waste material from the coconut shell. Future studies need to be carried out to enhance the properties of these CS concrete and further its applications in other types of ceramic matrix composites.

Acknowledgment

This work is supported by UPM under GP-IPS grant, 9647100 and Newton Fund, 6300896. The authors would like to express their gratitude and sincere appreciation to the Department of Aerospace Engineering, Faculty of Engineering, Universiti Putra Malaysia and Laboratory of Biocomposite Technology, Institute of Tropical Forestry and Forest Products (INTROP), Universiti Putra Malaysia (HiCOE) for the close collaboration in this research.

References

[1] Shah A U M, Sultan M T H, Cardona F, Jawaid M, Talib A R A and Yidris N 2017 Thermal analysis of bamboo fibre and its composites *BioRes.* **12** 2394-2406.

[2] Leman A S, Shahidan S, Nasir A J, Senin M J, Zuki S S M, Ibrahim W M H, Deraman R, Khalid F S and Azhar A T S 2017 Properties of concrete containing coconut shell powder (CSP) as a filler *IOP Conf. Ser. Mater. Sci. Eng.* **271** 012006

[3] Utsev and Taku J K 2012 Coconut shell ash as partial replacement of ordinary portland cement in concrete production *Int. J. Sci. Tech. Res.* **18** 86–89

[4] Salmah H, Koay S C and Hakimah O 2012 Surface modification of coconut shell powder filled polylactic acid biocomposites *J. Thermoplast. Compos. Mater.* **26** (6) 809–819

[5] Mostafa N H, Ismarrubie Z N, Sapuan S M and Sultan M T H 2017 Fibre prestressed composites: Theoretical and numerical modelling of unidirectional and plain-weave fibre reinforcement forms *Compos. Struct.* **159** 410-423.

[6] Gunasekaran K, Kumar P S and Lakshmipathy M 2010 Compatibility studies on the coconut shell cement composites *ICT J.* **0972** (2998) 11-1

[7] Thanyakon S and Amnart S 2016 A study on coconut shell powder filled in epoxy resin: a remedy for electrical tree growth inhibition *Key Eng. Mater.* **718** 36-39

[8] Jacob O A, Ayogu H and Olawale M S 2014 Evaluation of mechanical properties of coconut shell fibres as reinforcement material in epoxy matrix *Int. J. Eng. Res. Tech. (IJERT).* **3** (2) 2337-2348

[9] Mostafa N H, Ismarrubie Z N, Sapuan S M and Sultan M T H 2016 Effect of equi-biaxially fabric prestressing on the tensile performance of woven E-glass/polyester reinforced composites *J. Reinf. Plast. Comp.* **35** 1093-103.

[10] Chary G H M and Ahmed K S 2018 Development and testing of coconut shell particle reinforced epoxy composite for power transmission applications *J. Eng. Res. App.* 26-35.

[11] Apeksha K and Sarvesh K J 2017 Performance of coconut shell as coarse aggregate in concrete *Const. Build. Mater.* **140** 150-156

[12] Leman A S, Shahidan S, Senin M S and Hannan N I R R 2016 A preliminary study on chemical and physical properties of coconut shell powder as a filler in concrete *IOP Conf. Ser.: Mater. Sci. Eng.* **160** 012059

[13] Lavanya B A, Sunitha M S, Umesha P and Chethan H H 2018 Experimental study of partial replacement of cement and coarse aggregate with fly ash and coconut shell *Int. Res. J. Eng. Tech.* **5** (4) 194-198

[14] Alok S, Savita S and Aditya K 2013 Study of mechanical properties and absorption behavior of coconut shell powder-epoxy composites *Int. J. Mater. Sci. App.* **2** (5) 157-161

[15] Salman S D, Leman Z, Sultan M T H, Ishak M R and Cardona F 2017 Effect of kenaf fibers on trauma penetration depth and ballistic impact resistance for laminated composites *Text. Res. J.* **87** 2051-2065.

[16] Salman S D, Sharba M J, Leman Z, Sultan M T H, Ishak M R and Cardona F 2016 Tension-compression fatigue behavior of plain woven kenaf/kevlar hybrid composites *BioRes.* **11** 3575-3586

[17] Salman S D, Leman Z, Sultan M T H, Ishak M R and Cardona F 2016 Influence of fiber content on mechanical and morphological properties of woven kenaf reinforced PVB film produced using a hot press technique *Int. J. Polym. Sci.* **2016** 1-11

[18] Karolína H, Petr V, Miroslav M, Robert D A and Alessandro R 2019 Experimental description of the aging of the coconut shell powder epoxy composite *ICMEM 2018, LNME*, 456–464

[19] Gunasekaran K, Annadurai R, Prakash C S and Anandh S 2017 Study for the relevance of coconut shell aggregate concrete non-pressure pipe *Ain Shams Eng. J.* **8** 523–530

[20] Amarnath Y and Ramachandrudu C 2012 Properties of concrete with coconut shells as aggregate replacement *Int. J. Eng. Inv.* **1** (6) 21-31

[21] Palak P Arora Dr N K and Shraddha R V 2015 Experiments on partial replacement of coconut shell as coarse aggregate in concrete *Int. J. Inno. Res. Sci. Tech.* **2** (1) 78-87

[22] Sourabh Er S and Anish Er T 2017 A review on using coconut shell as coarse aggregate in concrete *Int. J. Adv. Res. Sci. Eng.* **2319** 8354

[23] Shambharkar R D, Aditya S, Kunal R and Dolly W 2018 Study on light weight characteristics of self compacting concrete using fine pumice powder and coconut shell *Int. Res. J. Eng. Tech.* **5** (3) 1379-1382

[24] Dodda N 2017 An experimental study on partial replacement of coconut shell in coarse aggregate *Int. J. Prof. Eng. Stud.* **8** (5) 36-46

[25] Parag S K and Sandhya R M 2014 Application of coconut shell as coarse aggregate in concrete a technical review *J. Eng. Res. App.* **4** (3) 498-501

[26] Dewanshu A and Kalurkar L G 2014 Coconut shell as partial replacement of coarse aggregate in concrete *IOSR J. Mech. Civ. Eng.* **1** (4) 100-103

[27] Mohan K, Abhishek B, Prithviraj M, Raghavendra and Vinay P 2018 Study on the effect of varying volume fraction on mechanical properties of coconut shell powder reinforced epoxy matrix composites *IOP Conf. Ser.: Mater. Sci. Eng.* **376** 012097

IOP Conf. Series: Materials Science and Engineering **670** (2019) 012028 doi:10.1088/1757-899X/670/1/012028

A short review on the extraction of kenaf fibers and the mechanical properties of kenaf powder composites

F S Shahar[1], M T H Sultan[1,2,3], A U M Shah[1,2] and S N A Safri[2]

[1]Department of Aerospace Engineering, Faculty of Engineering, Universiti Putra Malaysia, 43400 UPM Serdang, Selangor Darul Ehsan, Malaysia
[2]Laboratory of Biocomposite Technology, Institute of Tropical Forestry and Forest Products (INTROP), Universiti Putra Malaysia, 43400 UPM Serdang, Selangor Darul Ehsan, Malaysia
[3]Aerospace Malaysia Innovation Centre (944751-A), Prime Minister's Department, MIGHT Partnership Hub, Jalan Impact, 63000 Cyberjaya, Selangor Darul Ehsan, Malaysia

E-mail: *thariq@upm.edu.my

Abstract. Kenaf fibers are rapidly growing in the global market and currently being used as a composite to replace other expensive composite materials available in the market. Kenaf powder has the potential to replace other traditional fillers due to its low cost, eco-friendliness, and renewability. The aim of this paper is to review on the mechanical properties of Kenaf fibers, and its powder composites. This paper discusses on the Kenaf plants itself, extraction process, physical properties of kenaf fibers, and finally the mechanical properties of the kenaf powder composites. It can be concluded that Kenaf powder composite has more opportunity for further growth with new methods and technologies which keeps on evolving.

1. Introduction

Since ancient civilisation, natural composites had been used to build up structures or products. One such example, is the use of straws and clays to build up walls. However, the use of natural fiber started to decline when the age of industrialisation started replacing these natural composites with higher strength materials. The downside of these materials is that they are too expensive. Thus, researchers started to find other possibilities that could replace this type of material which could reduce the cost while still retaining the high strength [1–5]. One of the solutions is replacing with a natural fiber composites. Natural fiber composites does not only have the high strength to weight ratio property, it is also eco-friendly, renewable, and biodegradable. Out of all available natural composites, a lot of attention are being given to Kenaf composites due to its low cultivation needs with high amount of harvest within half a month.

Kenaf could grow as high as 6 meters within 6 to 8 months of plantation and produce around 6 to 10 tonnes of dry fiber each year. Kenaf markets are expanding rapidly locally and globally especially with the introduction of new technologies and applications that could potentially improve Kenaf fibers mechanical and chemical properties and replace other materials in the coming future [6]. This paper will be reviewing specifically on Kenaf extraction, its powdered form, and the mechanical properties of Kenaf powder composites.

2. Kenaf extraction

The extraction process of Kenaf fiber are essential to determine the quality of extracted fiber. The harvesting and extraction of Kenaf fibers has always depend on the location, equipment, method of processing and functions of the final products [7]. Mainly, there are two methods of harvesting fibers, which are hand harvesting, and using the whole stalk harvesters.

In hand harvesting, the process starts from harvesting manually by hand. The plant will be cut near the ground level with a sickle or a curved blade. This harvesting process has been used since 6000 years ago, and therefore it is the most traditional way of harvesting. This method requires human labour and time-consuming. The other method of harvesting and extraction is by using Kenaf Whole Stalk Harvesters. In this method, harvesting could be summarised into three steps which starts with pulling and windrowing, followed by retting, and finally balling parallel stalks. Currently, there are two type of harvesters available in the market which are the Sugarcane-type harvesters and forage-type harvesters. In both type of harvesters, researchers were focussing more on adapting with existing machines instead of innovating a new type of harvesters specifically for kenaf. Sugarcane-type harvesters are unmodified or slightly modified type of sugarcane harvesters. This harvester uses rotating blades or circular cutting blades to cut off the base of the kenaf stalks and also separate the low fiber from foliage and top portions of the plants. The long stalks will then laid down in a long windrow on the field to dry via the harvesting machine. Unlike Sugarcane-type harvesters which will harvest the long stalks vertically, forage-type harvesters will harvest kenaf by chopping the crops into small pieces. After that, a balling equipment will be used to turn the chopped kenaf into different sizes of square bales or large round bales.

Figure 1. Kenaf decorticator machine.

Figure 2. Kenaf whole stalk harvester machine.

After harvesting, it will undergoes retting process to extract the fibers. Currently, there are four types of retting process that has been used which are biological retting (dew retting, water retting, enzyme retting), mechanical retting (using decorticator), chemical retting, and physical retting [8].

In biological retting, the dew retting were carried out by spreading out the stalks on the field and expose them to the rain, sun and dew for several weeks until the stalks begin to separate from the fiber naturally [8]. Due to the long exposure to the natural conditions, fibers tends to lose its original colour and turns brown. Furthermore, this method reduces the fibers strength and quality. Water retting involves tying several bundles of stalks and immersed them in streams or water of at least 60 cm in depth. This process usually took around 2 weeks until the bast fibers naturally separates from the core fibers. These bundles were then taken out of the water and left to dry. Finally, enzyme retting the safest and fastest process of retting, where the core fibers can be taken out from the bast fibers after several hours. Some examples of enzymes that has been used are pectinase, and xylanase.

The second type of retting process is mechanical retting which uses Kenaf Decorticators [8]. A decorticator is basically a machine which could separate the bast and core fibers for further processing. It could ensure that the fiber will not be damage while being extracted. Some of the component of this machines are the frame (to hold all the components), knife (to cut the stalks), knife plate (to hold the knife in place), feeding mouth and security cover (to place the stalks into the machine with safety), and much more. This decorticator machine could reduce the amount of labor needed while still maintaining harvest at a faster rate. In the latest research made by Makanjuola et al., they evaluated the performance of a Kenaf decorticator [9]. The decorticator machine works by feeding the stalks into feeding mouth, then the stalks will be sent to the beaters to be crushed. The crushed stalk that contain both bast fiber and core fiber will be pushed into the delivery plate. Finally, the bast fiber will be separated from the crushed core fiber manually.

The third process which is chemical retting is a process where the stalks were treated with acid and alkalis to remove the core fibers easily from the bast fibers [8]. In these method, the stalks are boiled with chemicals such as sodium hydroxide, sodium benzoate, and hydrogen peroxide, at a specific temperature for several hours, and then washed with clean water. The fibers obtained from this method usually felt coarser and rough compared to other methods.

The last process for fiber extraction is Physical retting [8]. This process involves two steps which starts with modifying the surface/volume ratio of fibers through chemical pre-treatment. Then, the stalks will be subjected to steam explosion where the fibers will be blown apart and separated. In this method, the water will be vaporized completely and increased in volume and it requires small amount of processing time. However, this method could only produce short fibers. Generally, the fibers extracted from these harvesting methods will result with properties as shown in Table 1, and the range of each properties could vary depending on the method used to harvest and extract the fibers.

Table 1. Properties and cost of synthetic and natural fibers [10, 11].

Fibers	Density (g/cm^3)	Tensile Strength (MPa)	Young's Modulus (GPa)	Elongation to failure (%)	Moisture Content (%)	Cost (RM/kg)
Carbon	1.40	400	230 – 240	1.4 – 1.8	-	33.48 – 58.60
E-Glass	2.55	3.0 – 3.5	63 – 67	2.5	-	8.37
S-Glass	2.50	2000 – 3500	70 – 73	1.8 – 3.2	-	8.37
Aramid	1.44	3000	124	2.5	-	-
Kenaf	1.45	284 – 930	21 – 60	1.6	6.2 – 20	1.58
Hemp	1.48	550 – 900	70	1.6 – 4.0	8	6.49
Flax	1.40	800 – 1500	60 – 80	1.2 – 1.6	7	13.02
Jute	1.30 – 1.48	493 – 800	13 – 26.5	1.16 – 1.8	12	3.88

In comparison, the density of synthetic fibers and natural fibers listed in Table 1 are almost the same. E-glass fiber possesses the highest density while Jute having the lowest, whereas Kenaf fibers is in the middle range density. The density of these materials will determine how light the final product will be and especially important when making sports equipment that involves speed. Table 1 also shows that the tensile strength of natural fibers are comparable with synthetic fiber with the exception of S-Glass fiber and Aramid fiber which have an extremely high tensile strength which is 2000-3500 MPa and 3000 MPa respectively. As for Kenaf fibers, it could reach as high as 930 MPa, which suggests it as a medium strength fibres compared to other fibers. Natural fibers stiffness could only reach as high as 80 GPa whereas synthetic fibers could reach as high as 240 GPa. The elongation to failure of natural fibers are also comparable to synthetic fibers, which lies between 1.2-2.5 %, with Kenaf fiber having 1.6% of elongation of failure. Unlike synthetic fibers, natural fibers has higher moisture content, which is one of its drawbacks, with Kenaf fiber within the range of high moisture content. Finally, the cost of natural fiber are extremely low in comparison with synthetic fibers. Kenaf has the lowest raw material cost leading to the production of low cost composite [10].

Kenaf fiber has the potential to replace synthetic fibers due to its low density, medium range tensile strength and stiffness, and extremely low cost of raw materials [12]. After the fibers has been extracted and separated, Kenaf powder was processed from the inner fiber (core) of the Kenaf stalk. The core fiber will be machined or crushed until it turns into a powder form as shown in Figure 3. This form were usually used as fillers when making a composite and has the highest possibility of replacing the traditional fillers such as talc and calcium carbonate.

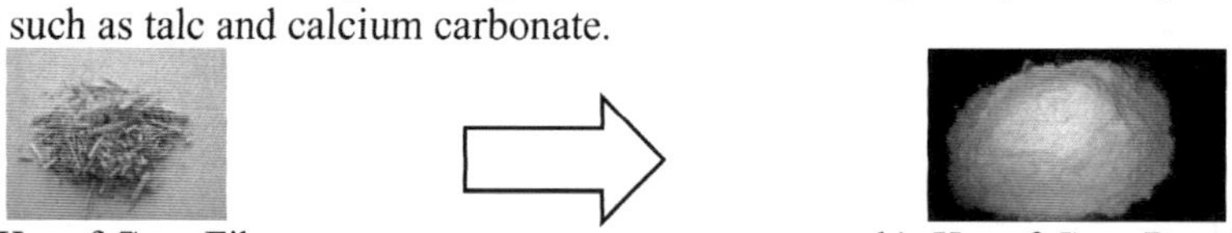

a) Kenaf Core Fiber b) Kenaf Core Powder

Figure 3. The core fiber (a) will be crushed and machined into powder form (b) Kenaf core powder

3. Mechanical properties of Kenaf powder composite

Composites are materials made up of reinforcements (fibers) and matrix (thermosets, thermoplastics, etc.) [13]. Composites which are made from a powdered form usually have lower tensile strength although it has lighter weight. Sometimes, in order to increase the strength of the composite, most fibers will be treated chemically. The chemical treatment can enhance the interfacial bonding, thus reduce the tendency of fibre pull out within the composites, which leads to failure [14–16]. The mechanical properties of kenaf powder composites vary depending on the type of treatment, fiber loading, and types of hybrids being used. Table 2 shows the mechanical properties of several types of Kenaf powder composites (KCP) reported previously.

Each of the work involved the use of varied amount of fibers starting from 0% up to 20-30%. The addition of Kenaf powder decreases the strength, but increases the stiffness of the composites [14–18]. Abdul Majid et al had conducted two experimental studies in 2016 [15] and 2018 [14]. Both of these experiments had used 70% Polyvinyl Chloride (PVC) and 30% Epoxidised natural rubber (ENR) with different Kenaf fiber loadings (0% - 20%). Both studies differ in terms of chemical treatment used which is Benzoyl Chloride (BC) [14] and Sodium dodecyl sulphate (SDS) [15]. From their research, it was found that the experiment used SDS has higher tensile strength compared from using BC which is 4 MPa and 3.5 MPa respectively.

Another research conducted by Mustaffa et al [16] also added the fiber loading from 0% up to 20% with 70% Polypropylene (PP) and 30% Virgin Acrylonitrile Butadiene Rubber (vNBR). When compared to Alias and Ismail's research [17], the tensile strength and Young's modulus of the composites had decreased significantly probably due to low percentage of PP. They also compared with presence of BC treatment. After being treated with BC, it was found that the tensile strength and Young's modulus had shown an increment. Alias and Ismail [17] conducted the experiment using 90% Polylactic Acid (PLA) and 10% Natural rubber (NR) with different fiber loadings. It was found that the higher the amount of fiber loading, the lower the tensile strength of the composite. When the fiber loading was the highest (20%) the tensile strength had been reduced from 41.39 MPa (0% KCP) to 34.21 MPa (20% KCP). However, the Young's Modulus increased from 1.459 GPa (0% KCP) to 2.578 GPa (20% KCP).

In different studies, Pang and Ismail [18] used 70% PP and 30% Waste tire dust (WTD) with variation in the amount of fiber loading and the form of Kenaf fibre. Similar with the previous researches, the higher the amount of fiber loading, the lower the tensile strength of the composite whereas the composite's Young's modulus kept increasing. It can also be seen that different kenaf form will give different results. In this case, the composite using short fibers will have better tensile strength and Young's Modulus compared to composites using powder form.

Table 2. Mechanical properties of Kenaf powder composite from several researches.

Composites	Tensile Strength (MPa)	Young's Modulus (GPa)	References
20% KCP + 70% PVC + 30% ENR + BC	3.50	0.013	[14]
20% KCP + 70% PVC + 30% ENR + SDS	4.00	0.010	[15]
30% KCP + 70% PP + 30% vNBR + BC	8.00	0.450	[16]
20% KCP + 90% PLA + 10% NR	34.21	2.578	[17]
20% KCP (powder) + 70% PP + 30% WTD	10.00	0.750	[18]

4. Conclusion

The global market of Kenaf are expanding rapidly over the past several years. Currently, industries are manufacturing more products by using Kenaf composites. By comparing several other fibers (carbon, glass, Kevlar, hemp, jute, flax) with kenaf, it was found that Kenaf fibers are cheaper and better strength among other natural fibers. Kenaf fibers can be extracted using three different methods which is hand harvest and retting, using decorticator machine, and whole stalks harvesters. In order to get kenaf powder, the core fiber of kenaf will be crushed completely into powder form. Several researches had been done in order to develop a composite using kenaf powder. The mechanical strength of kenaf powder composite are commonly enhanced using chemical treatments. Thus, with the rapid growth of research being carried out, it can be predicted that these kenaf powder composites could be strengthen to resemble a carbon fiber composite and act as a replacement.

Acknowledgments
This work is supported by UPM under GP-IPS grant, 9647100 and Newton Fund, 6300896. The authors would like to express their gratitude and sincere appreciation to the Department of Aerospace Engineering, Faculty of Engineering, Universiti Putra Malaysia and Laboratory of Biocomposite Technology, Institute of Tropical Forestry and Forest Products (INTROP), Universiti Putra Malaysia (HiCOE) for the close collaboration in this research.

References

[1] Mostafa N H, Ismarrubie Z N, Sapuan S M and Sultan M T H 2017 Fibre prestressed composites: Theoretical and numerical modelling of unidirectional and plain-weave fibre reinforcement forms. *Compos. Struct.* **159** 410–423
[2] Mostafa N H, Ismarrubie Z N, Sapuan S M and Sultan M T H 2016 Effect of equi-biaxially fabric prestressing on the tensile performance of woven E-glass/polyester reinforced composites. *J Reinf. Plast. Compos.* **35** 1093–1103
[3] Salman S D, Leman Z, Sultan M T H, Ishak M R and Cardona F 2016 Influence of fiber content on mechanical and morphological properties of woven kenaf reinforced PVB film produced using a hot press technique. *Int. J. Polym. Sci.* **2016** 1–11
[4] Nor A F M, Sultan M T H, Jawaid M, Azmi A M R and Shah A U M 2019 Analysing impact properties of CNT filled bamboo/glass hybrid nanocomposites through drop-weight impact testing, UWPI and compression-after-impact behaviour *Compos. Part B Eng.* **168** 166–174
[5] Ismail K I, Sultan M T H, Shah A U M, Jawaid M and Safri S N A 2019 Low velocity impact and compression after impact properties of hybrid bio-composites modified with multi-walled carbon nanotubes *Compos. Part B Eng.* **163** 455–463
[6] Salman S D, Sharba M J, Leman Z, Sultan M T H, Ishak M R and Cardona F 2016 Tension-compression fatigue behavior of plain woven kenaf/kevlar hybrid composites *BioRes.* **11** 3575–3586
[7] Kamal I, Thirmizir M Z, Beyer G, Saad M J, Rashid N A A and Kadir Y A 2014 Kenaf for biocomposite: an overview *J Sci. Technol.* **6** 41–66
[8] Aisyah G S and Tajuddin R M 2014 Trends in natural fibre production and its future *5th Brunei Int.Conf. Eng. and Tech.* DOI: 10.1049/cp.2014.1080
[9] Makanjuola G A, Ayorinde T A, Aluko O B, Owolarafe O K and Sanni L A 2019 Performance evaluation of a kenaf decorticator *Agric. Eng. Int. CIGR J* **21** 192–202
[10] Mochane M J, Mokhena T C, Mokhothu T H, Mtibe A, Sadiku E R, Ray S S, Ibrahim I D and Daramola O O 2019 Recent progress on natural fiber hybrid composites for advanced applications: A review. *Express. Polym. Lett.* **13** 159–198
[11] Lakshumu N A, Jagadeesh V and Raju Bahubalendruni M V A 2017 A Review on chemical and physical properties of natural fiber reinforced composites *Int. J. Adv. Res. Eng. Technol.* **8** 56–68
[12] Salman S D, Leman Z, Sultan M T H, Ishak M R and Cardona F 2017 Effect of kenaf fibers on trauma penetration depth and ballistic impact resistance for laminated composites *Text. Res. J.* **87** 2051–2065
[13] Shah A U M, Sultan M T H, Cardona F, Jawaid M, Talib A R A and Yidris N 2017 Thermal analysis of bamboo fibre and its composites *BioRes.* **12** 2394–2406
[14] Abdul Majid R, Ismail H and Mat Taib R 2018 Processing, tensile, and thermal studies of poly(vinyl chloride)/epoxidized natural rubber/kenaf core powder composites with benzoyl chloride treatment *Polym. - Plast. Technol. Eng.* **57** 1507–1517
[15] Abdul Majid R, Ismail H and Taib R M 2016 Sodium dodecyl sulfate (SDS) as a filler treatment: effects on mechanical, morphological and swelling properties of poly (vinyl chloride) (PVC)/epoxidized natural rubber (ENR)/kenaf core powder composites *Mater. Res. Innov.* **20** 504–511
[16] Mustaffa Z, Ragunathan S, Othman N S, et al. 2018 Thermoplastic elastomer composite using benzyl chloride treatment on kenaf core powder mixing with polypropylene and virgin

Acrylonitrile Butadiene Rubber *IOP Conf. Ser. Mater. Sci. Eng.* DOI: 10.1088/1757-899X/429/1/012013.

[17] Alias N F and Ismail H 2019 Processing torque, tensile properties and morphological study of PLA/NR/kenaf biocomposite *AIP Conf. Proc.* DOI: 10.1063/1.5089383.

[18] Pang A L and Ismail H 2014 Influence of kenaf form and loading on the properties of kenaf-filled polypropylene/waste tire dust composites: A comparison study *J. Appl. Polym. Sci.* DOI: 10.1002/app.40877.

IOP Conf. Series: Materials Science and Engineering **670** (2019) 012029 doi:10.1088/1757-899X/670/1/012029

Impact response performance of pineapple leaf fibre (PALF)/carbon reinforced hybrid composite

K H Khor, M S Abdul Majid, M J M Ridzuan, and E M Cheng

School of Mechatronics Engineering, Universiti Malaysia Perlis, 02600 Arau, Perlis, Malaysia

E-mail: shukry@unimap.edu.my

Abstract. The number of fossil fuels is decreasing dramatically, whereas the amount of waste around the world is increasing due to the non-biodegradable petroleum-based product such as plastics. Therefore, impact response performance of pineapple leaf fibre/carbon reinforced hybrid epoxy composite was analysed with different layering sequences to replace non-biodegradable materials. The fibres laminates used for the hybrid composites were fabricated using interwoven between pineapple leaf fibre(PALF) and carbon fibre reinforced epoxy resin. The composites were made by using vacuum infusion method with a weight factor of 30% fibre and 70% epoxy resin. Other than the pure PALF reinforced epoxy composite, the PALF/carbon reinforced hybrid composites were using weight factor of 24% PALF and 6% carbon fibre. 5 types of layering sequences composites were fabricated that were, pure PALF, carbon/PALF/carbon, PALF/carbon/PALF, alternative PALF with carbon, and alternative carbon with PALF. Charpy impact test was used to obtain the impact strength of the hybrid composites. From the result obtained shows that layering sequences of PALF/carbon/PALF exhibited the highest impact strength that is around 100kJ/m^2 compare with others, followed by alternatively PALF with carbon, alternatively carbon with PALF, carbon/PALF/carbon and lastly pure PALF.

1. Introduction

This research project aims to study the impact response performance of pineapple leaf fibre with carbon reinforced epoxy composite. The increase of petroleum-based products, such as plastics causes many negative impacts on the environment [1], [2]. One of the solutions is using natural fibre as a substitute for synthetic fibres since they are eco-friendly and having excellent mechanical and physical properties. There are different types of fibres, for example, bast fibres, pineapple leaf fibres, core fibres, grass and reed fibres [3], [4]. Pineapple has become the 2nd important tropical fruits in the world in 2002, according to the Food Agricultural Organization of the United Nations. In 2018, there are about 24.8 million tons of pineapple fruits being produced annually in the whole world [5]. Among all the natural fibres, pineapple leaf fibre with epoxy composite shows a 40% better of mechanical properties such as impact, tensile and flexural strength compared to others. Therefore, pineapple leaf fibre composite can be a potential composite for further development and applicable for manufacture [6]. In recent year, hybrid composites are being used worldwide due to their low density, low price, unlimited or sustainable available, and most importantly, environment-friendly [7]–[10]. However, there are a few factors that affect the strength of a hybrid composite, such as properties of fibres, fibre to matrix interface bonding, the arrangement of the fibres and the failure strain of each fibre. The ideal or best hybrid composite can be created when the fibres are having high strain compatible[11].

2. Materials and methods

2.1. Materials

The woven PALF and woven carbon fibre are purchased from a local supplier in Malaysia. The PALF and carbon fibre are used as a reinforcement material in the woven form with a dimension of 300 mm × 300 mm, as shown in Figure 1(a) and (b) respectively. The epoxy resin of EpoxAmite 100 series resin is being used as the polymer matrix for the composite together with a slow hardener, named Epoxamite 103 slow hardener. The epoxy resin is mixed with the slow hardener with a weight ratio of 100g: 28.4g according to the instruction of the supplier.

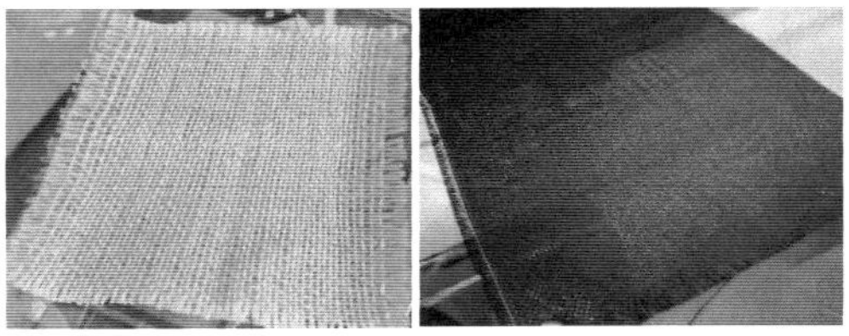

(a) (b)

Figure 1. (a) PALF and (b) carbon fibre dimension of 300mm x 300mm.

2.2. Composites Fabrication

The hybrid composites are prepared by using a PALF with a weight content of 24% ± 2% and carbon fibre weight content of 6% ± 2% having different types of layering sequence as listed in Table 1. Vacuum infusion method is used to fabricate the hybrid PALF/carbon fibre reinforced composites is illustrated in Figure 2.

Table 1. Abbreviation for various types of layering sequences of composites prepared.

Type of layering sequences	**Symbol**
a. Pineapple leaf/carbon/pineapple leaf	PCP
b. Carbon/pineapple leaf/carbon	CPC
c. Pineapple leaf/carbon/pineapple leaf/carbon	PCPC
d. Carbon/pineapple leaf/carbon/pineapple leaf	CPCP
e. Pure pineapple leaf	-
f. Pure carbon	-

(a) (b)

Figure 2. (a) The infusion vacuum machine used for fabrication (b) shows the fabrication process of the composites.

2.3. Charpy impact test

LS-22006-50 Charpy Impact tester is used for testing the specimens' impact strength. The specimens are placed in the Charpy Impact machine horizontally on the anvil for testing according to standard ASTM D6110. The pendulum is raised to an angle of 120° and lock to the height to prepare for impact testing. When all setup is ready, the handle is pulled as to release the impactor from the locking and knock on the specimens. After the specimens are broken, the angle of the impactor after impact is recorded, and the impact resistance is calculated.

3. Results and discussion

The impact resistance or absorbed impact energy per unit meter width of the composite is determined through the difference of potential energies before and after impact divided by the width of the composite. Table 2 shows the specimens before and after the impact with different layering sequences and the SEM image of each specimen.

From the result of figure 3, PCP has the highest average impact resistance. The PCP has the average impact resistance of 1.642kJ/m, which is 548% of pure PALF composites of 0.296kJ/m. For both the layering sequences of CPC and CPCP, the average impact strength of the hybrid composite doubled that of the pure PALF composite to around 0.5797kJ/m and 0.4891kJ/m respectively. Whereas for the layering sequences of PCPC, the average impact strength increased three times to 0.899kJ/m. Whereas for the impact strength of the composites shown in table 3 are determined from the difference of potential energies before and after the impact divided by the cross-sectional area of the composites' impact surface. The performance of impact strength concerning to difference layering sequences of composites is very alike to impact resistance, with layering sequence of PCP hybrid composite having the highest impact strength of 100.3 kJ/m^2. This is followed by PCPC hybrid composite with the impact strength of 61.09 kJ/m^2. The CPCP and CPC hybrid composites found to yield similar impact strength of 31.4 kJ/m^2 and 30.6 kJ/m^2, respectively. Neat PALF composite yields the lowest impact strength of 15.7436kJ/m^2. From the research of N. Agarwal, it shows that 30% wt of carbon fibre reinforced epoxy composite has an impact strength of 12kJ/m^2 [12]. This shows that carbon fibre has a lower impact strength compared to PALF; having a superior impact, and energy absorption compared to other natural fibre, for example of Kenaf fibre [13]. Therefore when carbon fibre layer used as the skin of hybrid composite result in lower impact strength or energy absorption compared with hybrid composite with pineapple leaf fibre layer as skin [14]–[16].

From the SEM result for pure PALF composite, it shows that the debonding between PALF and the epoxy matrix and the fibre pullout from the epoxy matrix. This shows that there was a weak interface between fibre and matrix, causing weak adhesive bonding and created a concentrated deformation region [17], [18]. For CPC, the result shows there are potholes in the composite which caused by entrapped moisture molecules and also river markings oriented in a variety of angle across the fibre surface will lead to overexposure of fibre and localised microscopic areas of fracture. The large area of void in the composite after the impact test observed is due to the pull out of fibres. This is mainly caused by the weak interface and adhesive bonding between fibre and matrix [17]. For the PCPC, the morphology suggests that the overall distribution of the epoxy matrix has fully covered the composite with little voids. The figures show that there is weak adhesive bonding among carbon fibre, PALF and epoxy with cause this layering sequence having the low impact response performance. This will cause the composite easily to debonds, and fibre easily pulls out from the matrix. For the CPCP, overall, there are quite some voids in the composites which are caused by the impurities and bursting moisture molecule

Table 2. Specimens of different layering sequences before and after impact.

	Before impact	After impact	SEM Result
Pure PALF		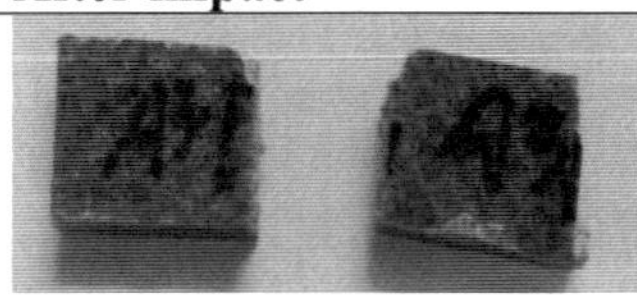	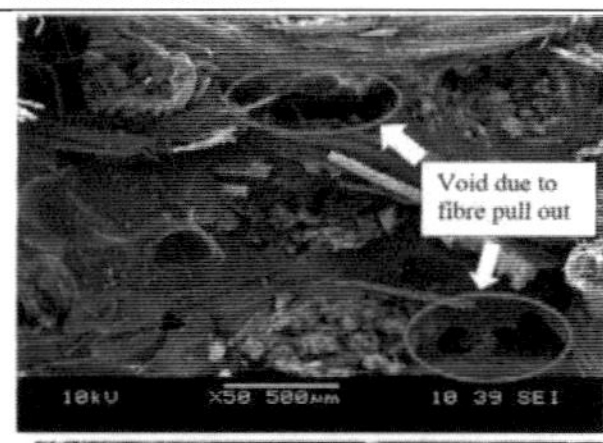
CPC			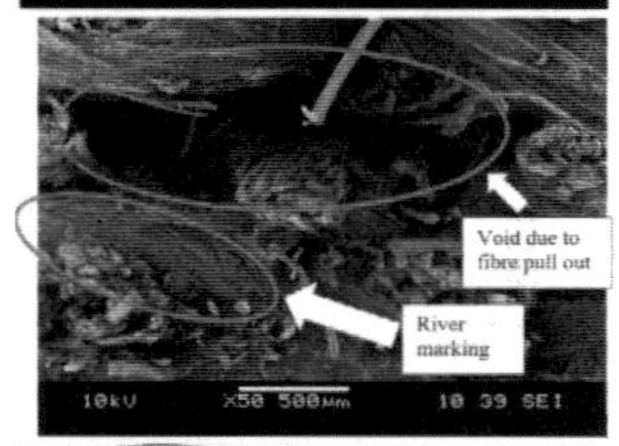
PCP	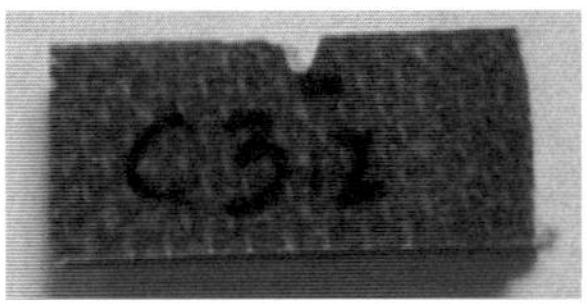		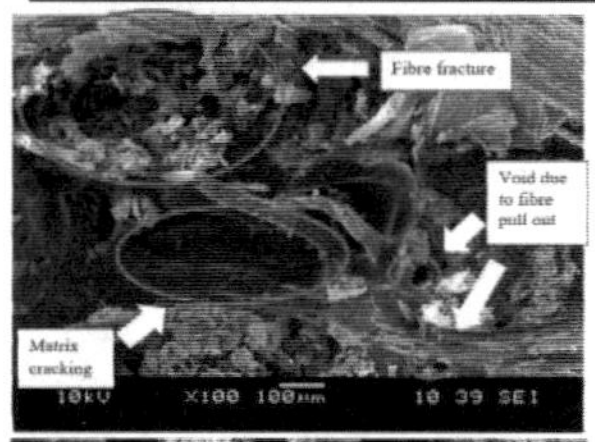
PCPC			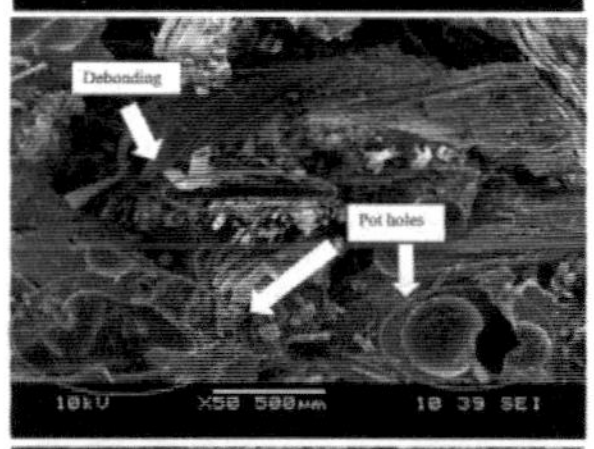
CPCP			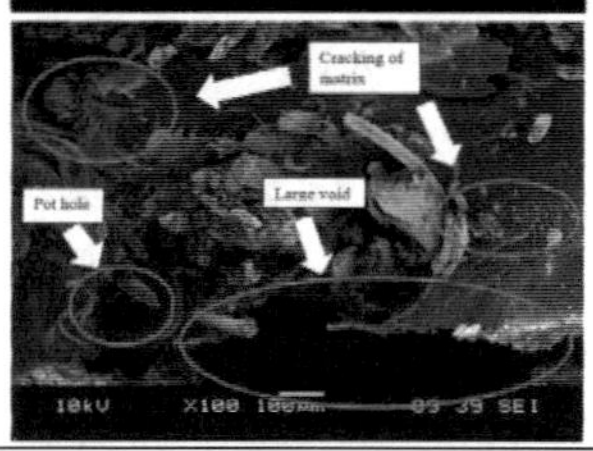

Table3. Average impact resistance and average impact strength of each layering sequences.

Type of layering sequences	*Average impact resistance, kJ/m*	*Average impact strength, kJ/m²*
Pure PALF	*0.2960 ± 0.0685*	*15.7436 ± 1.9407*
CPC	*0.5797 ± 0.264*	*30.6439 ± 2.8693*
PCP	*1.6419 ± 0.1447*	*100.2997 ± 5.4824*
PCPC	*0.8990 ± 0.0312*	*61.0910 ± 9.2268*
CPCP	*0.4891 ± 0.0309*	*31.3828 ± 1.6209*

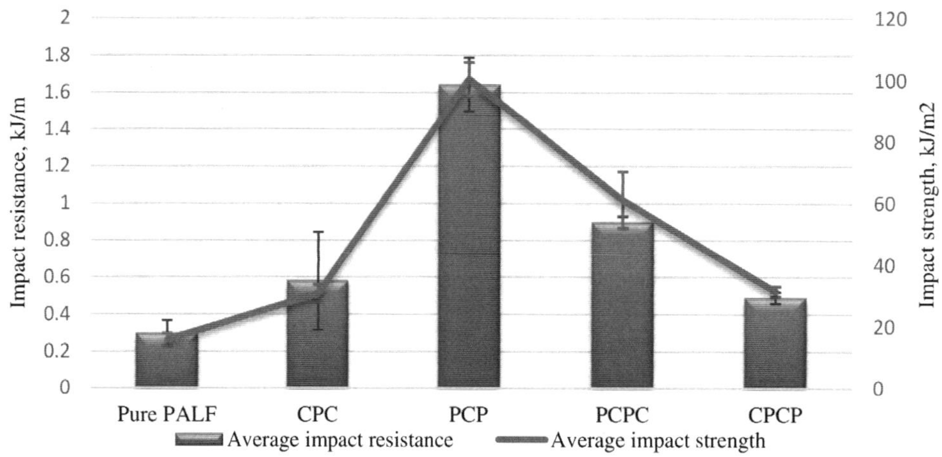

Figure 3. Graph of average impact resistance and average impact strength with different layering sequences.

Moreover, the interface of carbon fibre and the epoxy matrix is weak and cause a lot of the carbon fibre pulled out after the impact test process. All of these weak points decrease the impact strength of the composite and cause the composite fracture easily. Lastly, for PCP, the epoxy matrix is well filled across the composite, and the interface of fibres with the matrix is excellent, very little of fibres pull out after the impact process.

4. Conclusion

From the result obtained in this research, the impact properties of the hybrid composite are stronger than of PALF composites alone. The incorporation of carbon fibre in the hybrid composite increased the impact resistance and impact strength of the composite and made them stronger and better resistant to breaking. It shows that hybrid PALF/carbon fibre reinforced composite has the potential to replace most of the use of plastic products. This can help to preserve the number of fossil fuels and also to reduce the waste around the world as natural fibre composite is biodegradable. Other than that, this also creates a new method as to dispose of the residual of the pineapple plants by turning them into a more useful composite rather than bury them under the ground. Different layering sequences of fibres created composites with different impact properties. In conclusion, with the same weight factor of pineapple leaf fibre, carbon fibre and epoxy but different layering sequences, the one using natural fibre as the surface of the hybrid composite such as PCP layering sequence exhibit the best impact response performance of 100.3kJ/m^2 which is almost nine times stronger than of pure PALF. As from the result of morphological analysis, the interface of fibres with matrix plays an essential role in impact response performance of a composite. The stronger the adhesive bonding of fibre with matrix and fibre with fibre, the more energy is needed to break the composite. In order to fabricate a stable hybrid composite, the weakness of bad interface of carbon with epoxy needs to be solved by using a chemical treatment to remove the impurities and the moisture molecules in fibres.

Acknowledgements
The authors would like to acknowledge the Ministry of Education (MOE), Malaysia, for financial support via the Fundamental Research Grant Scheme (Ref: FRGS/1/2019/TK03/UNIMAP/02/4).

References
[1] "Tons of waste dumped - globally, this year," *The World Counts*. [Online]. Available: http://www.theworldcounts.com/counters/shocking_environmental_facts_and_statistics/world_waste_facts. [Accessed: 17-Nov-2018].
[2] Ridzuan K, Abdul Majid M J M, Afendi M S, Rahman M, Azduwin Y M N 2016 Thermal, composition and morphological properties of untreated and alkali-treated Napier fibres *J. Eng. Appl. Sci.* **11** (11) 2485–2489.
[3] "Advantages of using natural fibre applications in composites," *Pond*. [Online]. Available: https://pond.global/advantages-of-using-natural-fibre-applications-in-composites/. [Accessed: 17-Nov-2018].
[4] Fahmi I, Abdul Majid M S, Afendi M, Helmi E A and Haameem J A M 2016 Low-velocity impact responses of Napier fibre/polyester composites *Int. J. Automot. Mech. Eng.* **13** 3226–3237.
[5] Leao A L, Souza S F, Cherian B M, Frollini E, Thomas S, Pothan L A and Kottaisamy M 2010 Pineapple leaf fibers for composites and cellulose *Mol. Cryst. Liq. Cryst.* **522** 36–41.
[6] Odusote J K and Oyewo A T 2016 Mechanical properties of pineapple leaf fiber reinforced polymer composites for application as a prosthetic socket *J. Eng. Technol.* **7** (1) 2180–3811.
[7] Yuanjian T and Isaac D H 2007 Impact and fatigue behaviour of hemp fibre composites *Compos. Sci. Technol.* **67** (15–16) 3300–3307.
[8] Kommula V P, Kanchireddy O R, Shukla M and Marwala T 2013 Study on impact strength of untreated and alkali treated napier grass fiber strands reinforced epoxy composites *Int. J. Res. Eng. Technol.* **2** (3) 101–106.
[9] Abdul Majid M S, Gibson A G, Hekman M, Afendi M and Amin N A M 2014 Strain response and damage modelling of glass/epoxy pipes under various stress ratios *Plast. Rubber Compos.* **43** (9) 290–299.
[10] Fartini M S, Abdul Majid M S, Ridzuan M J M, Amin N A M and Gibson A G 2016 Compressive properties of napier (pennisetum purpureum) filled polyester composites *Plast. Rubber Compos.* **45** (3) 136–146.
[11] Puglia D, Biagiotti J and Kenny J M 2005 A review on natural fibre- based composites — part II - application of natural reinforcements in composite materials for automotive industry *J. Nat. Fibers* **1** (3) 23–65.
[12] Agarwal N and Manish Bhargava 2017 Mechanical and fracture toughness analysis of woven carbon fibre reinforced epoxy composites *Int. J. Sci. Res. Eng. Technol.* **6** (1). 17–22.
[13] Asim M, Jawaid M, Abdan K, Ishak M R and Alothman O Y 2017 Effect of hybridization on the mechanical properties of pineapple leaf fiber/kenaf phenolic hybrid composites *J. Renew. Mater.* **6** (1) 38–46.
[14] Yahaya R, Sapuan S M, Jawaid M, Leman Z and Zainudin E S 2015 Effect of layering sequence and chemical treatment on the mechanical properties of woven kenaf–aramid hybrid laminated composites *Mater. Des.* **67** 173–179.
[15] Venkateshwaran N and ElayaPerumal A 2012 Mechanical and water absorption properties of woven jute/banana hybrid composites *Fibers Polym.* **13** (7) 907–914

IOP Conf. Series: Materials Science and Engineering **670** (2019) 012029 doi:10.1088/1757-899X/670/1/012029

[16] Ridzuan M J M, Abdul Majid M S, Afendi M, Azduwin K, Kanafiah S N A and Dan-mallam Y 2015 The effects of the alkaline treatment's soaking exposure on the tensile strength of napier fibre," *Procedia Manuf.* **2** 353–358.
[17] Poursartip A, Ashby M and Beaumont P 1985 Matrix cracking of composites inside a scanning electron microscope *J. Mater. Sci. Lett.* **4** (7) 845–847.
[18] Rout J, Misra M, Mohanty A K, Nayak S K and Tripathy S S 2003 Journal of Reinforced Plastics and Composites SEM Observations of the Fractured *J. Reinf. Plast. Compos.* 22 (12) 1083–1100.

IOP Conf. Series: Materials Science and Engineering **670** (2019) 012030 doi:10.1088/1757-899X/670/1/012030

Thermal properties of pineapple leaf/kenaf fibre reinforced vinyl ester hybrid composites

A A Mazlan[1], M T H Sultan[1,2,3], A U M Shah[1,2] and S N A Safri[2]

[1]Department of Aerospace Engineering, Faculty of Engineering, Universiti Putra Malaysia, 43400 UPM Serdang, Selangor Darul Ehsan, Malaysia
[2]Laboratory of Biocomposite Technology, Institute of Tropical Forestry and Forest Products (INTROP), Universiti Putra Malaysia, 43400 UPM Serdang, Selangor Darul Ehsan, Malaysia
[3]Aerospace Malaysia Innovation Centre (944751-A), Prime Minister's Department, MIGHT Partnership Hub, Jalan Impact, 63000 Cyberjaya, Selangor Darul Ehsan, Malaysia

*Corresponding author: thariq@upm.edu.my

Abstract. In this modern technology era, the usage of natural fibres has been widely used in various field of applications such as aerospace, maritime, military, automotive as well as structural building. This is due to the reason that they are lightweight, cheap, good sustainability and biodegradable. Only a small amount of research work being conducted on the hybridization of pineapple leaf and kenaf fibres reinforced with vinyl ester (PALF/KF/VE) hybrid composites. Thus, in this present work, the PALF/KF/VE hybrid composites are presented. The thermal properties of PALF, KF, VE as well as hybrid composites were investigated based on their structural components. PALF was in long fibre form while kenaf fibre was in woven form fabricated to form hybrid composites by using the hand lay-up and hot compression moulding. TGA testing was conducted to determine their respective thermal properties. From the experimental testing, the results showed that the thermal stability of hybrid composites enhanced rather than fibres themselves. The thermal degradation of PALF and KF occurred at the range temperature of 250-370 °C while neat VE, PALF/VE and PALF/KF/VE occurred at 300-450 °C. Thus, these composites can be used for various application such as automotive, aerospace, structural and others.

1. Introduction

Recently, natural fibres become one of hot topic issues discussed in the composite's application [1]. The research work on the natural fibres have been attracted the researchers to explore them. This is due to the reason that they have more advantages over the synthetic fibres such as biodegradable, lightweight, cheap in cost and high availability [2]. The natural fibres can be used as a reinforcement but only in the low-stress application. Pineapple, ananas comosus is one of the largest tropical fruits in Malaysia. Several research works had been conducted to study the properties of pineapple leaf fibre (PALF), mechanically, thermally and chemically [3], [4], [5]. TGA dan DSC testing being worked by the researchers to investigated the thermal properties of PALF [6]. Another study showed that PALF composites recorded the onset temperature approximately at 270 °C for thermal degradation [7].

Kenaf is well known as a hardy plant with a fibrous stalk which resistant to insect and ability to grow under climatic condition that required less water, pesticides and fertilizers. Globally, kenaf is one of the suitable biological resources that can replace the fossil fuels because of its physical properties. A lot of research work being conducted on the kenaf fibres (KF) and its composites [8], [9], [10], [11]. Besides, the thermal degradation behaviour of kenaf/PALF hybrid composites were investigated [12] with respect to the fibre content and length. Vinyl ester (VE) resin is a thermosetting polymer that produced by the mixture of epoxy and unsaturated polyester resin. VE has high toughness, strength and chemical resistance. Therefore, VE possessed high thermal and electrical insulation properties. A lot of research works being worked on to investigate the properties of VE [13], [14]. A research work conducted on the properties of betel nut husk reinforced with VE composites [15]. Since there is no previous research work that examined the thermal properties of PALF/KF/VE hybrid composites, the present research work examined the thermal properties of PALF, kenaf fibre, neat VE, PALF/VE and PALF/KF/VE hybrid composites. The discussion is done based on the component of the composites. The hybridization of PALF with KF enhanced the thermal stability of composites.

2. Materials and Methodology

2.1 Materials

PALF, in the roving form of a thin yarn, with the average thickness of 1 mm, was used as reinforcement material. Prior to reinforcement, the PALF was chopped into 30 mm length, by using a chopper gun. Woven KF was used to be hybridized with PALF. In this study, the resin used was MFE-11, a premium standard Bisphenol-A type epoxy VE resin supplied by Sino Polymer Co., LTD. The hardener that was used was methyl ethyl ketone peroxide (MEKP).

2.2 Methodology

In this study, PALF, kenaf fibres, neat VE, PALF/VE and PALF/KF/VE hybrid composites were fabricated. The woven kenaf fibre were used as addition to the present research work which focusing on the hybridization of PALF and kenaf fibre. The hand lay-up and hot compress method was used to prepare the composites. The ratio used for fibres to matrix resin was 30:70. The temperature used for the 40 tonnes hot compression machine was 70°C at 3 minutes. After that, the composites were left for 3 hours before it is hardened completely. Thermal behaviours composites were examined using a thermogravimetric analyser (TGA DTA; Universal V3.9A TA instruments) in the Institute of Tropical Forestry and Forest Products (INTROP), Universiti Putra Malaysia (UPM). The heating rate of the sample has been set to 10°C/min.

3. Results and Discussion

Thermogravimetric analysis or thermal gravimetric analysis (TGA) were studied to determine the thermal properties of the PALF, kenaf fibre, VE resin and PALF/KF/VE hybrid composites.

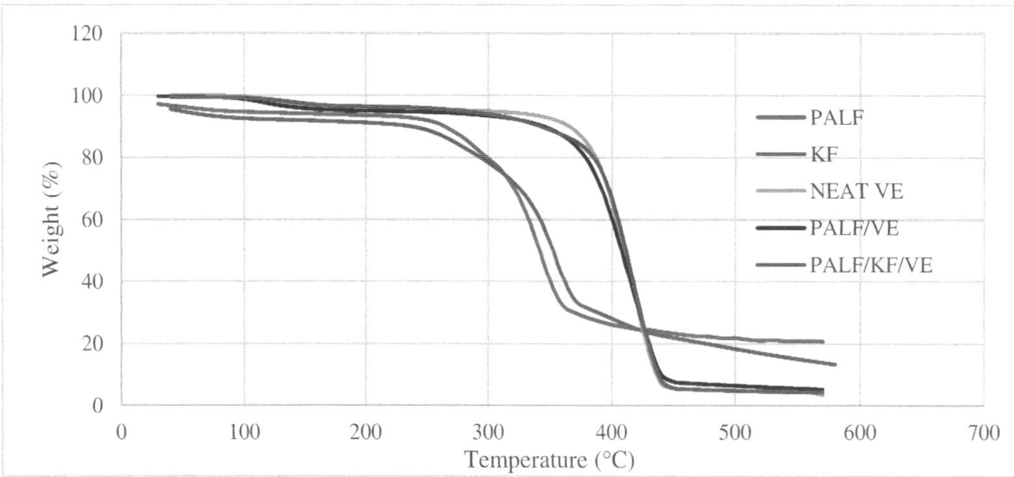

Figure 1. TGA curve of PALF, KF, Neat VE, PALF/VE and PALF/KF/VE hybrid composites.

Based on Figure 1, it can be observed that the thermal degradation of PALF and KF occurred at early stage. There were several steps of degradation of the composites and their respective components [16]. From 50 °C to 250 °C, both of the fibres started to degrade slowly at the first stage. The first stage indicated the weight losses due to the evaporation or vaporisation of water molecules in the composites [17]. This is due to the facts that PALF and KF are the natural resources that obviously have high content of moisture. However, in the range of temperature of 250-370 °C, as the temperature increase, there was major weight loss occurred in the stage due to the decomposition of hemicellulose of the PALF [18]. The final stage of degradation of PALF and KF can be observed at temperature range from 370-580 °C because of the decomposition of lignin, ash and cellulose content in the fibres [19], [20]. The percentage of weight of PALF, KF, neat VE, PALF/VE and PALF/KF/VE hybrid composites were illustrated as in table 1.

Table 1. Data for Weight Loss at Selected Temperature.

Sample	Weight Loss (%)								
	$T_{60°C}$	$T_{100°C}$	$T_{180°C}$	$T_{260°C}$	$T_{300°C}$	$T_{370°C}$	$T_{450°C}$	$T_{530°C}$	$T_{570°C}$
PALF	4.18	5.3	6.18	9.16	20.53	70.11	76.71	78.84	79.31
KF	5.87	7.43	8.46	12.52	21.75	66.13	78.14	83.71	86.12
Neat VE	0.03	0.4	4.07	4.55	5.04	11.31	94.36	95.46	96.51
PALF/VE	0.51	1.17	4.77	5.49	6.51	16.08	92.21	94.06	94.68
PALF/KF/VE	0.6	0.78	3.41	4.3	5.99	14.68	94.27	95.42	95.63

The neat VE, PALF/VE and PALF/KF/VE hybrid composites losses its water at first stage (100-300 °C). As the temperature increase from 300 to 450 °C, these composites loss their weight significantly as degradation of hemicellulose occurred. Last but not least, at the final stage of thermal degradation, the degradation of lignin, ash and cellulose occurred [21], [22], [23]. In the case of composites, the thermal degradation of neat VE, PALF/VE and hybrid composites showed difference to each other. The thermal degradation of neat VE and PALF/KF/VE hybrid composites illustrated almost the same trend to each other while for the thermal degradation of PALF/VE, the trend differed. The neat VE showed better thermal stability because VE is one of the synthetic thermoset matrixes which expected to produce better thermal stability. The reinforcement of natural fibres with VE enhanced the performance of the fibre composites. It can be observed from the figure 1 that showed that the degradation of PALF/VE occurred faster compared to the neat VE and hybrid composites. From 100 °C to 300 °C, the PALF/KF/VE hybrid composites presented lower degradation compared to PALF/VE composites. This is due to the fact that the hybrid composites composed of PALF and KF rather than the composites that composed of PALF only. The hybrid composites will always show better performance than an individual fibre composite. The lower degradation that occurred explained that the hybrid composites has better thermal stability compared to PALF/VE composites. When the rapid degradation occurred from 300 °C to 450 °C, the graph still showed that hybrid composites possessed lower degradation compared to PALF/VE composites due to the components existed in the composites. However, as the temperature continued to increase, the residue of PALF/VE was higher than PALF/KF/VE hybrid composites.

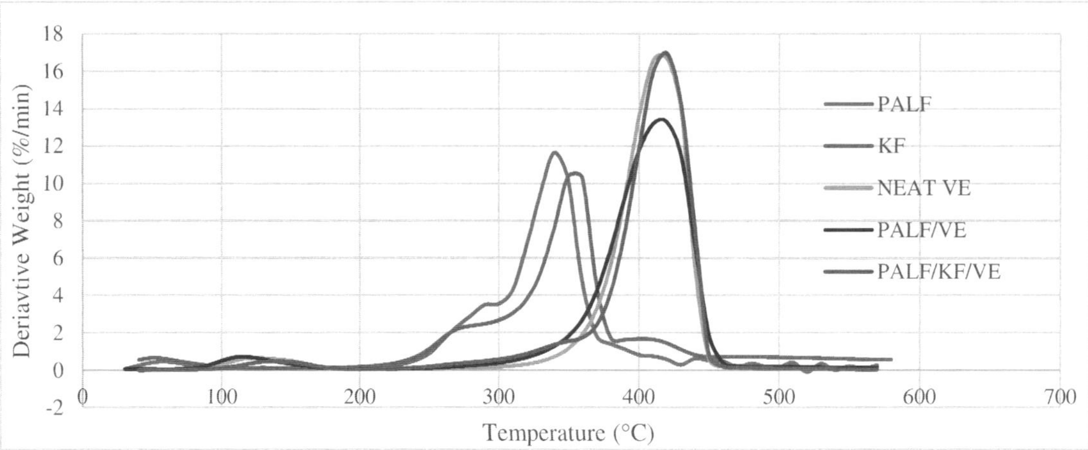

Figure 2. DTG curve of PALF, KF, Neat VE, PALF/VE and PALF/KF/VE hybrid composites.

Figure 2 showed the DTG of PALF, KF, neat VE, PALF/VE and PALF/KF/VE hybrid composites. The graph presented the rate of percentage of weight loss as the temperature keeps increasing. From the figure 2, in comparing the fibres with the composites, the degradation of weight of PALF and KF occurred earlier than the composites. From 300 °C to 390 °C, PALF showed higher weight loss compared to KF. For the case of neat VE, PALF/VE and PALF/KF/VE hybrid composites, the rate of degradation of these composites showed higher compared to the PALF's and KF's. The rapid degradation started to occur from 300 °C to 390 °C. Among these three composites, neat VE and hybrid composites illustrated almost the same rate of degradation while the PALF/VE composites showed lower rate of degradation. After 450 °C, the rate of degradation of these composites started to show almost the same trend to each other. Based on the analysis that has been done, it can be concluded that the hybridization of PALF and KF improved the thermal stability of the composites. PALF/KF/VE hybrid composites showed better thermal stability compared to PALF/VE composites due to the extra component added to the composites which is KF.

The TGA and DTG graph clearly depicts the thermal stability of the PALF/KF/VE hybrid composites, revealing that the hybrid composite is a highly resistant material, even without hybridizing it with other synthetic materials. This statement is in strong agreement with other established findings, reporting that the hemicellulose is responsible for the thermal degradation of fibres, while the existence of lignin in the lignocellulose fibres improves the thermal stability [24].

4. Conclusion

From the TGA testing, it can be noticed that the degradation of PALF and KF occurred at 250-370 °C while the degradation of neat VE, PALF/VE and PALF/KF hybrid composites occurred at 300-450 °C due to the properties of VE. In the DTG graph, the rate of degradation of neat VE, PALF/VE and hybrid composites is higher compared to the PALF and KF and rapid degradation occurred between 300 to 390 °C. In comparing the PALF/VE and PALF/KF/VE hybrid composites, it shows that the hybridization of PALF and KF improved the thermal stability compared to PALF/VE composites. The thermal degradation of PALF and KF showed a good potential to be used as a reinforcement in the various application such as automotive, aerospace, structural and others.

Acknowledgement

This work is supported by UPM under HICoE grant, 6369107 and Newton Fund, 6300896. The authors would like to express their gratitude and sincere appreciation to the Department of Aerospace Engineering, Faculty of Engineering, Universiti Putra Malaysia and Laboratory of Biocomposite Technology, Institute of Tropical Forestry and Forest Products (INTROP), Universiti Putra Malaysia (HiCOE) for the close collaboration in this research.

IOP Conf. Series: Materials Science and Engineering **670** (2019) 012030 doi:10.1088/1757-899X/670/1/012030

References

[1] Shah A U M, Sultan M T H, Cardona F, Jawaid M, Talib A R A and Yidris N 2017 Thermal analysis of bamboo fibre and its composites *BioRes.* **12** 2394-2406.

[2] Nor A F, Sultan M T H, Jawaid M, Azmi A M and Shah A U M 2019 Analysing impact properties of CNT filled bamboo/glass hybrid nanocomposites through drop-weight impact testing, UWPI and compression-after-impact behaviour *Compos. Part B Eng.* **168** 166-174.

[3] Mohamed A R and Khalina A 2014 Mechanical and thermal properties of josapine pineapple leaf fiber (PALF) and PALF-reinforced vinyl ester composites *Fiber Polym.* **15** 1035–1041.

[4] Siregar J P, Salit M S, Ab Rahman M Z and Mohd Dahlan K Z H 2011 Thermogravimetric analysis (TGA) and differential scanning calometric (DSC) analysis of pineapple leaf fibre (PALF) reinforced high impact polystyrene (HIPS) composites *Pertanika J. Sci. Technol.* **19** 161–170.

[5] Siakeng R, Jawaid M, Ariffin H and Sapuan S M 2018 Thermal properties of coir and pineapple leaf fibre reinforced polylactic acid hybrid composites *IOP Conf. Ser. Mater. Sci. Eng.* **368**.

[6] Wan Nadirah W O, Jawaid M, Al Masri A A, Abdul Khalil H P S, Suhaily S S and Mohamed A R 2012 Cell wall morphology, chemical and thermal analysis of cultivated pineapple leaf fibres for industrial applications *J. Polym. Environ.* **20** 404–411.

[7] Oliveira G *et al.* 2014 Effect of surface treatment on performance of pineapple leaf fiber-polycarbonate composites *Polym. Test.* **5** 628–632.

[8] Salman S D, Leman Z, Sultan M T H, Ishak M R and Cardona F 2017 Effect of kenaf fibers on trauma penetration depth and ballistic impact resistance for laminated composites *Text. Res. J.* **87** 2051-2065.

[9] Salman S D, Sharba M J, Leman Z, Sultan M T H, Ishak M R and Cardona F 2016 Tension-compression fatigue behavior of plain woven kenaf/kevlar hybrid composites *BioResources* **11** 3575-3586.

[10] Salman S D, Leman Z, Sultan M T H, Ishak M R and Cardona F 2016 Influence of fiber content on mechanical and morphological properties of woven kenaf reinforced PVB film produced using a hot press technique *Int. J. Polym. Sci.* **2016** 1-11

[11] Pang C, Shanks R A and Daver F 2015 Characterization of kenaf fiber composites prepared with tributyl citrate plasticized cellulose acetate *Compos. Part A Appl. Sci. Manuf.* **70** 52–58.

[12] Aji I S, Zainudin E S, Khalina A, Sapuan S M and Khairul M D 2012 Thermal property determination of hybridized kenaf/PALF reinforced HDPE composite by thermogravimetric analysis *J. Therm. Anal. Calorim.* **109** 893–900.

[13] Jofre-reche A, Su J C and Alia C 2018 Characterization of the chemical structure of vinyl ester resin in a climate chamber under different conditions of degradation *Polym. Degrad. And Stab.* **153** 88–99.

[14] Alhuthali A, Low I M and Dong C 2012 Characterisation of the water absorption, mechanical and thermal properties of recycled cellulose fibre reinforced vinyl-ester eco-nanocomposites *Compos. Part B Eng.* **43** 2772–2781.

[15] Yusriah L and Sapuan S M 2018 Properties of betel nut husk reinforced vinyl ester composites. Elsevier Ltd.

[16] Jumaidin R, Sapuan S M, Jawaid M, Ishak M R and Sahari J 2017 Thermal, mechanical, and physical properties of seaweed/sugar palm fibre reinforced thermoplastic sugar palm Starch/Agar hybrid composites *Int. J. Biol. Macromol.* **97** 606–615.

[17] Sanyang M L, Sapuan S M, Jawaid M, Ishak M R and Sahari J 2016 Effect of plasticizer type and concentration on physical properties of biodegradable films based on sugar palm (arenga pinnata) starch for food packaging *J. Food Sci. Technol.* **53** 326–336.

[18] Asim M *et al.* 2018 Thermal, physical properties and flammability of silane treated kenaf/pineapple leaf fibres phenolic hybrid composites *Compos. Struct.* **202** 1330-1338.

[19] Fung K L, Xing X S, Li R K Y, Tjong S C and Mai Y W 2003 An investigation on the processing of sisal fibre reinforced polypropylene composites *Compos. Sci. Technol.* **63**, 1255–1258.

IOP Conf. Series: Materials Science and Engineering **670** (2019) 012030 doi:10.1088/1757-899X/670/1/012030

[20] El-Shekeil Y A, Sapuan S M, Zainudin E S and Khalina A 2011 Optimizing processing parameters and fiber size for kenaf fiber reinforced thermoplastic polyurethane composite *Key Eng. Mater.* **471–472**, 297–302.
[21] Elkhaoulani A, Arrakhiz F Z, Benmoussa K, Bouhfid R and Qaiss A 2013 Mechanical and thermal properties of polymer composite based on natural fibers: Moroccan hemp fibers/polypropylene *Mater. Des.* **49** 203–208.
[22] El-Shekeil Y A, Sapuan S M, Jawaid M and Al-Shuja'a O M 2014 Influence of fiber content on mechanical, morphological and thermal properties of kenaf fibers reinforced poly(vinyl chloride)/thermoplastic polyurethane poly-blend composites *Mater. Des.* **58**, 130–135.
[23] Abdel Hakim A A, Nassar M and Sultan M 2011 Preparation and characterization of rigid polyurethane foam prepared from sugar-cane bagasse polyol *Mater. Chem. Phys.* **129**, 301–307.
[24] Manfredi L B, Rodríguez E S, Wladyka-Przybylak M and Vázquez A 2006 Thermal degradation and fire resistance of unsaturated polyester, modified acrylic resins and their composites with natural fibres *Polym. Degrad. Stab.* **91** 255–261.

IOP Conf. Series: Materials Science and Engineering **670** (2019) 012031 doi:10.1088/1757-899X/670/1/012031

Charpy and Izod impact properties of natural fibre composites

N M Z Abidin[2], M T H Sultan[1,2,3], A U M Shah[1,2] and S N A Safri[2]

[1]Laboratory of Biocomposite Technology, Institute of Tropical Forestry and Forest Products (INTROP), Universiti Putra Malaysia, 43400 UPM Serdang, Selangor Darul Ehsan, Malaysia
[2]Department of Aerospace Engineering, Faculty of Engineering, Universiti Putra Malaysia, 43400 UPM Serdang, Selangor Darul Ehsan, Malaysia
[3]Aerospace Malaysia Innovation Centre (944751-A), Prime Minister's Department, MIGHT Partnership Hub, Jalan Impact, 63000 Cyberjaya, Selangor Darul Ehsan, Malaysia

*Corresponding author: thariq@upm.edu.my

Abstract
Low velocity impact can be divided into several types of evaluation. Charpy and Izod impact testing are two types of testing that mostly used by the industry as it was known as an economical quality control method. In this short review, previous studies on the charpy and izod impact testing on natural fibre composites will be discussed. It was found that the percentage fibre loading is the mostly factors being analysed, besides the treatment applied and size of fibre. It was suggested that the range of 10-30 % fibre loading were the mostly applied in short fibre natural composites.

1. Introduction

Nowadays, research on impact properties of composites are vastly growing to investigate the strength properties of composites for industrial application purposes. Impact is a collision of two or more bodies, either interaction between elastic, plastic, fluid or any combination of these. Generally, there are four types of impact velocity which are low velocity, high velocity, ballistic and hypervelocity [1-3]. The Charpy and Izod impact test were categorized in low velocity impact, which determine the impact strength and toughness of a material. An impact blow is delivered to a test specimen by means of a pendulum-type hammer and the impact value is determined from the energy required to break the specimen. Results of impact tests are expressed in term of impact energy (J/m) or impact strength (kJ/m^2). Figure 1 illustrates the experimental set up for Izod and Charpy impact test.

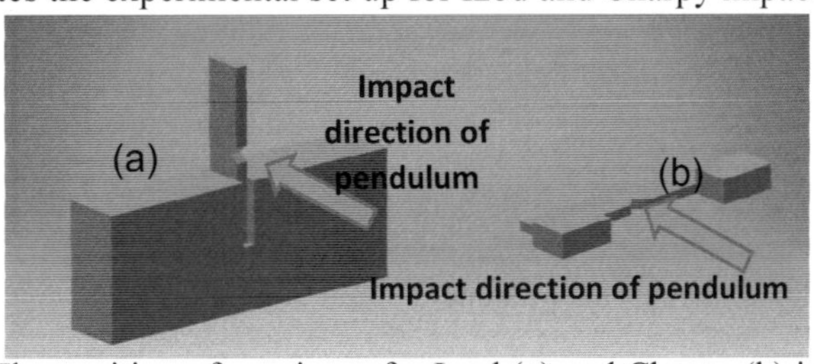

Figure 1. The position of specimen for Izod (a) and Charpy (b) impact test.

It was shown that both Charpy and Izod impact test applied almost the same principle. Only several differences, such as the shape and position of the specimen may differ the type of impact test. Table 1 summarize the differences between Charpy and Izod impact test.

Table 1. The comparison between Charpy and Izod impact test. [4]

	Charpy Impact Test	**Izod Impact Test**
Position of specimen	Horizontal	Vertical
Type of notch	V-notch & U-notch	V-notch
Direction of notch to the striker	Facing toward the pendulum	Facing away from the pendulum
Type of hammer	Ball pin hammer	Farming hammer
Striking point	Centre of specimen	Upper tip of specimen
Materials tested	Metals	Plastics & Metals
Common specimen dimensions	55 x 10 x 10 mm	64 x 12.7 x 3.2 mm (plastic) 127 x 11.43 mm round bar (metal)
Common standards	ASTM E23, ASTM D6110, ASTM D256, ISO 179, EN 10045-1	ASTM D256, ASTM E23, ISO 180

2. Charpy Impact Properties

In industrial scale, Charpy impact can be described as an economical quality control method to determine the impact toughness and notch sensitivity of materials. Besides having a very low cost, the simple procedure in preparing samples and obtaining results quickly has made this testing a favourable standard in various industries. The example of applications, which used Charpy impact test in their industries, are the pipeline, the construction of pressure vessels and the bridges. These industry has push the development of tough, strong, and weldable material. In the research and development stage, various studies had been conducted on Charpy impact test. Different types of material with various factors affecting the impact properties, such as the application of chemical treatment, percentage of fibre loading and the fibre length, had been analysed. Table 2 shows the detailed research on Charpy impact test of natural fibre composites covered by previous researchers.

Table 2. Reported research on Charpy impact test of natural fibre composites.

Material	**Variation**	**Standard**	**Impact Energy (J/m)**	**Impact Strength (kJ/m²)**		**Ref.**
				kJ/m²	**Composites**	
Kenaf bast fiber reinforced metal laminate system	*Treated / untreated *Fibre loading (%): 50/60/70 *Fibre length (cm): 3,6,9	ASTM E-23	-	Max: 157.04 (Min: 71.62)	Treated 70% 9cm	[5]
Giant Bamboo fiber reinforced epoxy	*Fibre loading (%): 0/10/20/30	ASTM D6110	Max: 72 ± 14	-	30%	[6]

Untreated mercerized mallow fiber reinforced epoxy	*Treated / untreated *Fibre loading (%): 0/10/20/30	ASTM D6110	Max: 905.49 ± 94.56 (Min: 96.07 ± 21.15)	-	Treated 30%	[7]
Buriti fibers reinforced epoxy	*Fibre loading (%): 0/10/20/30	ASTM D6110	Max: 114.17 ± 14.37 (Min: 13.77 ± 2.07)	-	30%	[8]
Continuous curaua fiber reinforced polyester	*Fibre loading (%): 0/10/20/30	ASTM D256	Max: 335.71 ± 35.67 (Min: 22.91 ± 9.77)	-	30%	[9]

It was seen in Table 2, percentage of fibre loading is the main variables being analysed in Charpy impact properties. In all studies listed in the tables, the highest percentage of fibre loading results in the maximum value of impact energy or impact strength comparing to other variation of fibre loading amount was tested. Tough materials absorb a lot of energy whilst brittle materials tend to absorb a little energy prior to fracture. The factors that affect material toughness include low temperatures, high strain rates and stress concentrators such as notches, cracks and voids. The review also suggested the range of percentage fibre loading for short fibres reinforced composites is between 0 to 30 %. The use of chemical treatment on natural fibres can enhanced the impact strength of composites due to the improvement of fibre-matrix interfacial bonding. Besides that, the length of short fibres could also contribute to the impact properties of composites. D.Sivakumar [5] claims that the impact energy absorbed was decreased by the shorter fibre length specimen due to the poor stress transfer upon impact.

3. Izod Impact Properties

In the industry, Izod impact test is used in medical, automotive, and plastic industries. The Izod test most commonly used to evaluate the relative toughness and impact toughness of the materials and often used in quality control application where it is a fast and economical test. The research study of the Izod testing are shown in Table 3 below.

Table 3. The research detail of Izod impact test for composite material.

Material	Variation	Standard	Impact Energy (J/m)	Impact Strength (kJ/m²)		Reference
				kJ/m²	Composites	
Hybrid Composite Materials	*Sisal+epoxy / *Jute+epoxy / *Sisal+Glass+epoxy / *Sisal+Jute+epoxy	ASTM D256		Max: 6593 (Min: 2418)	Sisal+Glass+Epoxy	[10]
Fique Fibre Reinforced Polyester Composites	*Fibre loading (%): 0/10/20/30	ASTM D256	Max: 260 (Min: 25)		30%	[11]

Hybrid Composite Material-Sisal & Coir	*Treated / untreated *Sisal (25%)+coir (15%) *Sisal (20%)+Coir (20%) *Fibre length (mm): 3,5	ASTM D256		Max: 44200 (Min: 29500)	Treated Sisal 20% + Coir 20% 3mm	[12]
Polyester matrix composites reinforced with blanket of the malva and jute fibers	Layer of blanket: 0,1,2	ASTM D256	Max: 104.18±6.19 (Min: 35.06±2.96)		0 layer blanket Jute 60%+Malva 40%	[13]
Polyester matrix composites reinforced with banana fibers	Fibre loading (%): 0/10/20/30	ASTM D256	Max: 265 (Min: 30)		30%	[14]
Palf reinforced polymeric composites	Fibre loading (%); 0/10/20/30	ASTM D256	Max: 503±116.22 (Min: 18±1.8)		30%	[15]
Epoxy Composite Reinforced with Mallow fibers	Fibre loading (%); 0/10	ASTM D256	Max: 113.21 (Min: 6.3)		10%	[16]

Based on Table 3, Izod impact test properties in term of toughness also will be increased by inclusion the high amount of fiber loading which are similar to charpy impact test result. In the table also shows that fibre's orientation and number of layers giving a different result of impact strength which increasing the composite strength. It has been proof based on the research by the Ygor Macabu de Moreas [13] where there is a remarkable increase in the notch toughness by increasing layers of malva blankets and jute.

4. Conclusion

A widely used testing in the case of impact behaviour study is the Charpy and Izod impact test. It is a simple analysis method for the measurement of the stiffness by breaking a random geometry bar. For homogenous and isotropic materials, these methods are being used and also useful for comparing the resistance to impact loading of different materials or the same material in different processing conditions.

Acknowledgement

This work is supported by UPM under GP-IPS grant, 9486400 and Newton Fund, 6300896. The authors would like to express their gratitude and sincere appreciation to the Department of Aerospace

Engineering, Faculty of Engineering, Universiti Putra Malaysia and Laboratory of Biocomposite Technology, Institute of Tropical Forestry and Forest Products (INTROP), Universiti Putra Malaysia (HiCOE) for the close collaboration in this research.

References

[1] Salman S D, Leman Z, Sultan M T H, Ishak M R and Cardona F 2017 Effect of kenaf fibers on trauma penetration depth and ballistic impact resistance for laminated composites *Text. Res. J.* **87** 2051-2065
[2] Nor A F, Sultan M T H, Jawaid M, Azmi A M and Shah A U M 2019 Analysing impact properties of CNT filled bamboo/glass hybrid nanocomposites through drop-weight impact testing, UWPI and compression-after-impact behaviour *Compos. Part B Eng* **168** 166-174
[3] Ismail K I, Sultan M T H, Shah A U M, Jawaid M and Safri S N A 2019 Low velocity impact and compression after impact properties of hybrid bio-composites modified with multi-walled carbon nanotubes *Compos. Part B Eng* Apr **163** 455-63
[4] Element Material Technology, https: www. element.com/nucleus/charpy-vs-izod-impact
[5] Sivakumar D, Khatiravan S, Ng L F, Ali M B, Selamat M Z, Sivaraos S and Bapokutty O 2018 Experimental Investigation on Charpy Impact Response of Kenaf Bast Fiber Reinforced Metal Laminate System *ARPN Journal of Engineering and Applied Sciences* **13** 824-826
[6] Gloria G O, Margaem F M, Ribeiro C G D, deMoraes Y M, da Cruz R B, Silva F de A and Monteiro S N 2015 Charpy Impact Tests of Epoxy Composites Reinforced with Giant Bamboo Fibers *Mater. Res.* **18** 178-184
[7] Nascimento L F C, Neves S, Monteiro, Louro L H L, da Luz F S, dos Santos J L, Braga F d O and Marcal R L S B 2018 Charpy Impact Test of Epoxy Composites Reinforced with Untreated and Mercerized Mallow Fibers *J Mater Res Technol* **7** (4) 520-527
[8] Barbosa A de P, Oliveira M P, Altoe G R, Margem F M and Monteiro S N 2014 Charpy Impact Test of Epoxy Matrix Composites Reinforced with Buriti Fibers *Materials Science Forum* **775-776** 296-301
[9] Barcelos M A, Simonassi N T, Monteiro S N and Margem F M 2014 Impact Tests in Polyester Matrix Composites Reinforced with Continuous Curaua Fiber *Congresso Brasileiro de Engenharia e Ciencia dos Materiais*
[10] Sathish S, Ganapathy T and Bhoopathy T 2014 Experimental Testing on Hybrid Composite Materials *Applied Mechanics and Materials* **592-594** 339-343
[11] Artur C P, Sergio N M, Foluke S A and Henry A C 2017 Izod Impact Tests in Polyester Matrix Composites Reinforced with Fique Fabric *Proceedings of the 3rd Pan American Materials Congress*
[12] Velmurugan G, Venkatesan S P, Prakash P V, Sathishkumar N and Vijayakumar N 2014 Mechanical Testing of Hybrid Composite Material (Sisal & Coir) *Int J of Scientific & Research Publications 4*
[13] Moraes Y M de, Ribeiro C G D, Margem F M, Monteiro S N, Margem J I and Vasconcelos J B 2017 Izod Impact Test in Polyester Matrix Composites Reinforced with Blanket of the Malva and Jute Fibers *The Minerals, Metals & Materials Series*
[14] Monteiro S N M, Margem F M, Assis F S de, Loiola R L and Oliveira M P 2014 Izod Impact Test in Polyester Matrix Composites Reinforced with Banana Fibers *Materials Science Forum* **775-776** 261-265
[15] Gomes M de A, Altoe G R, Teles M C, Gloria G O, Margem F M and Monteiro S N 2015 Izod Impact Test of Palf Reinforced Polymeric Composites *70 Congress Annual ABM*
[16] Sivasankaraiah T, Srinivas V K, Umashankar C H and Reddy D G 2017 Impact Characteristics of Polymer Matrix Composites *Int J Engineering Sciences & Research Technology*

Static analysis on Malaysian Yankee's pineapple leaf fiber/epoxy composite

M I Najeeb[1], M T H Sultan[1,2,3*], A U M Shah[1,2] and S N A Safri[2]

[1]Department of Aerospace Engineering, Faculty of Engineering, Universiti Putra Malaysia, 43400 UPM Serdang, Selangor Darul Ehsan, Malaysia
[2]Laboratory of Biocomposite Technology, Institute of Tropical Forestry and Forest Products (INTROP), Universiti Putra Malaysia, 43400 UPM Serdang, Selangor Darul Ehsan, Malaysia
[3]Aerospace Malaysia Innovation Centre (944751-A), Prime Minister's Department, MIGHT Partnership Hub, Jalan Impact, 63000 Cyberjaya, Selangor Darul Ehsan, Malaysia

*Corresponding author: thariq@upm.edu.my

Abstract. The static analysis on pineapple leaf fibre (PALF) reinforced epoxy composites was carried out in this study. The effect of fiber alignment in X and Y axis were studied while maintaining total fiber loading of 10% by weight. The bending and compression strength were analyzed by performing flexural and compression test. The composite properties was strongly influence by the fiber orientation where Y axis shows improvement about 97% in bending and 5.3% in compressive strength compared to neat epoxy. However, as for X orientation bending test shows lower bending strength compared to neat epoxy but improve 4% in compressive strength. It can be concludedthat the fiber alignment in Y-direction is more suitable to be utilized for automotive interior or household applications that demand high strength properties.

1. Introduction

In 2016, natural fiber composite market size was valued at USD 4.46 billion worldwide. It is expected to grow 11.8 % of compound annual grow rate (CAGR) from 2016-2024 due awareness on environment pollution and high demand for light weight component from automotive industry [1]. An example of automotive interior component produced from natural materials is door panels that are made from flax/sisal reinforce thermoset resin [2]. Among all natural fibers, pineapple leaf fibers (PALF) is the best candidate to be used as reinforcement in composite as it shows superior mechanical strength due to high cellulose content [3]. In Malaysia alone, there are more than 10 varieties of pineapple namely Josapine, Sarawak, Moris Gajah, Yankee, Maspine, N36, etc which proves that there are abundant of leaf waste that could be convert into sustainable materials. Previous study on Josapine's PALF on 30mm and 200mm fiber length reinforce composite shows that longer fiber had higher bending strength [4]. Fiber length of 2,4 and 6 cm and loading of 50,60 and 70% of Josapine's PALF reinforce starch composites properties had been studied in term of tensile, flexural and hardness [5]. Extend to this study, the author had varied the starch (75, 100 and 250 μm) size reinforce with three different cultivars namely Josapine, Moris Gajah and Sarawak [6]. On top of that, in different studies evaluation of 10% fiber (Moris Gajah, Maspine, Josapine and N36) loading reinforced Polylactic acid composites had been done on tensile and flexural properties [7]. Besides, Josapine's

PALF reinforced polypropylene composites with fixed 100mm length and different weight percentage (30,40,50,60,70%wt) had investigated it's tensile and hardness strength [8]. Author [9] concluded that PALF had the credibility to be utilized in bio fuel and bio composites. However, there are none study on Yankee's PALF natural fiber composite. Previous studies had shown that different variety of plants yield different properties. However, the potentials of the Yankee Pineapple variety in the composite industry remains untapped to date. No studies had been carried out on the PALF natural fiber composite drawn from the Yankee variant. This study is aimed at investigating the mechanical properties of the leaf fiber of the new Malaysian pineapple variety in composite applications.

2. Material Preparation

The natural fiber employed in this study was pineapple leaf's fiber (PALF) of Yankee variety collected at Teluk Panglima Garang, Selangor, Malaysia. The pineapple leaf is collected after the pineapple harvesting process and the age of the collected leaf is aged around 1-2 years old. The collected leaf was then extracted mechanically at Universiti Tun Hussein Onn and the extracted fiber was called as PALF. The PALF are then sun dried for 2 days to remove its water content. The dried PALF are comb and stored in closed compartment to prevent absorption of water and contamination from surrounding such as dust. Based on chemical composition analysis on this fiber, it was found out there were cellulose (47.74%), hemicellulose (15.98%) and lignin (2.44%). The natural fiber composite was fabricated by aligned the fiber continuously in one direction as shown in Figure 1. The total fiber loading reinforcing in the matrix and thickness was fixed at 10% by weight and 4mm respectively. A pure epoxy plate was also fabricated as control samples. Before the fabrication process start, the surface of mould was wax for ease of removal. The ratio used for epoxy to hardener was 3:1 and was stir uniformly for 5 min which then was used to impregnated with the fiber in a mould. The mould was cured by using hot press at 120° for 10minutes with a constant 250 bars. Directly after hot press continue with cold press for 5 minutes with constant 250 bars. The composite was then cut according to X and Y fiber alignment as shown in the Figure 1. Compression test was carried as per ASTM D695 with dimension 12.7 x 12.7 x 4 mm and the cross speed 1.3mm/min. Flexural test was carried out according to ASTM D7264 with standard span to thickness ratio of 32:1, width 13 mm through three-point bending test using Universal Testing Machine (UTM). The rate of displacement was set at 2mm/min. Morphology of tested composite sample were studied using scanning electron microscope (SEM, Hitachi S3400N). The samples were sputter coated with gold for better visualization.

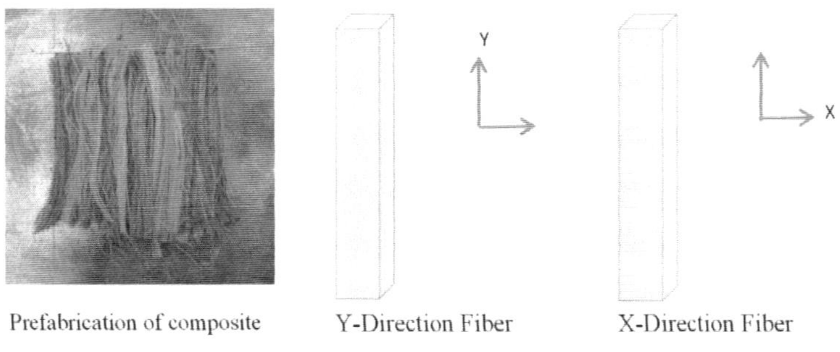

Figure 1. Fiber alignment condition in fabricated specimen.

3. Results and Discussions

3.1 Flexural Analysis

Figure 2 shows flexural strength and flexural modulus for all the tested samples.

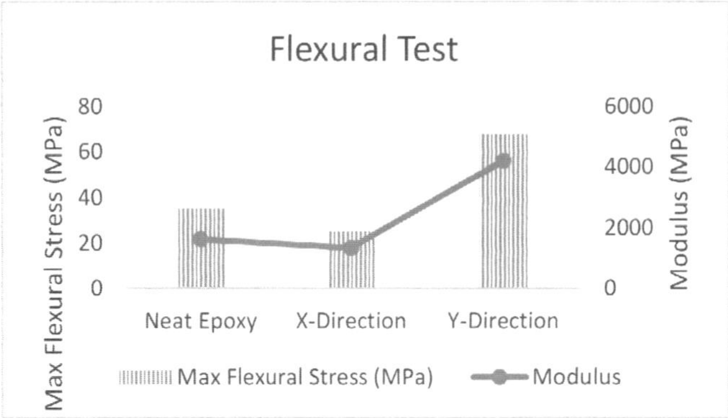

Figure 2. Flexural strength of composite.

Composite made from fibre in Y-direction shows 93.7% improvement in flexural maximum stress compared to neat epoxy. This shows that, PALF helps in transferring the load more effectively in Y-axis throughout the composite, thus contribute to the total strength of the composite. Besides, the Y-direction fiber composite shows the highest modulus indicate that it had the highest stiffness. However, composite made from PALF in X-direction experience reduction 28.3% in strength to neat epoxy maybe due to short distance stress distribution throughout the fiber end causing it to fail at lower stress. Other reason could be contributed by low interfacial bonding between PALF as discussed in morphology analysis section. The results of this research show that composite made from Y-direction fiber alignment are comparable with other PALF composite as per Table 1 even though, the fiber loading used is 10%. This shows that, the fiber alignment plays a huge role in contributing to the overall composite strength. By considering both mentioned parameters in a composite fabrication, fiber loading percentage can be optimized so that it is possible to achieve higher natural fiber composite strength compared to synthetic composite.

Table 1. PALF Composite.

Fiber condition	Layering method	Matrix	Flexural strength (MPa)	Reference
10%; continuous fiber Yankee's PALF	Y-direction	Epoxy	68.13	Current research
30%; length 30mm PALF(P)	PPPPPP	Polyester	107	[10]
40%; length 0.8-1mm Indonesia PALF	Short fiber	Phenolic	65	[11]
30%; length 30mm Josapine's PALF	Short fiber	PLA	32	[4]
30%; length 200mm Josapine's PALF	Long Fiber	PLA	52	[4]

3.2 Compression Analysis

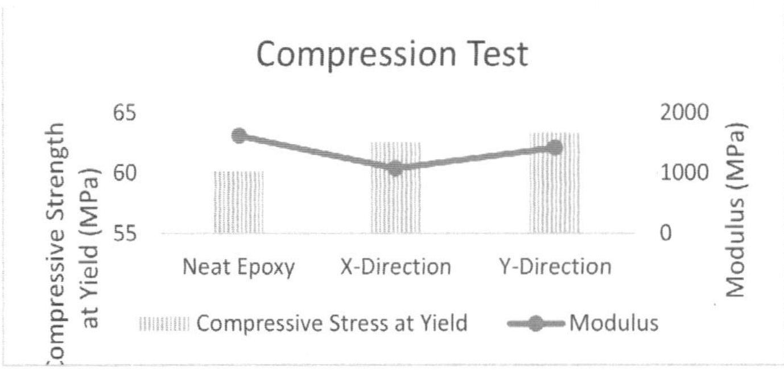

Figure 3. Compressive strength of composite.

Figure 3 shows that by adding 10% of Yankees PALF into the neat epoxy composite, helps to improve the compressive strength up to 5.3% regardless of the fiber orientation. It also shows that the fiber orientation in the matrix plays important role as it contribute to the matrix strength as well. Composite made from fiber in Y direction shows 1.3% higher compressive strength compared to composite made from fiber in X direction as shown in Figure 3. However, even though Y-direction fiber composite had the highest compressive strength, the stiffness of the composite reduced compared to neat epoxy. Previous study [12] shows that composite made from long pineapple fiber hybridised with woven glass fiber ($G/P_Y/P_X/P_Y/G$) epoxy composites had compressive strength of 64.4 MPa. The different of compressive strength compared to composite made from fiber in Y direction in current study was only 2% lower. This shows that, the compressive strength is comparable even the current study did not do any hybridisation process and without using any synthetic fiber. This shows that having right orientation according to the load direction will help the stress distribution more effectively hence increasing the strength of the composite.

3.3 Morphology Analysis
The morphology of fails flexural testing specimen were taken randomly and studied using SEM. From the cross-section surface, Figure 4 shows there are small gap between the fiber-matrix (which as shown by the arrow) and fiber pull out. This mean there are some part of fiber that are not properly merge with the matrix or it is not strong enough to hold the load causing it to shear hence creating the gap. The gap shows there were limited bonding between fiber and matrix as fiber had hydrophilic nature properties while the matrix had hydrophobic properties resulting to poor interfacial bonding [13-17].

Figure 4. Cross section crack from flexural test.

Conclusion
Generally, the addition of PALF in epoxy matrix help increasing the composite strength. The orientation of fiber in matrix plays an important role in absorbing the external stress. Composite made from Y-direction fiber orientation were the best configuration in this study where the maximum bending and compressive stress that it can hold before it fails were 68.13 MPa and 63.3 MPa respectively. From morphology analysis by SEM, it can be concluded that the interfacial bonding fiber-matrix play a vital role in contributing to the composite mechanical properties.

Acknowledgment
This work is supported by UPM under HICoE grant, 6369107 and Newton Fund, 6300896. The authors would like to express their gratitude and sincere appreciation to the Department of Aerospace Engineering, Faculty of Engineering, Universiti Putra Malaysia and Laboratory of Biocomposite Technology, Institute of Tropical Forestry and Forest Products (INTROP), Universiti Putra Malaysia (HiCOE) for the close collaboration in this research.

References

[1] Natural Fiber Composites (NFC) Market Size, Share & Trends Analysis Report By Raw Material, By Matrix, By Technology (Injection Molding, Compression Molding, Pultrusion), By Application, And Segment Forecasts, 2018 - 24

[2] Holbery J and Houston D 2006 Natural-fiber-reinforced polymer composites in automotive applications *Jom* **58** (11) 80-86

[3] Mukherjee P S and Satyanarayana K G 1986 An empirical evaluation of structure-property relationships in natural fibres and their fracture behaviour *J. Mater. Sci.* **21** (12) 4162-68

[4] Ramli S N R, Fadzullah S H S M and Mustafa Z 2017 The effect of alkaline treatment and fiber length on pineapple leaf fiber reinforced poly lactic acid biocomposites *Jurnal Teknologi* **79** (5-2)

[5] Selamat M Z, Razi M, Kasim A N, Dharmalingam S, Putra A, Yaakob M Y, and Daud M A M 2016 Mechanical properties of starch composite reinforced by pineapple leaf fiber (PLF) from josapine cultivar *ARPN J. Eng.and Appl. Sci.* **11** (16)

[6] Selamat M Z, Tahir M S Z, Kasim A N, Dharmalingam S, Putra A, Yaakob M Y and Daud M A M 2018 Effect of starch sizes particle as binder on short pineapple leaf fiber composite mechanical properties In *MATEC Web of Conf.* **150** 04008

[7] Munawar R F, Jamil N H, Shahril M K, Rahim S A, Muhammad S, Abidin Z and Lau K T 2015 Development of green composite: pineapple leaf fibers (palf) reinforced polylactide (pla) In *Appl. Mech. and Mater.* **761** 520-525

[8] Kasim A N, Selamat M Z, Daud M A M, Yaakob M Y, Putra A and Sivakumar D 2016 Mechanical properties of polypropylene composites reinforced with alkaline treated pineapple leaf fibre from Josapine cultivar *Int. J. Auto. and Mech. Eng.* **13** 3157

[9] Alwani M S, Khalil A, Islam M N, Nadirah W W and Dungani R 2014 Fundamental approaches for the application of pineapple leaf fiber in high performance reinforced composites *Polimery* **59** (11-12) 39

[10] Nayak S Y, Heckadka S S, Amin N M, Sadanand R V and Thomas L G 2018 Effect of Hybridization on the Mechanical Properties of Chopped Strand Mat/Pineapple Leaf Fibre Reinforced Polyester Composites in *MATEC Web of Conf.* **153** 01006

[11] Asim M, Jawaid M, Abdan K and Ishak M R 2018 The effect of silane treated fibre loading on mechanical properties of pineapple leaf/kenaf fibre filler phenolic composites *J Polym Environ* **26** (4) 1520-1527

[12] Manickavasagam V M, Ramnath B V, Elanchezhian C, Jenish J, Jayavel S and Muthukumar V 2014 Investigation on impact and compression Properties of Pineapple Reinforced Polymer Composite in *Appl. Mech. and Mater.* **591** 116-119

[13] Mostafa N H, Ismarrubie Z N, Sapuan S M and Sultan M T H 2017 Fibre prestressed composites: Theoretical and numerical modelling of unidirectional and plain-weave fibre reinforcement forms *Compos. Struct.* **159** 410-423.

[14] Mostafa N H, Ismarrubie Z N, Sapuan S M and Sultan M T H 2016 Effect of equi-biaxially fabric prestressing on the tensile performance of woven E-glass/polyester reinforced composites *J Reinf. Plast. Comp.* **35** 1093-103.

[15] Salman S D, Sharba M J, Leman Z, Sultan M T H, Ishak M R and Cardona F 2016 Tension-compression fatigue behavior of plain woven kenaf/kevlar hybrid composites *BioRes.* **11** 3575-3586.

[16] Salman S D, Leman Z, Sultan M T H, Ishak M R and Cardona F 2016 Influence of fiber content on mechanical and morphological properties of woven kenaf reinforced PVB film produced using a hot press technique *Int. J. Polym. Sci.* **2016** 1-11.

[17] Salman S D, Leman Z, Sultan M T H, Ishak M R and Cardona F 2017 Effect of kenaf fibers on trauma penetration depth and ballistic impact resistance for laminated composites *Text. Res. J.* **87** 2051-2065.

IOP Conf. Series: Materials Science and Engineering **670** (2019) 012033 doi:10.1088/1757-899X/670/1/012033

Optimal control of a tetrahedral configured reaction wheels for Quaternion rotation based on rigid satellite model

T Vil Cherd[1], R Daud[1] and S Yaacob[2]
[1] School of Mechatronic Engineering, Universiti Malaysia Perlis, Arau, Perlis, Malaysia
[2] Electrical, Electronics and Automation Section, UniKL Malaysia Spanish Institute, Hi-Tech Park, Kulim, Kedah, Malaysia

Abstract. Reaction wheel is the most popular actuator for the attitude manoeuvring of a satellite system due to its compact size, reliability and able to produce precise torque. However, for redundancy purposes, the reaction wheels are assembled in a tetrahedral configuration that is able to generate torque in any direction even when one of the reaction wheel fails. The tetrahedral configured reaction wheel causes the maximum torque generation to be uneven in any direction. Hence, the quaternion rotation is completed at different rate in different direction. In order to optimize the rotation rate, an optimal control using iterative method via GPOPS toolbox. It is then compared with the traditional Eigen-axis Quaternion Feedback control. The improvement shown by the optimal control to be between 3.49% to 25.11% improvement in manoeuvre time depending on the direction of manoeuvre. The optimal control is able to outperform the traditional control method.

1. Introduction

Since 1957, artificial satellites have been placed into the Earth's orbit. These satellites play important in various purposes, namely, military intelligence and defence, communications, space explorations and the scientific study [1].

The orientation of satellites need to be changed on every mission to achieve desired outcome. Hence, attitude control is needed to alter the satellite attitude in the desired direction. Generally, the control methods are classified as active or passive. Passive method has been utilized in [2]. Some of the active methods include generalized predictive control method [3], control method based on Lyapunov [4] and nonlinear control based on linear matrix inequality [5].

The main purpose of this paper is to introduce a control function on each of the three satellite's kinetic equations. These control functions are within the framework of an optimal control problem via PMP. Also, we show that these controls are able to drive the satellite's attitude to desired direction at minimum time compared to the traditional control method.

2. The reaction wheel

The attitude control of a satellite system involves the cooperation between a controller and actuator where the controller provides the desired input while the actuator generates the input to the overall system based on the controller. A typical satellite system is equipped with four or more wheels of Reaction Wheel (RW) for the purpose of redundancy in case when one or more RW fails [6]. A attitude manoeuvring actuator system with a configuration of four RW is shown in Figure 1 in reference to RazakSAT® satellite [7]. The tetrahedral configuration is crucial for redundancy purpose because if one of the RW fails, the other three RW is still able to generate torque in any of the three axis.

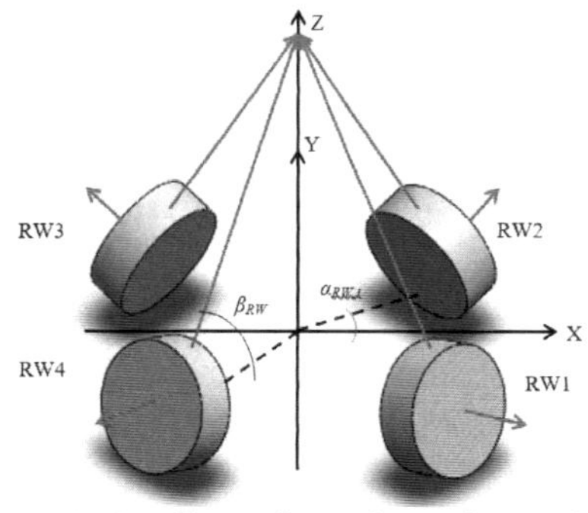

Figure 1. Configuration of reactions wheels.

3. The equation of motion

The momentum of the RW is defined as the product of RW's inertia and angular velocity as shown in Equation 1.

$$\boldsymbol{h}_{r,j} = \boldsymbol{I}_{r,j}\boldsymbol{\Omega}_j \tag{1}$$

where $I_{r,i}$ and Ω_i are the moment of inertia and the angular velocity of the RW respectively. $h_{r,j}$ is the angular momentum with reference to the satellite body frame. The torque with respect to the wheel frame, $\tau_{r,i}$ is the time derivative of Equation 2.

$$\boldsymbol{\tau}_{r,j} = \frac{d\boldsymbol{h}_{r,j}}{dt} = \boldsymbol{I}_{r,j}\dot{\boldsymbol{\Omega}}_j \tag{2}$$

Hence, the torque generated by the RW, $\boldsymbol{\tau}_{out,i}$ is

$$\boldsymbol{\tau}_{out,i} = \frac{d\boldsymbol{h}_{out,j}}{dt} = {}^{C}\boldsymbol{R}^{w}\boldsymbol{I}_{r,j}\dot{\boldsymbol{\Omega}}_j \tag{3}$$

h_{out} is the angular momentum with reference to the satellite body frame and ${}^{C}\boldsymbol{R}^{w}$ is the transformation matrix for the RW orientation with respect to the body frame. By Euler's transport theorem adopted from [8], the inertial Equation 4 is obtained.

$$\boldsymbol{\tau}_{out,j} = {}^{C}\boldsymbol{R}^{w}\boldsymbol{\tau}_{r,j} + \dot{\boldsymbol{\theta}}_s \times \boldsymbol{I}_{r,j}\boldsymbol{\Omega}_j \tag{4}$$

Hence, the RW array generated a torque that is equivalent but in opposite direction in the body frame due to the moment exchange between the RW and the satellite body, as shown in Equation 5.

$$-\begin{bmatrix}\tau_X\\ \tau_Y\\ \tau_Z\end{bmatrix} = \boldsymbol{\tau}^{N}_{out,j} = {}^{C}\boldsymbol{R}^{w}\boldsymbol{\tau}_{r,j} + \dot{\boldsymbol{\theta}}_s \times \boldsymbol{I}_{r,j}\boldsymbol{\Omega}_j \tag{5}$$

4. The Eigen-axis Quaternion Feedback

The common controller applied in satellite system is the Eigen-axis Quaternion Feedback (EQF). The controller uses an Eigen-axis manoeuvre so that the attitude manoeuvring problem forms a simple single angle rotation. The model obtained from the EQF control logic often used in satellite attitude control system is shown in Equation 6 [9].

$$\boldsymbol{\tau}'_{req} = -\mathrm{k}_c \mathbf{I}_s \mathbf{q}_{123} - \mathrm{c}_c \mathbf{I}_s \dot{\boldsymbol{\theta}}_s + \dot{\boldsymbol{\theta}}_s \times \mathbf{I}_s \dot{\boldsymbol{\theta}}_s \tag{6}$$

The torque $\boldsymbol{\tau}'_{req}$ is what the EQF controller is trying to control. Equations 6 contain the gyroscopic coupling term which is not desired in solving a linear motion along an arc. $\mathbf{I}_s$ is the total moment of inertia. The nonlinearity can be removed by simply adding the gyroscopic term into the EQF equation to remove the nonlinearity that exists in the control Equation 6. Hence, Equation 7 is obtained.

$$\begin{aligned} \boldsymbol{\tau}_{req} &= \boldsymbol{\tau}'_{req} - \dot{\boldsymbol{\theta}}_s \times \mathbf{I}_s \dot{\boldsymbol{\theta}}_s \\ &= -\mathrm{k}_c \mathbf{I}_s \mathbf{q}_{123} - \mathrm{c}_c \mathbf{I}_s \dot{\boldsymbol{\theta}}_s \\ &= \mathbf{I}_s \dot{\boldsymbol{\theta}}_s \end{aligned} \tag{7}$$

The control torque $\boldsymbol{\tau}_{req}$ is effective control of the satellite body frame. Equation 8 can be rearranged as Equation 8.

$$\begin{aligned} \boldsymbol{\tau}_{req} + \mathrm{k}_c \mathbf{I}_s \mathbf{q}_{123} + \mathrm{c}_c \mathbf{I}_s \dot{\boldsymbol{\theta}}_s = 0 \\ \mathbf{I}_s \dot{\boldsymbol{\theta}}_s + \mathrm{k}_c \mathbf{I}_s \mathbf{q}_{123} + \mathrm{c}_c \mathbf{I}_s \dot{\boldsymbol{\theta}}_s = 0 \end{aligned} \tag{8}$$

The inertia is a constant, hence it is taken out to obtain Equation 9.

$$\ddot{\boldsymbol{\theta}}_s = -\mathrm{k}_c \mathbf{q}_{123} - \mathrm{c}_c \dot{\boldsymbol{\theta}}_s \tag{9}$$

The satellite attitude is measured in three-dimensional space. However, by applying the Eigen Axis which simplifies the rotation into a single axis, the trajectory of the rotation can also be reduced to single angle, while the body rate of the satellite can also be measured as a single angular rate, as shown in Equation 10.

$$\begin{aligned} \theta_E &= \theta_s - \theta_d \\ \ddot{\boldsymbol{\theta}}_E &= -\mathrm{k}_c(\theta_s - \theta_d) - \mathrm{c}_c(\dot{\theta}_s - \dot{\theta}_d) \end{aligned} \tag{10}$$

Assume that a desired final attitude and rate are known. The controller will manoeuvre the error from initial state to zero. For a satellite to change its attitude, it is considered a rest to rest motion. Hence, the desired rate is zero, and the Equation 11 is obtained.

$$\dot{\theta}_d = 0 \qquad \frac{\mathrm{u}}{\mathrm{I}_{tot}} = \ddot{\theta}_E = -\mathrm{k}_c(\theta_s - \theta_d) - \mathrm{c}_c \dot{\theta}_s \tag{11}$$

As shown in Equation 11, the EQF controller equation is similar to the PD control system.

5. The Optimal Control

The minimum time optimal control is done using GPOPS for developing the ideal control function with the shortest time to complete desired attitude rotation. The optimal control problem is obtained by compiling the dynamic models and states of the system from Equations 1 to 11 shown in Equation 12.

$$\text{Given} \qquad \boldsymbol{x}^{\boldsymbol{T}} = [q_1, q_2, q_3, q_4, \dot{\theta}_x, \dot{\theta}_Y, \dot{\theta}_Z, \Omega_1 \dots, \Omega_4] \tag{12}$$

IOP Conf. Series: Materials Science and Engineering **670** (2019) 012033 doi:10.1088/1757-899X/670/1/012033

$$\boldsymbol{u}^T = [\tau_{r,1}, \dots, \tau_{r,4}]$$

Minimize $J = t_f$

Subject to $\ddot{\boldsymbol{\theta}}_s = \frac{\boldsymbol{\tau} - \dot{\boldsymbol{\theta}}_s \times \boldsymbol{I}_s \dot{\boldsymbol{\theta}}_s}{\boldsymbol{I}_s}$ $\quad -\boldsymbol{\tau} = \boldsymbol{\tau}^N_{out,j} = {}^C\boldsymbol{R}^W \boldsymbol{\tau}_{r,j}$

$$\boldsymbol{x}_o^T = [q_{0,1}, q_{0,2}, q_{0,3}, q_{0,4}, 0, 0, 0, 0, 0, 0, 0]$$

$$\boldsymbol{x}_f^T = [q_{f,1}, q_{f,2}, q_{f,3}, q_{f,4}, 0, 0, 0, 0, 0, 0,]$$

$$-\tau_{max} < \boldsymbol{\tau}_{r,j} < \tau_{max} - \Omega_{max} < \boldsymbol{\Omega}_i < \Omega_{max}$$

Where, x^T refers to the states, x_o^T is the initial state and x_f^T is the final state. u^T is control variable, while J is the cost function.

6. The Simulation

The parameters applied for the attitude manoeuvring simulation is shown in Table 1.

Table 1. Simulation parameters.

Variable	Value
$\boldsymbol{I}_s$	$\begin{bmatrix} 200 & 0 & 0 \\ 0 & 200 & 0 \\ 0 & 0 & 400 \end{bmatrix} kg \cdot m^3$
τ_{max}	$0.403\ N \cdot m$
Ω_{max}	$314\ rad/s$

The results from the attitude maneuvering parameters in quaternion are shown in Figure 2.

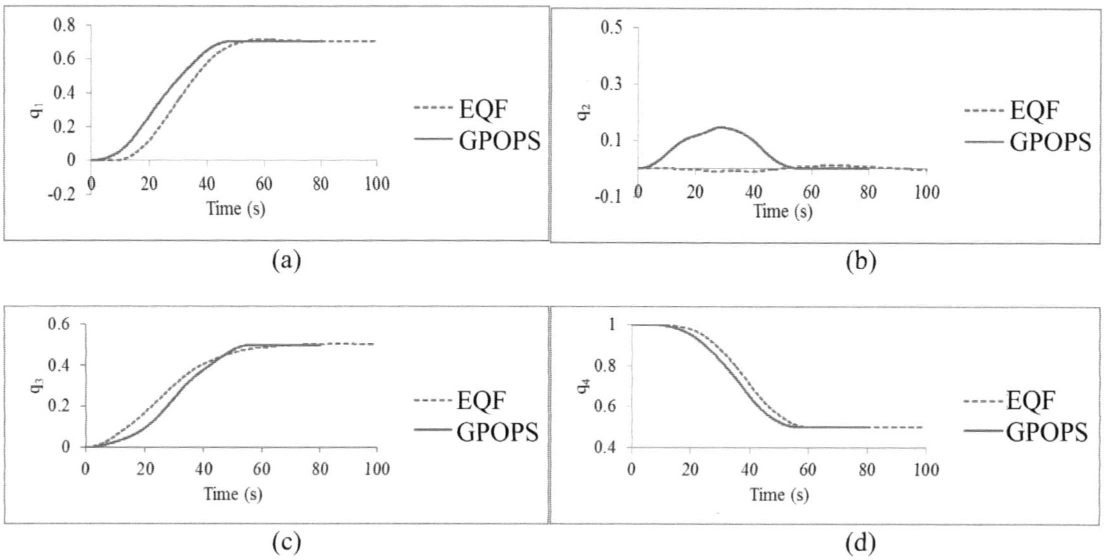

Figure 2. Attitude rotation of satellite in Quaternion (a) q_1 (b) q_2 (c) q_3 (d) q_4

The results from Figure 2 are translated into Euler angles as shown in Figure 3. The figure shows that the GPOPS solution takes 55.45 seconds to complete the manoeuvring which is 25.11% faster than the EQF manoeuvre at 74.04 second. Regardless of the axis of rotation, the GPOPS which generates time optimal manoeuvre will certainly be as fast as or faster than the Eigen axis manoeuvre.

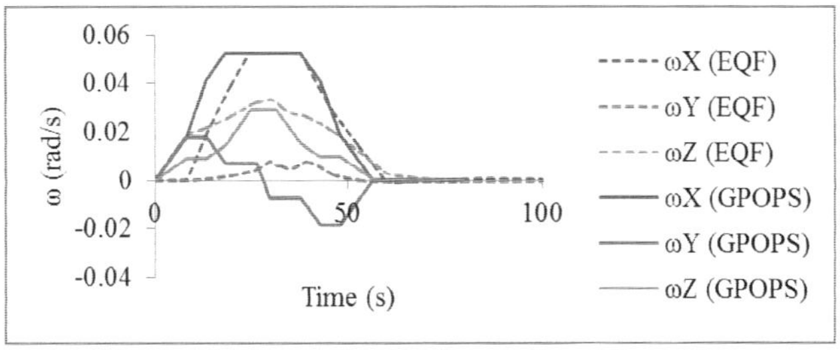

Figure 3. Rate of attitude rotation of satellite.

The optimal solution manoeuvre is able to achieve the final desired attitude much faster for two main reasons. Firstly, the optimized solution is able to take advantage of Off-Eigen axis motion. Hence, this enables the satellite to manoeuvre at a greater magnitude in both torque and momentum. Secondly, the manoeuvre rate of the EQF is confined by the pseudo-inverse restriction. On the other hand, the optimal solution is able to achieve momentum and torque at the full physical capacity of the RW. Table 2 presents a comparison of a manoeuvre-time between the EQF and GPOPS about a 120° manoeuvres on different Eigen axis. Each case moves from the initial origin [0, 0, 0, 1] quaternion to the designated attitude. Interestingly, the time-optimal maximum body rate consistently outperforms the EQF, this accounts for much off-Eigen Axis movement.

Table 2. Manoeuvring performance at difference Quaternion.

q	Eigen Axis	GPOPS (s)	EQF (s)	Improvement (%)
[0, 0 , 1, 0.5]	[0, 0, 1.155]	63.6371	65.94	3.49
[0.5, 0.5, -0.5, 0.5]	[0.577, 0.577, -0.577]	55.5748	61.18	9.16
[-0.612, -0.433, -0.433, 0.5]	[0.7071, 0.5, 0.5]	53.3832	60.83	12.24
[0.707, 0, 0.5, 0.5]	[0.8165, 0, 0.577]	56.7654	65.11	12.82

7. Conclusion

Time optimal control is performed to obtain the control which will generate desired attitude change at the minimum. This is done via the application of GPOPS toolbox while verification of the optimality is also done. The works presented shows a 3.49% to 25.11% improvement in manoeuvre time. The performance increase is due to two major factors. Firstly, the optimal control is able to fully utilize the all the capacity of the RW while the EQF is plagued by the pseudo-inverse limitation. Secondly, the application of optimal control allows the trajectory to deviate from the effective Eigen axis to achieve faster manoeuvre by utilizing the torque that is unavailable to the effective Eigen axis manoeuvre.

Acknowledgments

Authors wishing to acknowledge assistance or encouragement from colleagues, and financial support from School of Mechatronic Engineering, UniMAP.

References

[1] Niknam M R, Kheiri H and Heydari A 2014 Three-axis optimal control of satellite attitude based on Ponteryagin maximum principle *Int. J. Ind. Math.* **8** (1) 41–48.
[2] Rawashdeh S A 2010 *Passive attude Stabilization for Small Satellites*. Kentucky: University of Kentucky.
[3] Kemih K and Liu W-Y 2007 *Constrained generalized predictive control of chaotic Lu system* **1**.
[4] Fragopoulos D and Innocenti M 2004 Stability considerations in quaternion attitude control using discontinuous Lyapunov functions *IEE Proc. - Control Theory Appl.* **151** (3) 253–258.

[5] Shou H-N and Jan Y-W 2005 Nonlinear H∞ for spacecraft attitude control *IFAC Proc. Vol.* **38** (1) 49–54.
[6] Ismail Z and Varatharajoo R 2010 A study of reaction wheel configurations for a 3-axis satellite attitude control *Adv. Sp. Res.* **45** (6) 750–759.
[7] Vil Cherd T, Bakar S A, Yaacob S and Hamzah N H 2015 Flexible dynamic model of attitude maneuvering of RAZAKSAT® satellite *J. Teknol.* **76** (12).
[8] Crews S R 2013 *Increasing slew performance of reaction wheel attitude control systems* Naval Postgraduate School, Monterey, California.
[9] Lee H, Ziad T and Khairon R 2002 *MACSAT Critical Design Review* Kuala Lumpur.

IOP Conf. Series: Materials Science and Engineering **670** (2019) 012034 doi:10.1088/1757-899X/670/1/012034

Flexural properties of hybrid synthetic/Napier fibres reinforced epoxy composites

M J M Ridzuan[1*], M S Abdul Majid[1], A Khasri[2], L Tajul[1], and E M Cheng[1]

[1]School of Mechatronic Engineering, Universiti Malaysia Perlis, Pauh Putra Campus, 02600 Arau, Perlis, Malaysia.
[2]Faculty of Engineering Technology, Universiti Malaysia Perlis, Unicity Alam Campus, 02100 Padang Besar, Perlis, Malaysia

* denotes corresponding author
Authors email addresses: *ridzuanjamir@unimap.edu.my, shukry@unimap.edu.my, azduwin@unimap.edu.my, liyanatajul@unimap.edu.my, emcheng@unimap.edu.my

Abstract. This paper present the flexural properties of hybrid synthetic/Napier fibres reinforced epoxy composites. The synthetic materials used were carbon and glass fibres. The composites were fabricated using the vacuum infusion method. Napier grass and synthetic fibres were used as reinforcement materials would be mixed with the epoxy resin. The samples were differed by volume fractions. The flexural tests has been conducted using a universal testing machine according with ASTM standard D790. The flexural strength and modulus of the hybrid synthetic/Napier fibre reinforced epoxy composites increased as the content of the synthetic fibre content increased. The flexural properties of the hybrid carbon/Napier reinforced epoxy composites were observed higher than the hybrid glass/Napier reinforced epoxy composites.

Introduction

Hybrid composite is a fabricated material where it is a combination of common matrix with two different type of fibres. Thus, it also have more extensive scope of potential application to be compared with other composite reinforcement. As resulted in previous researched, the hybrid composites which contained 5% sodium hydroxide with Napier grass fibre exhibited great stiffness and flexural strength which lower percentages of void contents of the composite [1]. The properties of the hybrid composites are dependent on the fibre content, fibre orientation, and fibre length, bonding between the matrix and fibre, and the arrangement of the fibres within the laminates. Previous studies on natural-synthetic fibre hybrid composites have primarily focused on reducing the use of synthetic fibres [2–4]. The study has described the potential advantages associated with natural-synthetic fibre hybridisation [5]. The goal of this composite where there is an adequate viewpoint proportion which is providing reinforcement function at least in one directions. Regular substance used to fabricate the composite is using the epoxy resin which have great flexural properties of material [6]. Recently, we have examined the potential use of Napier fibre in polymer composites [7–9]. Therefore, in this work, we have compared the flexural properties of hybrid carbon/Napier fibres reinforced epoxy composites and hybrid glass/Napier fibres reinforced epoxy composites.

IOP Conf. Series: Materials Science and Engineering **670** (2019) 012034 doi:10.1088/1757-899X/670/1/012034

Materials and Method

1.1. Extraction of Napier Grass Fibre

Initially, the stems are cleaned and crushed with a mallet before immersed under tap water for approximately three to four weeks, for separation and decomposition process as shown in Figure 1(a). The fibres are then manually extracted from the stem internodes as illustrates in Figure 1(b) up to the removing the excessive cellulose from the fibre in Figure 1 (e). Next, the fibres were cleaned with a distilled water and sun- dried for a couple of days to remove the excess moisture as shown in Figure 1(f). Meanwhile, the fibres extracted are measured within 170-200 mm in length. Finally, the extracted raw fibres are sun-dried for a few days to remove excess moisture and the fibres are further dried in an oven at 55°C about 10 minutes before for the vacuum infusion specimen preparation.

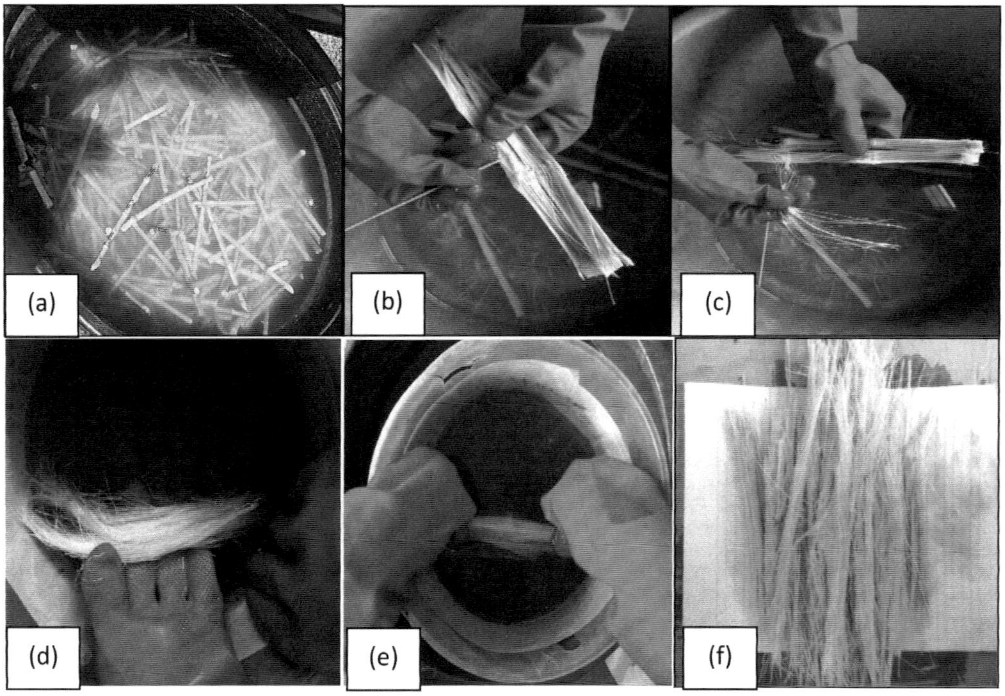

Figure 1. Extraction process.

1.2. Hybrid Composite Fabrication

The composites are the combining of Napier fibre and synthetic fibre which being prepared as a sandwich structure in dimensions of 320 x 300 x 0.1 mm per layer, thus fabricated using the vacuum infusion method [10]. There are two types of structural composite which are laminar composite and sandwich panel [11], where the special part of making this composite material in this experiment is the arrangement of the composite laminate which is clearly known as sandwich composite. Where some of the middle layers of the laminate are replaced by a very lightweight core material or fibre, typically being foam, or natural fibre which is low in density, such as Napier grass fibre, which is prepared and sorted unidirectional direction and formed as a mat or ply in Figure 2. In addition, sandwich panel is also a kind of layered composite which consist of two faces at the outer sheets and one core at the middle of the two faces. Figure 1 shows the schematic structure for the sandwich panels of the structural composites. Besides, sandwich composites are very attractive because of the mechanical properties of the structures, which are very lightweight with a very high bending stiffness thus usually have great impact and fatigue resistance.

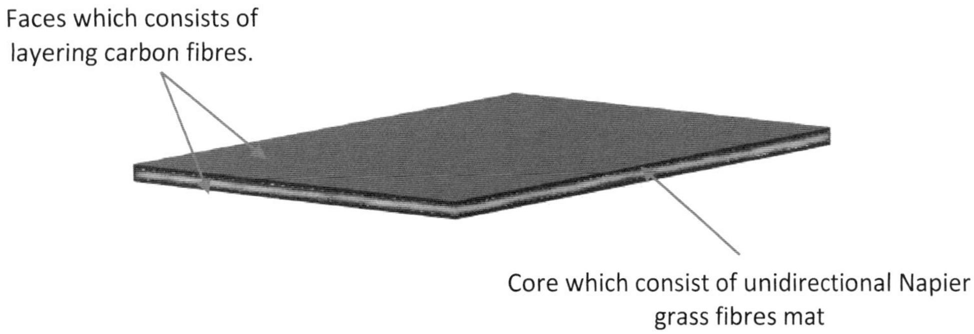

Figure 2. The schematic sandwich structure of hybrid carbon/Napier epoxy composites.

1.3. Flexural Test

Three point flexural tests are conducted on the composite specimens, according to ASTM D790 standards using a universal micro tester (INSTRON 5848), with a load cell of 2kN. Thus, the span length to depth ratio is set as 16:1. The specimens are machined to dimensions of 125 x 13 x (3.2 ± 0.4) mm^3 and being tested at a crosshead speed of 2.5 mm/min. Five specimens are utilized for each of the sample, thus result for flexural strength and modulus were recorded for analysis.

Result

1.4. Flexural Strength

Ability of the composites to withstand a bending load and deformation before it reached the breaking point can be described from the flexural strength. The flexural strength of hybrid carbon/Napier epoxy composites and hybrid glass/Napier epoxy composites are presented in Figure 1. The flexural strength of the Napier fibres reinforced epoxy composites (So) increased from 47.31 to 186.68 MPa with an addition of carbon fibre at 6% of volume fraction (S1). The flexural strength of the hybrid carbon/Napier epoxy composites increased as the volume content of the synthetic fibre increased. Thus resulting the flexural strength by 229, 301 and 456 MPa for S2, S3, and S4 samples, respectively. This strength degradation was measured and represented equivalent strength reduction when comparing the results of both hybrid composites for each sample, whereas yielding increment by 212, 211, 200, and 265% for S1, S2, S3 and S4 samples, respectively. Therefore, the level of degradation reduces as the content of synthetic fibres increased, thus yielding a greater strength of hybrid composites.

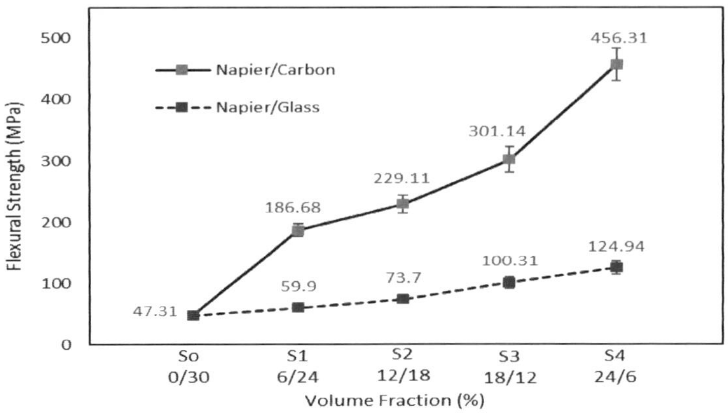

Figure 3. Tensile strength of hybrid synthetic/Napier reinforced epoxy composites

1.5. Flexural Modulus

Figure 3 shows the flexural modulus of Napier fibres reinforced epoxy composites (So) increased from 2.5 to 6.8 GPa with a content of 6% volume fraction of glass (S1), thus yielding an increment about 28% from 6.8 to 8.7 GPa with 6% content of carbon fibre for S1 sample. As the carbon fibre content increased, the flexural modulus of the hybrid carbon/Napier epoxy composites steadily increased to 14.75, 18.65 and 25.76 GPa for S2, S3, and S4 samples, respectively. The increment of the flexural modulus clearly shows that the hybridization of hybrid carbon/Napier epoxy composites was stronger and more rigid, in comparison with hybrid glass/Napier epoxy composites, thus yielding a better flexural performance. The flexural properties of the composites were greatly influenced by the stiffness of both synthetic and natural fibres. Thus, the flexural modulus of hybrid glass/Napier epoxy composites for S1 sample was 6.77 GPa, representing a reduction of over 22%. Furthermore, the strength reduction for S2, S3 and S4 samples were measured by 41, 34, and 49%, respectively.

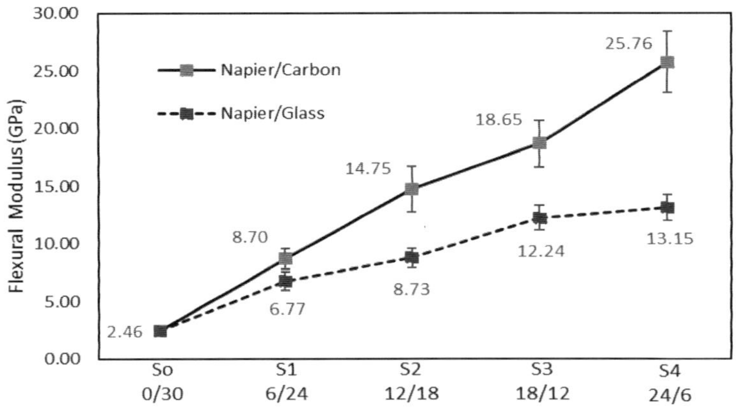

Figure 4. Flexural modulus of hybrid synthetic/Napier reinforced epoxy composites

Conclusion

The flexural properties of hybrid carbon/Napier epoxy composites achieved an excellent hybridization compared to hybrid glass/Napier epoxy composites. The higher content of carbon and glass fibres in the reinforcement of the hybrid composites increased the strength and modulus of the hybrid composites. A good interfacial bonding between the adhesions of the fibre with the matrix, resulted a higher strength and modulus of the hybrid composites.

Acknowledgement

This research was financed by the Ministry of Education, Malaysia, through the Fundamental Research Grant Scheme; (Ref: FRGS/1/2019/STG07/UNIMAP/02/2). The authors would also like to acknowledge the School of Mechatronic Engineering, Universiti Malaysia Perlis (UniMAP) for the equipment and technical assistance. The authors are much gratitude to the dedicated staff for the fruitful discussions and input to the project.

References

[1] Ridzuan M J M, Abdul Majid M S, Hafis S M and Azduwin K 2017 The effects of alkali treatment on the mechanical and morphological properties of Pennisetum purpureum/glass-reinforced epoxy hybrid composites *Plast Rubber Compos* **46** (10) 421–30. doi:10.1080/14658011.2017.1368966.

[2] Joshi S V, Drzal L T, Mohanty a K and Arora S 2004 Are natural fiber composites environmentally superior to glass fiber reinforced composites? *Compos Part A Appl Sci Manuf* **35** 371–6. doi:10.1016/j.compositesa.2003.09.016.

[3] Assarar M, Zouari W, Sabhi H, Ayad R and Berthelot J-M 2015 Evaluation of the damping of hybrid carbon–flax reinforced composites *Compos Struct* **132** 148–54. doi:10.1016/j.compstruct.2015.05.016.

[4] Kumar C S, Arumugam V, Dhakal H N and John R 2015 Effect of temperature and hybridisation on the low velocity impact behavior of hemp-basalt/epoxy composites *Compos Struct* **125** 407–16. doi:http://dx.doi.org/10.1016/j.compstruct.2015.01.037.

[5] Jawaid M and Abdul Khalil H P S 2011 Cellulosic/synthetic fibre reinforced polymer hybrid composites: A review *Carbohydr Polym* **86** 1–18. doi:http://dx.doi.org/10.1016/j.carbpol.2011.04.043.

[6] Mohammed L, Ansari M N M, Pua G, Jawaid M and Islam M S 2015 A Review on Natural Fiber Reinforced Polymer Composite and Its Applications *Int J Polym Sci* **2015** 1–15. doi:10.1155/2015/243947.

[7] Fartini M S, Abdul Majid M S, Ridzuan M J M, Amin N A M and Gibson A G 2016 Compressive properties of Napier (Pennisetum Purpureum) filled polyester composites *Plast Rubber Compos* **45** 136–46. doi:10.1080/14658011.2016.1149911.

[8] Revati R, Majid M S A, Ridzuan M J M, Basaruddin K S, Rahman Y M N, Cheng E M, et al. 2017 In vitro degradation of a 3D porous Pennisetum purpureum/PLA biocomposite scaffold *J Mech Behav Biomed Mater* **74**. doi:10.1016/j.jmbbm.2017.06.035.

[9] Ridzuan M J M, Majid M S A, Afendi M, Rahman M N, Azduwin Y K 2016 Thermal, composition and morphological properties of untreated and alkali-treated napier grass fibres *J Eng Appl Sci* **11** 2485–9. doi:10.3923/jeasci.2016.2485.2489.

[10] Ridzuan M J M, Abdul Majid M S, Khasri A, Basaruddin K S and Gibson A G 2019 Effect of moisture exposure and elevated temperatures on impact response of Pennisetum purpureum/glass-reinforced epoxy (PGRE) hybrid composites. *Compos Part B Eng* **160** 84–93. doi:10.1016/j.compositesb.2018.10.029.

[11] Summerscales J and Short D 1978 Carbon fibre and glass fibre hybrid reinforced plastics *Composites* **9** 157–66. doi:10.1016/0010-4361(78)90341-5.

IOP Conf. Series: Materials Science and Engineering **670** (2019) 012035 doi:10.1088/1757-899X/670/1/012035

Mechanical properties evaluation of nonwoven industrial cotton waste produced by needle punching method

S N H Husain[1], A H A Rashid[1], M F Shaari[1], M N R[1], N N M Hassan[1], M R Ahmad[2], and R A Rashid[1]

[1]Innovative Manufacturing Technology Focus Group, Advanced Technology Centre, Faculty of Engineering Technology, Universiti Tun Hussein Onn Malaysia, Campus (Pagoh Branch), Pagoh Higher Education Hub, Pagoh, Muar, Johor, Malaysia
[2]Textile Research Group, Faculty of Applied Sciences, Universiti Teknologi Mara (UiTM) Universiti Teknologi Mara (UiTM), Shah Alam, Selangor, Malaysia

Abstract. This paper describes the fabrication method and mechanical characterization of nonwoven fabrics from industrial fabric wastes by different machine variables. Cotton waste fibres underwent the mechanical recycling methods, needle punching process and worked into a continuous web of nonwoven fabrics. The measurement method of full factorial design of experiment was implemented in this study as a systematic and efficient way to distinguish between the interaction of more than one factor which are the fibre feeder speed and the number of stacking layers, respectively. Three different fibre feeder speed of 1.8 m/s, 2.2 m/s and 2.8 m/s and three levels of stacking layer (4, 5 and 6 layers) were examined. Analysis of variance (ANOVA) was used to determine the significant factors that influenced the mechanical properties of the web nonwoven fabrics. The mechanical properties of nonwoven web were measured by bursting strength test and puncture resistance test. The results displayed that the difference in fibre feeder speed and number of stacking layer significantly affect the puncture resistance performance. S26L web resisted the highest load during puncture resistance test which is 70.5 N while S14L resisted the lowest load of 25.5 N. Whereas, bursting strength performance was only affected by the fibre feeder speed. Lower fibre feeder speed produced stronger nonwoven fabric. Hence, the faster the fibre feeder speed and the higher number of stacking layer give good puncture resistance properties of nonwoven fabrics while the bursting properties of nonwoven fabrics shows the opposites results.

1. Introduction

The demand of new textile design keeps growing every year making the growth of the textile production inflated. The growth of the world's population in line with the rising living standards has resulted in the continued increase in global fibre utilizations to contribute to the increase in textile waste [1, 2]. Maintaining statistics, at current rates of demographic and economic growth, global annual consumption of fibre will reach 110 million tonnes in 2020 [3, 4]. Combating the situation, it is projected that global fibre supplies will increase from 87 million tonnes to 240 million tonnes by 2050 [5]. It shows that the global fibre supply is not in line with the use of fibre. More production leads to more waste and greater environmental impact [6].

"Fast fashion" phenomenon is the right term in which describes the production of low cost clothing in high volumes intended for short lifetime and made to be disposed [7] has been such a great influence on the trend changes. This resultant in higher amounts of the fabric waste in the landfills. Fibres are basically used in many other application such as the production of clothing, interior, technical textiles and household [8, 9]. Manufacturing of textile products causes a large quantity of wastes, commonly disposed into landfill or used for energy recovering [10]. Burying and land filling the textile waste had been the major concerns for the environment [11]. Morley *et al.* [12] reported that, a total amount of 7.3 million tonnes of apparel waste was generated each year in Europe and 50% of them end up being at the landfills. Landfilling is a preferred choice to dispose waste as it is cost-effective and simpler than focusing on recycling the waste into usable products [13]. Therefore, it was absolutely necessary to establish a method to handle the constantly increasing levels of textile waste which is recycling method.

According to Björquist *et al.* [14]; Leonas [15]; Vadicherla & Saravanan [16] there are two types of textile recycling methods which are chemical and mechanical recycling methods. For natural fibres such as cotton, the main recycling method is mechanical recycling such as weaving, nonwoven and knitting method. In mechanical recycling, fibres that are obtained by defibrillation are used for the production of new yarns, nonwovens and other textiles, upholstery stuffing, insulation and roofing felt, carpet components and lower quality blankets [17, 18]. From this standpoint the aims of this study are to produce cotton waste nonwoven web and evaluate the mechanical properties by means of mechanical recycling. Currently, less research was done using the nonwoven needle punching method to re-process the cotton waste. This research leads to a new exploration of properties and potential products from cotton wastes.

2. Materials and methodology

The cotton fabric wastes were supplied by Golden Gate Technologies Sdn. Bhd., Melaka, Malaysia. The needle punching process were made by the conversion of fibres/filaments into webs and bonded together at different ways such as subsequent bonding by chemical, thermal or mechanical means. In this study, the webs were produced by three different fibre feeder speed (1.8 m/s, 2.2 m/s and 2.8 m/s) and three different types of stacking layer were undergone the punching repetition. The different stack up web of four, five and six layers experienced the repeated punching process. This process was done to ensure the complete compactness of the cotton waste nonwoven web. The cotton waste nonwoven web underwent two types of mechanical properties test which are bursting strength and puncture strength test. The bursting strength was tested using a bursting strength test machine which followed ASTM D3786 testing method. The average of five readings were taken for each of the samples. The puncture strength performance was measured by using a Universal Testing Machine which followed ASTM D4833-07.

Factorial design is applied because it is the most effective approach in the experiment involving more than one factor. The factors or parameters in this project need to be identified, followed by selection the levels at which each factor will be examined [19]. This study has focused on two factors which are the fibre feeder speed and the number of stacking layer. Therefore, the full factorial design with 3 x 2 designs was used. The statistical analysis was carried out using one way analysis of variance (one way ANOVA) to examine the significant effects of each factor and level. It is stated that a p-value of less than 0.05 was considered significant [20]. Samples coding and formulations are presented in Table 1.

Table 1. Samples coding and formulation.

Machine speed	Repeat punching by number of layers	Samples coding
S1 (1.8 m/s)	4	S14L
	5	S15L
	6	S16L
S2 (2.2 m/s)	4	S24L
	5	S25L
	6	S26L
S3 (2.8 m/s)	4	S34L
	5	S35L
	6	S36L

3. Results and discussions

The results of bursting and puncture were calculated to the mean grouping using analysis of variance (ANOVA). Table 2 shows summarization of the statistical analysis (ANOVA) results of the mechanical properties at difference samples of cotton waste nonwoven web. Fibre feeder speed shows significant effects ($p < 0.05$) for bursting strength and interactions of factors were also present. In contrast, ANOVA results for bursting shows that the factors of stacking layer did not affect the response. The p -value denotes more than 0.05. However, R^2 value was recorded as 70.92%, which indicated the data is stable with intermediate degree of variability. Main effect plot of mean data and interaction plot are displayed in Figure 1(a) and 2(a). The machine speed, S1 and S2 produced slightly similar values of bursting strength, while S3 yielded lower bursting strength value. This is because the lower fibre feeder speed, the higher punching density at the same spot on the web which resulting in higher fibre interlocking activities. Thus, this result shows that higher bursting strength is required to break the bonding between fibres in the web. Das *et al.* [21] reported that higher punching density will not only resulted in higher mechanical performances, but also in greater amount of fibre damage and breakage. It is also proved that some research stated that the strength of a nonwoven web increase proportionally to the punch density [22].

Table 2. ANOVA test for; (a) bursting test; (b) puncture test.

Sources	*Df*	*SS*	*MS*	*F*	*P*
(a) Bursting test					
Fibre feeder speed	2	1.0783	0.53915	16.44	**0.000**
Stacking layer	2	0.1057	0.05284	1.61	0.214
Fibre feeder speed*Number of stacking layer	4	1.6959	0.42397	12.93	**0.000**
Error	36	1.1806	0.03280		
Total	44	4.0605			
(b) Puncture test					
Fibre feeder speed	2	6978.6	3489.28	148.71	**0.000**
Stacking layer	2	1191.8	595.90	25.40	**0.000**
Fibre feeder speed* Number of stacking layer	4	320.4	80.11	3.41	**0.018**
Error	36	844.7	23.46		
Total	44	9335.5			

Df: degree of freedom; SS: sum of squares; MS: mean square; F: F-test; p: p-Value.

In Table 2, the puncture test result shows that both of the fibre feeder speed and stacking layer were proven to affect significantly the puncture resistance of the nonwoven webs ($p < 0.05$). The analysis also showed the presence of factor of interaction. Figure 1(b) and 2(b) displayed main effects plots and interaction plots for puncture resistance of mean data. S2 produced higher puncture resistance values and gradually drop as the speed increases. This is due to the same reason as stated by Das *et al.* [21]. When needling density is too high (which is related to the speed of the fibre feed), the bonding structure of the nonwoven tend to damage and the puncture resistance of nonwoven will decrease accordingly [23, 24]. This reason proved that S3 produces lower puncture resistance as compared to S2. As for the number of stacking layers, the interaction also shows that higher stacking layers gives nonwoven web with higher puncture resistance value. Figure 2 and 3 show the highest mean of bursting strength observed at S14L while the highest mean of puncture resistance is S26L, respectively. However, from the results of the thickness proved that the optimum speed to be used was 2.2 m/s.

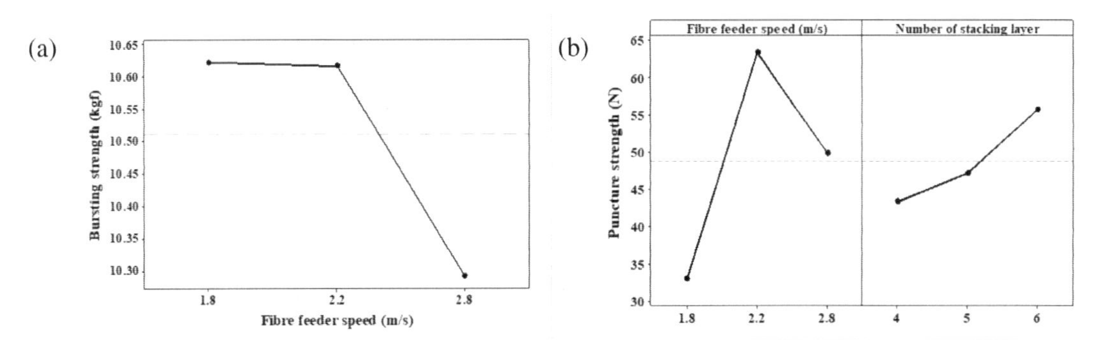

Figure 1. Main effects plot: (a) bursting strength and (b) puncture strength.

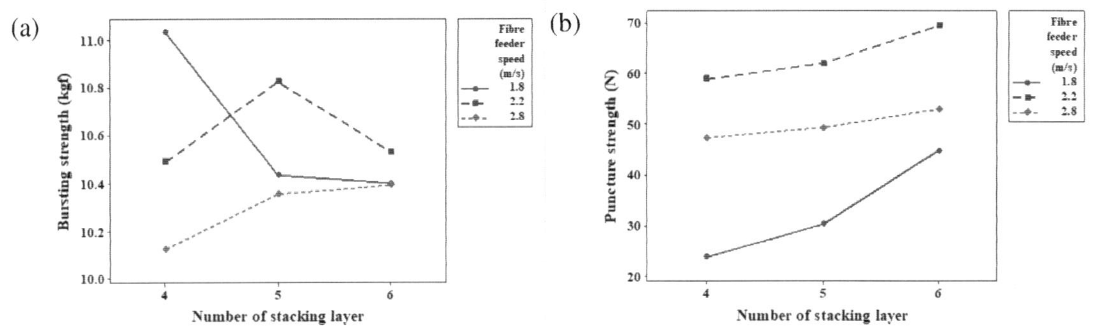

Figure 2. Interaction plot: (a) bursting strength and (b) puncture strength.

4. Conclusions

In this study, the cotton waste nonwoven web was successfully measured and prepared by using needle punching method. The processing parameters of the fabrics affected the mechanical properties of the fabrics. Both fibre feeder speed and number of stacking layers were found statistically significant ($p < 0.05$) based on puncture strength test while the only significant results for bursting test is fibre feeder speed. The increasing number of stacking layers exhibited the higher of puncture properties but oppositely to the bursting properties. As a conclusion, the optimum speed is 2.2 m/s and the optimum number of stacking layer is six layers of cotton waste nonwoven web.

Acknowledgements

This research was supported by Universiti Tun Hussein Onn Malaysia under TIER 1 Research Grant, Vote Number H086. The authors also wanted to express their sincere gratitude to Universiti Tun Hussein Onn Malaysia, for providing the facilities and instruments throughout this research.

References

[1] Lin S D 2012 Recycled fibre industry facing opportunities and challenges in the new situation *Resource Recy* **11** 44–47.

[2] Lu J J and Hamouda H 2014 Current status of fiber waste recycling and its future *Adv Mater Res* **878** 122–131. https://doi.org/10.4028/www.scientific.net/amr.878.122

[3] Vončina B, Volmajer Valh J, Vajnhandl S, Majcen le Marechal A, Aneja A and Lobnik A 2016 A new circular economy concept: From textile waste towards chemical and textile industries feedstock *Proceedings of the 16th AUTEX World Textile Conference* Ljubljana.

[4] Broda J, Przybyło S, Gawłowski A, Grzybowska-Pietras J, Sarna E, Rom M and Laszczak R 2018 Utilisation of textile wastes for the production of geotextiles designed for erosion protection *J Text Ins* **5000** 1–9. https://doi.org/10.1080/00405000.2018.1486684

[5] Björquist S, Aronsson J, Henriksson G and Persson A 2018 Textile qualities of regenerated cellulose fibers from cotton waste pulp *Text Res J* **88** (21) 2485–2492. https://doi.org/10.1177/0040517517723021
[6] Altun Ş 2012 Prediction of textile waste profile and recycling opportunities in Turkey *Fibres and Textiles in Eastern Europe* **94** (5) 16–20.
[7] Asaadi S, Hummel M, Hellsten S, Härkäsalmi T, Ma Y, Michud A and Sixta H 2016 Renewable high-performance fibers from the chemical recycling of cotton waste utilizing an ionic liquid *ChemSusChem* **9** (22) 3250–3258. https://doi.org/10.1002/cssc.201600680
[8] Mohd Rus A Z and Mat Hassan N N 2014 Influence of multilayer textile biopolymer foam doped with titanium dioxide for sound absorption materials *Eng Mater* **595-595** 750-754.
[9] Mat Hassan N N and Mohd Rus A Z 2013 Influences of thickness and fabric for sound absorption of biopolymer composite *Appl Mech Mater* **393** 102-107.
[10] Wazna EL M, Gounni A, Bouari EL A, Alami EL M and Cherkaoui O 2018 Development, characterization and thermal performance of insulating nonwoen fabrics made from textile waste *J Ind Text* **48** (7) 1167-1183. https://10.1177/1528083718757526
[11] Mishra R, Behera B and Militky J 2014 Recycling of textile waste into green composites: performance characterization *Polym Polym Compos* **16** (2) 101–113. https://doi.org/10.1002/pc
[12] Morley N, Slater S, Russell S, Tipper M and Ward D G 2006 Recycling of low grade clothing waste *Oakdene Hollins, Nonwovens Innovation & Research Institute Ltd* (September).
[13] Moh Y C and Abd Manaf L 2014 Overview of household solid waste recycling policy status and challenges in Malaysia *Resources, Cons Recy* **82** 50–61. https://doi.org/10.1016/j.resconrec.2013.11.004
[14] Björquist S, Aronsson J, Henriksson G and Persson A 2017 Textile qualities of regenerated cellulose fibers from cotton waste pulp *Text. Res. J* **88** (21) 2485-2492. https://doi.org/10.1177/0040517517723021
[15] Leonas K K 2017 The use of recycled fibers in fashion and home products *Text Clothing Sus* 55–77. https://doi.org/10.1007/978-981-10-2474-0
[16] Vadicherla T and Saravanan D 2015 Roadmap to sustainable textiles and clothing 139–160. https://doi.org/10.1007/978-981-287-164-0
[17] Preda C, Cîmpeanu C G, Harpa R and Racu C 2006 Geotextiles containing fibers recovered from reusable textile materials. Part I. Fibrous blends with fibers reclaimed from textile wastes *Ind Textila* **57** 104–110.
[18] Bhatia D, Sharma A and Malhotra U 2014 Recycled fibers: An overview *Int J. Fiber Text Res* **4** 77–82.
[19] Azrin Hani A, Ahmad R, Mariatti M, Hashim M and Lim T 2015 Impact behaviour of woven coir-epoxy composite: Effects of woven density and woven fabric treatment *Proceedings of the Institution of Mechanical Engineers, Part L: J Mater: Des Appl* **230** (1) 240–251. https://doi.org/10.1177/1464420714567744
[20] Ryan B, Joiner B and Cryer J 2012 Minitab handbook: Updated for release 16 sixth edition. *Books/Cole, Cengage Learning*
[21] Das D, Nonwoven Technology, http://nptel.ac.in/courses/ 116102014
[22] Gnaba I, Soulat D, Jaouachi B, Vroman P, Ferreira M, Wang P and Omrani F 2018 Mechanical behavior of flax/polypropylene commingled nonwoven at dry scale: Influence of process parameters *Text Res J* **89** (5) 791-800. https://doi.org/10.1177/0040517518755789
[23] Alan G and Tercan M 2017 A preliminary study on puncture resistances of top and bottom layers of multi-layered needlepunched nonwoven geotextiles *IOP Conference Series: Mater Sci Eng* **254** (19). https://doi.org/10.1088/1757-899X/254/19/192002
[24] Li T T, Wang R, Lou C W, Lin J Y and Lin J H 2014 Effects of needle-punching and thermo-bonding on mechanical and EMI shielding properties of puncture-resisting composites reinforced with fabrics *Fibers Polym* **15** (2) 315–321. https://doi.org/10.1007/s12221-014-0315-8

IOP Conf. Series: Materials Science and Engineering **670** (2019) 012036 doi:10.1088/1757-899X/670/1/012036

Performance of composite sand cement brick containing paddy husk

M H Ahmad, M N Arib, F Mat and U Kassim

School of Mechatronic Engineering, Universiti Malaysia Perlis, Pauh Putra Campus 02600 Arau, Perlis, Malaysia.

Corresponding author: *mnarib@unimap.edu.my

Abstract. This research concentrates on the performance of composite sand cement brick containing paddy husk also known as rice husk. The main objective of this research is to determine the mechanical properties possess by the composite sand cement brick containing rice husk, such as water absorption and compressive strength. The bricks were designed to contain 0- 4% of RH. The brick specimens were naturally sun dried and be cured for 28-days in protected space. The percentage of water absorbed increases from controlled specimen to 1% of RH and slightly decreases at 2% of RH.The value then increases significantly at 3% and 4% of RH. The optimum amount of the RH percentage in this research would be at 2% correspond to the water absorption of specimens. For the compressive strength, the results show steep decrease from controlled specimen to 1% of RH and began to increase from 2% till 3% of RH and significantly decrease again. The optimum amount of the RH percentage in this research would be at 3% of RH correspond to the compressive strength of specimens. Thus implies that the addition of rice husk into the brick causing the brick to have lower density, subsequently increase water absorption and decrease the compressive strength of the brick

1. Introduction

The oldest brick manufactured in ancient method by a process known as soft mud process where the condition of the brick was relatively moist and pressed into the moulds by hand [1]. Currently making the clay brick as the oldest and most durable construction technique invented by human [2]. The clay brick available in form of burnt or sun dried has been long ago been around in earlier civilization approximately 10 thousand years ago and considered as easy to produce and light weight than natural stone or rubble, easy to fix into shape and function as wall which is durable and strong against fire. Nowadays, all around the world many brickswere used as combination with other masonry material such as stone, roof or floors [2].

Block or brick is considered as very essential for every construction in the industry is generally used and requested exceptionally. While white brick, unfired brick also known as cement brick which is widely uses when constructing and remodel of houses and other small construction in Malaysia because of it handy to procure and not as high as other brick in term of pricing [3]. Currently, world agricultural sector growth are undergoing a very fast pace. However, the active expansion ofthe agro-based industry is accompanied by wastes. By salvaging fibers from predisposed leaves and husk to be integrate in masonry material can helps reducing wasteful landfill and harmful open incineration[4]. This research aims to determine the strength of the brick with respect to compressive strength and the water absorbed as the brick containing reinforcing material compared to normal brick.

2. Methodology

The mixtures were moulded into a mould with the size of 212 mm in length, 90 mm in width and 63 mm in depth. The mix design ratio of sand cement brick used is 1:4 of sand cement. The process of moulding the brick started after the all materials were mixed including water, then placing the mixture into the mould in 3 layers and been tampered for 25 times for each layer. At 24-hours period the brick specimens were removed from the mould to be labelled according to the mixture proportion and to be cured. The process of fabricating the bricks repeated for remaining bricks proportions. The process of curing takes place for 28 days after the bricks were removed from the mould. There were 15 fabricated specimensfor compression test and 5 specimens for water absorption test, respectively.

The objective of the compression and water absorption tests were to define the bricks mechanical properties. For compression test, 3 specimens were prepared according to the percentages of 0%, 1%, 2%, 3% and 4% rice husk with ratio of 1:4 while for the water absorption test, only 1 specimen prepared for each percentage of rice husk. The test conducted on the specimen was in duration of 28 days curing. The procedure for testing the compression strength test was consistent with MS EN 12390-3:2012 "Testing Hardened Concrete". The procedure for water absorption test followed BS 3921:1985.

3. Result and discussion

20 specimens were used to conduct the experimental test and elaborated from tabulated data and graphical form.

3.1. Water Absorption

This test was conducted to determine the permeability of the specimens. The percentage of water absorbed was calculated based on the mass weighed before and after the immersion in the water for 72-hours. The quantity of water absorbed by the specimens or the absorptivity of the specimens was determined by the water content ratio to the dry mass of the specimens. The amount of water absorbed by the specimens was influenced based on the percentages of rice husk present in the brick because the added rice husk forming the void of volume which the water sinks into. Both amount of water absorbed, and the void volume give effect to density.

From results in Figure 1, it depicted that the water absorption of the specimens containing rice husk increased significantly as the rice husk percentage added for all mixtures. The amounts increased for lowest and highest water absorption of specimen with rice husk mixture were almost 1.89 and 3.14 times the controlled specimen. However, at rice husk percentage of 2% the water absorbed decreased about 1.776% and begin to increase again about 2.169% at rice husk of 3% and eventually absorbing water for the most at 17.901% when the bricks specimen containing 4% of rice husk. The controlled specimen was obviously less permeable compared to other specimen, but among the specimen having a mixture of rice husk the specimen of 2% showed the less permeable as to another specimen. From the results it can be concluded that the present of rice husk in the constituent of brick was greatly affected the water absorption characteristic of a brick.

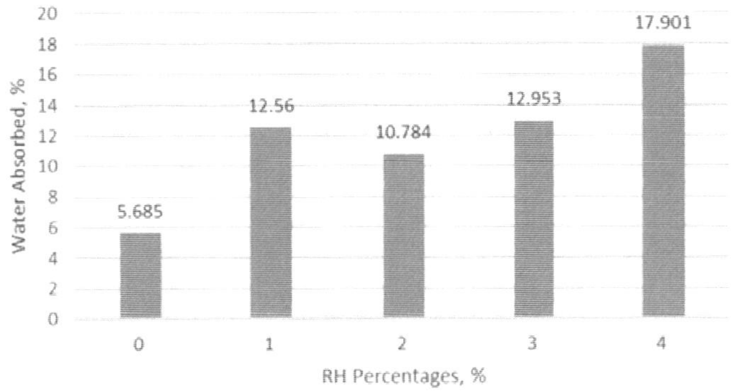

Figure1. Water absorbed against RH percentage.

3.2. Compressive Strength

This test was conducted to determine the strength of the specimen brick to withstand the compression under load with intend to failure. The load applied to the specimen was applied with no shock and increased continually until the failure and the maximum load was recorded. The compressive strength was obtained by dividing the maximum force with cross-sectional area. According to British Standard [5] the bricks should have load carrying capacity of more than 7 N/mm^2. The results of compressive strength of bricks specimens were depicted in Figure 2.From the figure, it can be seen that the brick specimens containing rice husk were inferior to the controlled brick, as the highest compressive strength of the composite brick specimens was 5.82 MPa which 2.15 times and about 63.2% less than strengths of the controlled brick with 15.802 MPa. Among the brick specimens containing rice husk, the brick with 4% was the lowest in term of strength compared to other specimens. The compressive strength of composite brick began to increase slightly from 1% RH brick with 4.384 MPa to 5.097 MPa of 2% RH brick and continually increased to 5.82 MPa of 3% RH brick which is the highest and from then it began to significantly decrease to 2.555 MPa at 4% RH brick.

The outcome from the experiment was unexpectedly low compared to controlled specimen. It is concluded that with addition of rice husk fiber to replace small amount of fine aggregate of sand of the 28-days cured white brick would lowered the compressive strength of the bricks. Low in compressive strength with presence of rice husk was attributed to the lack of rice husk to strengthen or enhance the cement paste as many large void spaces formed up and did not construct cementation links among the sand particle. In this case, the compressive strength decreased and no improvement in the cementation matrix [6,7,8].

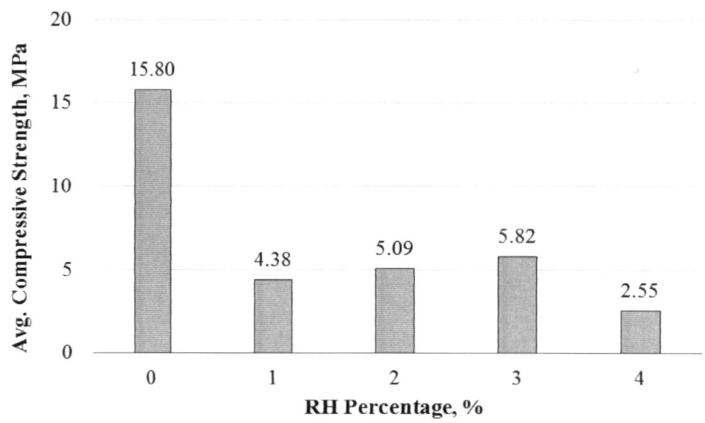

Figure 2. Average compressive strength against RH percentages.

4. Conclusion

In this research, the bricks specimen was fabricated using mould prepared to have the size of the brick available in the industry. The fabrication processes involving preparation of the material according to the mix design ratio, casting the bricks and cured for 28-days to obtain the constant mass. The rice husk used in the research was raw material without any treatment.

In this research, there are two primary objectives to be achieved which one of them was to determine the compressive strength of sand cement brick containing rice husk. The objective was successfully accomplished although the result was unexpectedly low compared to the controlled brick. The highest compressive strength recorded among the brick specimens was 5.82 MPa of 3% RH while the lowest was 2.55 MPa of 4% RH. The compressive strength was very low because of the density, shape, size and distribution of the rice husk in the cementation was coarse and not evenly distributed. Uneven distribution of the rice husk created the lumps of cement and leaves the matrix a void space and porous thus weaken the bricks. As the density decreased the compressive strength also decreased. The compressive strength shows a steep decrease from controlled specimen to 1% of RH and began to increase from 2% till 3% of RH and significantly decrease again. The optimum amount of the RH percentage in this investigation would be at 3% of RH regarding the compressive strength of specimens with RH.

The second objective was to determine water absorption of sand cement brick containing rice husk. The objective was successfully accomplished within the expectation. The highest recorded water absorption among the specimens with rice husk was 17.901% of 4% RH while the lowest was 10.784% of 2% RH. The overall result based on the graph of the water absorption. The percentage of water absorbed began to increase from controlled specimen to 1% of RH and slightly decreases at 2% of RH and increases significantly at 3% and 4% of RH. The water absorption was high because of the permeability of the rice husk and the pore of the brick and their void volume was larger sufficiently to absorb the water. The optimum amount of the RH percentage in this investigation would be at 2% of RH corresponding to the water absorption of specimens with RH.

References

[1] Shakir A A and Mohammed A A 2013 Manufacturing of bricks in the past, in the present and in the future: a state of the art review *Int. J. Adv. Appl. Sci.* **2** (3) 145–156.
[2] Lourenço P B, Fernandes F M and Castro F 2010 Handmade clay bricks: chemical, physical and mechanical properties *Int. J. Archit. Herit.* **4** (1) 38–58.
[3] Kubissa W, Jaskulski R, Koper A and Szpetulski J 2015 Properties of concretes with natural aggregate improved by RCA addition *Procedia Eng.* **108** 30–38.
[4] Chan C 2011 Effect of natural fibres inclusion in clay bricks : physico-mechanical properties *World Acad. Sci. Eng. Technol.* **5** (1) 7–13.
[5] British Standards Institution, "Specification for clay bricks," *Bs:En:3921*, no. November, 350, 1985.
[6] Gidde M R and Jivani A P 2007 Potential of rice husk in india a literature review *Waste to Wealth* 586–590.
[7] Kumar A, Mohanta K, Kumar D and Parkash O 2012 Properties and industrial applications of rice husk : a review *Int. J. Emerg. Technol. Adv. Eng.* **2** (10) 86–90.
[8] Yadav J and Singh P B R 2015 Study on comparison of boiler efficiency using husk and coal as fuel in rice mill *SAMRIDDHI A J. Phys. Sci. Eng. Technol.* **2**.

IOP Conf. Series: Materials Science and Engineering **670** (2019) 012037 doi:10.1088/1757-899X/670/1/012037

Thermal polymer composites of hybrid fillers

S S Samsudin[1], M S Abdul Majid[2], M J M Ridzuan[2] and A F Osman[1]

[1]School of Materials Engineering, Universiti Malaysia Perlis, Jejawi, 02600, Perlis
[2]School of Mechatronic Engineering, Universiti Malaysia Perlis, Pauh Putra Main Campus, 02600, Perlis

Abstract. Thermal polymer composite materials for heat dissipation purposes have been widely researched over the previous decade due to their ease of processing, electric insulating, low cost and lightweight. These materials, which embedded with high thermal conductive fillers in the polymer matrix, exhibit high electrical insulation and thermal conductivity properties. These materials, however, are very hard to manage because usually it contains more than 70% volume of filler in order to be able to rapidly dissipate heat. Several approaches have been conducted by different research groups to enhance heat transfer through polymers while decreasing its filler contents. This paper presents a systematic review on several approaches which are used for enhancing the thermal properties of polymer composites by incorporating various types of fillers. The primary goal of this review is to identify any future trends in thermal polymer composites that regulate and dictate the method of thermal conduction.

1. Introduction

Since the past half-century, the drive for quicker, cheaper computing and its long-related requirements of growing device density and progressive device miniaturisation have push researchers, scientists and engineers to continuously develop fresh and ever-improving materials, instruments, procedures and design methodologies. Today, as electronic products are chasing light, pocket-sized and large-scale integration, more troubles are raised by heat issues that occur within products.

Outstanding thermal conduction is a must in electronic package and countless industrial processes that involved with heat dissipation using heat sink. To efficiently dissipate heat, thermal composite materials must have excellent thermal conductivity through which heat transfer happens with low contact resistance. The use of thermally conductive material would only be cost-consuming if excellent thermal contacts failed to be produced.

Majority of polymer resins are considered as thermal insulators. Regardless of this disadvantage, polymers are favoured for use in multiple industrial applications including circuit boards in power electronics, heat exchangers, electronic appliances and machinery. This justifies the latest important attempts to overcome the constraints of traditional polymers in studies of thermally conductive composite materials from polymer based.

2. Thermal conductivity of polymers

Polymer usually have thermal heat transfer limitations owing to polymer conductivity properties that depending on many factors, such as the crystallinity, temperature, macromolecular orientation, bonding strength, side groups molecular weight, type and defects or structural defects and processing conditions [1-2]. Polymer crystallinity has a strong effect on thermal conductivity, which means the thermal conductivity of semi-crystalline polymer was reported to increases with the crystallinity while on the other hand, the amorphous polymer will demonstrate the lowest thermal conductivity. Table 1 lists the typical values for thermal conductivity of common polymers.

Table 1. Thermal conductivity of commercial polymers [2-4].

Polymer	Thermal Conductivity at 25°C (W/mK)
Low density polyethylene (LDPE)	0.28 – 0.32
High density polyethylene (HDPE)	0.38 – 0.58
Polypropylene (PP)	0.18 – 0.24
Polystyrene (PS)	0.13
Polyvinyl chloride (PVC)	0.21
Polymethyl methacrylate (PMMA)	0.25
Nylon (N6,6)	0.30
Polyether ether ketone (PEEK)	0.30
Polytetrafluoroethylene (PTFE)	0.27
Phenol resin	0.24 – 0.29
Epoxy resin	0.17 – 0.21

A new, simple and cost-effective way of improving thermal dissipation from heat source is provided by thermally conductive polymer composites. As regards excellent corrosion resistance, light and processability, thermal polymer composites possess a great deal of potential as compared to traditional thermal conductive metals [5].

3. Thermal conductivity of conductive fillers

Hence the production of a high thermal conductivity polymer-based material involves the introduction of high thermal conductivity filler particles embedded into the polymer matrix. The thermal conductivity of fillers is primarily determined by the way how heat transfers in filler particles. Table 2 lists the thermal conductivity of frequently used thermally conductive fillers.

Table 2. Thermal conductivities of some thermally conductive fillers [6].

Fillers	Category	Thermal Conductivity (W/(mK)
Aluminum	Metal	234
Copper	Metal	386
Silver	Metal	427
Carbon nanotube (CNT)	Carbon	1000–4000
Carbon fibre	Carbon	300–1000
Graphene	Carbon	2000–6000
Graphite	Carbon	100–400
β-Silicon nitride (β-Si_3N_4)	Ceramics	103–200
Hexagonal boron nitride (h-BN)	Ceramics	185–300
Aluminum nitride (AlN)	Ceramics	100–300
Diamond	Ceramics	1000
β-Silicon carbide (β-SiC)	Ceramics	120
α-Alumina (α-Al_2O_3)	Ceramics	30

High filler loadings (>70 vol. %) are normally required to attain the suitable value of thermal conductivity in thermally conductive polymer composites, which represents a significant processing challenge. Indeed, the processing capabilities such as the possible extrusion and injection moulding often limit the quantity of fillers in the formulation and thus the efficiency of thermal conductivity [2]. The usual restriction of composite materials is the maximum packing fraction, which mainly caused problems in terms of processability, mechanical performance characteristics and the case of thermally conductive composite materials is the ability to dissipate heat rapidly and successfully.

4. Thermal polymer composites of hybrid fillers

Therefore, having hybrids compositions of filler is essential for the preparation of the next generation of heat dissipation materials since hybrid fillers system can increase the packing density, lower the system viscosity as well as enhancing the thermal conductivity of the composites. Hybrid filler can be tailored by adding several mixtures of filler with different sizes, shapes, types and dimensionality (1D, 2D or 3D) into polymer-based composites. The effectiveness of hybrid filler consisting of different shapes, types, sizes and dimensional was reported by earlier researchers in their project [7-30] and lists in Table 3.

Table 3. Summary results in hybrid filler for thermal conductivity polymer composites.

Filler	Matrix	Filler Fraction	TC (W/mK)	Year	Ref
MWCNT/ nano SiC	Epoxy	30 vol%	2.000	2010	[7]
MWCNT/ micro SiC	Epoxy	5.0/55 wt%	7.000	2010	[8]
MWCNT@SiO_2	Epoxy	1.0 wt%	0.213	2011	[9]
BN/MWCNT	Epoxy	30/1.0 vol%	1.912	2011	[10]
MWCNT/MGP	Epoxy	0.1/0.9 wt%	0.320	2011	[11]
GO/MWCNT	Epoxy	49.36/0.64 wt%	4.700	2012	[12]
MWCNT@MgO	Epoxy	2.0 wt%	0.290	2013	[13]
RGO@SiO_2	Epoxy	1.0 wt%	0.322	2013	[14]
MWCNT/GO	PEEK	0.45/1.0 wt%	0.440	2013	[15]
MWCNT/GNP	Epoxy	5.0 wt%	0.500	2013	[16]
Al_2O_3@GSs	PVDF	40 wt%	0.586	2013	[17]
GNP/micro SiC	Epoxy	5.0/55 wt%	7.060	2013	[18]
SiO_2@RGO	Epoxy	30 wt%	0.452	2014	[19]
SiO_2@MWCNT	PU	1.0 wt%	0.310	2014	[20]
BN@MWCNT	PI	3.0 wt%	0.388	2014	[21]
BN/C nano fiber	Epoxy	6.0/8.0 wt%	0.600	2015	[22]
Graphene/Al_2O_3	Silicone	1.0 wt-%/63 vol%	3.450	2015	[23]
SiC nanowires/GSs	PI	7.0 wt%	0.580	2015	[24]
Alumina fiber/GNP	Epoxy	50 vol%	1.620	2015	[25]
BN/MWCNT	PE	50/1.0 wt%	1.641	2016	[26]
Al_2O_3@GNP	Epoxy	12 wt%	1.490	2016	[27]
BNNs/ionic liquid	PMMA	34 vol%	7.300	2017	[28]
Al_2O_3/BT	PP	10 vol%	0.900	2018	[29]
Graphite/GNP	Epoxy	0.16/ 6.0 vol%	~2.500	2019	[30]

Generally, almost all the investigations performed prepared these composites via solution mixing [7-25,27-28] as well as moulding for thermoplastic matrix [26,29] and high-speed mixing by planetary centrifugal mixer [30]. On the other hand, these researches [7-21,23-25,27-28] were favoured by functionalized fillers and surface modifications on the fillers, with the exclusion of these researches [22,26,29-30], where the fillers as received from the supply were used. Most scientists have utilized surfactants such as coupling agents for a better interaction of fillers and matrix in order to reduce thermal interfacial resistance between matrix and fillers that will enhance thermal conductivity [12]. Interface thermal resistance seems to decrease and suppressed phonon scattering when surface treated fillers are used. Comparing the results between composites with and without surface treated particles, the former usually demonstrate greater thermal conductivity than the latter. Most of the modified fillers with coupling agent [7-8,10,17-19,25] showed that chemical surface treatment is conducive to the enhancement of thermal conductivity of polymer composites.

For carbon fillers; multiwalled carbon nanotubes (MWCNT) [7-13, 15, 16, 20-21] and graphene nanoplatelets/nanosheets (GNP/GNS) [15], it was reported that the functionalization of the fillers increases interfacial heat transport between the epoxy matrix and the fillers compare to the non-functionalization fillers and facilitates better dispersion of fillers in the matrix, which is beneficial to the enhancement of the thermal conductivity of the composites. In the event of the carbon nanotubes (CNT), Zhou [8] in his study revealed that multiwalled carbon nanotubes (MWCNT) demonstrated the greatest potential for effective thermal conductivity improvement of epoxy compared with single-walled carbon nanotubes (SWCNT) and double-walled carbon nanotubes (DWCNT) due to their relatively lower surface area, higher aspect ratio and better dispersion in an epoxy matrix.

Some of the hybrid fillers [9,13,17,19-21,27-28], however, were designed to thermally conduct but electrically insulate by adapting the coating technique. The highly thermally conductive filler (reduced graphene oxide (RGO), MWCNT and GNP/GNS) will be coated with high electrical insulation filler (silicon dioxide (SiO_2), magnesium oxide (MgO), aluminum oxide (Al_2O_3) and boron nitride (BN). This excellent characteristic makes coating technique a powerful candidate for thermal management in electronic packaging and electronic devices. According to Zhao et al. [20], abundant hydroxyl (-OH) groups on the surface of the SiO_2 layer create outstanding compatibility with the polyurethane (PU) matrix, enhancing the interfacial interaction between PU and MWCNTs and decreasing its thermal resistance.

As studied by several researchers [10-12, 24-25,30], the utilization of high aspect ratio filler in tubes, flakes, platelets, sheets, wires and fibres structures also contribute to enhance the thermal conductivity of the composites by increasing the number of heat transfer pathways. The pathways were construct by a long and tortuous bridge that resulting in an increased contact surface area between the hybrid fillers and the matrix. This long bridge will inhibit any filler aggregation leading to reducing the phonon scattering effect at the interface by building a great thermal transfer pathway and effectively reduce interfacial thermal resistance in the composites.

According to Ren et al. [26] and Yao et al. [29], their studies indicated that morphology and fillers size are an important factor in the formation of connected filler networks, which affects the thermal conductivity of the composites. Morphology analysis of the fillers and composites discovered that the variance in size of the 0D fillers is more conducive than 2D fillers towards the formation of effective thermally conductive pathways. 2D fillers was more easily intertwined with other filler and created more compact thermally conducive networks. This enhancement was achieved by the powerful synergistic work of fillers of varying sizes, their compact bond and directional arrangement, which enable elevated thermal conductivity with low filler loading.

5. Conclusions

In summary, thermally conductive polymer composites by using hybrid fillers have been reviewed in this paper. Various fillers produce synergistic impacts, which enhance the efficient thermal conductivity above what models would expect. The synergistic impacts are most frequently based on variation in filler size, which can assist to form percolating networks and enhance their packing ratios, or variation in aspect ratios wherein high aspect ratio fillers will created long bridge networks. The thermal conductivity of hybrid polymer composites is significantly higher than that of polymer composites using any single filling scheme; hybrid filling systems can therefore simultaneously achieve a synergistic impact and reduce costs.

Acknowledgements

The authors would like to acknowledge the Ministry of Education (MOE), Malaysia, for financial support via the Fundamental Research Grant Scheme (Ref: FRGS/1/2019/TK05/UNIMAP/02/11).

References

[1] Li A, Zhang C and Zhang Y F 2017 Thermal conductivity of graphene-polymer composites: Mechanisms, properties, and applications *Polymers* **9** (9) 437.

[2] Han Z and Fina A 2011 Thermal conductivity of carbon nanotubes and their polymer nanocomposites: A review *Progress in polymer science* **36** (7) 914-44.

[3] Tsekmes I A, Kochetov R, Morshuis P H and Smit J J 2013 Thermal conductivity of polymeric composites: A review *2013 IEEE International Conference on Solid Dielectrics (ICSD)* 678-681.

[4] Vadivelu M A, Kumar C R and Joshi G M 2016 Polymer composites for thermal management: a review *Composite Interfaces* **23** (9) 847-72.

[5] Hong J H, Park D W and Shim S E 2010 A review on thermal conductivity of polymer composites using carbon-based fillers: carbon nanotubes and carbon fibers *Carbon letters* **11** (4) 347-56.

[6] Chen H, Ginzburg V V, Yang J, Yang Y, Liu W, Huang Y, Du L and Chen B 2016 Thermal conductivity of polymer-based composites: Fundamentals and applications *Progress in Polymer Science* **59** 41-85.

[7] Yang K and Gu M 2010 Enhanced thermal conductivity of epoxy nanocomposites filled with hybrid filler system of triethylenetetramine-functionalized multi-walled carbon nanotube/silane-modified nano-sized silicon carbide *Composites Part A: Applied Science and Manufacturing* **41** (2) 215-21.

[8] Zhou T, Wang X, Liu X and Xiong D 2010 Improved thermal conductivity of epoxy composites using a hybrid multi-walled carbon nanotube/micro-SiC filler *Carbon* **48** (4) 1171-6.

[9] Cui W, Du F, Zhao J, Zhang W, Yang Y, Xie X and Mai Y W 2011 Improving thermal conductivity while retaining high electrical resistivity of epoxy composites by incorporating silica-coated multi-walled carbon nanotubes *Carbon* **49** (2) 495-500.

[10] Teng C C, Ma C C, Chiou K C, Lee T M and Shih Y F 2011 Synergetic effect of hybrid boron nitride and multi-walled carbon nanotubes on the thermal conductivity of epoxy composites *Materials Chemistry and Physics* **126** (3) 722-8.

[11] Yang S Y, Lin W N, Huang Y L, Tien H W, Wang J Y, Ma C C, Li S M and Wang Y S 2011 Synergetic effects of graphene platelets and carbon nanotubes on the mechanical and thermal properties of epoxy composites *Carbon* **49** (3) 793-803.

[12] Im H and Kim J 2012 Thermal conductivity of a graphene oxide–carbon nanotube hybrid/epoxy composite *Carbon* **50** (15) 5429-40.

[13] Du F P, Tang H and Huang D Y 2013 Thermal conductivity of epoxy resin reinforced with magnesium oxide coated multiwalled carbon nanotubes *International Journal of Polymer Science* **2013**.

[14] Hsiao M C, Ma C C, Chiang J C, Ho K K, Chou T Y, Xie X, Tsai C H, Chang L H and Hsieh C K 2013 Thermally conductive and electrically insulating epoxy nanocomposites with thermally reduced graphene oxide–silica hybrid nanosheets *Nanoscale* **5** (13) 5863-71.

[15] Hwang Y, Kim M and Kim J 2013 Improvement of the mechanical properties and thermal conductivity of poly (ether-ether-ketone) with the addition of graphene oxide-carbon nanotube hybrid fillers *Composites Part A: Applied Science and Manufacturing* **55** 195-202.

[16] Safdari M and Al-Haik M S 2013 Synergistic electrical and thermal transport properties of hybrid polymeric nanocomposites based on carbon nanotubes and graphite nanoplatelets *Carbon* **64** 111-21.

[17] Qian R, Yu J, Wu C, Zhai X and Jiang P 2013 Alumina-coated graphene sheet hybrids for electrically insulating polymer composites with high thermal conductivity *Rsc Advances* **3** (38) 17373-9.

[18] Zhou T, Wang X, Cheng P, Wang T, Xiong D and Wang X 2013 Improving the thermal conductivity of epoxy resin by the addition of a mixture of graphite nanoplatelets and silicon carbide microparticles *Express Polym Lett* **7** (7) 585-94.

[19] Huang L, Zhu P, Li G, Lu D D, Sun R and Wong C 2014 Core–shell SiO_2@ RGO hybrids for epoxy composites with low percolation threshold and enhanced thermo-mechanical properties *Journal of Materials Chemistry* A **2** (43) 18246-55.
[20] Zhao J, Du F, Cui W, Zhu P, Zhou X and Xie X 2014 Effect of silica coating thickness on the thermal conductivity of polyurethane/SiO_2 coated multiwalled carbon nanotube composites *Composites Part A: Applied Science and Manufacturing* **58** 1-6.
[21] Yan W, Zhang Y, Sun H, Liu S, Chi Z, Chen X and Xu J 2014 Polyimide nanocomposites with boron nitride-coated multi-walled carbon nanotubes for enhanced thermal conductivity and electrical insulation *Journal of Materials Chemistry* A **2** (48) 20958-65.
[22] Raza M A, Westwood A V, Stirling C and Ahmad R 2015 Effect of boron nitride addition on properties of vapour grown carbon nanofiber/rubbery epoxy composites for thermal interface applications *Composites Science and Technology* **120** 9-16.
[23] Yu W, Xie H, Yin L, Zhao J, Xia L and Chen L 2015 Exceptionally high thermal conductivity of thermal grease: synergistic effects of graphene and alumina *International Journal of Thermal Sciences* **91** 76-82.
[24] Dai W, Yu J, Liu Z, Wang Y, Song Y, Lyu J, Bai H, Nishimura K and Jiang N 2015 Enhanced thermal conductivity and retained electrical insulation for polyimide composites with SiC nanowires grown on graphene hybrid fillers *Composites Part A: Applied Science and Manufacturing* **76** 73-81.
[25] Zha J W, Zhu T X, Wu Y H, Wang S J, Li R K and Dang Z M 2015 Tuning of thermal and dielectric properties for epoxy composites filled with electrospun alumina fibers and graphene nanoplatelets through hybridization *Journal of Materials Chemistry* C **3** (27) 7195-202.
[26] Ren P G, Si X H, Sun Z F, Ren F, Pei L and Hou S Y 2016 Synergistic effect of BN and MWCNT hybrid fillers on thermal conductivity and thermal stability of ultra-high-molecular-weight polyethylene composites with a segregated structure *Journal of Polymer Research* **23** (2) 21.
[27] Sun R, Yao H, Zhang H B, Li Y, Mai Y W and Yu Z Z 2016 Decoration of defect-free graphene nanoplatelets with alumina for thermally conductive and electrically insulating epoxy composites *Composites Science and Technology* **137** 16-23.
[28] Morishita T and Takahashi N 2017 Highly thermally conductive and electrically insulating polymer nanocomposites with boron nitride nanosheet/ionic liquid complexes *RSC Advances* **7** (58) 36450-9.
[29] Yao J, Hu L, Zhou M, You F, Jiang X, Gao L, Wang Q, Sun Z and Wang J 2018 Synergistic Enhancement of Thermal Conductivity and Dielectric Properties in Al_2O_3/$BaTiO_3$/PP Composites *Materials* **11** (9) 1536.
[30] Levy I, Wormser E M, Varenik M, Buzaglo M, Nadiv R and Regev O 2019 Graphene–graphite hybrid epoxy composites with controllable workability for thermal management *Beilstein journal of nanotechnology* **10** (1) 95-104.

IOP Conf. Series: Materials Science and Engineering **670** (2019) 012038 doi:10.1088/1757-899X/670/1/012038

Receding height of water droplet in the film boiling regime

S Illias[1,a], S Hussain[1], N A Rosman[1], N S Abdullah[1], A I M Shaiful[1], M N B Omar[1], K A Ismail[1] and H Ani[2]

[1]School of Manufacturing Engineering, Universiti Malaysia Perlis, Pauh Putra Campus, 02600 Arau, Perlis, Malaysia.
[2]Department of Manufacturing and Materials, Kulliyyah of Engineering, International Islamic University Malaysia, PO Box 10, 50728 Kuala Lumpur, Malaysia.

Abstract. The purpose of this research was to study and investigate the receding height of water droplet after it impinged on hot aluminum surface. On top of that, we also investigated the relationship between this receding phenomenon and residence time of water droplet. Degassed and distilled water was used as a droplet. The droplet diameter was maintained at 4.0 mm and the impact height was set to be around 65.0 mm corresponding to impact velocity of 1.129 m/s. The liquid temperature was kept around 16 °C corresponding to liquid subcooling of ΔT_{sub} = 84 K. The material selected in this experiment was aluminum surface. The surface had been polished until it became a mirror polished surface. The dimensions of the material surface were 30.0 mm in diameter and 8.0 mm in height. The phenomena were recorded at 10,000 frames per second using a high speed video camera. As a result, it was observed that the receding height during film boiling regime was almost similar to each other even the surface temperature was different. Furthermore, the residence time during the experiment was also close to the theoretical calculation.

1. Introduction

Drop impact research is widely conducted around the world as an experimental work to study the complicated evaporation characteristics and boiling phenomena. For decades, through droplet impact [1-2] and boiling research, a lot of new data and findings have been successfully gained by engineers and scientist that led to an improvement in thermal engineering system and application. On top of that, boiling studies also are important for heavy industrial area such as nuclear power plant [3], boiler, industrial chillers and other thermal application system that needs an accurate reading of temperature, stable temperature fluctuation and stable cooling performance. Basically, there are three (3) types of boiling mode in heat transfer. The first one is nucleate boiling [4] where generation of steam is actively being generated during liquid-solid contact. In nuclear power plant, this steam generation studies is important to enhance the capability of steam generator in order to spin the turbine. The second one is transition boiling [5-6] where it is well known as the most difficult topics in boiling heat transfer study. During transition boiling regime, the surface temperature increases while the cooling performance reduces. Meanwhile, when the boiling conditions shift to a much higher surface temperature, the transition boiling will shift to film boiling [7-8]. During film boiling, the cooling process is at the lowest rate due to the vapor film formation during the liquid solid contact. This vapor film [9] hinders the heat transfer from the solid to the liquid. Therefore, in order to increase our understanding regarding film boiling, we need to further our studies especially in maximum spreading [10], receding height, residence time, bouncing phenomena, effects of surface roughness and many others. There are a few numbers of researchers that give their commitment in order to study the boiling and evaporation phenomena. For instance, Fujimoto et al. [11] investigated the collision of single water droplets with a hot Inconel 625 alloy surface examined by a two-directional flash photography technique using two digital still cameras and three flash units. The experiments were conducted under the following conditions: the pre-impact diameters of the droplets ranged from 0.53 to 0.60 mm, the impact velocities ranged from 1.7 m/s to 4.1 m/s, and the solid surface temperatures ranged from 170 °C to 500 °C. As a result, they found that at surface temperatures of less than or equal to 300 °C, the blowout of vapor bubbles occurred at early stages for a large droplet. At a surface temperature of 500 °C, the two dimensionless deformation behaviors of the droplets were very similar to each other. Meanwhile, Kim et al. [12] studied the effect of surface roughness height, wettability and nano-porosity on Leidenfrost phenomena. Their study provides interesting and very meaningful findings for other drop impact research. Wang et al. [13] conducted experimental and theoretical investigations regarding maximum spreading factor of an impacting droplet onto solid surfaces with contact angle hysteresis. Experiments were conducted with deionized water on six surfaces with different wettability. Other researcher such as Illias et al. [14-15] also still continue their work in order to study this complicated phenomena especially critical heat flux and Leidenfrost phenomena.

In this paper, we focus our study in the film boiling regime. For the test liquid and heated surface, distilled water and aluminum material were selected during the experimental work. The material dimensions were 30.0 mm in diameter and 8.0 mm in height. We investigated the receding height of water droplet after it impinges with hot aluminum surface in the film boiling regime. On top of that, we also investigated the residence time of the water droplet and compared the experimental result with theoretical calculation. The experimental work was recorded using a high speed video camera at the frame rate of 10,000 fps. As a result, the receding pattern of water droplet seemed to have a similarity among each other even though the temperature is different. Furthermore, the residence time agreed closely with the theoretical calculation.

2. Experimental apparatus and procedure

Figure 1 illustrates a schematic diagram of the experimental apparatus. The experimental apparatus consisted of high speed imaging system, thermal recording device, droplet dispenser and heated surface. The heated surface is a material disk made of aluminum which has a diameter of 30.0 mm and 8.0 mm in height. The material was retained levelly and was heated from its periphery by an electric cartridge-type heater. Degassed and distilled water was used as a droplet. The water droplet diameter was retained 4.0 mm, and falling height was kept at 65.0 mm. Theoretically, the impact speed is approximately 1.129 m/s. This height is in the range that the droplet itself does not disintegrate by the collision energy of the droplet. The temperature of the droplet was kept at 16 ºC by the circulation of the tab water in the circumference of the nozzle. The temperature of the heating surface was measured by attaching the thermocouple with ceramic adhesive at two points on the surface. The droplet impact, receding phenomena and other boiling behavior was photographed in real time and was recorded by a frame rate of 10,000 fps.

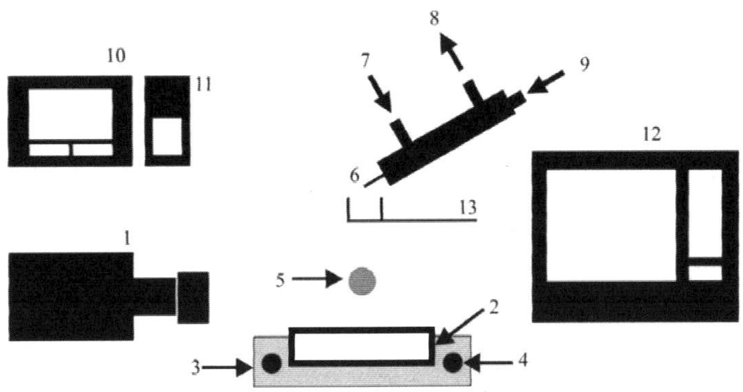

Figure 1: Experimental apparatus

1: High speed video camera 2: Material disc (aluminum) 3. Copper block 4. Cartridge heater 5. Water droplet 6. Droplet dispenser 7. Water-in 8. Water-out 9. Degassed and distilled water 10. TV Monitor 11. Monitor controller 12. Pen recorder 13. Shutter

3. Related formula

The related formula for residence time is shown in Eq. (1) where ρ, v, D_o and σ are the liquid density, impact velocity of droplet, initial diameter, and the surface tension force, respectively. From the theoretical calculation, the residence time τ_r, for this experiment was approximately 23.0 ms.

$$\tau_r = \left(\frac{\pi}{4}\right)\sqrt{\frac{\rho Do^3}{\sigma}} \qquad (1)$$

4. Result and discussions

Figure 2 shows high speed images of water droplet during impact on hot aluminum surface at the surface temperature of T_w = 320 °C in the film boiling regime. The numerical value under each photograph was the elapsed time after the droplet collision. The initial height for the droplet was approximately 4.0 mm. From Figure 2, it is observed that when the droplet impinged on the hot surface, the droplet height started to decrease slowly as time increased. As shown in Figure 2 (b) at 2.0 ms, the droplet height decreased and it formed a hat shape like on the surface. Interestingly, there were no liquid dispersion detected during the liquid-solid contact and this phenomena represented the film boiling regime. It was seemed the water droplet kept spreading on the hot surface until it reached its maximum spreading diameter at approximately 6.0 ms . Then, the water droplet started to recede back and close to its original diameter. The high speed images of this receding process are shown in Figure 2 (g) – (k). Finally, the water droplet seemed to bounce from the hot surface as shown in Figure 2 (m) at approximately 23.0 ms. This unique bouncing phenomenon normally occurred only in the film boiling regime at a very high surface temperature. From the scientific point of views, during the liquid-solid contact, a very thick vapor cushion or vapor film was form between the liquid and solid surface. This vapor film hindered the heat transfer from the solid to the liquid. This was the main reason why the cooling performances drop dramatically during film boiling regime. The bouncing phenomenon on aluminum heated surface is clearly shown in Figure 2 (m) – (p).

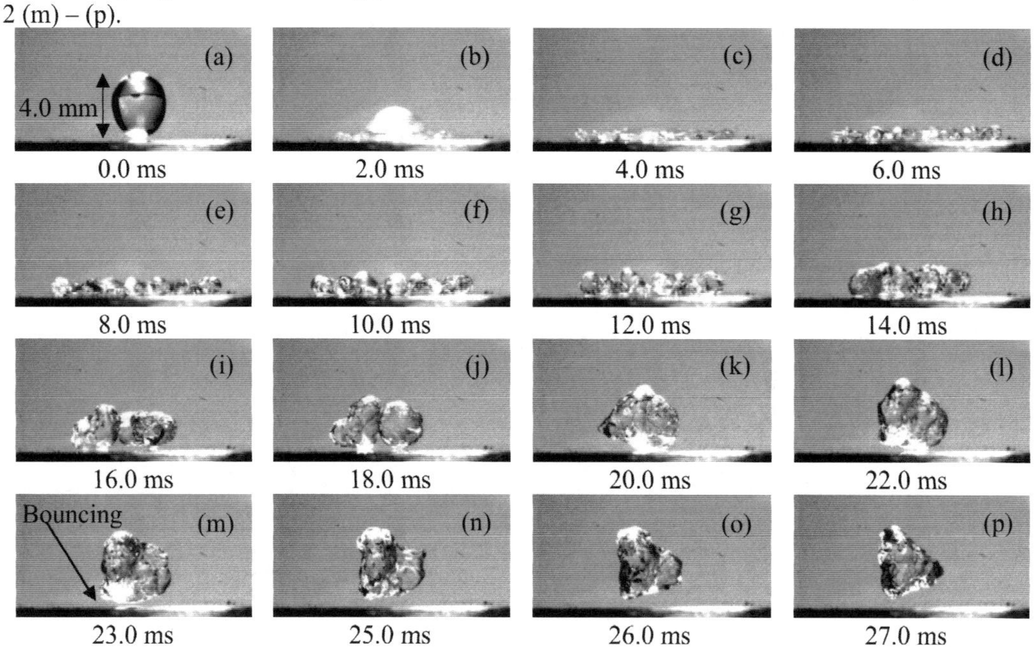

Figure 2: Film boiling phenomena at the surface temperature of T_w = 320°C

Meanwhile, Figure 3 shows the relationship between receding heights at three (3) different surface temperatures in the film boiling regime (T_w = 320, 340 and 360 °C). From Figure 3, it was observed that the receding height of a water droplet seemed to have a similar pattern even the surface temperature was different. Upon impact on the hot surface, the droplet height seemed to reduce slowly and there were no liquid dispersion occurred during the liquid-solid contact. During the first impact, a thick vapor layer was form between the liquid and the hot surface. As the surface temperature of T_w = 320 °C, droplet height seemed to reach it minimum reading of 0.51mm at approximately 5.0ms. Meanwhile, the minimum droplet height for surface temperature of T_w = 340 and 360 °C were approximately 0.59mm and 0.56mm, respectively. Interestingly, all minimum height occurred at same time of 5.0 ms even the surface temperature was different. On top of that, all data showed that the minimum height occur almost the same period when the maximum spreading took place on the hot surface. After the droplet reached the minimum height, the droplet height increased again and the pattern could be seen in Figure 3. For surface temperature of T_w = 320 °C, the receding height of droplet seemed to increase again to its original diameter before it was bounced off from the hot surface as shown in Figure 2. The bouncing phenomena took place at approximately 23.0 ms. This bouncing phenomena occurred due to the vapor pressure that was released from the hot surface. On the other hand, for surface temperature of T_w = 340 and 360 °C, the droplet height seemed to increase dramatically. The image of droplet became out of range and cannot be used for scientific calculation. Therefore, the black arrows in Figure 3 shows the time when the droplet became out of range during the high speed image analysis. From Figure 3 also, it is observed that the residence time recorded was 23.0, 22.8 and 23.6 ms for surface temperature of T_w= 320, 340 and 360 °C, respectively. For easy reference, the residence time for all surface temperatures was shown in Figure 3 by broken dashed line. From Figure 3, it can be concluded that the droplet receding height occurred almost at the same time of 5.0 ms even though the surface temperature was different. Thus, it can be said that the surface temperature did not give any effect to the receding phenomena of the droplet. The measured residence time also agreed closely with the theoretical calculation.

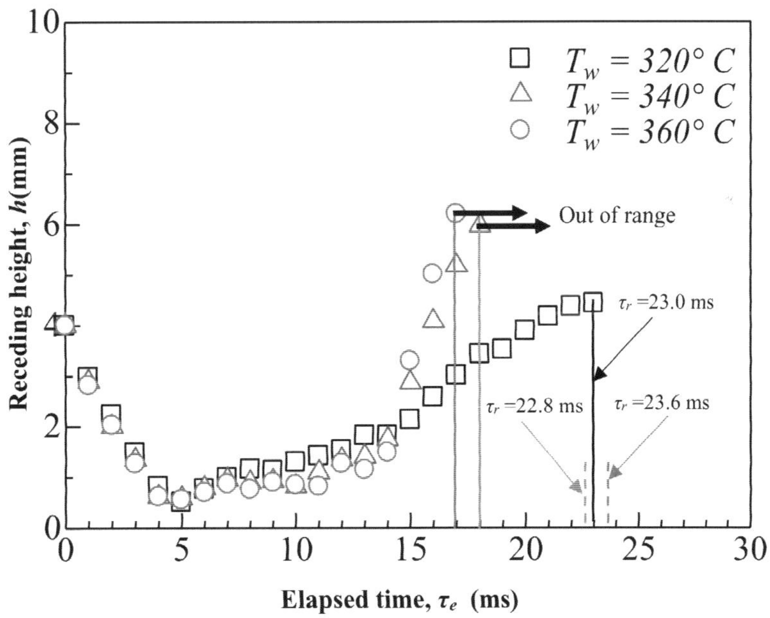

Figure 3: Relationship between receding height and elapsed time in the film boiling regime

5. Conclusion

From the experimental work, it could be concluded that the droplet receding height occurred almost at the same time which was about 5.0 ms even though the surface temperature was different. Thus, it could be said that the surface temperature did not give any effect to the receding phenomena of the droplet and the residence time agreed closely with the theoretical calculation.

Acknowledgments

The author would like to acknowledge the support from the Fundamental Research Grant Scheme (FRGS) under grant number FRGS/1/2018/TK03/UNIMAP/02/11.

References

[1] Guo C, Zhao D, Sun Y, Wang M and Liu Y 2018 Droplet impact on anisotropic superhydrophobic surfaces *Langmuir* **34** (11) 3533-3540.

[2] Abolghasemibizaki M and Mohammadi R 2018 Droplet impact on superhydrophobic surfaces fully decorated with cylindrical macrotextures *Journal of colloid and interface science* **509**, 422-431.

[3] Fujisawa K, Yamagata T and Fujisawa N 2018 Damping effect on impact pressure from liquid droplet impingement on wet wall *Annals of Nuclear Energy* **121** 260-268.

[4] Kim J M, Kong B, Wongwises S and Ahn H S 2018 Effect of h-BN coating on nucleate boiling heat transfer performance in pool boiling *Experimental Thermal and Fluid Science* **98** 12-19.

[5] Ma X and Cheng P 2019 Numerical simulation of complete pool boiling curves: from nucleation to critical heat flux through transition boiling to film boiling *Nuclear Science and Engineering* **193** (1-2) 1-13.

[6] Pandey V, Biswas G and Dalal A 2018 Effect of surface wettability and electric field on transition of film boiling to nucleate boiling *Numerical Heat Transfer, Part A: Applications* **74** (3) 1105-1120.

[7] Avedisian C T, Kuo W C, Tsang W and Lowery A 2018 A film boiling study of ethanol pyrolysis *Industrial & Engineering Chemistry Research* **57** (24) 8334-8340.

[8] Wang Q and Chen R 2018 Ultrahigh flux thin film boiling heat transfer through nanoporous membranes *Nano letters* **18** (5) 3096-3103.

[9] Beck F R, Virani N, Mohanta L, Sohag F A, Ray A and Cheung F B 2018 Study of vapor film dynamics and heat transfer through an image processing technique *International Journal of Heat and Mass Transfer* **125** 1310-1320.

[10] Gordillo J M, Riboux G and Quintero E S 2019 A theory on the spreading of impacting droplets *Journal of Fluid Mechanics* **866** 298-315.

[11] Fujimoto H, Oku Y, Ogihara T and Takuda H 2010 Hydrodynamics and boiling phenomena of water droplets impinging on hot solid *International Journal of Multiphase Flow* **36** (8) 620-642.

[12] Kim H, Truong B, Buongiorno J and Hu L W 2011 On the effect of surface roughness height, wettability, and nanoporosity on Leidenfrost phenomena *Applied Physics Letters* **98** (8) 083121.

[13] Wang F, Yang L, Wang L, Zhu Y and Fang T 2019 Maximum spread of droplet impacting onto solid surfaces with different wettabilities: Adopting a rim-lamella shape *Langmuir*.

[14] Illias S, Ishak M S A, Hussain S and Ismail K A 2016 High speed visualization and analysis of maximum spreading of water droplet during impact on hot horizontal surface *International Journal of Applied Engineering Research* **11** (22) 10832-10837.

[15] Illias S, Hasan M N, Mitsutake Y and Monde M 2014 High speed observation and measurement of surface temperature and surface heat flux during impact of a droplet on hot surface *Proc. of 15th International Heat Transfer Conference (IHTC-15)*, Kyoto, Japan.

IOP Conf. Series: Materials Science and Engineering **670** (2019) 012039 doi:10.1088/1757-899X/670/1/012039

Analysis of insertion and removal torque loading for miniscrew and miniplate skeletal anchorage

Z H Khor, R Daud*, N S Kamarrudin, I Ibrahim, F Mat and K S Basaruddin

Mechanical Engineering, School of Mechatronic Engineering, Universiti Malaysia Perlis, Arau, Perlis, Malaysia

ruslizam@unimap.edu.my

Abstract. Mini-Implant such as miniscrew and miniplate is used as a stabilizer in stable the crack occurs at the mandibular bone and maxillary bone for recovery. Reducing the screw loosening may reduce the trauma of insertion and removal that able to increase the rates of osseointegration. This paper presents the investigation of stress transfer between the interaction mini-implant and bone during applied in insertion and removal torque regards to stress distribution between mini-implant and bones. The finite element simulation involves three different materials mini-implant such as Titanium Alloy, (TiA), Stainless Steel, (SS) and Magnesium Alloy, (MgA). The result identified that the best material of mini-implant used to insert at mandibular bone is Titanium, (TiA) with the moment of 90 Nmm to perform high 0.70 STP value and average SEDTP of 2.60 compared to other materials. Whereas, for the material of mini-implant used to inserted at maxillary bone is Magnesium Alloy, (MgA) with the moment of 100 Nmm to perform high 1.11 STP value and average SEDTP of 1.20.

1. Introduction

Skeletal anchorage device that used in orthodontic treatment is to stabilize the moveable teeth, realign the teeth to correct position or connect the fracture in either maxillary and mandibular bones [1]. Fracture of maxillary and mandibular bone may cause by the force applied over the force that can withstand by the bone such as transport coincidence, fall from staircases, impact with a strong moving object such as stone [2]. The insertion torque and removal torque is showing the different moment applied will cause different stress distribution that affects the changes of mechanical stability between the mini-implant and the bones. Stability of the mini implant at the inserted site can be categorized into two aspects such as primary stability and secondary stability [3-4]. Primary stability which is known as mechanical stability of the mini-implant after inserted into the bone. While Secondary stability is defined as the biological stability in the regeneration and remodeling occurs in the bone after some period the insertion had done. In order to achieved high primary stability aspect after insertion, we have to identify the optimal position site for the insertion of mini-implant and the mechanical factors that cause the instability of inserted mini-implant parameters of the mini-implant and torque applied [5-7].

2. Finite Element Modeling

This paper presents the 3D simulation of mini-implant and bones. The full model of mini-implant and bones were generated by SolidWorks 2018 and the finite element analysis were conducted using

ANSYS Workbench v15.0. The stress transfer parameter (STP) and shear stress transfer parameter (SEDTP) were calculated by applied Wolff's law. Then, the von Mises stress on the screw thread and von Mises stress at the bone between miniscrews were evaluated. Figure 1 shows the detail of simulation works.

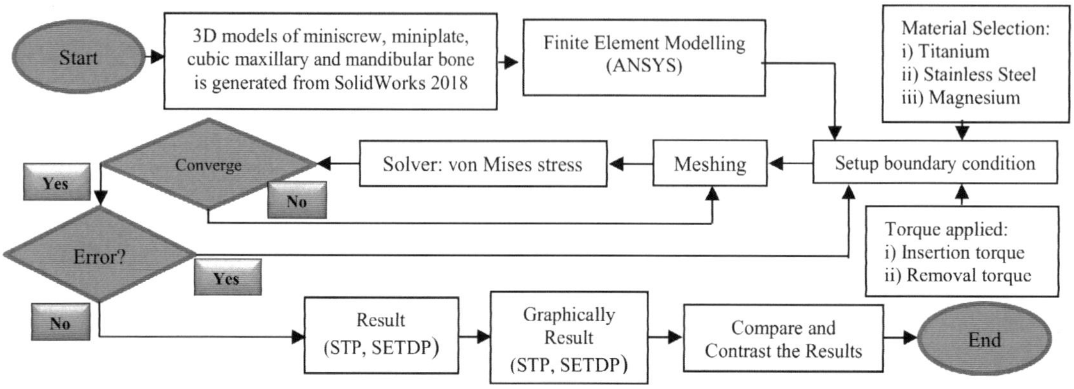

Figure 1. Research's flow chart

Three-dimensional models of miniscrews, miniplate, maxillary and mandibular bone were designed in SolidWorks 2018 software. The diameter of miniscrew for mandibular bone is 2.0 mm. In contrast, the diameter for miniscrew of maxillary bone is 1.5 mm. Both of them have a screw length of 8.0 mm [8]. The height of mandibular bone and the maxillary bone is 12 mm, cortical bone in mandibular and maxillary bone consist of 1 mm and 3 mm height and the rest in the bone is cancellous bone. Both wide and length were 40x20 mm and the bones modal was design according from the bones models used in Gautham Sivamurthy as shown in Figure 2 and Figure 3 [6]. The mechanical properties are shown in Table 1. Moreover, the contact surface of each body surface that contacts each other is automatically generated by ANSYS Workbench v15.0.

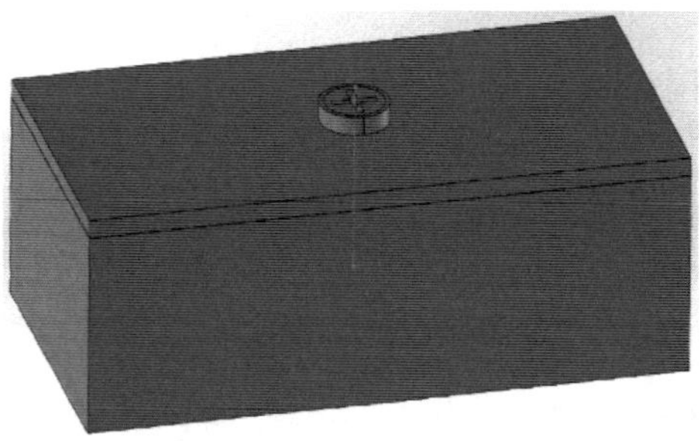

Figure 2. Assembly of miniscrew, miniplate and mandibular bone

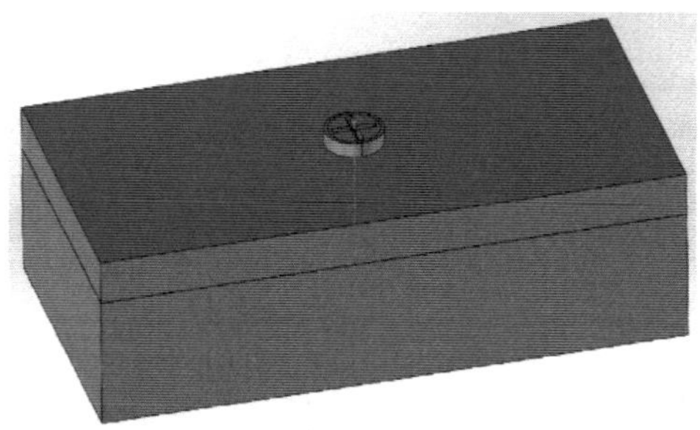

Figure 3. Assembly of miniscrew, miniplate and maxillary bone

Table 1. Mechanical properties of bone and mini-implants used [6], [9]

Material	Density, kg/m^3	Young's modulus(MPa)	Poisson's ratio(v)
Titanium, TiA	4506	107,000	0.34
Stainless Steel, SS	7700	207,000	0.30
Magnesium alloy	1800	45,000	0.29
Maxillary Cortical bone	641	15,000	0.30
Maxillary Cancellous bone	481	1,500	0.30
Mandibular Cortical bone	800	15,000	0.30
Mandibular Cancellous bone	640	1,500	0.30

In order to obtain less percentage error result, the sizing at the relevance center is being set fine and refinement surface is done on each of the surfaces to increase the elements of the meshing shown in Figure 4 and Figure 5.

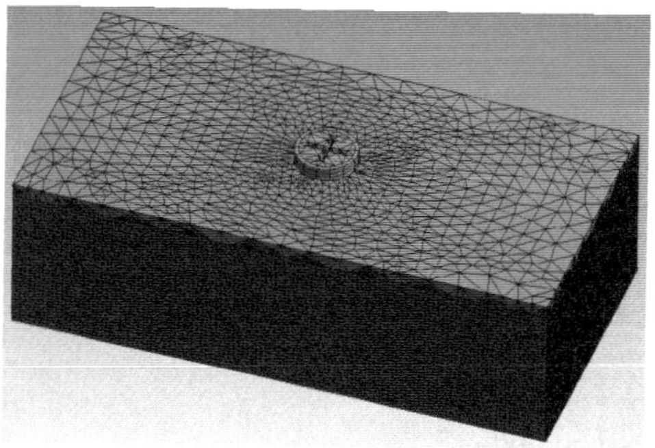

Figure 4. Coarse meshing

Figure 5. Fine meshing

Based on Haase and Rouhi [10], stress transfer parameters (STPs) were used to calculate the stress transfer between cortical bone and cancellous bone in the maxillary bone and mandibular and screw threads shown in Figure 6. The equation of stress transfer is expressed as

$$STP\ \alpha = \frac{\sigma_b}{\sigma_t} \tag{1}$$

$$STP\ \beta = \sum_{i=j=2}^{i=j=N} \frac{\sigma_{bi}}{\sigma_{tj}} \tag{2}$$

$$STP\ total = STP\ \alpha + STP\ \beta \tag{3}$$

According to the research of Kristina et al. state that the bone modeling process should consider both stress-strain energy density [11]. Hence, Strain energy density transfer parameters (SEDTPs) is to calculate the transfer of strain energy density to the bone.

$$SEDTP\ \alpha = \frac{\sigma_b \varepsilon_b}{\sigma_t \varepsilon_t} \tag{4}$$

$$SEDTP\ \beta = \sum_{i=j=2}^{i=j=N} \frac{\sigma_{bi}\varepsilon_{bi}}{\sigma_{tj}\varepsilon_{tj}} \tag{5}$$

$$SEDTP\ total = SEDTP\ \alpha + SEDTP\ \beta \tag{6}$$

3. Result and Discussion

Figure 6 shows the von Mises stress distribution under different torque loading. Figures 7-8 show the Magnesium Alloy, (MgA) has high STP and SEDTP values compare to another two materials Titanium Alloy, (TiA) and Stainless Steel, (SS) the applied moment of 100 Nmm for Magnesium Alloy, (MgA). Mini-implants in maxillary bone shows best performance of STP 1.11 and SEDTP 1.20. Figures 9-10 show that Magnesium Alloy, (MgA) with STP nearly to 1 which means that both mini-implant and bone share the same amount of stress in order to maintain the stress shielding. However, larger deformation of strain is found which is more than 3 times the strain that bone requires to withstand the loading. It's defined the bone layer get more damage compared to the miniscrew. In figures 10, shows the a big drop of SEDTP at moment 60 Nmm and 70 Nmm is due to the von Mises strain distribution is larger when the moment of 50 Nmm and 60 Nmm is applied at the adjacent of screw , N =1 compare to the moment applied of 70 Nmm. With less strain distribution, the miniscrew may get damage before it fully inserted into the bone. By comparing the STP and SEDTP with the moment applies at 60 Nmm and 90 Nmm for materials Stainless Steel, (SS) and Titanium Alloy, (TiA). The suitable materials used for mini-implant is Titanium Alloy, (TiA). This is due to higher STP of 0.70 and SEDTP of 2.60 compare to Stainless Steel, (SS) materials which have STP of 0.67 and SEDTP of 2.94.

Figure 6. Von Mises stress distribution

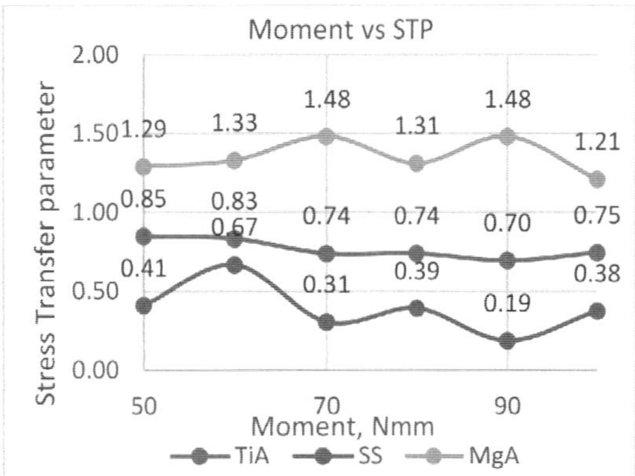

Figure 7. Moment vs STP graph of Mini-implant with mandibular bone

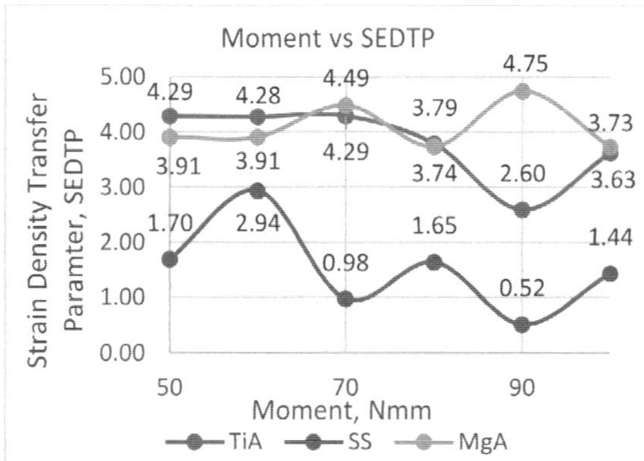

Figure 8. Moment vs SEDTP graph of Mini-implant with mandibular bone

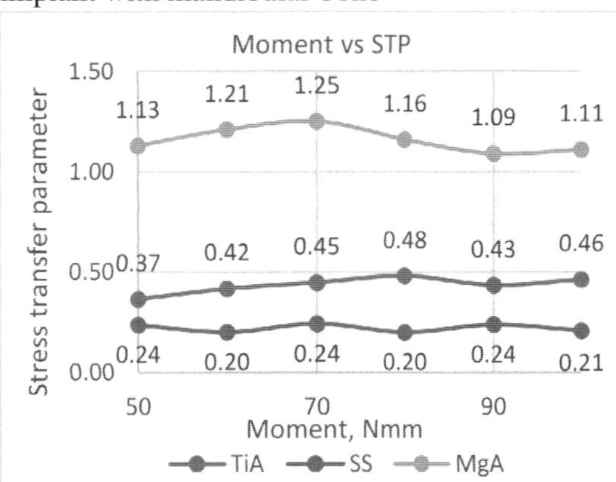

Figure 9. Moment vs STP graph of Mini-implant with maxillary bone

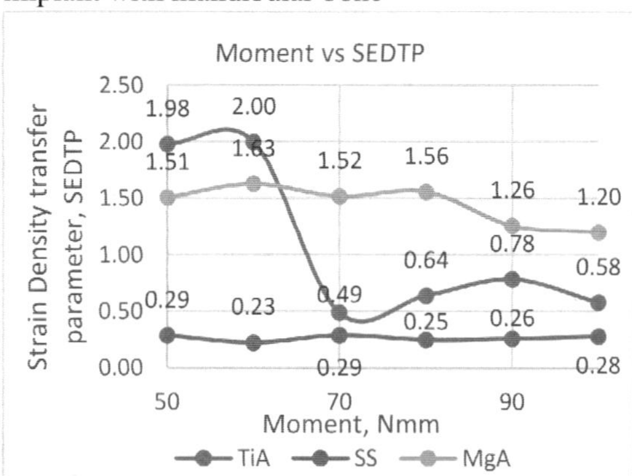

Figure 10. Moment vs SEDTP graph of Mini-implant with maxillary bone

4. Conclusion

The best torque applied is 90 Nmm and best materials used in mandibular bone is Titanium Alloy, (TiA) which produced 0.7 STP and 2.60 SEDTP, whereas the best torque applied based on the results was 100 Nmm and best materials used in maxillary bone is Magnesium Alloy,(MgA) which produced 1.11 STP and 1.2 SEDTP. It is concluded that the surgery biological stability after the insertion torque and the removal torque in range of applied moment is in the range of 10 Nmm to 150 Nmm.

Acknowledgments

The author would like to acknowledge the support from the Fundamental Research Grant Scheme (FRGS) under a grant number of FRGS/1/2018/TK03/UNIMAP/02/8 from the Ministry of Higher Education Malaysia.

References

[1] Liu Y J et al. 2015 Mechanical properties of temporary anchorage device *J. Dent. Sci.* **10** (1) 68-73.
[2] Most S P 2018 Fractures of the Jaw and Midface - Injuries and Poisoning - MSD Manual Consumer Version *Departments of Otolaryngology-Head & Neck Surgery (Plastic), Stanford University School of Medicine*, 2018. [Online]. Available: https://www.msdmanuals.com/home/injuries-and-poisoning/facial-injuries/fractures-of-the-jaw-and-midface. [Accessed: 22-Nov-2018].
[3] Ibrahim N F I, Daud R, Hassan M K A, Zain N A M, Kamaruddin N S, Azizan A F, Bajuri Y 2018 Effect of ring material and diameter on orthopedic implant stability: External fixation in Femur bone *International Journal of Engineering and Technology (UAE)* **7** (4) 190-193.
[4] Ibrahim N F I, Daud R, Hassan M K A, Zain N A M and Bajuri Y 2018 Half pins stress shielding interaction behavior at implant-bone interface *International Journal of Integrated Engineering* **10** (5) 204-208.
[5] Uesugi S, Kokai S, Kanno Z and Ono T 2018 Stability of secondarily inserted orthodontic miniscrews after failure of the primary insertion for maxillary anchorage: Maxillary buccal area vs midpalatal suture area *Am. J. Orthod. Dentofac. Orthop.* **153** (1) 54-60.
[6] Sivamurthy G and Sundari S 2016 Stress distribution patterns at mini-implant site during retraction and intrusion—a three-dimensional finite element study *Prog. Orthod.* **17** (4).
[7] Ausiello P, Franciosa P, Martorelli M and Watts D C 2012 Effects of thread features in osseo-integrated titanium implants using a statistics-based finite element method *Dent. Mater.* **28** (8) 919-927.
[8] Saeed K, Nadim M, Morcos S, Kyung H M and El-Kady A 2017 In vitro assessment of maximum insertion and removal torque with three different miniscrews on artificial maxilla and mandible *J. World Fed. Orthod.* **6** (3) 115-119.
[9] Callister W D 2006 *Materials Science and Engineering An Introduction*, Seventh Ed. Carl Wood of Utah State University and W. Roger Cannon of Rutgers University.
[10] Haase K and Rouhi G 2013 Prediction of stress shielding around an orthopedic screw: Using stress and strain energy density as mechanical stimuli *Comput. Biol. Med.* **43** (11) 1748-1757.
[11] Bhardwaj A, Gupta A, Jain P, Rani A and Yadav J 2015 Classification of human emotions from EEG signals using SVM and LDA Classifiers *2015 2nd International Conference on Signal Processing and Integrated Networks (SPIN)* 180–185.

IOP Conf. Series: Materials Science and Engineering **670** (2019) 012040 doi:10.1088/1757-899X/670/1/012040

Analysis of macro channel heat exchanger for fever relief cool cap using thermoelectric

C Y Goh[1], M N Y Rahman[1], Z M Razlan[1,*], S A Bakar[1], W K Wan[1], I Zunaidi[2], A Harun[3], M S M Hashim[1], M K Faizi[1], I Ibrahim[1], N S Kamarrudin[1] and M S Muhamad Azmi[1]

[1]School of Mechatronic Engineering, Universiti Malaysia Perlis, Pauh Putra Campus, 02600 Arau, Perlis Malaysia
[2]Faculty of Technology, University of Sunderland, St Peter's Campus, Sunderland, SR6 0DD, United Kingdom
[3]School of Microelectronic Engineering, Universiti Malaysia Perlis, Pauh Putra Campus, 02600 Arau, Perlis Malaysia

[*]Corresponding author: zuradzman@unimap.edu.my

Abstract. Fever is happened when the human having infection or other illness which the body temperature is higher than normal body temperature. The heat load released during the fever cause the uncomfortable condition to the human. Thermoelectric is used in the design of heat exchanger to provide cooling by using electricity which reliefs the fever uncomfortable. The thermal comfort zone for human and the analysis of the heat load from the human head surface are determined. The analysis focuses on the fever teenagers aged range from 15 to 25. The range of the thermal comfort zone is from 20 °C to 25 °C whereas the range for the total heat load from the head surface is from 10.56 W to 16.39 W. The findings from the analysis will be used in the design of heat exchanger that suites the requirement from the analysis result. The fluid flow effect of the working fluid and thermal effect of the heat exchanger are simulated by ANSYS Fluent. In addition, the different types of Peltier plate and fluid as well as the different materials of heat exchanger and tube have applied into heat exchanger design model for observing the effect of the fluid flow in heat exchanger. Using ANSYS Fluent, the temperature contours are compared and found that the combination of alumina Al_2O_3 nanofluid, copper heat exchanger have better performance due to high heat transfer rate while the tube materials show no effect. The best performance of cooling in the design of the heat exchanger is proposed. These design criteria include the selection of the material of heat exchanger, tube and fluid as well as the selection of the device such as Peltier plate, battery, cooler fan and pump.

1. Introduction

The range for human normal body temperature is between 36.5 °C and 37.5 °C, which the condition that over the range will be known as fever. Fever is happened when the human having infection or other illness, the thermostat in human body readjusts which results fever to fight the infections. However, a high body temperature more than 39.5 °C will result some serious consequences such as hyperthermia and heat stroke that may cause the brain damage or even cause death to the human beings. Thermoelectric uses the new technology to generate electricity by using a temperature difference and to provide heating or cooling by using electricity. Thermoelectric is used in medical profession to reduce human high fever body temperature [1]. It is believed that the design of the heat exchanger cooling systems can relief the fever uncomfortable and release the heat load from head surface to achieve thermal comfort by using thermoelectric. In the medical field, the thermoelectric Peltier elements are used in the cooling of living tissues [2].

The aims of this research are to study the human body behaviour on heat and to analyse the heat load on the human head surface. At the same time, the aims are to do heat transfer analysis in heat exchanger and propose the design of heat exchanger that suites the requirement. The analysis focuses on the fever teenagers aged range from 15 to 25 and focuses on the human head surface. The design of the heat exchanger is focused on the parameters such as the types and the materials of the parts. The analysis is done on the fluid flow effect and thermal effect of the heat exchanger which are simulated by ANSYS Fluent.

2. Methodology

2.1. Analysis of heat load on head surface

A group of the teenagers who aged from 15 to 25 and who had a fever are picked to record their measurement data. The measurement data needed are mass, height and the fever temperature. These measurement data help to determine a range of the heat load from the head surface. The measurement data is aimed to collect from 10 male teenagers and 10 female teenagers. The mass of the teenagers is measured by using a weight scale while the height of the teenagers is measured by using wall mounted height rods. Furthermore, the fever temperature of the teenagers is measured by using body non-contact infrared thermometer. Below are the formulae used in the analysis.

$$\dot{Q}_{rad} = \varepsilon\, \sigma\, A_s\, (T_s^4 - T_{sur}^4) \tag{1}$$

$$A_s = 0.202\, m^{0.425}\, h^{0.725} \tag{2}$$

$$\dot{Q}cond = k\, A\, \frac{\Delta T}{\Delta x} \tag{3}$$

$$\dot{Q}_{latent} = \dot{m}_{vapor}\, h_{fg} \tag{4}$$

2.2. The model of the heat exchanger

The brief idea of the design is presented in the schematic diagram in Figure 1. In the model, the heat is absorbed by the heat exchanger from the human head surface. The working fluid inside the heat exchanger transfer the heat by working fluid through the function of the pump which generates the mass flow. The heat is then passed to the water tank and the heat is absorbed by the Peltier plate. Heat is released through the hot side of the Peltier plate by the use of a heat sink and a fan. A real and more detailed of the whole model of the heat exchanger is displayed in the Figure 2. A power source is used to provide current to the Peltier plate, fan and pump. Furthermore, a microcontroller is used to control the performance of the Peltier plate, fan and pump by limiting or controlling the current direct to them.

2.3. Analysis and design of the heat exchanger

The model of the heat exchanger is used for thermal analysis in the ANSYS Fluent. The fluid flow and thermal analysis is done by changing the parameters to determine the highest heat transfer rate and the best performance for the design. The parameters concerned are the types of Peltier plate and fluid as well as the materials of heat exchanger and tube. There are five types of Peltier plates such as TM 71-1.0-3.0, TM 127-1.0-3.0, TM 71-1.0-4.0, TM 31-1.0-6.0 and TM 71-1.0-6.0 are considered in the design. The fluids considered are water and alumina nanofluid Al_2O_3 [3], [4]. For heat exchanger materials considered are aluminium and copper whereas for tube materials considered are plasticized polyvinyl chloride (PVC-P), polypropylene and the silicone rubber.

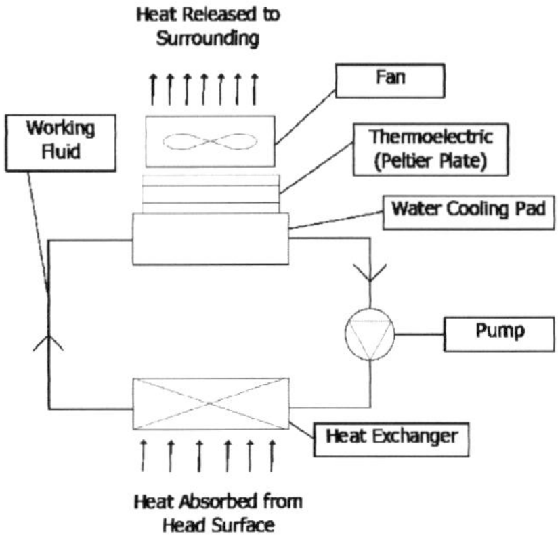

Figure 1. Schematic Diagram of the Heat Exchanger.

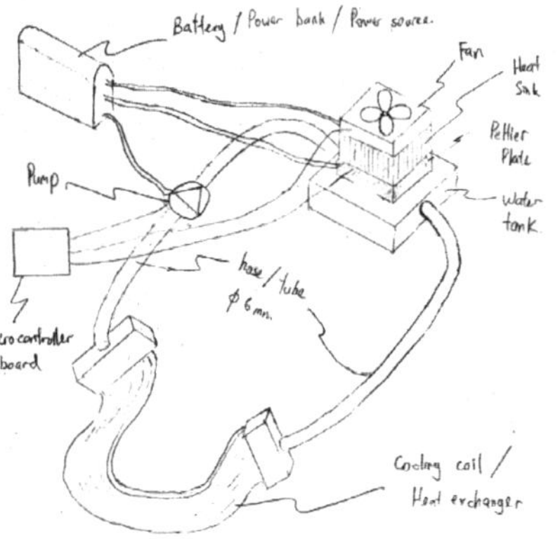

Figure 2. Whole Diagram of the Heat Exchanger.

3. Result and Discussion

3.1. Study of human body behaviour on heat

The topic for the human body behaviour on heat is studied and is explained in mechanical definition. A constant human core body temperature between 36.5 °C to 37.5 °C must be maintained, which the purpose is to prevent the high temperature damages the body systems; if above the temperature is called fever. Heat generation and heat dissipation are the types of heat transfer in human body. These heat transfer in human body is done by metabolic heat production, heat conduction, heat convection, heat radiation and evaporation.

In addition, the important part for the human body behaviour on heat is the thermal comfort for human which is the individual satisfaction thermal environment condition. A range from 20 °C to 25 °C is the thermal comfort range for human [5]. A lower than or higher than the range temperature will cause discomfort for the human. This range of thermal preferences is used as the preferable temperature in the design of heat exchanger for the human fever relief.

3.2. Determination of the heat load on the head surface

The data measurements for the 20 fever teenagers who aged range from 15 to 25. The heat load by conduction, radiation and evaporation are calculated and they are summed up to the total heat load from the head surface. The evaporation heat is assumed to be the same for all the teenagers. From the study, the range for the total heat load from the head surface is from 10.56 W to 16.39 W.

Table 1. Summary for the findings of the heat load on the human head surface.

	Minimum Heat Load	Maximum Heat Load
Male	13.26 W	16.39 W
Female	10.56 W	14.37 W

3.3. Thermal analysis of heat exchanger

The model of the heat exchanger is cut into symmetry since the both left side and right side are the same. By cutting the model into symmetry, the analysis will be easier and faster. Outside the model is the case for the cooling system, while inside the model is the fluid domain which will be determined the flow and the heat transfer rate.

There are total of 12 conditions by different materials, such as heat exchanger (aluminum or copper), tube (PVC-P, polypropylene or silicone rubber) and fluid (water or alumina nanofluid). The

IOP Conf. Series: Materials Science and Engineering **670** (2019) 012040 doi:10.1088/1757-899X/670/1/012040

boundary condition is set to the temperature of heat is 38 °C and the temperature of cooling is 5 °C. The tube materials show no effect on the heat transfer rate. For the heat exchanger materials show a little effect whereas the different fluid shows significant effect. Figure 3 shows the significant effect on the fluid flow and heat transfer rate by changing the working fluid and the heat exchanger materials.

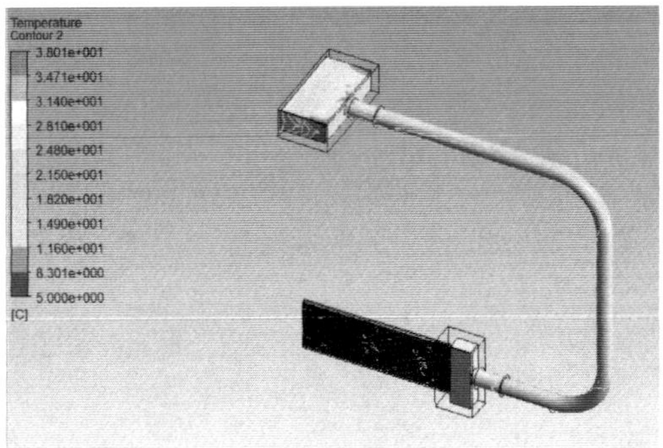

a)

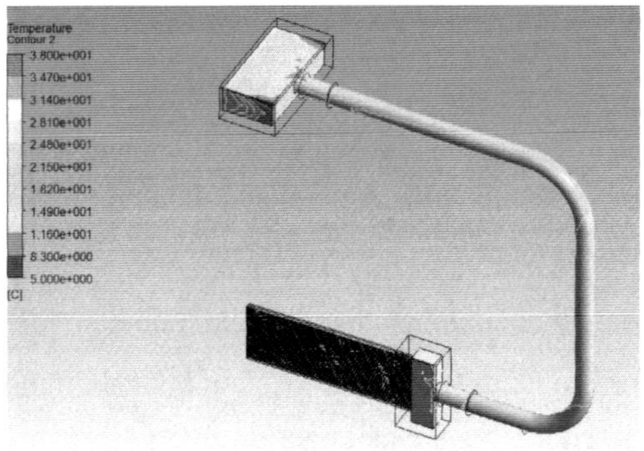

b)

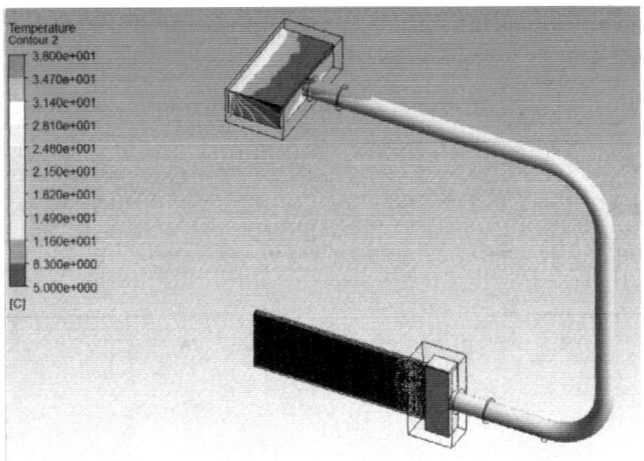

c)

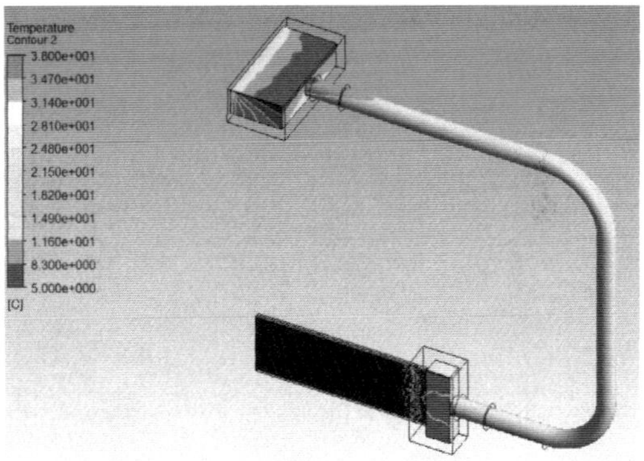

d)

Figure 3. Temperature Contour– a) Water and Aluminium Heat Exchanger, b) Water and Copper Heat Exchanger, c) Nanofluid and Aluminium Heat Exchanger, d) Nanofluid and Copper Heat Exchanger

3.4. Design of the heat exchanger

The design will consider Lithium Polymer (LiPo) battery pack or Power bank to provide power source to power the Peltier plate. When using the LiPo battery pack, the current is set to have 20 W of heat absorbed in Peltier plate, whereas the current in power bank is 2.1 A which determines the Peltier plate that has at least 20 W of cooling. Table 2 and Table 3 show the results of suitable Peltier plate.

Table 2. Values of Heat Load Absorbed by Peltier Plate using LiPo Battery Pack

Part Number	I (A)	V (V)	$\dot{Q}c$ (W)	P_{in} (W)
TM 127-1.0-3.0	2.0	10.0	20.0	20.00
TM 71-1.0-4.0	3.9	8.2	16.0	31.98
TM 71-1.0-6.0	3.8	5.8	20.0	22.04

IOP Conf. Series: Materials Science and Engineering **670** (2019) 012040 doi:10.1088/1757-899X/670/1/012040

Table 3. Values of Heat Load Absorbed by Peltier Plate using Power Bank

Part Number	I (A)	V (V)	$\dot{Q}c$ (W)	P_{in} (W)
TM 127-1.0-3.0	2.1	12.0	21.0	25.20
TM 71-1.0-4.0	2.1	5.0	9.0	10.50
TM 71-1.0-6.0	2.1	3.1	12.0	6.51

The design of the heat exchanger will be proposed to improve the performance. The proposal of the heat exchanger will include the selection of the material of heat exchanger, tube and fluid as well as the selection of the device such as Peltier plate, battery, DC cooler fan and pump. This proposal is proposed in the mechanical definition. Table 4 shows the proposal made based on the results.

Table 4. The selection for the design of heat exchanger with description.

Design Criteria	Selection Proposed	Description
Material of heat exchanger	Copper	Higher heat transfer rate
Material of tube	Silicone Rubber	Higher percentage of elongation, soft material
Fluid	Alumina nanofluid	Higher heat transfer rate
Types of Peltier plate	TM 127-1.0-3.0	Able to absorb 20 W of heat load with low power input
Battery	LiPo battery pack	Able to provide higher current and voltage
Fan	DC cooler fan	Able to provide the air flow rate of 0.1302 m^3/min
Pump	DC micro pump	Able to provide the water flow rate of 0.0249 x 10^{-3} m^3/min

4. Conclusion and Recommendation

In a nutshell, the human body behaviour on heat is studied and the analysis of the head surface heat load is done. The thermal behaviour of the heat exchanger cooling system is investigated by using ANSYS Fluent. The effect of changing the materials in the heat exchanger cooling system is analysed as well. Last but not least, the design of heat exchanger cooling system that having the best cooling performance is proposed. By summing up all the results, the conclusions are:

- The human body behaviour on heat is done by metabolic heat production, heat conduction, heat convection, heat radiation and evaporation. The range of the thermal comfort for human is from 20 °C to 25 °C.
- The analysis of the heat load on the human head surface shows there is a different heat load between genders. A range from 10.56 W to 16.39 W of the heat load is analysed from them.
- Based on the temperature contours, alumina nanofluid and copper heat exchanger have a higher heat transfer rate, whereas there is no effect on changing the tube materials.
- A combination of copper heat exchanger, silicone rubber tube, nanofluid, Peltier plate model TM 127-1.0-3.0, LiPo battery pack, DC cooler fan and DC micro pump is proposed for the design to boost the performance of cooling system.

In this research, the heat exchanger cooling system is analysed by changing the fluid as well as the materials for heat exchanger and tube. There is a significant effect on heat transfer by changing of fluid. However, a different dimension of the heat exchanger may affect the performance and the heat transfer rate. Hence, it is recommended to research and study on the effect of changing the dimension of the heat exchanger.

In addition, this research is studied and a best performance of the design for heat exchanger is proposed in the mechanical definition only. Therefore, this research of design best performance of the heat exchanger in term of mechatronic field or electrical field is suggested to study in the future. For example, the design of microcontroller to control the current passing through the Peltier plate, DC cooler fan and DC pump. A changing of the current in power source may affect the performance of the Peltier plate, DC cooler fan and DC pump. At the same time, this may affect the heat transfer rate in the heat exchanger as well which it is recommended to study the design of the heat exchanger in term of mechatronic field or electrical field.

Acknowledgments

The author would like to acknowledge the support from the Malaysian Institute of Road Safety Research (MIROS), Ministry of Transport Malaysia for their research grant scheme under Universiti Malaysia Perlis (UniMAP) project code number of 9002-00072.

References

[1] Baldry M, Timchenko V and Menictas C 2018 Thermal modelling of controlled scalp hypothermia using a thermoelectric cooling cap *J. Therm. Biol.* **76** 8–20.
[2] Uhlik M and Sladecek V 2018 Cooling Device with Peltier Element for Medical Applications *IFAC-PapersOnLine* **51** (6) 54–59.
[3] Ho C J, Liao J -C, Li C -H, Yan W -M and Amani M 2019 Experimental study of cooling performance of water-based alumina nanofluid in a minichannel heat sink with MEPCM layer embedded in its ceiling *Int. Commun. Heat Mass Transf.* **103** 1–6.
[4] Afifah A N, Syahrullail S and Nor Azwadi C S 2015 Natural convection of aluminium oxide-water nanofluid *J. Teknol.* **75** (11).
[5] Lechner N 2015 *Heating, cooling, lighting : sustainable design methods for architects.*

IOP Conf. Series: Materials Science and Engineering **670** (2019) 012041 doi:10.1088/1757-899X/670/1/012041

The effect of graphene loading on natural rubber latex/graphene stretchable conductive material

W M Che[1] *, P L Teh[1], C K Yeoh[1] and A J Jalilah[1]

[1] School of Material Engineering, Universiti Malaysia Perlis, Taman Muhibah, 02600 Jejawi, Perlis, Malaysia

Abstract. The demand for the soft and flexible stretchable electronic has been increased due to the flourish development of electronic field nowadays. This can be achieved by incorporating graphene nanoplatelet (GNP) into natural rubber latex (NRL). This paper was aimed to investigate the effect of the GNP content on the properties of the NRL/GNP composites. The incorporating of GNPs into NRL increased the density and modulus of the composite but reduced the tensile strength, elongation at break and crosslink density. On the other hand, the NRL/GNP has reached the percolation threshold at 7phr of GNP, at this content the composite became electrically conductive.

1. Introduction

The demand for the stretchable electronic has been increased due to the flourish development of electronic field. This became a big challenge for the material engineers since conductivity and stretchability were hard to coexist in a material due to stretching tend to lengthen the bonding and increased the resistivity of the material [1]. By the effort they paid, they successfully created few methods in order to made the stretchable electronic devices which able to withstand the repeat stretched, twisted and banded without compromising on the conductivity [2]. In this paper, graphene nano-platelet (GNP) was incorporated into natural rubber latex (NRL). Latex which acted as the matrix endued elasticity for the composite while GNP improved the mechanical properties of the composite materials. As percolation threshold achieved, the GNP particles contacted with each other and the conductive pathways were and the composite became conductive.

2. Materials and Method

Natural rubber latex (NRL) with TSC 60% was supplied by MG Color Sdn. Bhd. The 50% dispersive compounding additives such as zinc oxide (ZnO), sulfur, zinc diethyldithiocarbamate (ZDEC) and zinc 2-mercaptobenzothiazole (ZMBT) were purchased from Furben Technique (M) Sdn. Bhd. Furthermore, stearic acid (SA) was supplied by Acidchem International Sdn. Bhd while the stabilizer, potassium hydroxide (KOH) was purchased from Johchem Scientific & Instruments Sdn. Bhd. Lastly. graphene nano-platelet (0540DX) was purchased from SkySpring Nanomaterials Inc, United State.

First, the KOH solution, ZnO, SA, GNP and accelerators were added by followed the sequence and stirred until homogeneous by using the magnetic bar. Then, sulfur was added into the mixture after sonification and stir until homogeneous. The mixture was then degassed by using the vacuum bump and desiccator before cast in mold. Lastly, the casted composite material was let for maturation under room temperature for 48 hours and post cure in the oven. The steps were repeated with different GNP loading (0phr, 1phr, 3phr, 5phr, 7phr and 9phr).

3. Characterization

To determine the crosslink density of the composites, weighed samples (~0.2g) were immersed into the toluene until the sample weight reach equilibrium. Then, crosslink density of the composites was calculated by using the Flory-Rehner equation:

$$v = \frac{1}{M_{cs}} = -\left[\frac{V_2 + \chi V_2 + \ln(1-V_2)}{\rho_2 V_2 \left(V_2^{\frac{1}{3}} - \frac{1}{2}V_2\right)}\right] \quad (1)$$

Where v is the effective number of moles of crosslinked chains per gram of polymer in mol/g; M_{cs} is the molecular weight between crosslink in g/mol; V_2 is the volume fraction of polymer in the swollen mass; V_1 is the molar volume of the solvent in ml/mol; χ is the polymer-solvent interaction parameter while ρ_2 is the density of the polymer in g/cm^3. The density of the conductive materials was determined by using the pycnometer density analyzer with model Micromeritics AccuPycII 1340v under the flow of helium gas at room temperature. Furthermore, tensile properties were determined by using the Instron 5569 Universal Testing Machines (UTM) according to the ASTM D412. Then the electrical conductivity test, was tested according to the ASTM D257 by using the Fluke 8845A/8846A 6.5-digit precision multimeter in direct current mode with voltage supply of 5V at room temperature. Then, both the bulk resistivity and bulk conductivity were calculated by using the Equation 2.

$$\text{Bulk Conductivity, } \sigma = \frac{d}{r.\ A}. \quad (2)$$

Where r is resistance of resistor, A is the area of the specimens in cm^2 and d stand for the thickness of the specimens in cm.

4. Results and Discussion

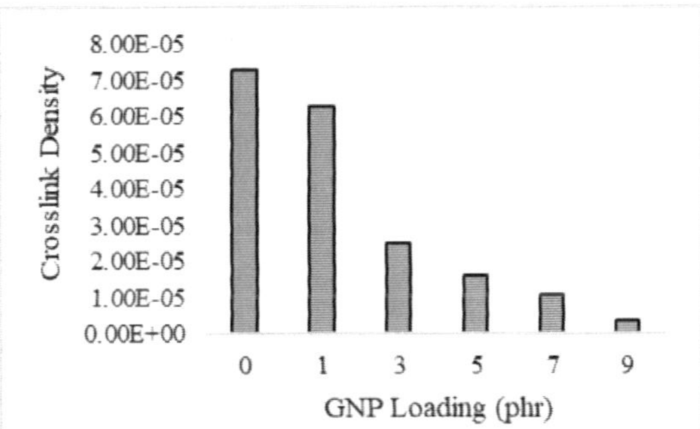

Figure 1: Crosslink density of composites with different GNP loading.

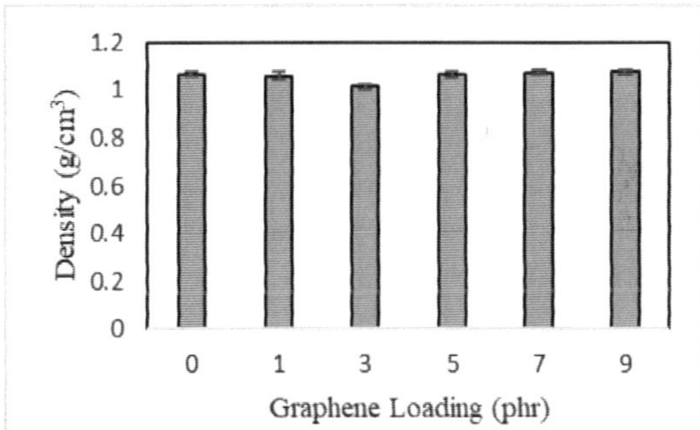

Figure 2: Density of composites with different GNP loading.

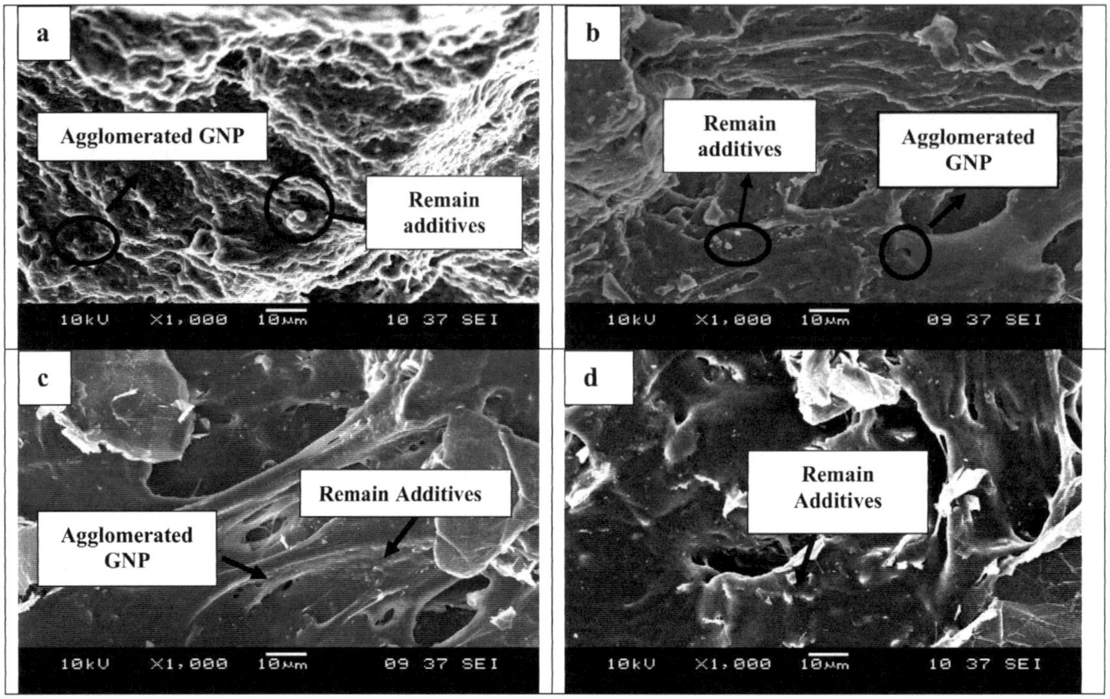

Figure 3: SEM of composites with different GNP loading (a) 0phr; (b) 1phr; (c) 7phr and (d) 9phr

Figure 1 shows crosslink density of the composite. The crosslink density decreased as the GNP loading increased from 0phr to 9phr. Localized of GNP between the interlayer of the NR macromolecular chains, restricted the formation of the C-S-C crosslinking between the sulphur and the rubber chains [3]. The crosslink density gave a significant effect on the physical properties such as solvent diffusion and tensile strength.

Then, the density of the conductive materials with different GNP contents was not significant as shown in ***Figure 2***. The density was decreased as GNP loading increased form 0phr to 3phr and then increased from 3phr to 9phr. There were some remained additives and agglomerated GNP within sample as shown in ***Figure 3***. The low interfacial adhesion between the matrix and the solid additive caused the formation of void around the interface resulting the reduction of the density. As more GNP added, the compounded mixture became denser. The increased of the overall mass caused the increased of the density as GNP loading increased from 3phr to 9phr.

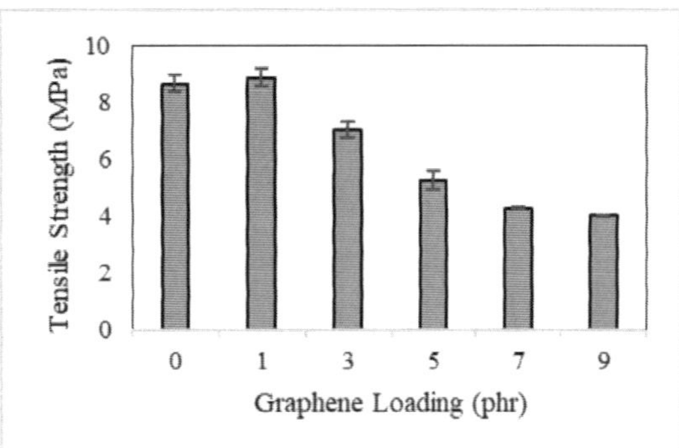

Figure 4: Tensile strength of composites with different GNP loading.

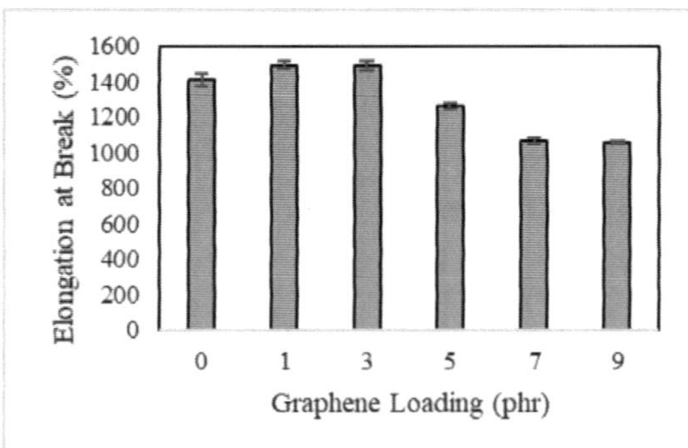

Figure 5: Elongation at break of composites with different GNP loading.

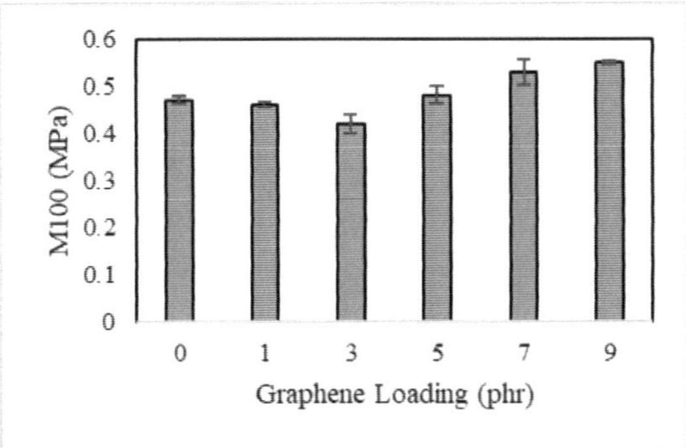

Figure 6: Modulus at 100% (M100) strain of composites with different GNP loading.

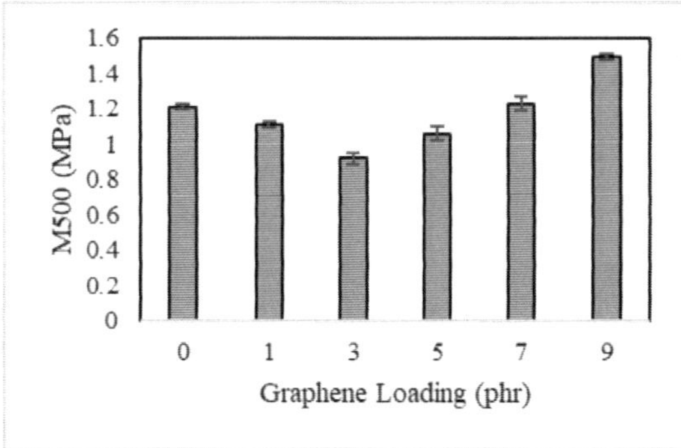

Figure 7: Modulus at 500% (M500) strain of composites with different GNP loading.

Figure 4 and ***Figure 5*** show the tensile strength and the elongation at break of the composites. Both increased from 0phr to 1phr GNP and then decreased as the subsequent increased of GNP loading. While for the modulus in ***Figure 6*** and ***Figure 7***, the M100 and M500 show a decreased trend from 0phr to 3phr GNP loading and then increased as the further increased of the GNP loading from 3phr to 9phr.

Better dispersion of GNP within the latex matrix allowed the GNP offer good stress transfer ability for the composite at 1phr GNP loading, resulting higher tensile strength [5]. As GNP loading increased, the strong cohesive forces between these carbonaceous particles tend to agglomerate with each other to form cluster form. The occurrence of filler agglomeration became serious as the GNP loading increased. since higher loading had a higher tendency to contact with each other during mixing and formed more GNP clusters. Then, The larger additive particles and the agglomerated GNP acted as the stress concentration point where the deformation initiated [4]. On the other hand, the high GNP content restrict the chain mobility the chains became unable to move freely, leading to the reduction of the elongation but increased in the modulus of the composites [6]

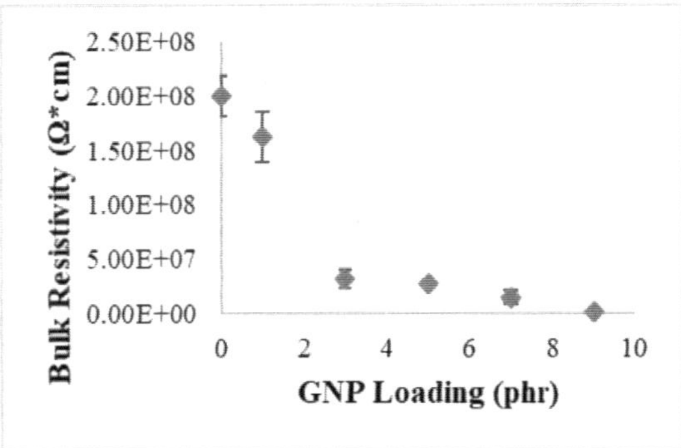

Figure 8: Bulk resistivity of composites with different GNP loading.

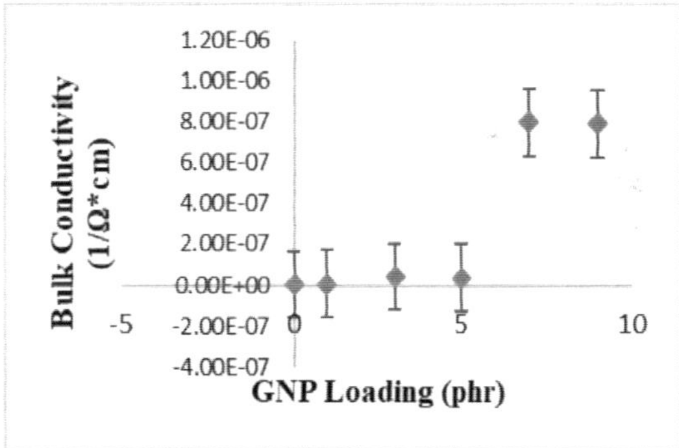

Figure 9: Bulk conductivity of composites with different GNP loading.

The GNP particles with high cohesive force tend to form agglomerate [7]. These GNP agglomerate structure help to improve the electrical conductivity of the composite [8]. When the GNP come contact with each other, the conductive pathway started to construct and allowed the electrons to move along the matrix and the composite started to conduct electricity once the percolation threshold achieved [7]. As the GNP added continuously, more conductive network formed, leading to a significant improvement of the conductivity. In this study, the percolation threshold was achieved at 7phr of GNP loading as shown in ***Figure 8*** and ***Figure 9*** where the bulk resistivity decreased significantly and there was a sudden increased of the conductivity.

5. Conclusion

The incorporation of the GNP particles into the NRL gave significant effect on the properties. The localization of the GNP particles between the NRL macromolecular chains restrict the formation of the C-S-C linkage, reduced the crosslink density which in turn reduced the tensile strength of the composites. Furthermore, the additives remained and the agglomerated GNP particles leading the reduction of density. As the GNP loading increased, the increased of the overall mass caused the increased of the density. Besides, the increased of the GNP loading restrict the mobility of the chains, resulting the reduced of the elongation at break of the composites but increased the modulus at 100%, and 500% strain. On the other hand, the GNP endued electrical conductivity to the NRL matrix and achieved the percolation threshold at 7phr.

Acknowledgement

The financial support of Fundamental Research Grant Scheme (FRGS) grant no: FRGS/1/2018/TK05/UNIMAP/02/13 is gratefully acknowledged.

References

[1] Spector D 2013 *This Stretchy Metal Will Change The Future Of Electronics.* Retrieved from https://www.businessinsider.com/highly-conductive-stretchable-material- flexible-electronics-2013-7/?IR=T
[2] Liang H -W, Guan Q -F, Zhu Z, Song L -T, Yao H -B, Lei X and Yu S -H 2012 Highly conductive and stretchable conductors fabricated from bacterial cellulose *NPG Asia Materials* **4** 1-6.
[3] Wang X, Xia Z, Yuan B, Zhou H, Li Z and Chen N 2013 Effect of curing temperature on the properties of conductive silicone rubber filled with carbonyl permalloy powder *Materials & Design* **51** 287–292.
[4] Ahmad M A 2017 *Approacheds to enhance filler-polymer interactions and cure properties of rubber compounds.*
[5] Fu S -Y, Feng X-Q, Lauke B and Mai Y -W 2008 Effects of particle size, particle/matrix interface adhesion and particle loading on mechanical properties of particulate–polymer composites *Composites Part B: Engineering* **39** (6) 933–961.
[6] Taylor P, Zhao F, Bi W and Zhao S 2012 Influence of crosslink density on mechanical properties of natural rubber vulcanizates influence of crosslink density on mechanical properties of natural rubber vulcanizates *J. Macromol. Sci, Part B Phys.* **50** 37–41.
[7] Santos R, Mould S, Formánek P, Paiva M and Covas J 2018 Effects of particle size and surface chemistry on the dispersion of graphite nanoplates in polypropylene composites *Polymers* **10** (2) 222.
[8] Potts J R, Shankar O, Du L and Ruoff R S 2012 Processing–morphology–property relationships and composite theory analysis of reduced graphene oxide/natural rubber nanocomposites *Macromolecules* **45** (15) 6045–6055.

Analysis on current flow style for vehicle alternator fault prediction

M A Abd Halim, M T A Rahman*, N Abdul Rahim, A Rahman, A F A Hamid and N A M Amin

School of Mechatronic Engineering, Universiti Malaysia Perlis (UniMAP), Kampus Pauh Putra, 02600 Arau, Perlis, Malaysia

*Corresponding author: tasyrif@unimap.edu.my

Abstract. Vehicle alternator is only seen as fragment piece in vehicle. This project will analyse the vehicle alternator current output flow style. A study on charging rate onto battery can be made based on this analyst. From this a prediction can be made on the vehicle alternator health and may prevent it from affecting other charging system component. Features extracted from the raw sample data are root mean square (RMS), waveform length (WL) and autoregressive (AR). These features will then go through normality test to find the sample is normally distributed or not. The normality test used in this experiment is Jarque-Bera (JB) test. After go through the normality test, it shows that need to continue with non-parametric test. Because JB test shows that p-value is less than 0.05 confidence level. Kruskal-Wallis is used as non-parametric data validation. In this test, the hypothesized value is the value is the median instead of the mean as in Analysis if Variance (ANOVA). The Kruskal-Wallis test evaluates for any significance difference in the population medians on a dependent variable across all levels of a factor. For classification K-Nearest Neighbour (KNN) is used to find the number of K to differentiate between classes. After that the K value is use in Holdout method for training and testing. Final result shows that the accuracy of this machine learning tools is 94%. This number is a good percentage to be able to called as vehicle alternator fault prediction.

1. Introduction

Data prediction plays a tremendous trend nowadays. For now, quite hype with Industrial Revolution 4.0 movement where we can have real-time monitoring data in a cloud[1]. With the help of Artificial Intelligence, it will help enhance the movement. Data prediction is a very useful added value in any system where from it may contribute to help in maintenance and research and development. Fault prediction on vehicle alternator will help user from go through such event such as car breakdown during travelling. In order to make prediction a study and analyse on alternator output pattern. Later on then can make prediction on fault state of vehicle alternator[2], [3].

2. Methodology

This experiment is done on a Proton Wire 1.3 engine setup. On this experiment only load from the engine is test without others vehicle electrical load. A current sensor which is ACS 756 embedded with Arduino Uno microcontroller is used[4]. The sensor was connected to the terminal B on the alternator. This terminal B is where the generated output from the alternator[5]. The collected raw data as in Figure 1 was then computed into Matlab software to continue with statistical analysis and fault

IOP Conf. Series: Materials Science and Engineering **670** (2019) 012042 doi:10.1088/1757-899X/670/1/012042

prediction. In statistical analysis, there are several tests that the raw data need to go through. First is Feature Extraction, here the feature that been extract is RMS, WL and AR. Then a normality test is done to find out the data is normally distributed or not. After that the feature data will go through data validation process which to find out which feature can be used for classification and which cannot[6]. Lastly classification and prediction, the data is then will be predict to find does it matching with the same class of it miss to other class.

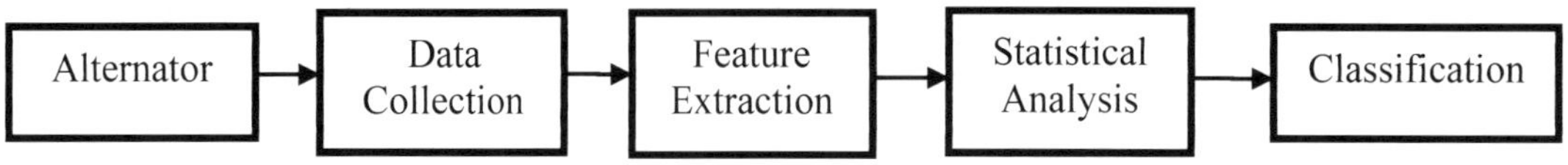

Figure 1. Block Diagram.

2.1. Data Collection
The data was collected in several RPM which is on idle, 2000 and 3000. There are 3 type of alternator used, Alternator A is a good alternator, Alternator B which is a bad alternator suspected with rectifier problem and Alternator C which is suspected with IC regulator problem.

2.2. Feature Extraction
The feature extraction process is done in the Matlab. The raw data is being segmented to 100 sample data in 5s time data window which equivalent to 20hz frequency. The extracted features are RMS, WL, and AR. RMS, a meaningful way of calculating the average of values over a period of time. WL, the horizontal distance between successive crests, troughs or other parts of a wave. AR, a time series model that uses observations from previous time steps as input to a regression equation to predict the value at the next time step.

2.3. Statistical Analysis
Succeeding the features extraction process is statistical analysis. Statistical analysis consists of normality test and data validation. Normality test is use to find the acquired features is normally distributed or not. After normality test then can know what type of classification tool need to be used in data validation which is either parametric or non-parametric test. For this normality test, Jarque-Bera test is the chosen one among many other normality tests. It is simply because is a test for skewness and kurtosis and very effective. Jarque-Bera test or known as JB test is where a null hypothesis is tested with vector from the normal distribution with an anonymous mean and variance[7].

$$JB = \frac{n}{6}(s^2 + \frac{(k-3)^2}{4}) \tag{1.1}$$

Where n is the sample size, s is the skewness and k are the sample kurtosis. For larger sample size, the test statistics has a chi-square distribution with two degree of freedom. This normality test is to measure the p-value of all these features to find out where the significance level when set to 0.05. When p-value<0.05, null hypothesis can be rejected. The hypothesis for normality test are as follows:

H_0: Data is normal distribution
H_1: Data is not normal distribution

From JB test, shows that it is accepted to reject the null hypothesis because the p-value<0.05 and shows the data is not normally distributed so need to continue the statistical analysis with non-

IOP Conf. Series: Materials Science and Engineering **670** (2019) 012042 doi:10.1088/1757-899X/670/1/012042

parametric test tools. All features are already analysed by Jarque-Bera test which is one of the time frequency domain analysis. For non-parametric data validation test, Kruskal-Wallis test is chosen. In this test, the hypothesized value is the value is the median instead of the mean (as in ANOVA). The Kruskal-Wallis test evaluates for any significance difference in the population medians on a dependent variable across all levels of a factor. If the requirements for a parametric test can be met, it is better to use the one-way independent measure ANOVA instead of Kruskal-Wallis since it is more powerful.

$$H = \frac{12}{N(N+1)}\left(\frac{R_1^2}{n_1}+\frac{R_2^2}{n_2}+\cdots+\frac{R_k^2}{n_k}\right) - 3(N+1) \tag{1.2}$$

Where n is the size of sample k, N is the total number of observations in combined samples, k is the number of samples or groups and R_k is the sum of ranks from sample k.

H_0: Median1 = Median2 = Median3
H_1: At least one group median is different from others

In Kruskal-Wallis test the to reject null hypothesis the p-value must less than 0.05. If it cannot reject null hypothesis means that the median for each group is the same and it may affect the classification accuracy later. Because it cannot differentiate what type of group is need to sort. For Kruskal-Wallis test, it shows that all features are rejecting null hypothesis except feature 6 which is AR4 feature where the p-value>0.05. Here a conclusion may arise which all features the median is the same between classes but for feature 6 at least one group median is different from others. An early prediction can be made where feature 6 cannot differentiate each class and it may affect the accuracy. So, a new dataset form where feature 6 is remove from it and from 7 features it become only 6 features on it. For the next classification method both datasets is done where a finding on the affection of feature 6 for this project.

2.4. Classification

In machine learning, KNN is used together with holdout method to do the prediction. KNN is a non-parametric, lazy learning algorithm. Its purpose is to use a database in which the data points are separated into several classes to predict the classification of a new sample point. KNN makes predictions just-in-time by calculating the similarity between an input sample and each training instance. The holdout method is the simplest kind of cross validation. The data set is separated into two sets, called the training set and the testing set. The function approximate fits a function using the training set only. Then the function approximate is asked to predict the output values for the data in the testing set[8]. The errors it makes are accumulated as before to give the mean absolute test set error, which is used to evaluate the model. The advantage of this method is that it is usually preferable to the residual method and takes no longer to compute. For this experiment, the holdout is used in 3 stages which is 60-40, 70-30 and 80-20.

3. Results and Discussions

In KNN the parameter to find classification is K value tested from K=1 to K=10. For holdout method, it is tested for 10 times and the average accuracy is then accumulated in Table 1.

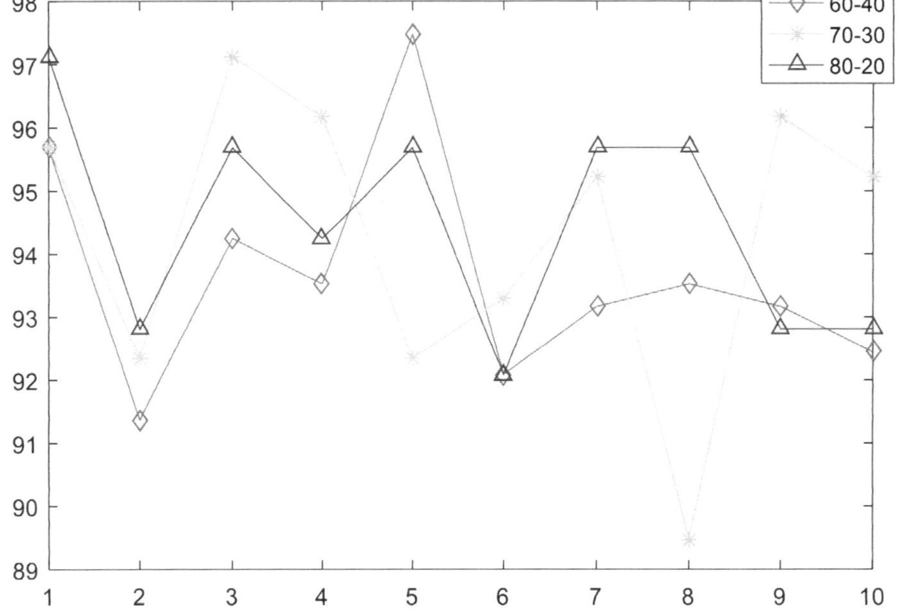

Figure 2. K-Nearest Neighbour

Figure.2 is the K value for KNN where it has been test from K=1 to K=10 and fine the best accuracy. Later then holdout method will come around to find out the distribution accuracy. The holdout method is done for 10 times then get the average accuracy.

Table 1. Holdout Method.

Type	Holdout 60-40	Holdout 70-30	Holdout 80-20
7 Features	94.3885	94.4019	94.8921
6 Features	94.3525	94.1627	95.0360

Based on Table 1 result we may assume the result is still approximately same. This step may lead to new approach with is reduce complexity where feature 6 is remove from the analysis but still lead to equivalent result. From the result, it shows that the accuracy for this model to do the prediction on the vehicle alternator fault is around 94%. Even though the feature 6 is removed it still resulting an equivalent result and this may help in reduced complexity.

4. Conclusions

Based on the normality test, the findings are that the data is not normally distributed. Jarque-Bera test is the chosen normality test. Therefore, it must go to non-parametric test. There are a lot of type of parametric test but the chosen one Kruskal-Wallis for the non-parametric data validation. The result from Kruskal-Wallis show that feature 6 does not reject the null hypothesis hence the median value for each class is approximately same. The sample data is then divided into 2 which one for all features and the other one with feature 6 exclude from it. This is because later in the experiment want to find did feature 6 affecting the accuracy. For non-parametric classification, K-Nearest Neighbour is chosen among other method. This is to find the number of K so the class can be differentiating. The test is done for 10 times and the number of K is trained by using cross validation technique which is Holdout method. For Holdout method there is 3 type of Holdout setting used. First with 60-40 holdout which means 60% training and 40% testing. Next with 70-30 holdout which means 70% training and 30% testing. Last but not least 80-20 holdout which means 80% training and 20% testing. All holdout setup was being run for 10 times and the average value is recorded. As a conclusion, the result from this method find that the accuracy for all 7 features intact were 94% and for only 6 features were also 94%. From this result can conclude that feature 6 which is AR4 does not affect the classification accuracy. So an addition conclusion that can be made that feature AR4 can be removed from the features extracted and this may result reduced complexity because it does not affect the accuracy much.

Acknowledgement

The authors would like to extend their sincere gratitude for the funding provided by the Universiti Malaysia Perlis (UniMAP) (Geran Penyelidikan Khas - No.9004-00064) which enable this research to be carried out.

References

[1] Hamid A F A et al. 2017 Connected car: Engines diagnostic via Internet of Things (IoT) *J. Phys. Conf. Ser.* **908** (1).
[2] Şimşir M, Bayir R and Uyaroılu Y 2016 Real-time monitoring and fault diagnosis of a low power hub motor using feedforward neural network *Comput. Intell. Neurosci.* **2016**.
[3] Bangalore P and Tjernberg L B 2015 An Artificial Neural Network Approach for Early Fault Detection of Gearbox Bearings *IEEE Trans. Smart Grid* **6** (2) 980–987.
[4] Dinio C J T et al. 2019 Automated water source scheduling system with flow control system *2018 IEEE 10th Int. Conf. Humanoid, Nanotechnology, Inf. Technol. Commun. Control. Environ. Manag. HNICEM 2018*.
[5] Mazlan R, Dan R and Zakaria M 2017 Experimental study on the effect of alternator speed to the car charging system *The 2nd International Conference on Automotive Innovation and Green Vehicle (AiGEV 2016)*.
[6] Sukor A S A, Zakaria A and Rahim N A 2018 Activity Recognition using Accelerometer Sensor and Machine Learning Classifiers *2018 IEEE 14th Int. Colloq. Signal Process. Its Appl.* 233–238.
[7] Reddy D N and Ravinder Y 2018 Novel receiver diversity combining methods for spectrum sensing using meta-analytic approach based on p-values *J. Eng. Sc. Tech* **13** (9) 2883–2897.
[8] Yadav S 2016 Analysis of k-fold cross-validation over hold-out validation on colossal datasets for quality classification *2016 IEEE 6th Int. Conf. Adv. Comput.* 78–83.

Characterisation of structural and physical properties of cellulose nanofibers from *Pennisetum purpureum*

R Revati[a], *M S Abdul Majid[b], M J M Ridzuan[c] and N F Mohd Nasir[d]

Universiti Malaysia Perlis (UniMAP), Kampus Tetap Pauh Putra, 02600, Pauh, Perlis

[a]*revark_1990@yahoo.com,** [b]***shukry@unimap.edu.my***,
[c]*ridzuanjamir@unimap.edu.my,* [d]*nashrul@unimap.edu.my*

Abstract
Recently, cellulose nanofiber (CNF) and its applications gain immense attraction in both research and industrial areas due to its attractive properties such as excellent mechanical properties, high surface area, rich hydroxyl groups for modification, and natural properties with 100% environmental friendliness. *Pennisetum purpureum,* also known as Napier grass fibre is a newly-identify plant which is highly sustainable throughout Malaysia. In this study, the typical extraction of cellulose nanofiber from *Pennisetum purpureum* method is summarised, in which the cellulose nanofiber extraction concerning alkali treatment and planetary ball milling is mainly introduced. Cellulose nanofiber from *Pennisetum purpureum* also has been extracted using different concentration acid hydrolysis, such as 20% and 30% to observe the effect towards the fibre bonding. The morphology, chemical structure and crystallinity of the fibre were obtained using scanning electron microscopy SEM, Fourier transform infrared spectroscopy FTIR, and X-ray diffraction XRD. SEM clarifies that the increment of acid hydrolysis lowers the fibre-fibre bonding upon freeze-drying of the cellulose nanofiber suspensions. The result of XRD and SEM shows that 30% of acid hydrolysis gives higher crystallinity and chemical effect towards the structure of cellulose nanofibers.

Keywords: *Pennisetum purpureum*, cellulose nanofiber, acid hydrolysis, ball milling

1. Introduction

Biomedical application using nanocellulose and nanocomposite is only recently discovered [1,2]. In order to enhance the physical and mechanical properties of the cellulose, most researchers aim to improve the extraction process. Many techniques were used to extract nanocellulose, including ball milling, ultrasonic technique and acid hydrolysis [1–3]. Ball milling is considered to be an attractive process that can be used to prepare nano-scale materials due to its simple operation and cost-effectiveness [4–6]. Thus, new bast fibres can be extracted into nanocellulose. The application potentials of *Pennisetum purpureum* (Napier grass) has yet to be explicitly explored on their applications in the bio-nanocomposites [7–10]. Cellulose obtained from this bast fibre can be potentially used as one of the nanomaterials to form a bio-nanocomposite since it possesses biocompatibility and biodegradability, in addition to an excellent physical property [11,12].

This study represents the extraction of nanocellulose from *Pennisetum purpureum* using acid hydrolysis together with the ball milling process. The production of nanocellulose from ball milling was supposed to increase the surface area and crystallinity of the cellulose. Based on several background research, acid hydrolysis has increased the crystallinity of the material, whereas ball milling is increased the surface area of cellulose. Characterisation including Scanning electron microscopy (SEM), Fourier transforms spectroscopy (FTIR), and X-ray diffraction (XRD) were conducted to investigate the structural and physical properties of the prepared cellulose nanofiber from *Pennisetum purpureum.*

2. Materials and Methods

2.1. Materials

Pennisetum purpureum fibres were purchased from Bukit Kayu Hitam, Kedah, Malaysia. Sodium hydroxide (NaOH), and sulphuric acid (H_2SO_4) were obtained from Fisher Chemical Co. Throughout the experiment, distilled water (DI) was used as a medium for multiuse.

2.2. Preparation of cellulose nanofiber (CNF) from Pennisetum purpureum

The *Pennisetum purpureum* (PP) fibres were extracted from the bast fibres using the water retting process. The powder of *Pennisetum purpureum* was treated using 12% sodium hydroxide (w/v) water solution for 2 h to eliminate lignin and hemicellulose completely. The residual was further processed using acid hydrolysis technique with varying concentration; in this study, the fibres undergo hydrolysis process with 20% and 30% concentration, CNF-PP_{20} and CNF-PP_{30} respectively. At the end of hydrolysis, the residue was dialysed with distilled water for 3 days until the neutral pH was reached. Nanocellulose was prepared using planetary ball milling with distilled water. Initially, cellulose was placed in a 160 mL zirconia container with distilled water and zirconia balls (15 mm in diameter). The ball milling was performed for 3 h at 14 Hz. The milled product was further sonicated for 10 minutes before the cellulose nanofiber was freeze-dried.

2.3. Characterisation

The morphologies of the extracted cellulose nanofibers were studied using SEM (JEOL, JSM 6460-LA). Samples were coated with a layer of platinum, Pt using a sputter coater and further analysed at 3 kV.

The crystalline structure and crystallinity index of the cellulose nanofiber was investigated by XRD (D2 Phase company Bruker).

The FTIR spectra of cellulose were recorded in the spectral region between 400 to 4000 cm^{-1} using an FTIR spectrometer (Perkin-Elmer RX1, UK) to study the chemical changes occurred during the extraction process from *Pennisetum purpureum.*

3. Results and Discussions

3.1 SEM

The morphologies of freeze-dried samples from CNF suspensions are shown in Figure 3.1. The dried sample of cellulose nanofiber CNF from *Pennisetum purpureum* with 20% and 30% of acid hydrolysis and ball milling, formed fibrous and agglomerate particles. Researchers have reported that the agglomeration of cellulose nanofiber CNF may occur during the freezing stage of the drying process[13]. In the freezing stage, the non-freezing bound fluid is eliminated by heating the samples under vacuum. The water vapour hence diffuses and induces structural development and rearrangement of cellulose nanofiber CNF particles. Thus, the enormous length (up to a hundred micrometers) and width (tens to hundreds of micrometers) are the consequence of the lateral agglomeration of the cellulose nanofiber CNF. Based on Figure 3.1(a), scanning electron microscopy (SEM) of CNF-PP_{20} shows the formation of a finely blended and interconnected fibrous structure. Irregularly shaped particles were obtained from samples CNF-PP_{20} and CNF-PP_{30}. CNF-PP_{30} shows that the size of particles is smaller compared to CNF-PP_{20}. The interconnection structure bonding of CNF-PP_{30} are not strong, and the structure is easy to break. As a conclusion, smaller fibres appear in the SEM images as the concentration of the acid hydrolysis increases. This apparently clarifies that the increment of acid hydrolysis lowers the fibre-fibre bonding upon freeze-drying of the cellulose nanofibers suspensions. Thus, different concentration of acid hydrolysis influences the cellulose nanofiber suspensions with different particle sizes and structural morphologies.

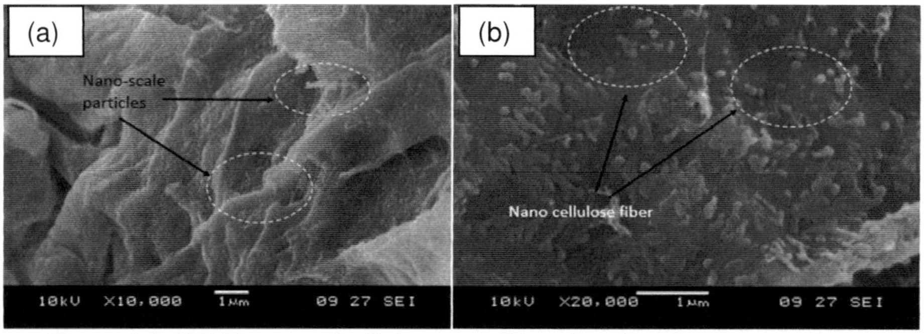

Figure 3.1: *SEM images of cellulose nanofibers from Pennisetum purpureum: (a) CNF-PP_{20} and (b) CNF-PP_{30}.*

3.2 XRD

The influence of acid hydrolysis and ball milling on the microstructure and morphology of isolated cellulose nanofiber from Pennisetum purpureum was analysed using X-Ray diffraction (XRD). Figure 3.2 (a) and (b) shows the XRD profile of developed cellulose nanofibers. Similar patterns are observed for both samples, which are CNF-PP_{20} and CNF-$PP_{30,}$ respectively. Different concentration of acid used will give different effect towards the crystallinity of the cellulose nanofibers Pennisetum purpureum fibres.

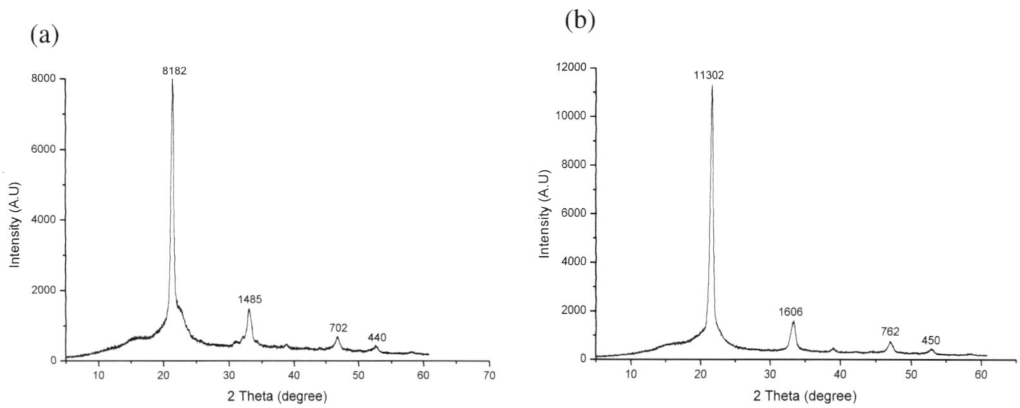

Figure 3.2: *X-Ray Diffraction pattern of (a) CNF-PP_{20} and (b) CNF-PP_{30}.*

Figure 3.2 (a) and (b) show the X-ray diffractograms of CNF-PP_{20} and CNF-$PP_{30.}$ Results of XRD studies show the main peak around the 2Θ value of 21.42° and 21.66° in CNF-PP_{20} and CNF-$PP_{30,}$ respectively. Interestingly, we observed that the intensity count for CNF-PP_{30} is higher compared with CNF-$PP_{20,}$ and this may occur due to an increment of concentration in acid hydrolysis. The XRD pattern of CNF-PP_{20} had a distinctive crystalline peak at 2Θ of 33.09°, 46.78°, and 52.63° which corresponded to the basal spacing of 2.71358Å, 1.94158Å, and 1.73537Å respectively. Meanwhile, the XRD pattern of CNF-PP_{30} had a very similar peak at 2Θ of 33.37°, 47.62°, and 52.79° which corresponded to the basal spacing of 2.69003Å, 1.939037Å, and 1.72909Å respectively. Based on the result of XRD for CNF-PP_{30}, the main peak pattern $2\Theta \approx 21.66^{\circ}$ are higher than Figure 3.2 (b), and the remaining peak of the graph are 33.37°, 47.62°, and 52.79°. Thus, the higher the acid concentration, the higher the crystallinity peak of cellulose nanofibers. This phenomenon proves that the role of chemical treatment successfully increased the cellulose fibre crystalline due to the removal of hemicellulose and lignin contents during the chemical treatment.

3.3 FTIR

The FTIR spectra of cellulose nanofiber from Pennisetum purpureum shows two distinctive peaks absorption region in the spectrums of 2800-3600cm-1 and 750-1750cm-1 were recognised in the spectra, as shown in Figure 3.3. Interestingly, the FTIR spectrum explains that significant changes between the cellulose nanofibers obtain under different acid hydrolysis were not detected. A broad band in the region 3335cm-1 observed from both samples; this indicates the O-H free stretching vibration of the CH2-OH structure on cellulose nanofiber and OH groups, which complement to inter and intramolecular hydrogen bond appear in absorbed water and cellulose. The bands noticed at 1316cm-1 are relevant to the C=C stretching or /and CH2 symmetric bending in an aromatic group of cellulose due to crystallinity band. The spectra of cellulose nanofibers achieved from CNF-PP_{30} shows a significant increment in the intensity of crystallinity functional group compared to the CNF-PP_{20}. This result proves that the concentration of acid hydrolysis might have increased the level of crystallinity of the cellulose nanofibers, as observed from the XRD result.

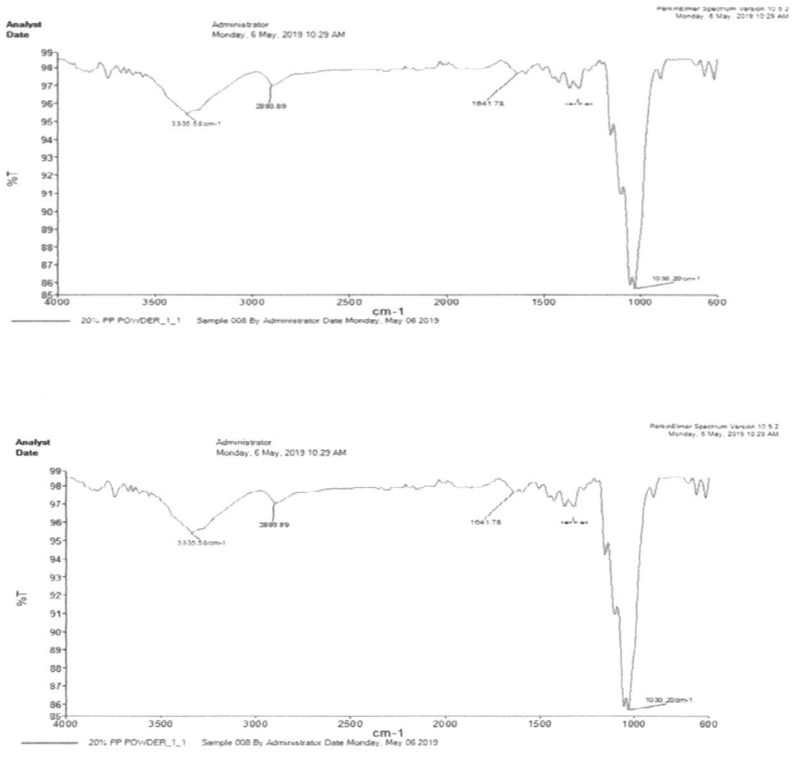

Figure 3.3: FTIR spectra (a) CNF-PP_{20} (b) CNF-PP_{30}

4. Conclusion

The main objective of this study was to extract and develop a cellulose nanofiber from Pennisetum purpureum fibre. A combination of few techniques was used for the extraction process of cellulose nanofiber from cellulosic materials, where acid hydrolysis with different concentration was used to weaken the fibrous structure, the extraction process was then continued with ball milling to crushed the weakened fibres. Those techniques were applied in this study to facilitate the conversion of cellulose into nano-sized fibres within the short term. SEM studies on the morphologies of dried samples of CNF-PP_{20} and CNF-PP_{30} shows that the higher concentration of acid hydrolysis used in this study is easier to weaken the interconnection or bonding of the fibre structure. The XRD graph showed that the highest crystallinity peak was observed at CNF-PP_{30} sample. There are almost 10% different of crystallinity percentage for both concentrations, which are 59.956% for CNF-PP_{30} sample, and 50.856% for CNF-PP_{20}. Thus, different chemical and mechanical method for the extraction process will affect the crystallinity structure. Meanwhile, the higher concentration used for the acid hydrolysis process, the crystallinity structure of cellulose nanofiber will increase. Next, characterization-using FTIR shows the result that the spectra of cellulose nanofibers achieved from CNF-PP_{30} show a significant increment in the intensity of crystallinity functional group compared to the CNF-PP_{20}. This result proves that the concentration of acid hydrolysis might have increased the level of crystallinity of the cellulose nanofibers, as observed from the XRD result.

Acknowledgements
The authors would like to acknowledge the Ministry of Education (MOE), Malaysia, for financial support via the Fundamental Research Grant Scheme (Ref: FRGS/1/2019/TK05/UNIMAP/02/11).

5. References

[1] Zhang Y, Nypelö T, Salas C, Arboleda J, Hoeger IC, Rojas OJ. Cellulose Nanofibrils. J Renew Mater 2013;1:195–211. doi:10.7569/jrm.2013.634115.
[2] Chirayil CJ, Mathew L, Thomas S. Review of recent research in nano cellulose preparation from different lignocellulosic fibers. Rev Adv Mater Sci 2014;37:20–8.
[3] Xie H, Du H, Yang X, Si C. Recent Strategies in Preparation of Cellulose Nanocrystals and Cellulose Nanofibrils Derived from Raw Cellulose Materials. Int J Polym Sci 2018;2018:1–25. doi:10.1155/2018/7923068.
[4] Suryanarayana C. Mechanical alloying and milling. Prog Mater Sci 2001;46:1–184. doi:10.1016/S0079-6425(99)00010-9.
[5] Baheti V, Abbasi R, Militky J. Ball milling of jute fibre wastes to prepare nanocellulose. World J Eng 2012;9:45–50. doi:10.1260/1708-5284.9.1.45.
[6] Posada P, Castro C, Serpa A, Gómez CH, Velásquez-cock J, Ga P. Influence of combined mechanical treatments on the morphology and structure of cellulose nanofibrils : Thermal and mechanical properties of the resulting films 2016;85:1–10. doi:10.1016/j.indcrop.2016.02.036.
[7] Ridzuan MJM, Abdul Majid MS, Khasri A, Basaruddin KS, Gibson AG. Effect of moisture exposure and elevated temperatures on impact response of Pennisetum purpureum/glass-reinforced epoxy (PGRE) hybrid composites. Compos Part B Eng 2019;160:84–93. doi:10.1016/j.compositesb.2018.10.029.
[8] Fartini MS, Abdul Majid MS, Ridzuan MJM, Amin NAM, Gibson AG. Compressive properties of Napier (Pennisetum Purpureum) filled polyester composites. Plast Rubber Compos 2016;45:136–46. doi:10.1080/14658011.2016.1149911.
[9] Revati R, Majid MSA, Ridzuan MJM, Basaruddin KS, Rahman Y. MN, Cheng EM, et al. In vitro degradation of a 3D porous Pennisetum purpureum/PLA biocomposite scaffold. J Mech Behav Biomed Mater 2017;74. doi:10.1016/j.jmbbm.2017.06.035.

[10] Ridzuan MJM, Majid MSA, Afendi M, Rahman MN, Azduwin YK. Thermal, composition and morphological properties of untreated and alkali-treated napier grass fibres. J Eng Appl Sci 2016;11:2485–9. doi:10.3923/jeasci.2016.2485.2489.

[11] Revati R, Abdul Majid MS, Ridzuan MJM, Normahira M, Mohd Nasir NF, Rahman Y. MN, et al. Mechanical, thermal and morphological characterisation of 3D porous Pennisetum purpureum/PLA biocomposites scaffold. Mater Sci Eng C 2017;75:752–9. doi:10.1016/j.msec.2017.02.127.

[12] Revati R, Majid MSA, Ridzuan MJM, Basaruddin KS, Rahman MNY, Cheng EM, et al. Journal of the Mechanical Behavior of Biomedical Materials In vitro degradation of a 3D porous Pennisetum purpureum / PLA biocomposite sca ff old. J Mech Behav Biomed Mater 2017;74:383–91. doi:10.1016/j.jmbbm.2017.06.035.

[13] Peng Y, Gardner DJ, Han Y. Drying cellulose nanofibrils: In search of a suitable method. Cellulose 2012;19:91–102. doi:10.1007/s10570-011-9630-z.

IOP Conf. Series: Materials Science and Engineering **670** (2019) 012044 doi:10.1088/1757-899X/670/1/012044

Isolation and characterisation of nanowhisker cellulose from Pennisetum purpureum

E F Sucinda, M S Abdul Majid*, M J M Ridzuan and N A M Amin

School of Mechatronic Engineering, Universiti Malaysia Perlis, Pauh Putra Campus, 02600 Arau, Perlis, Malaysia.

* denotes corresponding author
E-mail: *sucinda94@gmail.com,* shukry@unimap.edu.my, ridzuanjamir@unimap.edu.my, nasrulamri.mohdamin@unimap.edu.my*

Abstract.
Pennisetum Purpurerum or Napier fibre is a sustainable, renewable and biodegradable waste enriched with cellulose. The methodology to recover agriculture *Pennisetum Purpureum* residue is by acquiring nanowhiskers cellulose (NWCs) from microcrystalline cellulose (MCC). In this study, acid hydrolysis with different time exposure was performed to degrade the MCC derived from *Pennisetum Pupureum* into NWCs. The characterisations of the isolated NWCs was conducted through a Scanning Electron Microscope (SEM) and Fourier Transform Infrared Spectroscopy (FTIR). In SEM morphology study, the NWCs shape ribbon-like with nano scales thickness was observed, and from the SEM result shows that the thickness of the sample was in nano size. The isolated NWCs from *Pennisetum purpureum* derived MCC has a high potential to be used for the manufacture of nanocomposite for the implementation of pharmaceutical and biomedical fields.

1. Introduction

In the present study, nanocrystalline cellulose (NCC) has acquired considerable attention in many potential application due to their abundance of the amount which can be inherently renewed, low cost and environmental friendly [1–3]. Variety sources of nanocrystal cellulose as known as nanowhiskers cellulose has been studied to discover the superior properties for any potential application such as Roselle, sugar palm, cotton pulp, rice, oat husks, napier and many more [4–8]. NWCs features high mechanical properties refer to the inter/intramolecular hydrogen bond, high aspect ratios and large specific surface area that influenced by their low weight and morphological features [9]. The production of NWCs can be obtained by chemical treatment and with or without mechanical treatment. However, the material properties like thermostability and crystallinity or even final structure and morphology were highly dependent on the drying process. The spray-dried products had higher thermal stability and crystallinity index when the nanocellulose was applied in non-polar thermoplastic compared to air-drying, freeze drying and supercritical-drying [10]. Moreover, the morphological structures of air-drying, freeze drying and spray-drying showed solid packed NCCs, ribbon-like structures and irregularly shaped particles with external voids, respectively[11]. For further utilised in some application, NWCs have to undergo reinforcement in polymer matrices for possibilities of high-performance structural application [12].

2. Materials and methods

2.1. Materials and chemicals

The *Pennisetum purpureum* also known locally as Napier fibre, was acquired from a local plantation in Bukit Kayu Hitam, in northern Peninsular of Malaysia. Sulfuric acid (H2SO4) and Sodium hydroxide (NaOH) was purchased from Fisher Chemical Co. and dialysis tube was purchased from A.R. Alatan Sains Sdn. Bhd.

2.2. Extraction and preparation of Pennisetum purpureum

The leaves and the roots were eliminated from *Pennisetum purpureum* stems and underwent the water retting process. In addition, the soaked fibres were washed with running water to remove the unwanted impurities then sun-dried to expel overabundance dampness.

2.3. Alkaline treatment

The alkali treatment process was conducted to purify the Napier cellulose. The fibres were treated with 12 wt% concentration of aqueous NaOH solution. The fibres were then soaked in the solution for two hours to neutralise and remove the balance of lignin and hemicellulose.

2.4. Grinding and sieving

The dried sample that has been treated by alkali treatment was ground into a powder and sieved using 63-micron size vibrating sieve machine to obtain the fine Napier fibres powder.

2.5. Acid hydrolysis

Sulphuric acid of 64 wt% concentration was used to hydrolyse three different reaction time of 30 min, 45 min and 60 min, NCC-I, NCC-II and NCC-III respectively. It was then, followed by centrifugation to remove the excess sulphuric acid and it was dialysed for 6 days to raise the pH value to obtain a neutral pH value. Afterwards, ultra-sonication was carried out to disperse the NWC suspensions for 30 min (6 s on and 2 s off). Eventually, the supernatant NWC suspension was freeze-dried to further the process of characterisation.

2.6. Characterisation

2.6.1. Scanning electron microscopy (SEM) analysis

Scanning electron microscopy SEM was observed using, JEOL (JSM 6460-LA). Generally, material surface morphology, orientation and dispersion of nanowhiskers were determined by SEM. For this analysis, the samples were coated by platinum using auto fine coater JFC-1600.

2.6.2. Fourier transform infrared spectroscopy (FTIR) analysis

The FTIR allows the chemical structure to be characterised by identifying the functional groups from each sample. A Spectrum 65 FTIR spectrometer (Perkin Elmer) was used to record the spectra between 4000–450 cm^{-1}.

3. Results and Discussion

3.1. SEM analysis

It was noticeably that all three NWCs formed plate-like materials and have only slightly different in size even though they were treated in three different reaction times. All three NWCs undergo the same alkali treatment process; this will not affect significantly on their surface morphologies because the amorphous region or non-cellulosic compound was removed most likely at the same rate. As shown in figure 3.1, it is observed that all three NWC-I, NWC-II and NWC-III have the thickness of less than 1 micron which can be considered as nanoscale size. Nevertheless, the length of the freeze-dried NWCs cannot be seen due to the formation of the ice crystals during the freeze-drying process. The freeze drying process is conducted in two stages. The first stage involves the freezing process and followed by the second stage; drying process. During the drying process, the froze samples in the formed of ice crystals will be sublimated. Therefore, the morphology of the freeze-dried NWCs formed has a plate-like shape as ice crystal grows in the same direction and oriented in plate-like structure in a parallel direction.

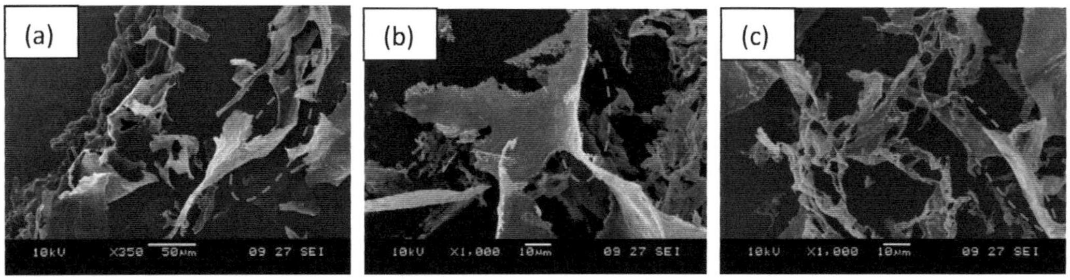

Figure 3.1. SEM micrographs of freeze-dried nanowhiskers cellulose with the treated hydrolysis reaction time (a) NCC-I, (b) NCC-II and (c) NCC-III.

3.2. FTIR analysis

The FTIR result in figure 3.2 showed similar spectra from three different acid hydrolysis time, which proved various time for acid hydrolysis process not affected the chemical structure on samples. The data for each peak at each different acid hydrolysis time was almost very similar. Figure 3.2 also showed at 3332.82 cm^{-1} that the stretching of O—H bonds was more intense with the increase of acid hydrolysis time [13]. At 2903.78 cm^{-1} (C—H symmetrical stretching) shows that prolonged reaction was affect the cellulose characteristics to be much visible [4]. The band around 1372 cm^{-1} indicated C-H asymmetric deformation. Furthermore, the hemicellulose content at 1740 cm^{-1} was also observed, but its gradually decrease as shown in figure 3.2 while the water absorption is at 1636.83 cm^{-1} (H—O—H stretching vibration) was found that the intensity is increasing[4].

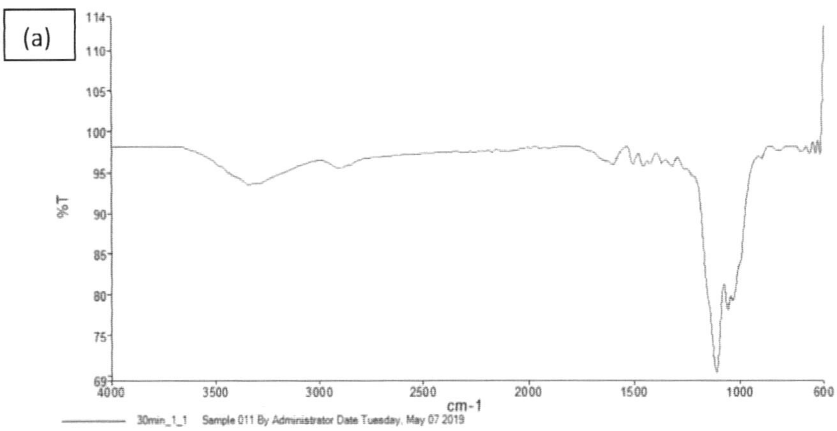

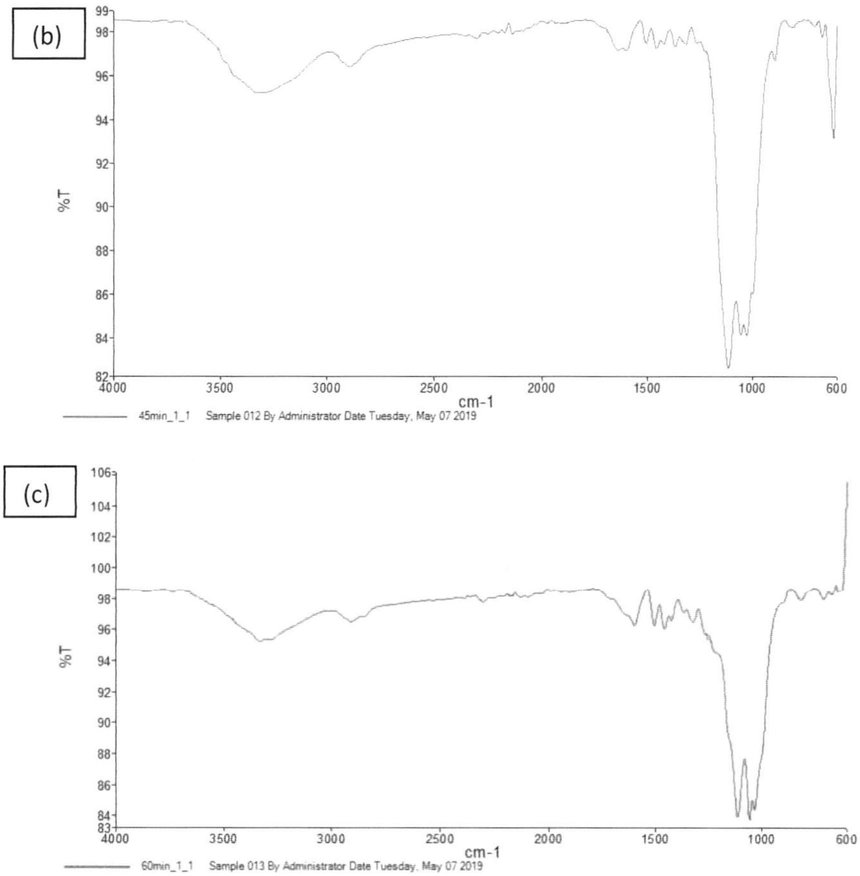

Figure 3.2. FTIR spectra (a) NWC-I, (b) NWC-II and (c) NWC-III.

4. Conclusion

The NWCs was successfully extracted and isolated by sulphuric acid through acid hydrolysis from *Pennisetum purpureum* fibres. NWCs exhibited a plate-like structure with nanoscale thickness; the morphological structure depended on final stage of the experiment which is the drying process. FTIR test which showed that the chemical structure does not affect by the time exposure of acid hydrolysis. It can be seen that the treated 64 wt% concentration acid hydrolysis of NWC-I, NWC-II and NWC-III do not have any changes in the chemical properties even though the time of acid hydrolysis treatment were different.

Acknowledgements

The authors would like to acknowledge the Ministry of Education (MOE), Malaysia, for financial support via the Fundamental Research Grant Scheme (Ref: FRGS/1/2019/TK05/UNIMAP/02/11).

References

[1] Habibi Y, Lucia L A, Rojas O J 2010 Cellulose nanocrystals: chemistry, self-assembly, and applications *Chem Rev* **110** 3479–500. doi:10.1021/cr900339w.

[2] Maciel M M Á D, Benini K C C de C, Voorwald H J C and Cioffi M O H 2019 Obtainment and characterization of nanocellulose from an unwoven industrial textile cotton waste: effect of acid hydrolysis conditions *Int J Biol Macromol* **126** 496–506. doi:10.1016/j.ijbiomac.2018.12.202.

[3] Revati R, Majid M S A, Ridzuan M J M, Basaruddin K S, Rahman Y M N, Cheng E M, et al 2017 In vitro degradation of a 3D porous Pennisetum purpureum/PLA biocomposite scaffold *J Mech Behav Biomed Mater* **74** 383-391. doi:10.1016/j.jmbbm.2017.06.035.

[4] Kian L K, Jawaid M, Ariffin H and Karim Z 2018 Isolation and characterization of nanocrystalline cellulose from roselle-derived microcrystalline cellulose *Int J Biol Macromol* **114** 54–63. doi:10.1016/j.ijbiomac.2018.03.065.

[5] Chen X Q, Pang G X, Shen W H, Tong X, Jia M Y 2019 Preparation and characterization of the ribbon-like cellulose nanocrystals by the cellulase enzymolysis of cotton pulp fibers *Carbohydr Polym* **207** 713–9. doi:10.1016/j.carbpol.2018.12.042.

[6] de Oliveira J P, Bruni G P, el Halal S L M, Bertoldi F C, Dias A R G and Zavareze E da R 2019 Cellulose nanocrystals from rice and oat husks and their application in aerogels for food packaging *Int J Biol Macromol* **124** 175–84. doi:10.1016/j.ijbiomac.2018.11.205.

[7] Fartini M S, Abdul Majid M S, Ridzuan M J M, Amin N A M and Gibson A G. Compressive properties of Napier (Pennisetum Purpureum) filled polyester composites *Plast Rubber Compos* **45** 136–46. doi:10.1080/14658011.2016.1149911.

[8] Ridzuan M J M, Majid M SA, Afendi M, Rahman M N and Azduwin Y K 2016 Thermal, composition and morphological properties of untreated and alkali-treated napier grass fibres *J Eng Appl Sci* **11** 2485–9. doi:10.3923/jeasci.2016.2485.2489.

[9] Abe K and Yano H 2011 Formation of hydrogels from cellulose nanofibers *Carbohydr Polym* **85** 733–7. doi:10.1016/j.carbpol.2011.03.028.

[10] Cai Z, Gardner D J, Tshabalala M A, Peng Y, Han Y and Kiziltas A 2013 Influence of drying method on the material properties of nanocellulose I: thermostability and crystallinity *Cellulose* **20** 2379–92. doi:10.1007/s10570-013-0019-z.

[11] Peng Y, Gardner D J, Han Y, Cai Z and Tshabalala M A 2013 Drying Cellulose Nanocrystal Suspensions.

[12] Revati R, Abdul Majid M S, Ridzuan M J M, Normahira M, Mohd Nasir N F, Rahman Y M N, et al. 2017 Mechanical, thermal and morphological characterisation of 3D porous Pennisetum purpureum/PLA biocomposites scaffold. *Mater Sci Eng C* **75** 752–9. doi:10.1016/j.msec.2017.02.127.

[13] An X, Wen Y, Cheng D, Zhu X and Ni Y 2016 Preparation of cellulose nano-crystals through a sequential process of cellulase pretreatment and acid hydrolysis *Cellulose* **23** 2409–20. doi:10.1007/s10570-016-0964-4.

IOP Conf. Series: Materials Science and Engineering **670** (2019) 012045 doi:10.1088/1757-899X/670/1/012045

Biomechanical analysis of patient-specific femur model of osteogenesis imperfecta with cortical and cancellous bone

L C Tan[1], M H Mat Som[1], K S Basaruddin[1], H Yazid[1], R Daud[1] and M S Awang[2]

[1]School of Mechatronic Engineering, Universiti Malaysia Perlis, Pauh Putra Campus, 02600, Arau, Perlis, Malaysia.
[2]Kuliyyah of Medicine, International Islamic University Malaysia, Jalan Sultan Ahmad Shah, Bandar Indera Mahkota, 25200 Kuantan, Pahang, Malaysia.

Abstract. Osteogenesis imperfecta (OI) is a fragile bone disease characterized by easy fractures. The femur consists of cortical and cancellous bone, each with different mechanical properties. Bone fractures often occur throughout patients' lifetime. However, doctors still have no quantitative method to predict fractures. This project's purpose is to investigate the mechanical behaviour of patient-specific OI femur from the finite element analysis. The fracture risk in daily activities (ADL) were examined. All the stress values were judged by the fracture criteria, assumed as 115 MPa. The exercises that exerted force more than 6 times of body weight could cause fractures. Cancellous bone was not affected in any case of ADL. The effects of force and stress on cancellous bone and its impact on fracture risk are negligible.

1. Introduction

Osteogenesis Imperfecta (OI), also known as "brittle bone disease" is a genetic disorder characterized by bones that break easily. It is caused by the mutation in the COL1A1 and COL1A2 genes that encode type I procollagen. Bones of OI patients break easily and they break for unknown reasons. They will experience bone fracture frequently throughout their lifetime. However, there is still no established clinical method to predict the bone fracture. Thus, further complicate the medical intervention, since the physicians are unable to predict when and where the fracture will happen in OI patients.

Fritz et al. [1] developed a standard femur model geometry and altered it to match the morphology of the type I OI subject to predict the fracture risk in type I OI patients. Wanna et al. [2–4] also developed ten deformed femur models with different deformed angle using ANSYS software to predict the femoral fracture risk during standing in OI patients. The specimen's mechanical properties of both studies were assumed isotropic and identified by nanoindentation (Young's modulus = 19 GPa and Poisson's ratio = 0.3). The fracture risk was examined via the fracture risk formula created by Keyak et al. [5] and predicted the fracture strength as 115 MPa.

The purpose of this study was to develop OI patient-specific femoral FE models using CT images. This patient-specific model was used to analyse the distribution of mechanical stress, fracture risk, fracture load under various loads, and the number of affected voxels in the cortical and cancellous bone. It is hypothesized that cortical bone is more affected by stress based on its mechanical properties. The results of FEA on the femur provided important predictive information for determining the risk of fracture in OI patients.

IOP Conf. Series: Materials Science and Engineering **670** (2019) 012045 doi:10.1088/1757-899X/670/1/012045

2. Methodology

A. Reconstruction of 3D Femoral Model

The 3D femoral finite element models were reconstructed using VOXELCON (Quint Corporation, Japan) by using type III OI images obtained from IIUM Medical Centre. The 3D femoral model was made up of 220 images of cortical bone and 220 images of cancellous bone. The cortical and cancellous bone were segmented from the image using global thresholding method [6]. By assigning 1 mm to width, 1 mm to height and 1mm to slice pitch of a pixel, a voxel with a volume of 1 mm^3 was created.

The mechanical properties were assigned based on the literature by Fan et al. on pediatric OI bone specimens [7], characterized by nanoindentation tests [1,8]. The femoral model is assumed to be isotropic, and Young's modulus of the cortical and cancellous bone are 19 GPa and 10 GPa, respectively, and the Poisson's ratio is 0.3.

B. Finite Element Analysis (FEA)

The 3D femoral model meshed for finite element analysis. The voxel element size of the femur model generated was based on the pixel size of original CT image. Therefore, the convergence test is not required. For all analysis cases, the voxel model was constrained at the distal femur as a fixed support to prevent motion. Figure 1 shows that the force, F was applied perpendicular to the proximal region of the sagittal femur, and the distal end of the femoral stem serves as a fixed support.

The patient was a 13 years old girl, and her weight of patient was assumed as 37.5 kg based on the literature of Graff et al. [9] in the OI children development chart. Five loading conditions were applied to the distal region of the femur model to simulate standing, walking and jumping and study their fracture risk. According to Bankoff's work [10], the femur needs to absorb approximately one-third of human body weight during standing (stance phase) and 3-7 times the body weight when walking. The proximal femur will absorb 15 to 20 times the bodyweight of the compressive forces while jumping. In order to estimate the optimal force that the femur can resist, 1000 N of force was applied to the femur to obtain the maximum von Mises stress. Next, the fracture load can be calculated using eq. (1).

$$\frac{1000\ \mathrm{N}}{\mathrm{F_{frac}}} = \frac{\mathrm{VMS_{1000}}}{115\ \mathrm{MPa}} \tag{1}$$

where VMS_{1000} is the maximum von Mises stress upon 1000 N of force and F_{frac} is the fracture load.

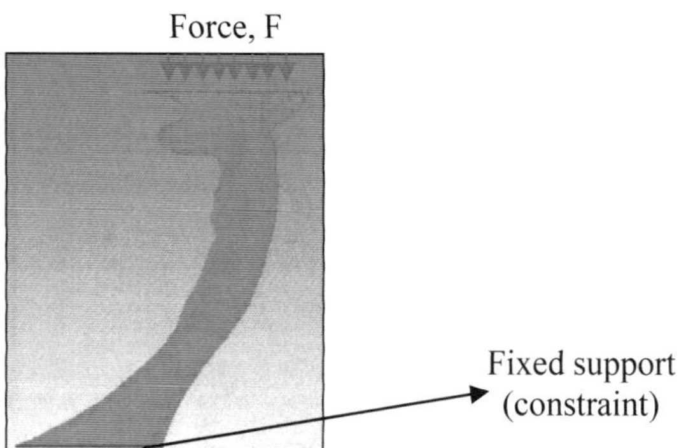

Figure 1: Surface force applied on the proximal femoral head with fixed support (constraint).

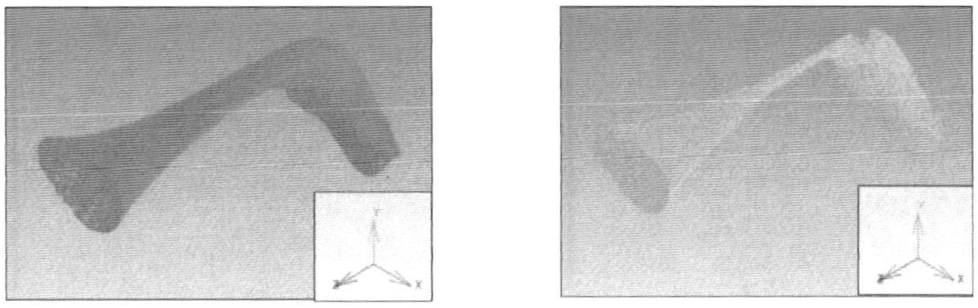

(a) (b)

Figure 2: xyz- plane: (a) Cortical bone component and (b) Cancellous bone component.

F. Fracture Risk

The fracture risk was calculated based on eq. (2), which was developed by Keyak et al [5]. The fracture strength of 115 MPa was assumed based on the nanoindentation test done by the previous study. When the fracture risk value is equal to or less than 1 (FR $\leq$1), the femur is considered fractured.

$$\mathrm{FR} = \frac{\text{Fracture strength}}{\text{von Mises stress}} \tag{2}$$

3. Results and Discussion

A. Reconstruction of 3D Femoral Model

Figure 2 shows the 3D femoral model that was reconstructed using VOXELCON software. This femur bone model consists of 220 cortical bone images and cancellous bone images. The whole femoral model has a total of 383,066 voxels elements made up from 190,955 voxels elements of cortical bone and 192,111 voxels elements of cancellous bone.

B. Finite Element Analysis (FEA)

Figure 3 shows the higher the load force is, the greater the area with the high-stress intensity (red). It was found that the von Mises stress of maximum walking loading and both of maximum and minimum jumping condition exceeded the fracture strength, 115 MPa.

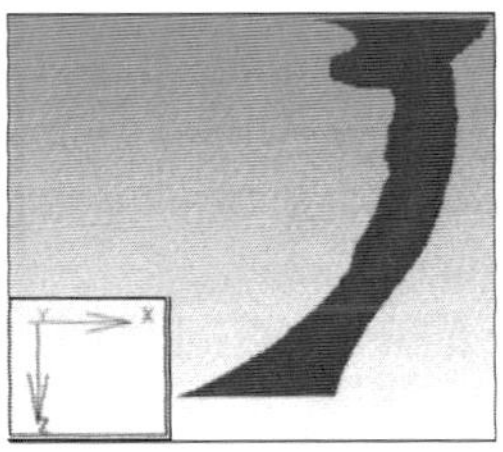

(a) Standing : 1/3 × BW
(7.7 MPa)

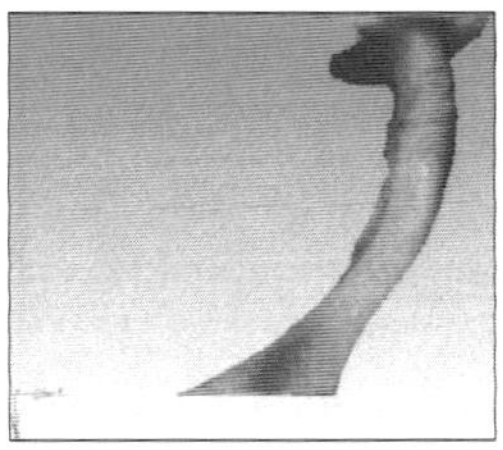

(b) Walking (min) : 3 × BW
(62.9 MPa)

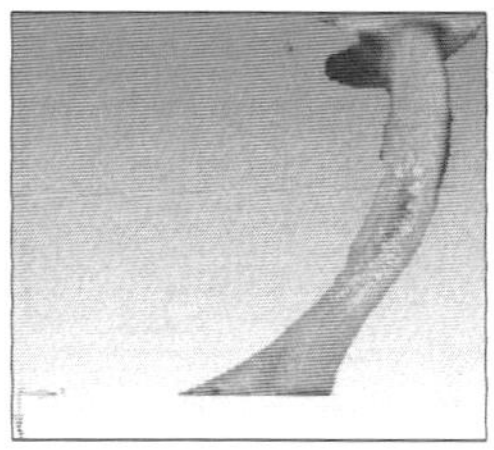

(c) Walking (max) : 7 × BW
(146.8 MPa)

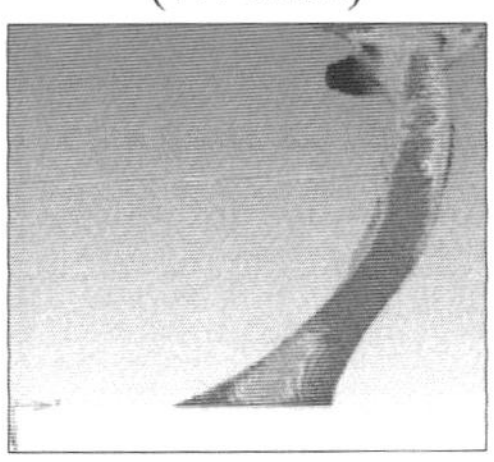

(d) Jumping (min) : 15 × BW
(314.5 MPa)

(e) Jumping (max) : 20 × BW
(419.3 MPa)

Figure 3: Von Mises stress distribution of femoral model under different loading conditions.

IOP Conf. Series: Materials Science and Engineering **670** (2019) 012045 doi:10.1088/1757-899X/670/1/012045

Figure 4 shows that the more intense the activity, the lower the value, the higher the risk of fracture. The standing and walking with 3 times the body weight force will not cause any fracture to OI patients due to their high value of fracture risk. When the proximal femur absorbs 7 times or more of body weight, the value of fracture risk drops below 1. This shows that these activities are dangerous to OI patients.

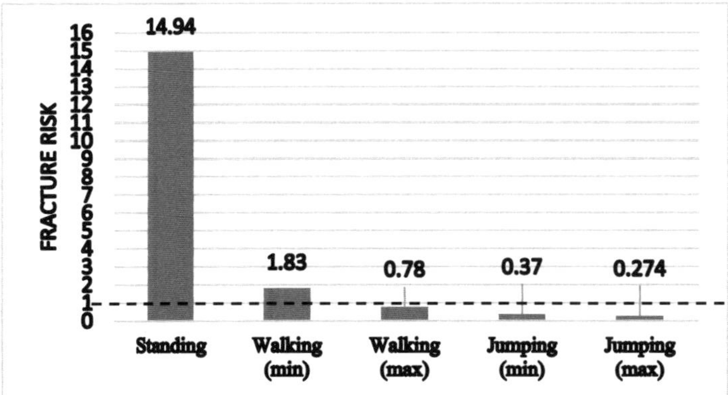

Figure 4: Fracture risk of the different daily activity. The lower the value, the higher the fracture risk.

This study reveals the feasibility of using the segmented cortical and cancellous images to create the geometric shape of the OI femur. The key parameters involved are the femur geometry, material properties and boundary conditions. Unlike normal bone tissue, the OI femur shows more isotropic material properties. This is considered to be the result of an abnormal collagen network [7]. The Young's modulus of cortical and cancellous bones is lower compared to normal bone. However, the cortical bone Young's modulus decreased more than cancellous bone. This is due to its higher collagen density, which leads to more sensitivity to collagen network deformation [7].

One of the goals of this study is to investigate the mechanical behaviour of OI femur from the finite element analysis. The stress distribution of von Mises stress indicates the starting point of fracture. The stress is concentrated in the deformed area, the diaphysis. If the force applied to the femur exceeded its stress limits, this critical site will become a fracture site. This is believed to be caused by the lateral curvature of OI femur. The lateral bending that causes more bone is away from the mechanical axis. Therefore, when standing, walking, or jumping, the compression load of the femur on the hip causes the bending moment to increase. This leads to an increase in the maximum stress on the lateral side of the bone. As compared to normal bone, OI femur diaphysis also narrower.

From the result, it reflects the fact that exercises that exert force more than 6 times of body weight can cause fractures. This shows the OI patient able to withstand their own weight during standing and normal walking. The finding of standing is similar to Wanna et al. [2].

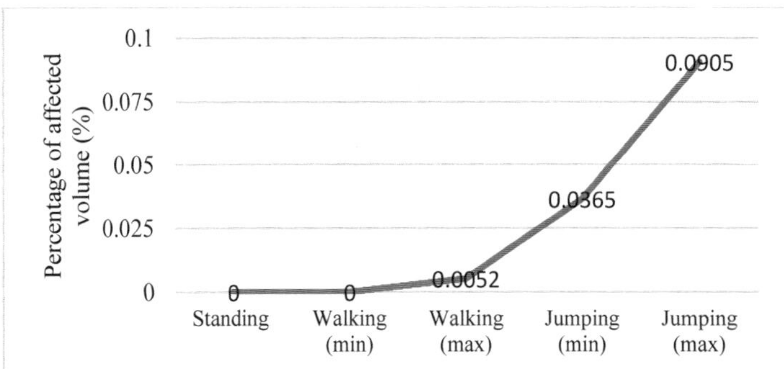

Figure 5: Percentage of affected volume against activities of daily living.

This study also indicates the affected voxel number of the cortical and cancellous bone based on the fracture criteria, where the fracture strength assumed as 115 MPa. Figure 5 shows percentage of volume affected increases drastically when the compression force exceeds 7 times or more of the body weight. No cancellous voxel was found to exceed the fracture strength. Most of force is absorbed by the cortical bone, and the effect of stress on the cancellous bone is negligible. The stress will be concentrated at the bowed area, as shown in the von Mises stress distribution of femur model and the fracture will be initiated at this site.

4. Conclusion

From the study, the OI patient can bear her weight during standing and normal walking. However, walk too fast and jumping are very risky to them. From the results, it was found that only cortical bone was affected when force was applied. Only when very high forces are applied can the cancellous bone be affected, which far exceeds the fracture load. It was found that the effects of force and stress on cancellous bone and its impact on fracture risk are negligible.

Acknowledgement

The authors would like to acknowledge the support from the Fundamental Research Grant Scheme (FRGS) under a grant number of FRGS/1/2016/TK03/UNIMAP/02/6 from the Ministry of Education Malaysia.

References

[1] Fritz J M, Guan Y, Wang M, Smith P A and Harris G F 2009 A fracture risk assessment model of the femur in children with osteogenesis imperfecta (OI) during gait *Med. Eng. Phys.* **31** 1043–8.

[2] Wanna S B C, Basaruddin K S, Mat Som M H, Mohamad Hashim M S, Daud R, Abdul Majid M S and Sulaiman A R 2017 Prediction on fracture risk of femur with Osteogenesis Imperfecta using finite element models: Preliminary study *Journal of Physics: Conference Series* **908** 012022.

[3] Wanna S B C, Basaruddin K S, Mat Som M H, Sulaiman A R, Shukrimi A, Khan S F, Abdul Majid M S and Ridzuan M J M 2018 Fracture risk prediction on children with Osteogenesis Imperfecta subjected to loads under activity of daily living *IOP Conference Series: Materials Science and Engineering* **429** 012004.

[4] Wanna S B C, Basaruddin K S, Som M H M, Salleh A F and Sulaiman A R 2018 Effect of loading direction on fracture of bone with osteogenesis imperfecta (OI) during standing *AIP Conference Proceedings* (2030) 020094.

[5] Keyak J H, Rossi S A, Jones K A and Skinner H B 1998 Prediction of femoral fracture load using automated finite element modeling *J. Biomech.* **31** 125–33.

[6] Mansor N, Beh Y M, Mat Som M H, Basaruddin K S, Mustafa N, Yazid H and Salleh A F 2018 Reconstruction of cortical and cancellous bone in tibia with osteogenesis imperfecta *J. Telecommun. Electron. Comput. Eng.* **10** 115–9.

[7] Fan Z, Smith P A, Eckstein E C and Harris G F 2006 Mechanical properties of OI type III bone tissue measured by nanoindentation *J. Biomed. Mater. Res. - Part A* **79** 71–7

[8] Sulong I, Basaruddin K S, Mat Som M H and Daud R 2018 Nanoindentation measurement on mechanical properties of Osteogenesis Imperfecta (OI): A review *ARPN J. Eng. Appl. Sci.* **13** 3435–40

[9] Graff K and Syczewska M 2017 Developmental charts for children with osteogenesis imperfecta, type I (body height, body weight and BMI). *Eur. J. Pediatr.* **176** 311–6

[10] Pria Bankoff A D 2012 Biomechanical Characteristics of the Bone *Human Musculoskeletal Biomechanics*

Optimization of micro-perforated sound absorber using Particle Swarm Optimization (PSO)

W H Tan[1, a], R Haslina[1], E A Lim[2] and H G Chuah[3]

[1] School of Mechatronic Engineering, Universiti Malaysia Perlis (UniMAP), Pauh Putra Campus, 02600 Arau, Perlis, Malaysia.
[2] Institute of Engineering Mathematics, Universiti Malaysia Perlis (UniMAP), Pauh Putra Campus, 02600 Arau, Perlis, Malaysia.
[3] Department of Engineering, Computing and Build Environment, KDU Penang College Univeristy, 32, Jalan Anson, George Town, 10400 George Town, Pulau Pinang, Malaysia.
[a] Correspondng Author: whtan@unimap.edu.my

Abstract. This study investigates the acoustic properties of micro-perforated panel (MPP) as an alternative to passive noise control. In order to obtain high sound absorption over broad frequency band, the analysis of sound absorption performance of MPP sound absorber should be conducted based on 3 major design parameters, such as perforation diameter, *d*, air cavity depth, *D*, and distance between perforations, *b*, of MPP. Next, the optimization of sound absorption coefficient for MPP sound absorber need to be done by using the Particle Swarm Optimization (PSO) algorithm and the optimum parameters of a single layer MPP sound absorber should be obtained for achieving higher sound absorption which the sound absorption coefficient, $\alpha > 0.5$ with a wider frequency band. The PSO is used in order to optimize the panel construction to provide maximum sound absorption coefficient in a determined wide band frequency range. Experiments are carried out at normal sound incidence and plane waves. A good agreement is achieved between the theory and the experiments.

1. Introduction

Micro-perforated panel (MPP) sound absorber is expected a significant sound absorption material [1–7]. The construction of MPP sound absorber is built from a perforated panel with an air space between the panel and rigid wall. The perforations on the panel are in sub-millimetre size in diameter. Figure 1 and 2 shows the construction of a MPP sound absorber. The construction of MPP sound absorber is considered relatively simple and uncomplex to be built. The acoustic performance of a MPP sound absorber is determined by perforation diameter (*d*), thickness of panel (*t*), distance between perforations (*b*), and the air gap depth (*D*) as depicted in Figure 1 and 2.

In 1975, Maa [1] proposed the first idea of MPP sound absorber. Based on Maa's theory, increase the density of perforation and decrease the perforation diameter able to achieve higher sound absorption for a wide frequency band [6]. In this study, optimization algorithm of particle swarm optimization (PSO) is applied to optimize the set of MPP sound absorber parameters for obtaining a better sound absorption coefficient.

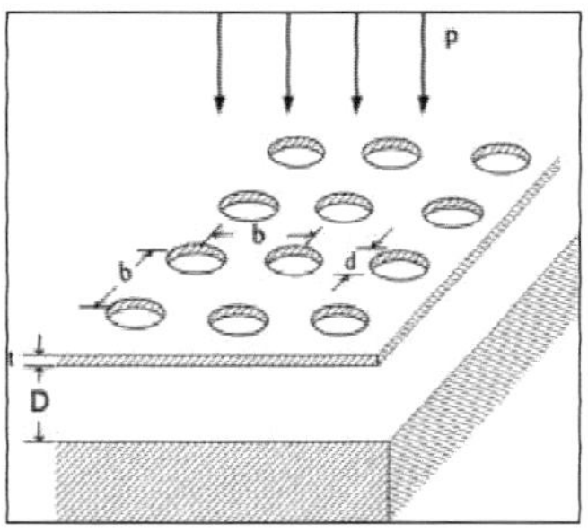

Figure 1. Layout of MPP sound absorber [5].

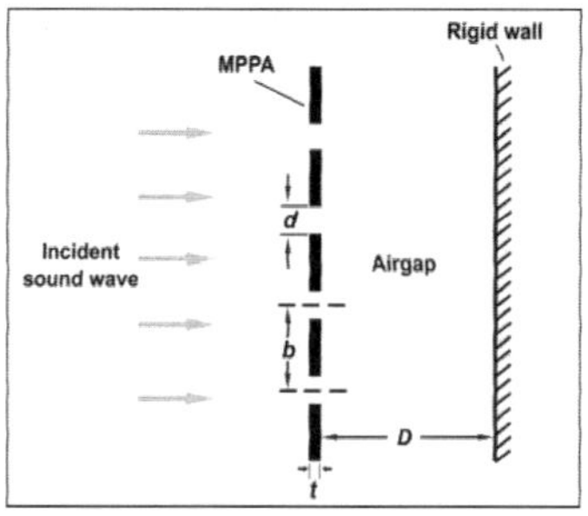

Figure 2. Schematic diagram of MPP backed with air gap [6]

2. Acoustical theory for MPP sound absorber

Acoustic performance of a micro-perforated panel (MPP) sound absorber is completely determined by perforation diameter (*d*), thickness of panel (*t*), distance between perforations (*b*), and the air gap depth (*D*), which was first put forward by Maa [1].

Maa observed the discontinuity between the two cases. He developed an approximation solution for the perforation in sub-millimetre size, which made *k* lie between 1 and 10. This structure is named as MPP sound absorber.

$$Z_1 = \frac{32\rho\mu t}{d^2}\sqrt{1+\frac{k^2}{32}} + j\omega pt\left(1+\frac{1}{\sqrt{3^2+\frac{k^2}{2}}}\right) \quad (1)$$

where $k=\sqrt{\left(\frac{\omega}{\mu}\right).\left(\frac{d}{2}\right)}$. According to Maa's theory, acoustic resistance of perforations becomes significant as the perforations size are very small without additional porous material when comparing with the conventional formula of perforated panel sound absorbers. By comparing with exact values, model proposed by Maa agrees well with the error less than 6% [1].

$$Z_{MPP} = \frac{Z_1}{\sigma\rho c} = r + j\omega m \quad (2)$$

With

$$r = \frac{32\mu}{\sigma c}\frac{t}{d^2}\left(\sqrt{1+\frac{k^2}{32}} + \frac{\sqrt{2}\,k}{8}\frac{d}{t}\right) \quad (3)$$

$$m = \frac{t}{\sigma c}\left(1+\frac{1}{\sqrt{3^2+\frac{k^2}{2}}} + 0.85\frac{d}{t}\right) \quad (4)$$

where σ is the perforation ratio, $\sigma = 78.5\ d^2/b^2$. If perforation is square lattices arrangement, *b*: distance between perforations, *c*: sound velocity, *r*: normalized specific acoustic resistance of MPP, and *m*: normalized acoustic mass of MPP. The normal specific acoustic impedance of the air gap behind MPP, normalized by ρc, becomes

$$Z_D = -j\cot\left(\frac{\omega D}{c}\right) \quad (5)$$

Normalized specific acoustic impedance of a MPP sound absorber can be determined by

$$z = Z_{MPP} + Z_D \quad (6)$$

For normal incidence, the sound absorption coefficient of MPP sound absorber, α is obtained by

$$\alpha = 1 - \left|\frac{z-1}{z+1}\right|^2 \quad (7)$$

3. Experiment Setup

In this study, two-microphone method was used for the micro-perforated panel (MPP) sound absorber sound absorption coefficient measurement. This measurement setup consists of four main equipment that are needed while conducting the experiment. There are impedance tube, microphones, data acquisition (DAQ) device and a computer. The MPP is placed at the end of tube and sample insertion is adjusted to form an air gap. The opposite end of the tube is connected with a loudspeaker to generate the sound source signal. On the tube, a pair of microphones is mounted with the inner wall. The mounted microphones are used to measure the incident and reflected sound signal generated by the loudspeaker in the tube. Figure 3 shows the two-microphone impedance tube measurement setup used in this study.

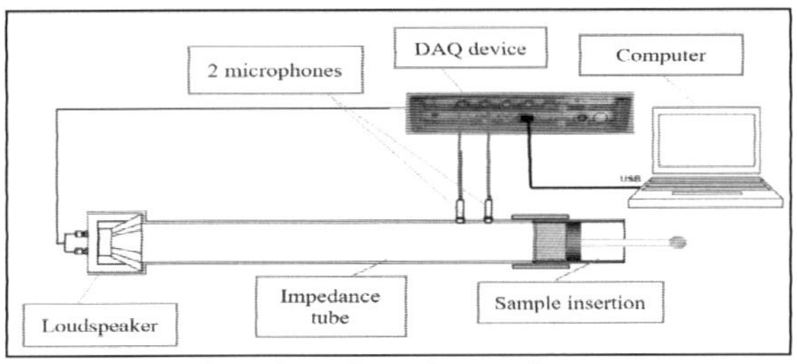

Figure 3. Two-microphone impedance tube measurement setup.

4. Analysis of Experimental Measurement

In this experiment, four samples of micro-perforated panel (MPP) sound absorber were prepared for sound absorption coefficient measurement. Perforation ratio, σ of the MPP sound absorber was calculated by the formula $\sigma = 78.5\ d^2/b^2$. The structural parameters of MPP are listed in Table 1.

Table 1. Structural parameters of MPP test specimen.

Specimen	Specimen diameter, *D* (mm)	Specimen thickness, *t* (mm)	Holes spacing, *b* (mm)	Holes diameter, *d* (mm)	Perforation ratio, σ (%)
A	108	1	3	0.5	2.18
B	108	1	7	0.5	0.40
C	108	1	5	0.9	2.54
D	108	1	5	0.3	0.28

4.1 Effect of air cavity, D

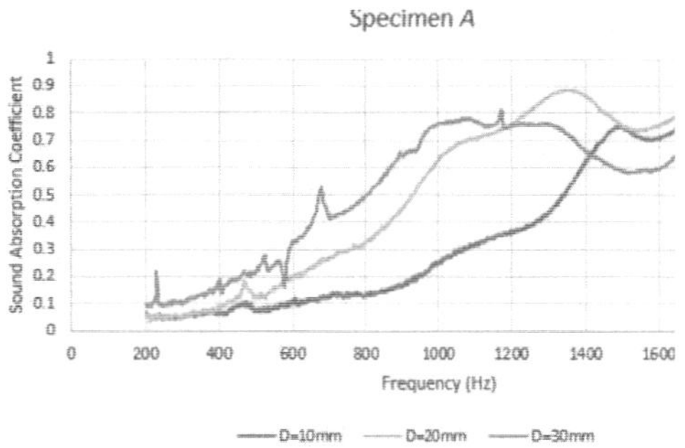

Figure 4. Sound absorption coefficient of Specimen *A* with multiple size air cavity

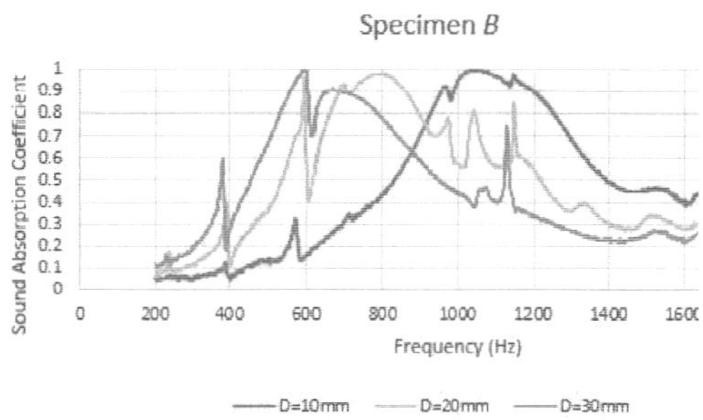

Figure 5. Sound absorption coefficient of Specimen *B* with multiple size air cavity.

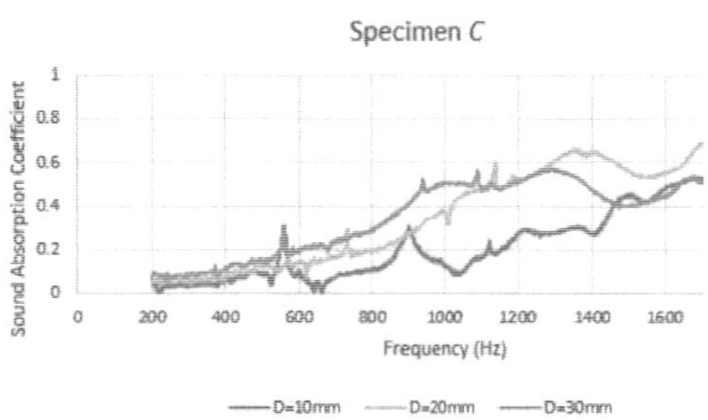

Figure 6. Sound absorption coefficient of Specimen C with multiple size air cavity

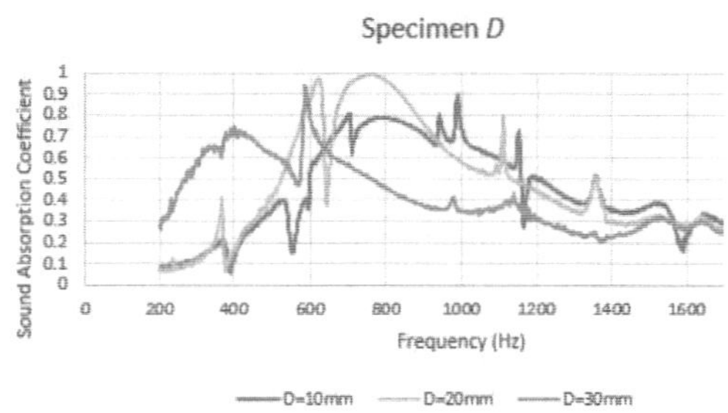

Figure 7. Sound absorption coefficient of Specimen *D* with multiple size air cavity

Figure 4 to 7 show the sound absorption coefficient (SAC) curves of four MPPs with different depth of air cavity (10mm, 20mm and 30mm). It is showed that all SAC figures have a similar variation regardless of perforation ratio.

It is observed that the width of maximum SAC increases when the depth of air cavity increasing. At the same time, the maximum SAC is found shifts to lower frequency range. This observation can be explained due to both open ends of hole acted as an acoustic mass, and the fluid in air cavity is acted as an acoustic spring. A spring mass resonance is created for this manner. SAC Peak occurs when the stiffness of air cavity cancels the mass of holes. However, the increasing of air cavity depth will reduce the stiffness of air cavity and SAC peak will be moved towards the lower frequencies, and the bandwidth of SAC peak becomes wider. In addition, the results show a smaller perforation ratio (Sample D, $\sigma = 0.28\%$, $b = 5$mm), the maximum SAC value increases simultaneously with the increasing of air cavity depth. On the other hand, maximum SAC decreases progressively as air cavity depth increases for higher perforation ratio (Sample *A* and Sample *C*). It can be noticed that the SAC peak of Sample *B* (σ = 0.4%, *b* = 7mm) is the highest and remains constant for the changing of air cavity depth.

In spite of this, for a higher perforation ratio (Sample *C*, σ = 2.54%, *b* = 5mm), the sound absorption capabilities are not satisfactory. Thus, the influencing parameters of thickness, perforation ratio and air cavity depth of MPP sound absorber should be combined in order to design a MPP sound absorber which able to obtain desired sound absorption capability. It is also alternate to attach porous material layer to wider the frequency bandwidth of SAC.

4.2 Effect of perforation ratio, σ.

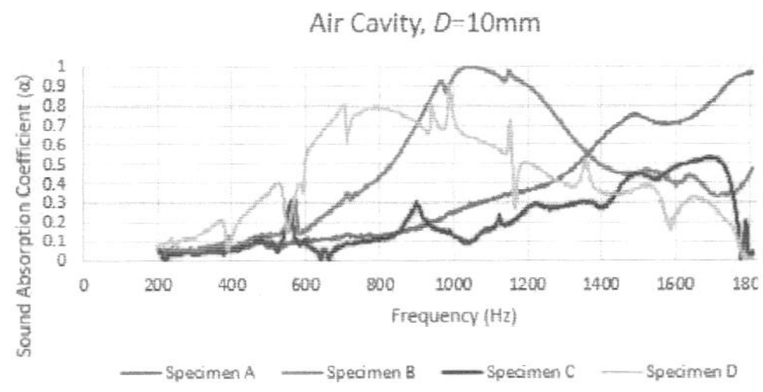

Figure 8. Sound absorption coefficient of MPP sound absorbers with air cavity, D = 10mm.

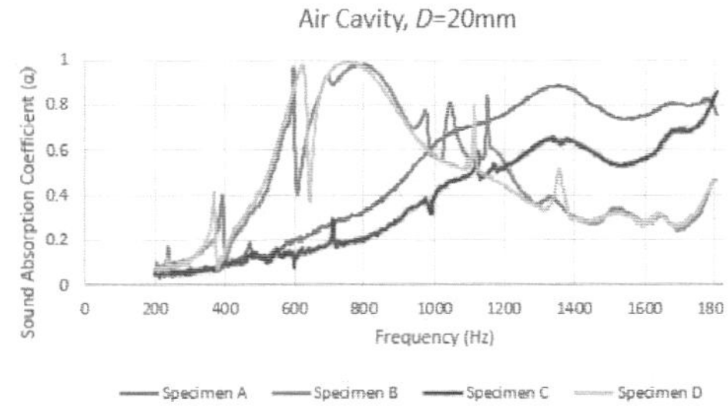

Figure 9. Sound absorption coefficient of MPP sound absorbers with air cavity, D = 20mm.

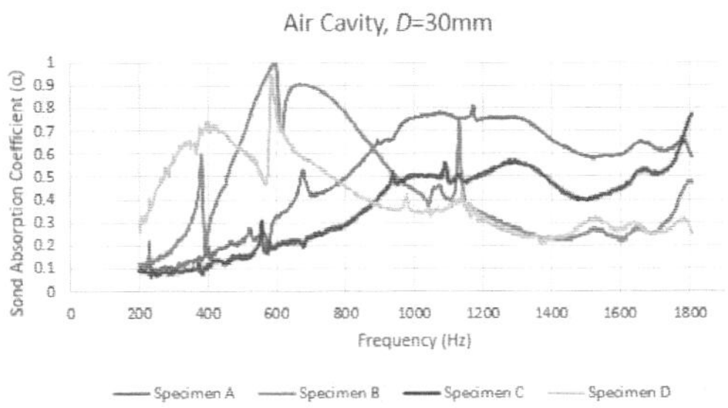

Figure 10. Sound absorption coefficient of MPP sound absorbers with air cavity, D = 30mm.

The effect of perforation ratio on sound absorption coefficient (SAC) of the four MPPs, when backed by an air cavity is shown in Figure 8 to10. In this study, the air cavity depths are 10mm, 20mm and 30mm.

Based on Figure 8 to 10, it is observed that increasing the perforation ratio of a MPP sound absorber yields a higher acoustic resonance frequency for the SAC peak. This observation is caused by increasing of perforation ratio results in the decrease of total acoustic mass for all of the perforations. Therefore, it increases the resonant frequency at which the SAC peak occurs. The outcomes also demonstrated that the SAC peak and the corresponding frequencies are similar for the case of an MPP sound absorber backed by an air cavity.

For the lower frequencies, MPP sound absorber with smaller perforation ratio will contribute a better SAC, whereas higher perforation ratio delivers a better SAC for the higher frequencies. It is necessary to adjust the perforation ratio of MPP sound absorber or combine with porous material layer to obtain the desired SAC peak frequency and a wider frequency range,.

4.3 Sound absorption of micro-perforated panel sound absorber by Particle Swarm Optimization (PSO).

In MATLAB particle swarm optimization (PSO) program, the range value of hole spacing, *b,* hole diameter, *d* and air cavity, *D* for Specimen *A*, Specimen *B*, Specimen *C* and Specimen *D* has been set. So, the exact best value which is the widest width of the graph for hole spacing, *b,* hole diameter, *d* and air cavity, *D* can be generated based on the widest frequency bandwidth of the graph of each specimen. Based on the result of the MATLAB for the PSO, the value for hole spacing, *b,* hole diameter, *d,* air cavity, *D* and the best value for the widest frequency bandwidth result from the graph are shown in Table 2.

Table 2. PSO result for MATLAB.

Result	Length
Hole spacing, *b*	6.6356 mm
Hole diameter, *d*	0.8755 mm
Air cavity, *D*	24.4078 mm
Widest frequency bandwidth	1999.2011 Hz

5. Conclusion

In this study, it is found that the acoustic resonant frequency depends on the depth of air cavity behind the micro-perforated panel (MPP). The acoustic resonant frequency of the corresponding value of sound absorption was reduced with increasing the depth of the air cavity behind the MPP inside the impedance tube. For the perforation ratio at lower frequencies, a smaller perforation ratio of the MPP layer gives a better sound absorption coefficient (SAC) whereas a higher perforation ratio gives a better SAC at higher frequencies. The effect of SAC for different holes diameter and constant holes spacing shows that reducing the size of the hole diameter can increase the SAC but narrow the frequency band in condition of uniform holes spacing and distance of the air cavity. Lastly, the optimization of SAC from the MPP sound absorber that is analysed by using particle swarm optimization (PSO) gives the result of the value for hole spacing, *b,* hole diameter, *d* and air cavity, *D* and the best value for the widest frequency bandwidth. Therefore, the present work analyzed the absorption properties of MPP sound absorber by using PSO, in order to obtain high sound absorption over a broad frequency band. The attention was focused on the influence of the hole spacing, *b,* hole diameter, *d* and air cavity, *D,* since varying this parameter represents an easy way to modify the sound absorption properties of the panel.

Acknowledgement

Authors wishing to acknowledge assistance or encouragement from colleagues, special grateful to Dr Haniza binti Yazid, lecturer from Biomedical Electronics Engineering Programme, UniMAP for guiding the MATLAB application. Author also would like to thank School of Mechatronic University Malaysia Perlis (UniMAP) for providing the equipment and environment to success the study.

References

[1] Maa D Y 1975 *Sci Sin.* **18** 55.
[2] Maa D Y 1987 *J. Noise Control Eng* **29** 77.
[3] Maa D Y 1998 *J Acoust Soc Am.* **104** 2861.
[4] Maa D Y 2018 *Shengxue Xuebao/Acta Acust.* **31** (6) 481.
[5] Maa D Y 2007 *Int J Acoust Vib.* **29** 3.
[6] Qian Y J, Kong D Y, Liu S M, Sun S M, Zhao Z 2013 *Appl Acoust.* **74** (7) 931.
[7] Liu Z, Zhan J, Fard M, Davy J L 2017 *Appl Acoust.* **121** 25.

IOP Conf. Series: Materials Science and Engineering **670** (2019) 012047 doi:10.1088/1757-899X/670/1/012047

Dielectric properties of kenaf filled epoxy composites

M J M Ridzuan[1*], L S S Liew[1], M S Abdul Majid[1], E M Cheng[1], A Khasri[2] and A Z Ahmad Firdaus[1]

[1]School of Mechatronic Engineering, Universiti Malaysia Perlis, Pauh Putra Campus, 02600 Arau, Perlis, Malaysia.
[2]Faculty of Engineering Technology, Universiti Malaysia Perlis, Unicity Alam Campus, 02100 Padang Besar, Perlis, Malaysia

* denotes corresponding author
Authors email addresses: ***ridzuanjamir@unimap.edu.my,** liewlss5303@gmail.com, shukry@unimap.edu.my, emcheng@unimap.edu.my, azduwin@unimap.edu.my, ahmadfirdaus@unimap.edu.my

Abstract. The dielectric properties of kenaf filled epoxy composites was investigated. Composites was fabricated using kenaf fibres and epoxy resin. The composites was manufactured using a cold press method; the volume fraction of the natural fibres were 7 and 14%. The dielectric test were conducted using Agilent/Keysight E8363B PNA Network Analyser. Result shows that the 14% Kenaf filled epoxy composites successfully decrease the dielectric constant between 12.4 GHz and 16.66 GHz. It shows that the dielectric constant increased with decreased on the fibre loading. Besides that, 14% Kenaf has a higher dissipation loss factor compared to 7% Kenaf filled epoxy composites and reached its peak at 17.89 GHz. In summary, the 14% Kenaf shows a lowest dielectric constant and permeability compared to 7% Kenaf filled epoxy composites.

1. Introduction

Nowadays, the composites formed by natural fibre/polymer has received big attention. Natural fibre/polymer-based composites with a great performance in mechanical properties, reasonable cost and low density had attracted the interest of the researchers [1]–[4]. Dielectric and electromagnetic materials with excellent dielectric and electromagnetic properties are of critical importance in various applications, such as electromagnetic interference (EMI), artificial muscles, drug release and energy storage capacitors [5]–[10]. The dielectric studies of minerals filled epoxy composite point out that the electric regular increased with the increased of filler loading [11]. The find out about of dielectric regular and dielectric loss as a feature of temperature and frequency is one of the most convenient and sensitive techniques of analysing the polymeric structure. Electric properties of pineapple reinforced polyethene composites have been studied by Jayamol et. al. [12]. They discovered that the expansion in the dielectric regular of the composite with fibre loading used to improve orientation and interfacial polarization. The electric properties of sisal fibre strengthened composite showed that the composite has electric powered anisotropic behaviour [13]. In this study, experimental were performed to evaluate the dielectric and electromagnetic properties of kenaf filled epoxy composites under dry condition.

IOP Conf. Series: Materials Science and Engineering **670** (2019) 012047 doi:10.1088/1757-899X/670/1/012047

2. Material and Methods

2.1 Materials
The composites consisted of kenaf fibres, and epoxy resin. Kenaf fibres were supplied from local supplier in Malaysia. EpoxAmite 100 series resin was used as a matrix, and mixed with a hardener at a ratio of 3:1.

2.2 Preparation of fillers
The Kenaf fibres was entered inside the grinding machine to produces fillers. Retsch sieve shaker was used to sieve the filler and separate it into various size. Filler size of the kenaf fibres is less than 500 μm.

2.3 Composites fabrication
Composites were fabricated using Kenaf fillers and epoxy resin. The volume fraction of kenaf fillers is 7 and 14%. The composites were manufactured using a cold press method. The fillers weighted by using Hanchen Analytical Balance JJ224BC high precision digital balance scale. The measurement can measure up to 200g and the tolerance is 0.0001g. Epoxy resin and hardener were measured using A&D SK-5001WP Washdown Scale.

2.4 Dielectric and Elctromagnetic test
The machine used to test the dielectric and electromagnetic properties of kenaf filled epoxy composites was Agilent/Keysight E8363B PNA Network Analyser and the frequency range used to conduct the experiment was P-band which between 12.4GHz and 18 GHz. In order to test the dielectric and electromagnetic properties of the kenaf fibre filled epoxy composites, the small piece of sample was inserted into the hollow middle holder. The result obtained was in both graphical and numerical form which included real and imaginary part of permittivity and permeability respectively.

3. Results and discussion

Figure 1 shows the effect of the Kenaf fibre loading on the dielectric constant for dry conditions. The 14% Kenaf filled with epoxy composites successfully decrease the dielectric constant between 12.4 GHz and 16.66 GHz. It shows that the dielectric constant increased with decreased on the fibre loading. The decrease in resistivity on the incorporation of fillers may arise from the presence of more polar groups, which facilitate the conducting process [14]. The incorporation of fillers into elastomers, generally decreases the volume resistivity and there is a critical loading of fibre, i.e., percolation threshold, at which the composites change from insulating to conducting material [15].

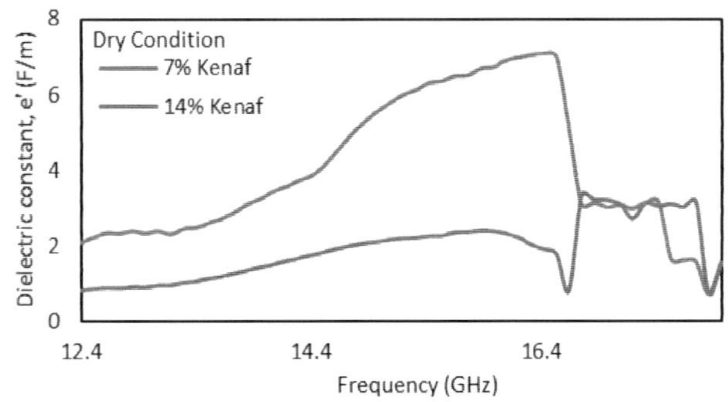

Figure 1: Dielectric constant of Kenaf filled epoxy composites

Figure 2 shows the effect of Kenaf fibre loading on the dissipation loss factor under dry conditions. According to Figure 2, the Kenaf fibre loading shows insignificant effect on the dissipation loss factor between 12.4 GHz and 17.22 GHz. However, 14% Kenaf has a higher dissipation loss factor compared to 7% Kenaf filled epoxy composites and reached its peak at 17.89 GHz. The peak of tan δ is associated with molecule movement and amount of energy dissipation. In composite material, energy dissipation is dependent on fibre/matrix interaction, a molecule moment in the polymer chain, strength of fibre, fibre breakage and crack propagation in composite material [16].

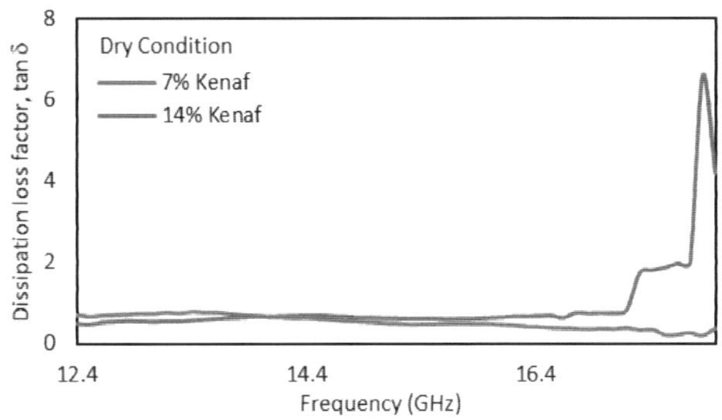

Figure 2: Dissipation loss factor of Kenaf filled epoxy composites

Figure 3 shows the effect of Kenaf fibre loading on the permeability for dry conditions. Based on the figure, the permeability for 14% Kenaf is the lowest compared to 7% Kenaf filled epoxy composites between 12.4 and 16.77 GHz. Magnetic permeability (μ) is the ability of a magnetic material to support magnetic field development. In other words, magnetic permeability is the constant in the proportionality between magnetic induction and magnetic field intensity. The greater the magnetic permeability of the material, the greater the conductivity for magnetic lines of force, and vice versa.

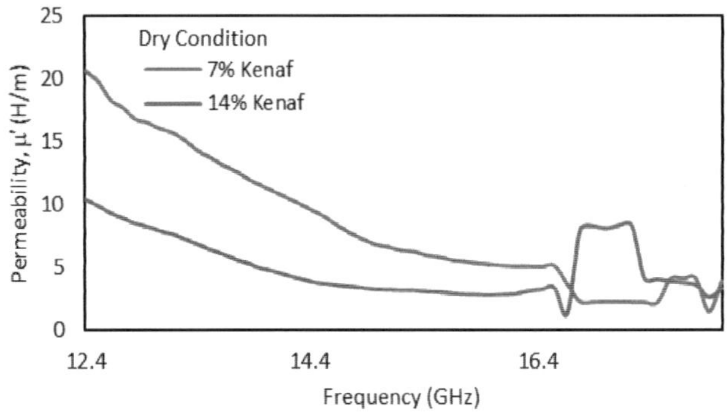

Figure 3: Permeability of Kenaf filled epoxy composites

4. Conclusion

In summary, the dielectric properties of kenaf filled epoxy composites was investigated. It shows that the dielectric constant increased with decreased on the fibre loading. the Kenaf fibre loading shows insignificant effect on the dissipation loss factor between 12.4 GHz and 17.22 GHz. However, 14% Kenaf has a higher dissipation loss factor compared to 7% Kenaf filled epoxy composites and reached its peak at 17.89 GHz. 14%. The Kenaf is the lowest compared to 7% Kenaf filled epoxy composites between 12.4 and 16.77 GHz.

Acknowledgement

This research was financed by the Ministry of Education, Malaysia, through the Fundamental Research Grant Scheme; (Ref: FRGS/1/2018/STG07/UNIMAP/02/1). Thc authors would also like to acknowledge the School of Mechatronic Engineering, Universiti Malaysia Perlis (UniMAP) for the equipment and technical assistance. The authors are much gratitude to the dedicated staff for the fruitful discussions and input to the project.

References

[1] Pathania D and Singh D 2009 A review on electrical properties of fiber reinforced polymer composites *Int. J. Theor. Appl. Sci.* **1** (2) 34–37.
[2] Fartini M S, Abdul Majid M S, Ridzuan M J M, Amin N A M and Gibson A G 2016 Compressive properties of Napier (Pennisetum Purpureum) filled polyester composites *Plast. Rubber Compos.* **45** (3) 136–146.
[3] Ridzuan M J M, Abdul Majid M S, Khasri A, Basaruddin K S and Gibson A G 2019 Effect of moisture exposure and elevated temperatures on impact response of Pennisetum purpureum/glass-reinforced epoxy (PGRE) hybrid composites *Compos. Part B Eng.* **160** 84–93.
[4] Ridzuan M J M, Majid M S A, Afendi M, Rahman M N and Azduwin Y K 2016 Thermal, composition and morphological properties of untreated and alkali-treated napier grass fibres *J. Eng. Appl. Sci.* **11** (11) 2485–2489.
[5] Cao M -S, Yang J, Song W -L, Zhang D -Q, Wen B, Jin H -B, Hou Z -L and Yuan J 2012 Ferroferric Oxide/Multiwalled Carbon Nanotube vs Polyaniline/Ferroferric Oxide/Multiwalled Carbon Nanotube Multiheterostructures for Highly Effective Microwave Absorption *ACS Appl. Mater. Interfaces* **4** (12) 6949–6956.
[6] Yousefi N, Sun X, Lin X, Shen X, Jia J, Zhang B, Tang B, Chan M and Kim J -K 2014 Highly Aligned Graphene/Polymer Nanocomposites with Excellent Dielectric Properties for High-Performance Electromagnetic Interference Shielding *Adv. Mater.* **26** (31) 5480–5487.
[7] Agnihotri N, Chakrabarti K and De A 2015 Highly efficient electromagnetic interference shielding using graphite nanoplatelet/poly(3,4-ethylenedioxythiophene)–poly(styrenesulfonate)

composites with enhanced thermal conductivity *RSC Adv.* **5** (54) 43765–43771.
[8] Zhang X, Liang G, Chang J, Gu A, Yuan L and Zhang W 2012 The origin of the electric and dielectric behavior of expanded graphite–carbon nanotube/cyanate ester composites with very high dielectric constant and low dielectric loss *Carbon N. Y.* **50** (14) 4995–5007.
[9] Li Y, Shi Y, Cai F, Xue J, Chen F and Fu Q 2015 Graphene sheets segregated by barium titanate for polyvinylidene fluoride composites with high dielectric constant and ultralow loss tangent *Compos. Part A Appl. Sci. Manuf.* **78** 318–326.
[10] Chen Y, Lin B, Zhang X, Wang J, Lai C, Sun Y, Liu Y and Yang H 2014 Enhanced dielectric properties of amino-modified-CNT/polyimide composite films with a sandwich structure *J. Mater. Chem. A* **2** (34) 14118.
[11] Wu S -L and Tung I -C 1995 Dielectric studies of mineral-filled epoxy *Polym. Compos.* **16** (3) 233–239.
[12] Jayamol G, Bhagawan S S and Thomas S 1997 Electrical Properties of Pineapple Fibre Reinforced Polyethylene Composites *J. Polym. Eng.* **17** (5) 383–404.
[13] Chand N and Jain D 2005 Effect of sisal fibre orientation on electrical properties of sisal fibre reinforced epoxy composites *Compos. Part A Appl. Sci. Manuf.* **36** (5) 594–602.
[14] "Dielectric Properties Results of this chapter have been accepted for publication in Journal of Polymer Engineering."
[15] Thomas S, Sinturel C and Thomas R *Micro- and nanostructured epoxy/rubber blends*.
[16] Jayamani E, Hamdan S, Rahman M R and Bin Bakri M K 2014 Comparative study of dielectric properties of hybrid natural fiber composites *Procedia Eng.* **97** 536–544.

IOP Conf. Series: Materials Science and Engineering **670** (2019) 012048 doi:10.1088/1757-899X/670/1/012048

Optimise 3D printing parameter on the mechanical performance of PLA-wood fused filament fabrication

***S F Khan[1,2], M M Zukhi, H Zakaria[1], M A M Saad[1]**

[1]School of Mechatronic Engineering, Universiti Malaysia Perlis, Pauh Putra Main Campus, 02600 Arau, Perlis, Malaysia.
[2]Additive Manufacturing and Design Lab, School of Mechatronic Engineering, Universiti Malaysia Perlis, Malaysia.

*corresponding author: sfkhan@unimap.edu.my

Abstract. The effect of printing parameter of tensile and compression specimens on production cost and mechanical performance of the material which is PLA with 20% wood content had been identified. The printing parameters selected are nozzle temperature, raster angle and printing speed. The raster angle and printing speed shows higher impact of tensile specimen on production cost while compression specimen only printing speed. Furthermore, for tensile experiment, raster angle gives highest impact than other parameter while for compression experiment each printing parameter giving an equal reaction to mechanical performance. By plotting the S/N ratio graph, higher nozzle temperature, lower raster angle and printing speed show the optimum printing parameters for tensile specimen. While optimum printing parameters for compression specimen are lower nozzle temperature, higher raster angle and printing speed.

1. Introduction

Additive manufacturing (AM) is the technology used to build 3D objects where the joining process of material was added layer by layer such in fused filament fabrication (FFF). There are many types of polymer that used in FFF which common material are ABS and PLA [1]. Thermoplastic is usually used in additive manufacturing because of the unique physicochemical properties and characteristic such as low cost, light weight, easy to fabricate and variability of compositions [2]. Yedige Tlegenov et al. [3], the lower nozzle temperature had tended to clog more than higher nozzle temperature. K. G. Jaya Christiyan et al. [4] state the tensile strength will be increase as well as the printing speed decrease. Lastly, Y. Wang et al. [5] state the higher raster angle will lower tensile strength. These three printing parameters gave highly impact for mechanical performance of the material. Based on literature reviews, there are many studies about pure PLA and ABS but still less study about composite material which mixed particularly with PLA or ABS. Most current study on literatures or experiments finding about how printing parameter affect to mechanical performance of pure PLA and ABS. Lastly, the literatures or experiments of the pass usually were about 3D printing to find on printing parameter effect in mechanical performance but not on production cost.

2. Methodology

This experiment will focus on characterizing tensile and compression strength of fused filament fabrication parts produced with PLA with 20% content of wood powder utilizing open source 3D printing software. The addition of 20% wood content increases the modulus of elasticity (MOE) value and reduce the density, any further addition of wood will result in lower MOE values [6]. The printing filament was obtained from Magma3DP and used as it.

2.1. Design of Experiment

To design the experiment, Taguchi method was chosen to simplify the experimental while analysis of variance (ANOVA) was used to determine the influence of each printing parameter on each mechanical property. The selected parameter was list in the table which shown in **Table 1** before conducting Taguchi method.

Table 1: Selected printing parameter

Printing Parameter	Value
Nozzle temperature (°C)	195, 205, 215
Raster angle (°)	0, 45, 60
Printing speed (mm/s)	30, 40, 50

Taguchi table being plotted after identify the selected printing parameters which is shown in **Table 2.** There are nine levels for each testing (tensile and compression) according to Taguchi L_9.

Table 2: Taguchi model for experiment

Level	Nozzle Temperature (°C)	Raster Angle (°)	Printing Speed (mm/s)	No. of specimens
1	195	0	30	5
2	195	45	40	5
3	195	60	50	5
4	205	0	40	5
5	205	45	50	5
6	205	60	30	5
7	215	0	50	5
8	215	45	30	5
9	215	60	40	5

2.2. 3D printing specimens

The 3D model be designed using CATIA V5 software by referring ASTM standard [7] [8] for tensile and compression dimension and testing specification. Then the CAD translated to STL file in order to slice the 3D printing software. The tensile and compression specimens were printed based on Taguchi table. The material that used is particle reinforced polymer composite which is PLA with 20% wood content.

2.3. Mechanical Testing and Scanning Electron Microscope (SEM)

Tensile and compression were tested experimentally using universal testing machine (UTM) which namely as SHIMADZU AG-100kN. An UTM machine has maximum 100kN of load and been set up constantly with 2 mm/min of speed for all tensile specimens while 5 mm/min for all compression specimens. SEM had been utilized during fracture surface analysis of 3D printed specimens to provide a vision to the failure mode by using SEM machine from HITACHI TM3000.

3. Result and Discussion

3.1. Tensile Strength

Based from **Table 3**, the highest average of ultimate tensile stress is level 8 which have 215°C of nozzle temperature, 45° of raster angle and 30 mm/s of printing speed. The circle plotted which shown in **Figure 1** is S/N ratio of larger the better as the optimum for each printing parameter. The optimum tensile specimen for printing parameter of nozzle temperature, raster angle and printing speed are 215°C, 0° and 30 mm/s.

3.2. Compression Strength

Based from **Table 4**, the highest average of maximum compression stress is level 3 which have 195°C of nozzle temperature, 60° of raster angle and 50 mm/s of printing speed. The circle plotted which shown in **Figure 2** is S/N ratio of larger the better as the optimum for each printing parameter. The optimum compression specimen for printing parameter of nozzle temperature, raster angle and printing speed are 195°C, 60° and 50 mm/s.

Table 3: Tabulated raw data of tensile test

Level	Ultimate Tensile Strength (MPa)					Average (MPa)	S/N ratio
	1	2	3	4	5		
1	12.92067	12.54519	11.71875	13.82212	13.52163	12.905672 (Middle)	22.2156
2	11.64375	13.22115	12.31971	11.64375	11.94399	12.15447	21.6947
3	9.390144	11.49351	12.54519	12.54519	12.99567	11.79394	21.4332
4	13.14615	13.29639	13.44639	12.84567	12.31971	13.01086	22.2861
5	13.14615	13.14615	13.14615	12.54519	12.92067	12.98086	22.2661
6	12.09447	11.56851	11.86899	11.79375	11.64375	11.79389	21.4330
7	12.16947	12.46995	13.22115	13.59688	12.99567	12.89062	22.2055
8	13.14615	13.29639	13.89736	13.07091	13.14615	13.31139 (Highest)	22.4845
9	11.79375	11.79375	12.09447	11.41827	11.49351	11.71875 (Lowest)	21.3776

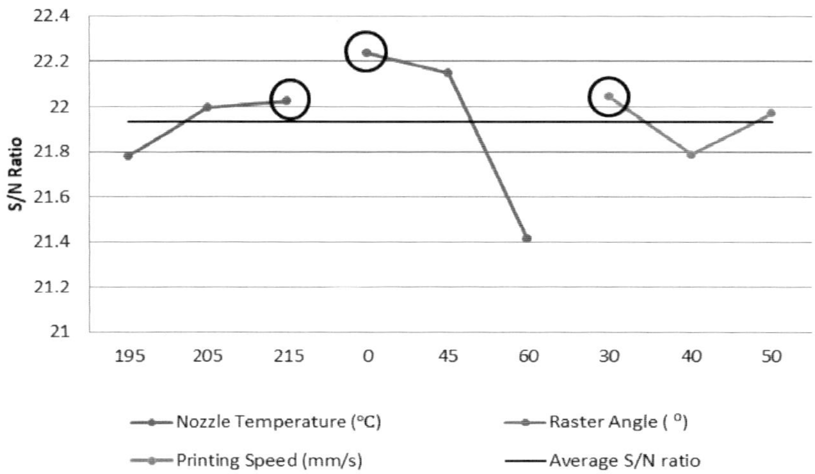

Figure 1: Tensile S/N plot for each printing parameter

3.3 Scanning Electron Microscope Result

As aforementioned, microstructure of specimens for tensile test and compression were inspected with the highest properties which is level 8.3 for tensile specimen and level 3.5 for compression specimen.

3.3.1 SEM of Tensile and Compression Specimens

In **Figure 3,** the breaking areas of every layer showed that the specimen had been made with 45°/45° of raster angle making the specimen stronger among other. Thus, the higher nozzle temperature and slower printing speed which are 215°C and 30 mm/s occur stronger bonding with layer by layer. This is because the contact area of each layer is bigger than other printing parameter.

Table 4: Tabulated raw data of compression test

Level	Maximum Compression Strength (MPa)					Average (MPa)	S/N ratio
	1	2	3	4	5		
1	96.5054	96.8014	96.0120	96.3574	96.4314	96.42152	39.6835
2	95.3953	96.1847	99.4164	96.1107	96.2587	96.67316	39.7061
3	100.650	96.3081	99.2437	98.2322	100.551	98.997 (Highest)	39.9124
4	95.4940	94.9019	91.3249	98.2076	96.3327	95.25222 (Lowest)	39.5775
5	95.9380	95.4940	95.8640	99.0957	95.5186	96.38206	39.6799
6	95.9380	95.1239	96.1600	97.6402	96.1354	96.1995 (middle)	39.6635
7	95.8640	95.3706	96.2587	96.8261	96.2834	96.12056	39.6563
8	94.5072	95.3706	96.0860	95.2473	95.9133	95.42488	39.5932
9	95.9133	95.6913	95.5186	97.6895	96.4561	96.25376	39.6684

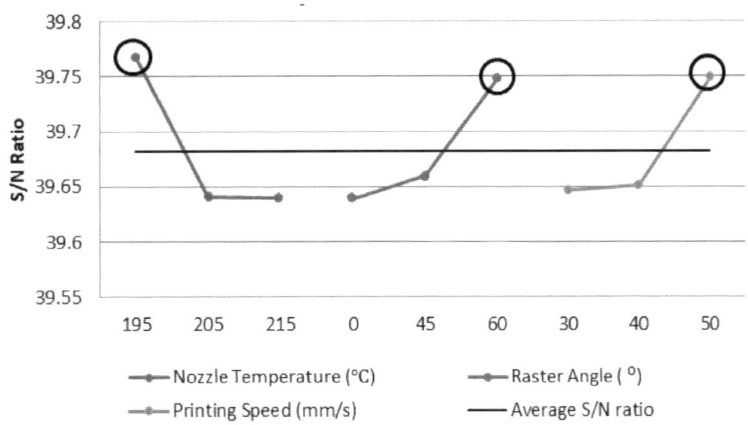

Figure 2: Compression S/N plot for each printing parameter

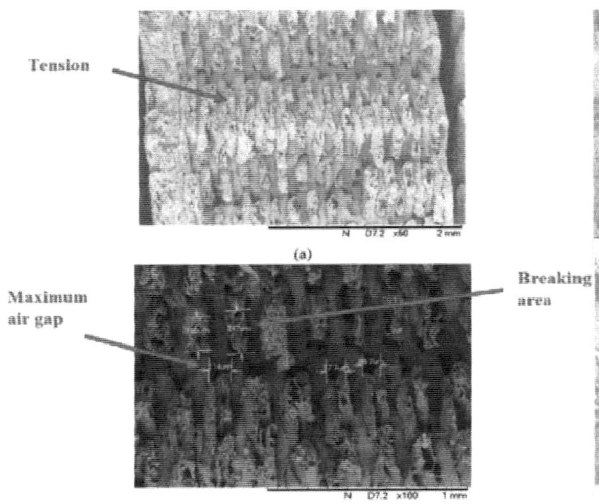

Figure 3: Highest strength of tensile specimen

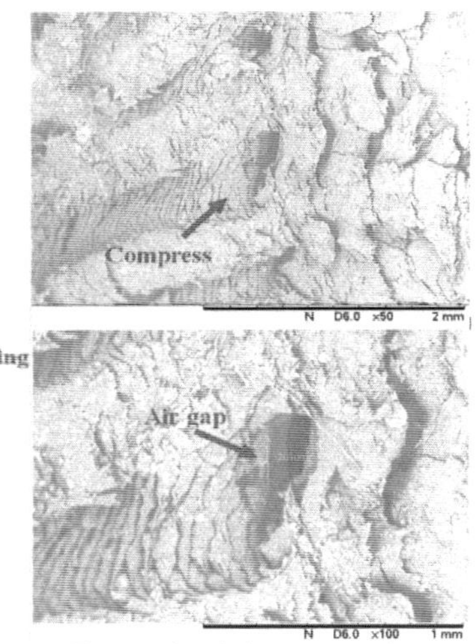

Figure 4: Highest strength of compression specimen

In **Figure 4**, the printing conditions gave the highest strength; however, tension break occurred due to failure of the strength that have a higher air gap which is maximum at 201μm. The higher size of air gap is proof that the stress occurred is higher stress.

4. Conclusion

The relationship between printing parameter and mechanical performance showed a good significant effect especially for tensile test where the raster angle is most influencing the printing parameter to the mechanical performance of specimen and also easily can be detected by using SEM for validate the result obtain. The research finding show consistency with others similar research [9-11]. For this research, the optimum printing parameters of PLA with wood content material for tensile strength are at 215°C of nozzle temperature, 0° of raster angle and 30 mm/s of printing speed. While the optimum printing parameters for compression strength are at 195°C of nozzle temperature, 60° of raster angle and 50 mm/s of printing speed. Thus, PLA with wood content can be used in scope of application area same as pure PLA such as medical tool, household item, architecture and engineering models, automotive prototype, and others.

Acknowledgement

The authors would like to express profound gratitude and deep appreciation to the School of Mechatronics Engineering, School of Manufacturing Engineering and Universiti Malaysia Perlis in providing researching platform well as providing the facilities and equipment.

References

[1] Bourell D *et al.* 2017 Materials for additive manufacturing *CIRP Ann. - Manuf. Technol.* **66** (2) 659–681.
[2] Yuan S, Shen F, Chua C K and Zhou K 2018 Polymeric composites for powder-based additive manufacturing: Materials and applications *Prog. Polym. Sci.* 1–28.
[3] Tlegenov Y, Hong G S and Lu W F 2018 Nozzle condition monitoring in 3D printing *Robot. Comput. Integr. Manuf.* **54** 45–55.
[4] Christiyan K G J, Chandrasekhar U and Venkateswarlu K 2016 A study on the influence of process parameters on the Mechanical Properties of 3D printed ABS composite *IOP Conf. Ser. Mater. Sci. Eng.* **114** (1).
[5] Wu W, Ye W, Wu Z, Geng P, Wang Y and Zhao J 2017 Influence of layer thickness, raster angle, deformation temperature and recovery temperature on the shape-memory effect of 3D-printed polylactic acid samples *Materials (Basel)* **10** (8).
[6] Kariz M, Sernek M, Obućina M and Kuzman M K 2018 Effect of wood content in FDM filament on properties of 3D printed parts *Materials Today Communications* **14** 135-140.
[7] ASTM D638 2015 *Standard test method for tensile properties of plastics ASTM Int.* **08** 46–58.
[8] van Apeldoorn J, Gilyén A, Gribling S and de Wolf R 2018 Convex optimization using quantum oracles **I** 1–8.
[9] Fernandes J, Deus A M, Reis L, Vaz M F and Leite M 2018 Study of the influence of 3D printing parameterson the mechanical properties of PLA *Proceedings of the 3rd International Conference on Progress in Additive Manufacturing (Pro-AM 2018)*, Singapore.
[10] Tymrak, Kreiger, and Pearce 2014 Mechanical properties of components fabricated with open-source 3- d printers under realistic environmental conditions *Materials & Design* **58** 242– 246.
[11] Wittbrodt and Pearce 2015 The effects of pla color on material properties of 3-d printed components *Additive Manufacturing* **8** 110–116.

IOP Conf. Series: Materials Science and Engineering **670** (2019) 012049 doi:10.1088/1757-899X/670/1/012049

Topology optimization of mandible fracture plate

J W C Teo[1] and *S F Khan[1,*]

[1] Additive Manufacturing and Design Lab, School of Mechatronic Engineering, Universiti Malaysia Perlis, Malaysia

[*]**corresponding author: sfkhan@unimap.edu.my**

Abstract. One of the most often fractured facial bones is mandible. It is also the most common fractured among the patient with multiple injuries. A mandible fracture plate is designed based on topology optimization to overcome the problem caused by standardized mandible fracture plate. The mandible fracture plate is design in Autodesk Inventor and it is analysed in shape generator to remove excessive material. The plate is simulated in ANSYS workbench in axial and multiple load. The plate is designed in three thickness that are 1.0mm, 1.5mm and 2.0mm. Each thickness is simulated using titanium alloy, stainless steel and polylactic acid (PLA). The mandible fracture plate is fabricated using Fused Filament Fabrication (FFF). By using topology optimization, the volume of mandible plate can be reduced up to 50%.

1. Introduction

The term "topology optimization" is generally used in structural optimization where the specification of design factors allows for a prediction of a material and load distribution [1]. Topology optimization is a structural optimization method. It recommends the best material distribution by continuous iterative calculations based on specified constraints and preserve regions in order to attain excellent structural performance [2]. The mandible is the largest and strongest bone of human face that forming lower jaws to hold lower teeth. Medical implants are devices or scaffolds that are located inside or on the body surface. Many implants are prosthetic for replacing missing parts of the body. Nowadays, customized implants have becoming more popular in craniofacial surgery due to saving operating time, proper preoperative planning and accurate adaptation to the broken or fracture part [3] due to the advancement in additive manufacturing and computer simulation technology. Constructive surgery using parts from 3D printing is increasingly popular. The purpose of this work is to determine the feasibility of fabricating topologically optimized conformal plate for the fixation of mandible fracture.

2. 3D Modelling

The mandible fracture miniplate was designed based on the captured data of the mandible model. The software used to create 3D model is Autodesk Inventor Professional 2019. Inventor is a computer aided drawing (CAD) software that has many features from sketching a profile, extruding a part and also assembly. The model was used in design visualization and simulation of the miniplate. The part created was saved in *.ipt* format and the assembly file in *.iam* format. The special feature in Inventor is shape generator. Shape generator was used to remove the excess part based on constraints specified; thus, saving material used. Fixed constraints with applied force were assigned initially onto the miniplate model before a set of preserve regions were defined in the model. Then, mass reduction

IOP Conf. Series: Materials Science and Engineering **670** (2019) 012049 doi:10.1088/1757-899X/670/1/012049

target and mesh size were set. The finer the mesh size, the higher the accuracy but the processing time will take longer. Then, a 3D mesh was produced as a guide for optimized design.

3. Simulation

ANSYS Workbench 18.1 was used to do the simulation in order to analyse the strength of mandible fracture miniplate. After 3D modelling, the file was imported into ANSYS for structural analysis. The maximum chewing force of an adult is 500N to 700N [2]. The force was applied to the miniplate before simulation was carried out. Besides, the maximum and minimum bending stress of fracture plate were also determined by using ANSYS. Sensitivity analysis was applied to reduce the computational time.

3.1 Simulation on Jaw Bone

The jaw model of human is shown in Figure 1. A 1mm gap was created to simulate the fracture of the mandible. The simulations focused on the effects of loading with two different reconstruction plates based on muscle forces on the mandible structures. Both the miniplates were simulated in ANSYS to check the equivalent (von-Mises) stress and total deformation. The miniplates were assigned with titanium alloy, stainless steel and polylactic acid (PLA) for comparison study as these are often used in fracture plate. The plates were attached onto the human fractured jaw to carry out simulation as shown in Figure 2. Meshmixer was used in generating the 3D model of the attachment on the mandible. Both the loads and miniplate were placed on the jaw model to carry out the simulation in ANSYS. Although bone is an example of natural composite material, its properties is different at every point. However; in this study, bone is assumed to be isotropic.

3.2 Loading condition

The model was designed specifically for mandible fracture related to masseter. When simulation was carried out, the loads were applied in various direction according to chewing force. This was due to the fracture plate will moves along with muscle due to chewing action. The loading condition is shown in Table 1. The loadings were applied to both side of the mandible as shown in Figure 1.

Table 1. Magnitude and direction of force at mandible [4]

Muscular forces	X	Y	Z
Masseter	50 N	-50 N	200 N
Medial pterygoid	0	-50	100
Temporalis	0	100	200

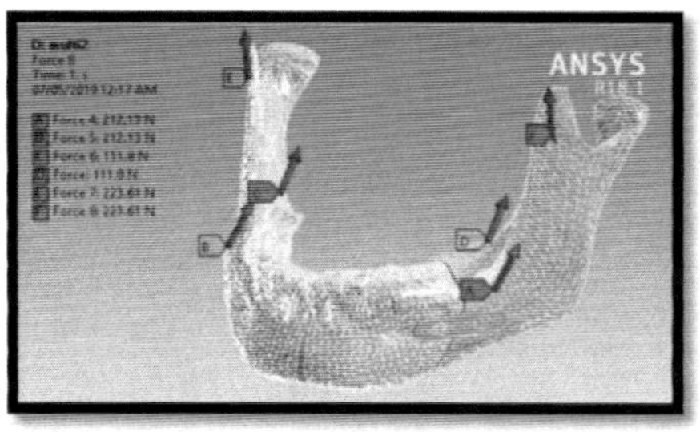

Figure 1. Loading on mandible

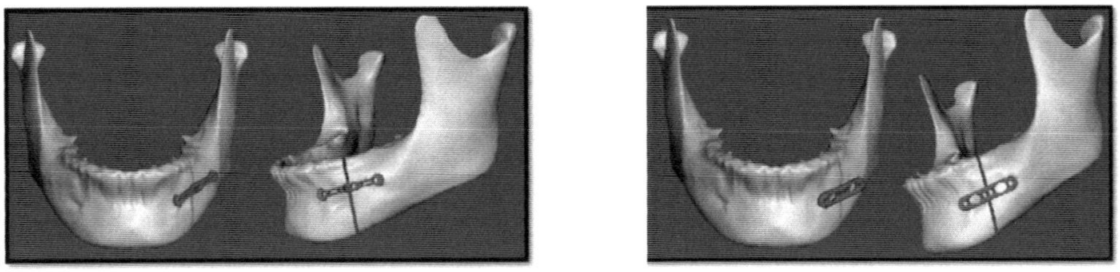

Figure 2. Miniplate attached on the mandible

4. Fabrication
The designed model was fabricated using additive manufacturing that build 3D object layer by layer from a CAD file. The fabricated model is compared with 3D model in term of geometry and dimensions. The purpose of comparison is to monitor the shrinkage of the printed model. A fitting test is performed to check the printed part. Lastly, finishing of the printed surface was carried out as surface finish is very important for an implant. To avoid stress shielding occurs, the chosen material needs to have similar properties with the surrounding bone. Table 2 shows the material properties of the different part. Titanium alloys (Ti6Al4V) and stainless steel are often used in fabricating implant because of their good biocompatibility, high chemical stability in the physiological environment and excellent mechanical properties. In addition, polylactic acid (PLA) is included in the simulation of miniplate as PLA coated with ceramics can be used to form composite in the manufacturing of mandible fracture plates.

Table 2. Material properties of different parts [5]

Type of material	Young's modulus (MPa)	Poisson's ratio	Strength (MPa)
Mandible (cortical bone)	13,700	0.30	85
Mandible (cancellous bone)	7930	0.30	0.22-10.44
Dentin	17,600	0.34	249
Ti6Al4V (scaffold)	105,000	0.30	897
Osteogenic material	7930	0.30	-

5. Results and Discussion

5.1 Volume reduction
Figure 3 illustrates the stages of the miniplate from original size to the final stage after applying axial load. Preliminary design of the plate is carried out to ensure this method is effective in volume reduction. Table 3 shows the volume changes from the original miniplate to the final stages for different thickness under axial loading. From original plate in stage 1, all of the three thicknesses experience almost 50 % of volume reduction. In order simulate actual loading of plate attached to fracture in mandible, multiple loading was applied. The excessive part is removed from the plate as shown in Figure 4 and Figure 5 after applying multiple load. The shape is similar when the thickness changes.

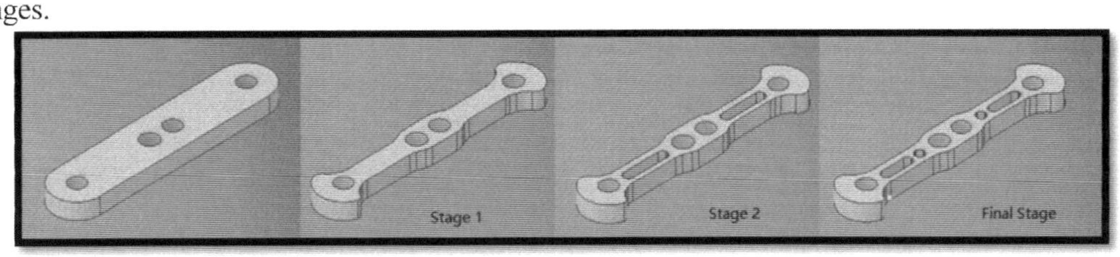

Figure 3. Removal stages after applying axial loading

IOP Conf. Series: Materials Science and Engineering **670** (2019) 012049 doi:10.1088/1757-899X/670/1/012049

Table 3 Volume of miniplates in different stages

Thickness	Original plate	Stage 1	Stage 2	Final Stage
1.0mm	0.147mm^3	0.076 mm^3	0.069 mm^3	0.070 mm^3
1.5mm	0.221mm^3	0.120 mm^3	0.109 mm^3	0.110 mm^3
2.0mm	0.295 mm^3	0.155 mm^3	0.131 mm^3	0.134 mm^3

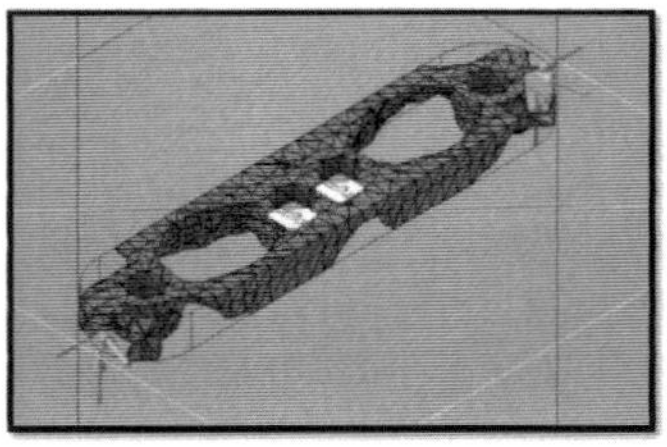

Figure 4. Design of plate with multiple loading

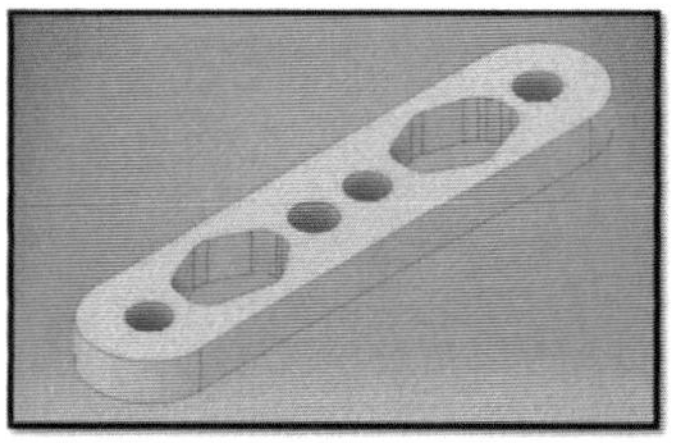

Figure 5. Shape generator result of multiple loading

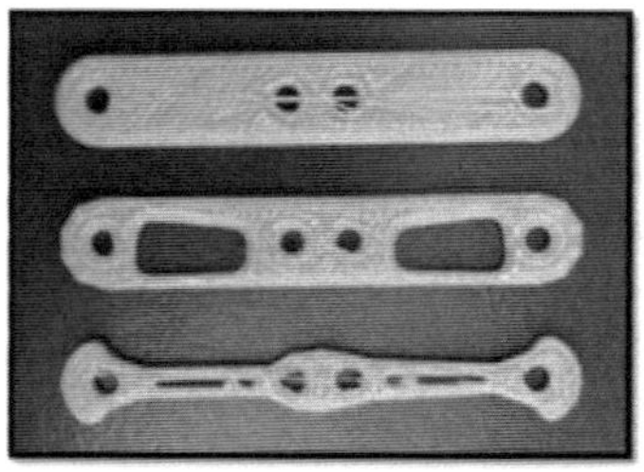

Figure 6. Printed mandible miniplates

5.2 Prototype

The miniplate were printed with Ender 3 Fused Filament Fabrication Modelling (FFF) using polylactic acid (PLA) as in Figure 6. At the top is the original plate and multiple loads miniplate is in the middle in Figure 6. While, the axial loading miniplate is at the bottom in Figure 6. Post-processing of the 3D printed miniplates were performed using XTC-3D brush-on coating and filling.

5.3 Simulation result

Figure 7 and Figure 8 illustrate the simulation result of multiple loading miniplate when attaching on bone for equivalent stress and total deformation. Table 4 show the outcome for both type of loading of the miniplate. From the result, the stress concentration is where the screw contacts the plate for fixation. The max label in red indicates the maximum von-Mises stress.

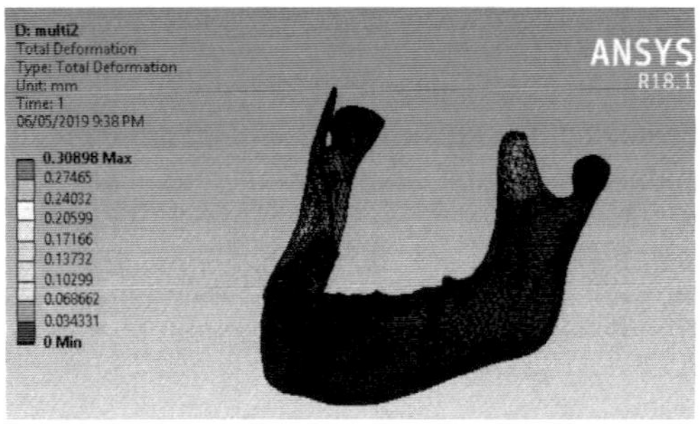

Figure 7. Equivalent stress of multiple loading miniplate attached on bone

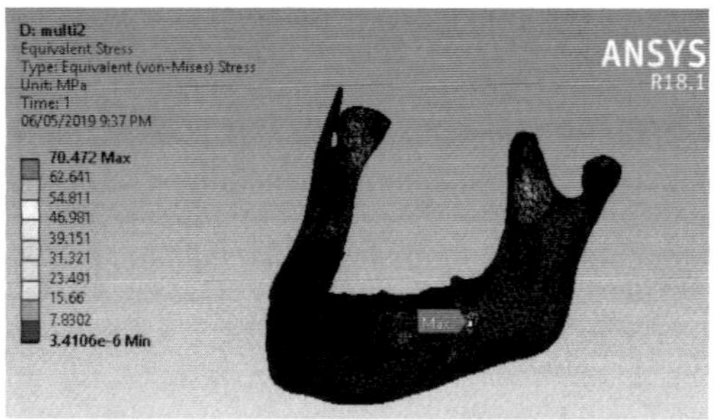

Figure 8. Total deformation of multiple loading miniplate attached on bone

Table 4. Simulation result of 2mm miniplate attached onto bone

Loading	Material	Equivalent (von-Mises) Stress		Total deformation	
		Max (MPa)	Min (MPa)	Max (mm)	Min (mm)
Axial	Titanium Alloy	68.106	2.6009e-11	0.00227	0
	Stainless Steel	60.402	5.8321e-12	0.00094	0
	PLA	85.731	3.8259e-11	0.00466	0
Multiple	Titanium Alloy	70.472	3.4106e-6	0.30898	0
	Stainless Steel	70.479	2.2967e-6	0.30895	0
	PLA	70.423	1.5504e-6	0.30966	0

6. Discussion

Preliminary design of the plate is carried out to ensure topology optimisation is an effective method in volume reduction. In the preliminary design, it shows that a 50% volume reduction can be achieved and be able to withstand the same loading condition as a non-topology optimised part. The purpose of preliminary design is to show the applicable of topology optimisation on volume reduction.

The miniplates in three thickness which is 1mm, 1.5mm and 2mm are simulated in ANSYS workbench with axial load and multiple load. Each thickness is simulated using three materials that are titanium alloy, stainless steel and polylactic acid (PLA). When same load is applied to the plate, the shape generator will form similar shape. By applying topology optimisation, the volume of the miniplate can be reduced up to 55%.

Axial loading plate done in three stages. The first stage shows the volume reduction from shape generator. Then, the part with minimum von-Mises stress is remove from the plate to form stage 2. Stage 3 includes the support that is designed to support the part with maximum von-Mises stress.

For 2mm axial miniplate attaching on bone, stainless steel has the lowest von-Mises stress and total deformation when compared to other two materials. The result for multiple loading miniplate is similar for the three material as the maximum von-Mises stress is located at the mandible bone.

The jaw model is cut into two pieces with 1mm gap to carry out simulation. This is to simulate the situation when the jaw is in fracture condition as the jaw model provided is a complete human jaw.

7. Conclusion

The shape generated from topology optimisation are often in organic form. With the combination of additive manufacturing technique, will enable the fabrication of optimised organic shape. Based on finite element analysis an appropriate miniplate for fractured mandible was optimised to produce conformal miniplate part at reduce material volume. This method is suitable in producing customised mandible fracture miniplate. It also achieved light weight structure with volume reduction up to 55% which is more than half. This method to design customized plate can be apply to fractures in other bone as it provides a stable environment for bone regrowth. Manufacturing using 3D printer is fast and effective for the treatment. Topology optimisation and additive manufacturing is suitable to be used in producing customised mandible facture miniplate. It enables more efficient materials utilisation in the design and reduce the time for fabrication of complex shape.

References

[1] Bendsøe M P 2008 Topology optimization *Encyclopedia of Optimization*, Boston, MA: Springer.

[2] Dai N, Zhu J F, Zhang M, Meng L Y, Yu X L, Zhang Y H, Liu B Y and Zhang S L 2018 Design of a maxillofacial prosthesis based on topology optimization *Journal of Mechanics in Medicine and Biology* **18** (3).

[3] Zenha H, Azevedo L, Rios L, Pinto A, Barroso M L, Cunha C and Costa H 2010 The application of 3-D biomodeling technology in complex mandibular reconstruction—experience of 47 clinical cases *European Journal of Plastic Surgery* **34** (4) 257–265.

IOP Conf. Series: Materials Science and Engineering **670** (2019) 012049 doi:10.1088/1757-899X/670/1/012049

[4] Al-Ahmari A, Nasr E A, Moiduddin K, Anwar S, Kindi M A and Kamrani A 2015 A comparative study on the customized design of mandibular reconstruction plates using finite element method *Advances in Mechanical Engineering* **7** (7) 168781401559389.
[5] Luo D, Rong Q and Chen Q 2017 Finite-element design and optimization of a three-dimensional tetrahedral porous titanium scaffold for the reconstruction of mandibular defects *Medical Engineering & Physics* **47** 176–183.

Impact response of Napier fibre reinforced nanomodified epoxy composites

W M A Wan Ramli, M S Abdul Majid, M J M Ridzuan and N A M Amin

Mechanical Engineering Programme, School of Mechatronics Engineering, Universiti Malaysia Perlis, Pauh Putra Campus, 02600 Arau, Perlis, Malaysia

Email addresses: adlan_wan@yahoo.com, shukry@unimap.edu.my, ridzuanjamir@unimap.edu.my, nasrulamri.mohdamin@unimap.edu.my

Abstract. The effects of nanoclay filler content on the impact properties of Napier fibre reinforced epoxy composites were investigated. The natural fibre based composites were fabricated using either untreated or 5% alkali-treated Napier fibre along with nanomodified epoxy resin which consists of various nanoclay filler contents which were 0, 3, 5 and 10 wt%. The composites were manufactured using cold compression moulding process. Morphological analysis was then conducted to further analyse the fractured specimens following the impact test. The maximum impact properties were recorded by the treated Napier fibre reinforced nanomodified epoxy with 3 wt% of nanoclay filler content. The SEM image of the Napier fibre reinforced nanomodified epoxy composites at the optimal nanofiller content also exhibited excellent interfacial adhesion between the epoxy matrix and the fibre reinforcement.

Keywords: Composites; nanoclay; alkali treatment; impact properties; natural fibre

1. Introduction

The development of natural fibre based polymer composites can be enhanced by means of fibres modification through chemical treatment or by incorporating interfacial additives within the polymer matrix [1]. This is important since many problems always occur at the interface between the hydrophilic natural fibres and the hydrophobic matrix polymer due to the incompatibility issues [2]. Commonly used chemical treatment is alkalinisation by using Sodium Hydroxide (NaOH) solution, which can reduce fibres' moisture absorption property and improve fibre-matrix interfacial adhesion characteristic. On the other hand, manipulation of desired properties on the matrix polymer can be achieved by adding a small percentage of nanofiller particles such as nanoclay and carbon nanotube (CNT). This most recent and advanced approach has been successfully proven in ameliorating the polymer matrix properties. It has been established that the introduction of nanoscale particles below than 10 wt% effectively modified the matrix epoxy with enhanced mechanical, morphological, thermal and electrical properties, comparing to traditional composites [3]. Therefore in this study, an investigation was implemented to analyse the effect of hybridisation between nanoclay filler particles and natural Napier fibre reinforced epoxy composites through an impact test analysis. Different amounts of nanoclay filler content were incorporated within the composites starting from 0 wt% then, 3, 5 and 10 wt% in order to determine the optimal value. Raw and 5% alkali-treated Napier fibres were also used in this experiment to differentiate their results towards the impact properties. By thoroughly elucidating the impact of properties of the cost-effective and environmental friendly of Napier fibre reinforced nanomodified epoxy composites, this study is profoundly expected to produce supportive manifestation for future development or application.

2. Materials and methods

2.1. Materials
The fabricated composites in this investigation consisted of Napier fibre, nanoclay filler additive and polymer epoxy. The Napier grass stems were supplied by a local farmer from Beseri, Perlis, Malaysia. Napier fibres were extracted from the internodes of the stem through a process called water retting [4–6]. During the water retting, the stems were fully immersed in water for almost a month before the extraction of the fibres was manually done. Some of the extracted Napier fibres were then treated with the alkali solution in order to prepare the treated fibres. The treated Napier fibres were formulated based on previous studies [7,8]. Hence, alkali solution with the optimal of 5% concentration was applied towards the natural fibres to remove the impurities and improve their surface. Next, the nanoclay filler used to modify the polymer epoxy is Montmorillonite type, which was supplied by the Sigma Aldrich Corporation. The surface of the supplied MMT clay has been modified with 25-30 wt% of trimethyl stearyl ammonium, making the nanofiller scale particles more compatible with the polymer epoxy. Lastly, the polymer epoxy employed in the development of the composites was obtained from Smooth-On Inc. Company.

2.2. Composites fabrication
In preparing the composites for the impact test, a mould was specially designed to produce laminates of size 300 mm × 300 mm with the thickness of 5mm. A thin layer of mould releasing agent was first applied to the whole surface of the mould several times. The weighted unidirectional Napier fibres were then arranged into the form of a mat in accordance with the size of the mould and placed into it. The process continues by mixing the weighted nanoclay filler particles into the required weight of the epoxy resin. Once the nanomodified epoxy was readily prepared, the hardener was then added into the mixture to form a completed polymer matrix system which then evenly poured into the mould containing the fibres mat. Then, the mould was closed before completely pressed by the compression machine. Finally, a series of rectangular plates, containing untreated and alkali-treated Napier fibres with 0, 3, 5 and 10 wt% nanoclay content were obtained for impact characterisation.

2.3. Impact test
Specimens in the form of rectangular shape with dimensions of 55 mm × 10 mm × 5 mm were cut from the produced composite plates using the 4000 Dremel cutter tool as preparation of the Charpy impact test. A 'v' notch was then made on every test specimen. This test was performed on the 50 J Charpy Impact Tester and conducted in accordance with the ASTM D6110 standard [9]. The energy absorbed and the impact strength of the composites are properties that were obtained and analysed from this impact test.

2.4. Morphological observation
In conducting an in-depth investigation of the fractured modes of the specimens following the impact test, therefore it was conducted using a Hitachi TM-3000 scanning electron microscope (SEM). In order to do the observation, small pieces from the fractured area of the composite specimens were firstly cut out as preparation in performing the SEM analysis. Then, they were uniformly coated using platinum prior to the scanning process.

3. Results and discussion
The results from the impact test were then summarised in the graph below. Figure 1 shows the energy absorbed and the impact strength of the test composites against nanoclay content. Clearly, from the graph, the energy absorbed and the impact strength was substantially displayed by the composites incorporated with 3 wt% of nanoclay content. Results show that 9.19 and 12.64 J of energy absorbed was revealed by the untreated and treated composite, respectively. Therefore, approximately 74 and 69% increment achieved by the untreated and treated composite respectively compared to when both of them were in the unfilled state. While the impact strength at the optimal nanoclay content obtained by the untreated and treated composites were 29.46 and 43.14 kJ/m^2, respectively. This means 59%

improvement stated by the untreated composite and close to 72% enhancement by the treated composite if compared to their similar composites when they were at 0 wt% of filler content. These were due to the development of the polymer composites from the hybridisation of natural fibre and nano-sized filler particles which reduced voids and improved their mechanical properties compared to the traditional natural fibre reinforced composites [10].

Still, increasing the nanoclay filler beyond the optimal value had resulted in an inconvenient situation where the impact properties of the developed composites abruptly dropped. The cause for the drop in both the energy absorbed and the impact strength of the composites can be attributed to the increment in the nanoclay loading. Introducing higher concentration of nanoclay filler content during fabrication of the composites yielded in the formation of agglomerated nanofiller particles, causing high-stress concentration zones where could act as the internal flaw within the composites [10,11]. This situation could lead to the embrittlement of the composites, weak interfacial interaction and improper adhesion between the nanofiller and the matrix. Hence, incorporating a higher amount of nanoclay fillers during the development of composites would prevent the nanofiller from functioning effectively. In fact, it reduced the energy absorption capability and impact strength of the developed composites.

Though, it can be seen from the figure that the alkali-treated fibre had a significant influence over the untreated fibre on increasing the energy absorbed from the impact test. This was due to the enhancement in the fibre-matrix interrelation. Indeed, at 0 wt% of nanoclay content, the treated fibre reinforced epoxy composite exhibited higher energy absorbed which was 7.49 J compared to the untreated composite at the same nanoclay content, 5.29 J which close to 42% improvement. Similar observation goes to the impact strength of the developed composites. The treated composites revealed higher impact strength at all nanoclay contents if compared to the untreated composites. From the results obtained, it can be suggested that the untreated composites had more inadequate fibre-matrix interfacial bonding as compared to the treated one. Unlike the treated fibres, raw fibres consist of hemicellulose, wax, oil, lignin and impurities that degenerated the interrelation between them and the epoxy polymer. The situation became worse due to the hydrophilic nature of the natural Napier fibres reduced the degree of compatibility between them and the hydrophobic epoxy polymer, which subsequently deteriorated the strength of the composites.

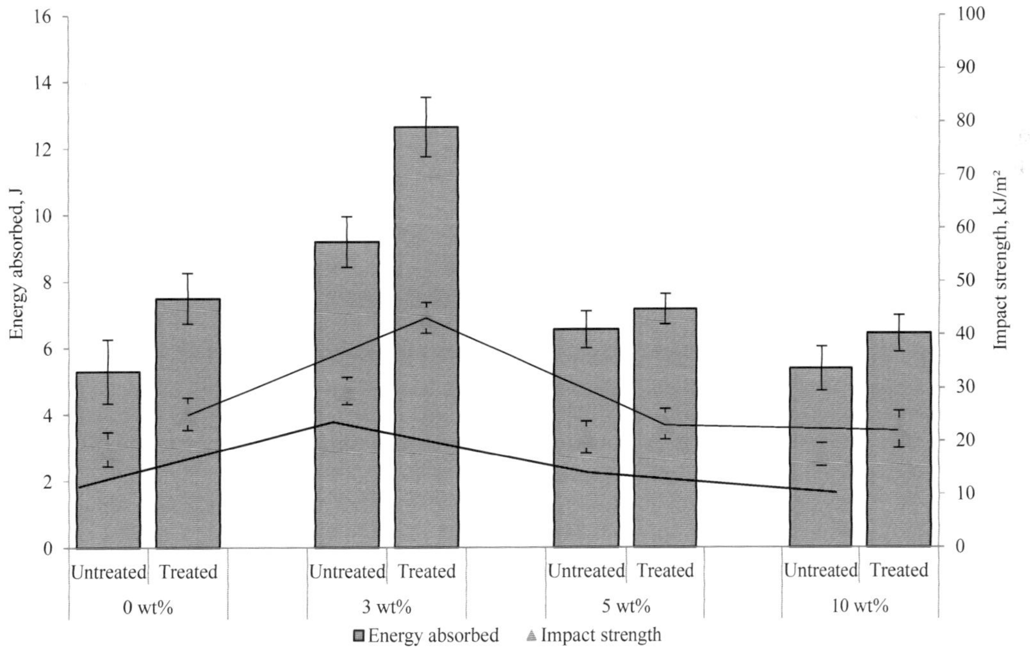

Figure 1. Impact properties of the developed composites against various nanoclay contents.

4. Fractured surface morphology

The effect of different nanoclay content can be observed by the dispersion of the nanoscale particles through the SEM analysis, as shown in Figure 2 below. Unfilled composites exhibited a smooth surface with the presence of void as shown in Figure 2 (a), indicating a brittle behaviour and poor resistance to impact force, thereby resulted in low impact strength of the developed composites [13]. Composites with 3 wt% of filler content, however, showed rough surface and reduced void as depicted in Figure 2 (b), implying better mechanical properties. This observation was in accordance with previously reported work [10]. The morphology of the composites with the optimal nanofiller content also demonstrated fibre bending and damaged end fibre, which reflected more impact energy consumed in order to break the composites. On the other hand, composites incorporated with 5 and 10 wt% of nanoclay content displayed relatively rougher surface and unevenly agglomerated nanoclay particles as can be observed in Figure 2 (c) and (d). More bubbles and voids were also observed in both of the developed composites due to the higher viscosity of the epoxy-clay mixture during their fabrication process, suggesting poor fracture resistance. In addition, pull out traces indicated the embrittlement of the composites owing to the higher concentration of nanoclay content.

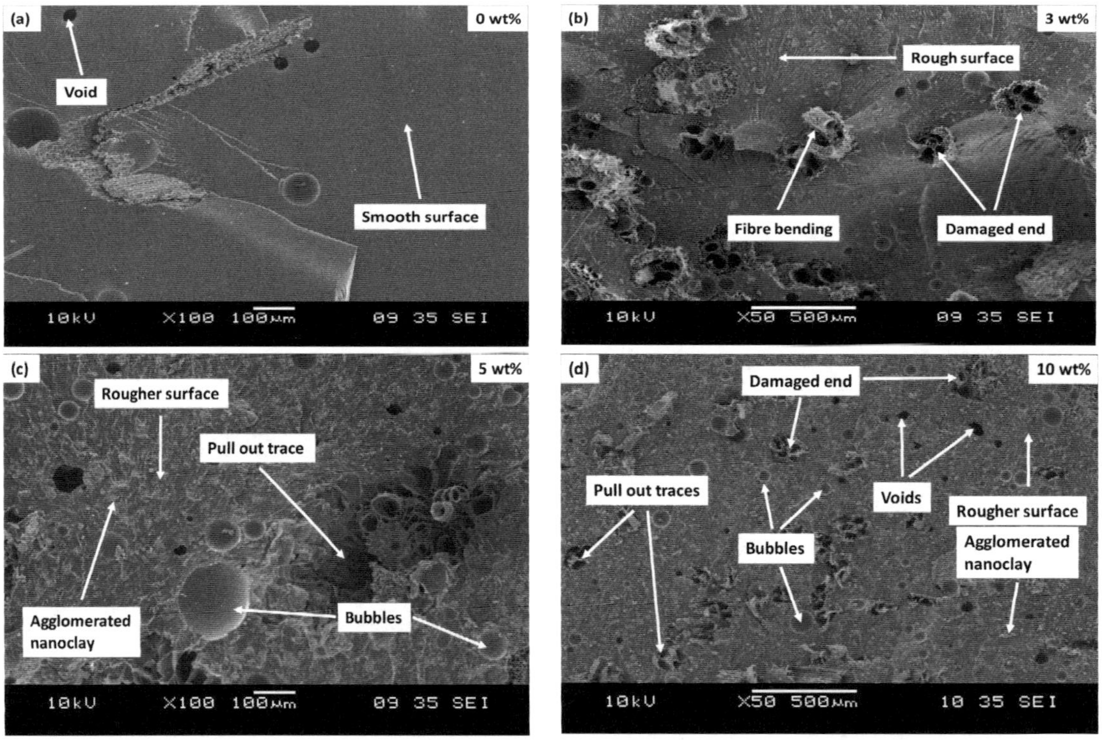

Figure 2. Fractured surface morphology of the developed composites.

5. Conclusion

The effects of various nanoclay content on the impact properties of untreated and 5% alkali-treated Napier fibre reinforced composites were investigated. The highest energy absorbed and impact strength of the developed composites was recorded when they were introduced with 3 wt% of nanoclay content. This was due to the homogenous dispersion of the nanoclay filler particles which successfully functioned in improving the impact resistance of the composites. Beyond the optimal content, however, observing the reduction of the impact properties of the composites. Treated composites yield the greatest performance in this investigation due to the better interrelation between the fibre and the epoxy matrix.

Acknowledgement
This research was financed by the Ministry of Education, Malaysia, through the Fundamental Research Grant Scheme; (Ref: FRGS/1/2019/TK05/UNIMAP/03/5). The authors would like to acknowledge the School of Mechatronic Engineering, Universiti Malaysia Perlis for the equipment and technical assistance.

References

[1] Saba N, Paridah M T and Jawaid M 2015 Mechanical properties of kenaf fibre reinforced polymer composite: A review *Constr. Build. Mater.* **76** 87–96.
[2] Fartini M S, Abdul Majid M S, Ridzuan M J M, Amin N A M and Gibson A G 2016 Compressive properties of Napier (Pennisetum Purpureum) filled polyester composites *Plast. Rubber Compos.* **45** 136–46.
[3] Saba N, Jawaid M, Alothman O Y, Paridah M and Hassan A 2016 Recent advances in epoxy resin, natural fiber-reinforced epoxy composites and their applications *J. Reinf. Plast. Compos.* **35** 447–70.
[4] Revati R, Majid M S A, Ridzuan M J M, Basaruddin K S, Rahman Y. M N, Cheng E M and Gibson A G 2017 In vitro degradation of a 3D porous Pennisetum purpureum/PLA biocomposite scaffold *J. Mech. Behav. Biomed. Mater.* **74**.
[5] Ridzuan M J M, Abdul Majid M S, Khasri A, Basaruddin K S and Gibson A G 2019 Effect of moisture exposure and elevated temperatures on impact response of Pennisetum purpureum/glass-reinforced epoxy (PGRE) hybrid composites *Compos. Part B Eng.* **160** 84–93.
[6] Ridzuan M J M, Majid M S A, Afendi M, Rahman M N and Azduwin Y K 2016 Thermal, composition and morphological properties of untreated and alkali-treated napier grass fibres *J. Eng. Appl. Sci.* **11** 2485–9.
[7] Kanchireddy O R and Muzenda E 2014 Effect of Fiber Surface Treatments on the Tensile Properties of Polycarbonate-Coated Napier Grass Fibers *2nd International Conference on Research in Science, Engineering and Technology (ICRSET'2014)* 194–7.
[8] Ridzuan M J M, Abdul Majid M S, Hafis S M and Azduwin K 2017 The effects of alkali treatment on the mechanical and morphological properties of Pennisetum purpureum/glass-reinforced epoxy hybrid composites *Plast. Rubber Compos.* **46**.
[9] D6110-15 2015 *Standard Test Method for Determining the Charpy Impact Resistance of Notched Specimens of Plastics* 1–17.
[10] Arpitha G R, Sanjay M R, Senthamaraikannan P, Barile C and Yogesha B 2017 Hybridization Effect of Sisal/Glass/Epoxy/Filler Based Woven Fabric Reinforced Composites *Exp. Tech.* **41** 577–84.
[11] Ranade A, D'Souza N A and Gnade B 2002 Exfoliated and intercalated polyamide-imide nanocomposites with montmorillonite *Polymer (Guildf)* **43** 3759–66.
[12] Lam C-K, Cheung H, Lau K, Zhou L, Ho M and Hui D 2005 Cluster size effect in hardness of nanoclay/epoxy composites *Compos. Part B Eng.* **36** 263–9.
[13] Singh K, Nanda T and Mehta R 2017 Addition of nanoclay and compatibilized EPDM rubber for improved impact strength of epoxy glass fiber composites *Compos. Part A Appl. Sci. Manuf.* **103** 263–71.

IOP Conf. Series: Materials Science and Engineering **670** (2019) 012051 doi:10.1088/1757-899X/670/1/012051

A review of motion capture systems for upper limb motion in throwing events: Inertial measurement unit

S F A Ahmad Noh[1], M E Baharudin[1], A. Mohd Nor[1], Mohd Sazli Saad[1] and Mohd Zakimi Zakaria[1]

[1]School of Manufacturing Engineering, Pauh Putra Campus, Universiti Malaysia Perlis, 02600 Arau, Perlis, Malaysia.

Abstract. Nowadays, the need of motion capture of an athlete's movement be among the most significance studies in sport performance analysis. The motion capture involves the use of sensing technology to track and store the movement. Thus, this review was done to present the studies on motion capture of the upper limbs movement in throwing events of the track and field sport. The databases were collected throughout the electronic media such as the Google Scholar, ScienceDirect, and Scopus. From literature, it was found that IMUs technology is not widely used in motion capture for upper limbs movement in throwing events. However, this device has a huge benefit to improvise the athlete's performances as well as their skills and techniques.

Keywords: motion capture, IMUs, upper limb motion, throwing events.

1. Introduction

By definition, motion capture of an athlete is a process of recording a movement in real time. In this context, the human motion capture was studies in sports research as it is significant to track and record the athlete's movement. In fact, it has been broadly utilized in numerous sports disciplines and activities for quantitative purposes such as the skills and competition assessment [1]. Moreover, access to quantitative performance variables can significantly improve overall athlete's performance, as well as sending information to the coaches, provides a real-time feedback thus recorded their performance over time. Recently, author in [2] assessed the latest growth in motion capture to analyse sports performance based on optical, non-optical and markerless technologies which subsequently categorized by biomechanical motion analysis, performance enhancement and new system validation.

Nowadays, the technologies in capturing the motion was based on the optical systems whereby these systems required multiple synchronized cameras and numerous of markers on the targeted subject to triangulate the three-dimensional (3D) positions [3]. In recent years, new optical motion capture systems were available in the consumer markets such as Kinect, which enabling to full body motion tracking. Besides using optical devices to profile the motion of the human body, the conservative wearable sensors was still desired to capture an accurate result as it can be used on site without requiring the huge space for camera positioning [4]. Nonetheless, it might be inconvenient to the user if too many cables attached to the body. Despite that, the motion capture based on video processing usually comes up with few restrictions as it may require numerous high speed cameras (usually more than 10) which must be structured and calibrated in a studio. Correspondingly, the subjects also need to wear an appropriate suit with retro reflective or any light emitting markers for huge amount of data to sort out, in which the real time will be difficult [5].

The rise of sensor signals processing techniques has cleared the way in utilizing the wearable sensors for screening the human status and performance. Amongst all the approaches, inertial sensors have been broadly discovered [6]. However, an accurate estimation with non-invasive sensors and non-limited workspace has becoming one of the major challenges in motion tracking. Author in [7] also stated that the ability of inertial sensors of sensing their own motion eventually makes them attractive for biomechanical analysis.

Thus, this paper described about the upper limb motion in throwing events and its motion capture systems. Besides that, inertial measurement unit (IMU) will become as a main focus in this paper. However, up to the author knowledge, the used of IMUs in simulating the throwing events was still limited and infrequent.

2. Throwing events

Historically, throwing events in track and field have been the subject of a number of studies [8,9]. In this review, the motion capture of throwing events of the track and field sports were focussed on the upper limbs movements of the athletes in discus, javelin, hammer and shot put throw, as shown in Figure 1. A classification of the throwing events made by [10] can be seen in Table 1.

Figure 1: Illustrations of the throwing events

Table 1. Classification of the throwing events **[10]**.

Throwing events	Implement	Form	Initial position
Discus	Light	Rotation	Backwards
Javelin	Light	Translation	Forward
Hammer	Heavy	Rotation	Backwards
Shot put	Heavy	Translation/ Rotation	Backwards

In discus throw, the athletes need to throw a metal disc weighing 2 kg, and 22 cm in diameter for men, 1kg for women and 18 cm in diameter for women, while be remain in a 2.5 m diameter circle. Author [11] stated that discus throw required high muscular power as it is an explosive track and field event. Nevertheless, author [12] stated that there was a high statistical correlation to throwing distance ($p<.05$) between the parameters hip velocity and shoulder velocity at touchdown of the brace leg. Another studies by [13] showed the relations between the release speed and release angle, and also the aerodynamic distance and release angle in which significantly affect the optimal release angle for the longest official distance. Meanwhile, javelin throw was technically demanding as the series of upper and lower extremity motions were important for throwing performances. Author in [14] stated that the optimum release angle, release speed and the height of javelin flight were the most vital criteria that defines the result of javelin throw according to the kinematic and dynamic technique characteristics.

Hammer throw demand a high power in the track and field event. As in turn phase, the resultant velocity of hammerhead was increased gradually with four turns and becoming more enlarged during double support phase. Past studies in hammer throw usually done by the lab based motion capture systems or using the standard 3D kinematic techniques such as the direct linear transform (DLT) [15]. Shot put throw can be classified as an open kinetic chain category with motivation to achieve a high rotational body speed and transfer energy to the shot by applying two putting styles known as the glide and spin techniques. Author [16] stated that the transfer of mechanical energy plays significant role in analyzing the athletes' movement. Author [17] concluded that the shot put performance depends on the release angle and speed whereby the result highly influenced by the athlete to shot with the possible combination of velocity, optimum angle, height and release distance.

3. Motion capture systems of the upper limbs movement

Motion capture system is a technology that enables the tracking of a moving object, which is rapidly growing research topic for human motion tracking. The upper limbs consist of two rigid segments (upper-arm and forearm) and two joints (shoulder joint and elbow joint) which then can be model as an articulated chain structure [18]. There are several motion capture systems currently available in the market. These systems can be divided into four categories as shown in Figure 2, which are visual tracking system, optoelectronic measurement system, inertial measurement unit and electromagnetic measurement system.

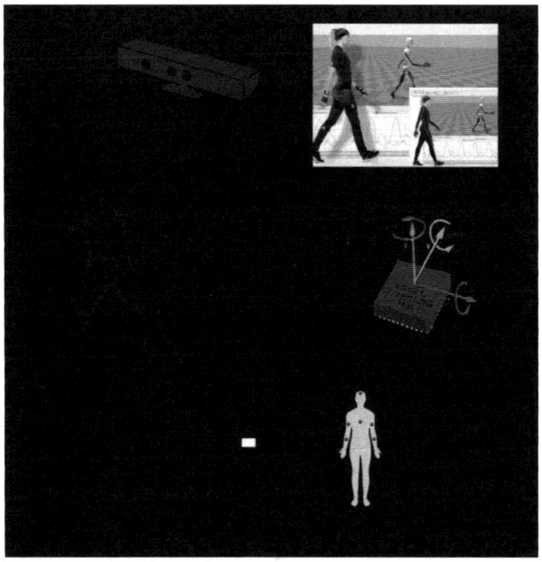

Figure 2: Motion capture system categories available in the market

Visual tracking system is a method of locating a target's position via cameras; whereby the cameras need to continually take frames from the target so that each frame is analysed to identify the object to be tracked in its position. These systems integrate more than one camera and use the method of triangulation to find the object in space. The function of this system broadly applied in motion based recognition, indexing video, human-computer interaction, navigating vehicles, and many more [19]. Meanwhile, optical measurement system is based on fixed cameras in the limited region. The precision depends on the number of the cameras, field of perspective and number of markers on the object. However, the amount of cameras leads to major problems in term of priced, portability, calibration, synchronization, labor and time to setup. It is vital to keep the distance between cameras and markers near the camera's focus distances during calibration to achieve for high accuracy. On the other side, there are two primary components in Electromagnetic Measurement systems: transmitters and sensors. The transmitter is an antenna producing a magnetic field in the three spatial directions, usually placed in a fixed location. Sensors are generally instruments that located to the object to be monitored. In the three spatial directions, these sensors obtain the magnetic field value so that location and orientation can be known. The electromagnetic measurement systems had showed the rise in medical applications [20]. Inertial measurement units (IMUs) is an electronic device consists of accelerometers, gyroscopes and magnetometers. The accelerometers measure the translational acceleration and acceleration due to gravity whereas the gyroscopes measure angular velocities. While magnetometer assist in direction orientation, it is useful in calibration and directing the data to the right position. Based on illustrated of the human arm in Figure 3, it can be seen that one IMU was placed on the upper arm, while another one on the forearm near the wrist. The IMUs measure accelerations and angular rates that are expressed in the sensor coordinate frame. Author [21] stated that the full body of human motion capture which can be classified as complex tasks also can be evaluate using the inertial sensors.

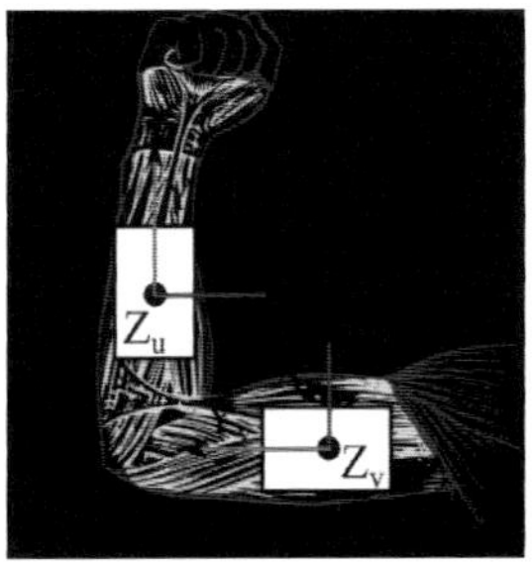

Figure 3: Illustration of IMUs placements on the human arms.

4. IMUs in throwing events

Since IMUs were light, small and minimally perceived by the athletes, it retains all the essential characteristics in evaluating the athlete's performance. A study by [22] applied the full body IMUs for a kinematic analysis and the valuation of its applicability whereas author in [23] examines the IMUs validity to measure data pertinent of the shoulder–pelvis separation angle, and torso-pelvis transverse plane orientation in discus throwing. In the case of hammer throw, a study by [24] built a wearable sensing system by using the IMUs to determine the dynamic vertical distances of the wrist and hip whereby the data can help in optimizing skills as it can verify the coordination between the lower and upper limbs. Following to the research, author [25] proposed to use IMU as a motion tracking sensor to replace the distance sensor as it is impossible to keep the distance sensor pointing vertically towards the ground during the hammer throw performance. Nevertheless, for the shot put, studies in [26] was done to determine the important kinematic variables whereby the athletes used rotational technique which been said to be very demanding from movement coordination point of view. Meanwhile, author in [27] studies the technology of inertial sensors (Moven system) to endorse a new motion analysis system. Regardless of restricting area of recording, condition of light, tracking and recognition of markers, the Moven system was said to be able to assess sport movement in outdoor sports too. However, to author's understanding and research restrictions, there are no studies of IMU's application in javelin throw till present year.

5. Conclusion

From the literature, it can be concluded that the application of IMU in the upper limb motion is still not widely applied. To author's concern, the studies of IMUs applicability in shot put throw could be broadening to more studies by means of parameters measured, kinematics and dynamics variable as there is still least valuable information on optimal angular position, angular acceleration, angle of throw and linear acceleration. The finding of the study should benefit the athletes with the aid of the mathematical models such as creating simulation to capture and analysis their motion in real time. In such way, not only the coaches could strategize for the best tactics, yet the athletes also can avoid any risks of injuries as they may now concern on the best angle and maximum put throw.

Acknowledgement

The author would like to acknowledge the support from the Fundamental Research Grant Scheme (FRGS) under a grant number of FRGS/1/2018/TK03/UNIMAP/02/19 from the Ministry of Education Malaysia.

References

[1] Molías L M, Ranilla J C and Cervera M G 2017 Pre-service physical education teachers' self management ability/: a training experience in 3D simulation environments *Retos Nuevas Tendencias En Educación Física, Deporte Y Recreación* **32** 30-34.

[2] Pueo B and Jimenez-Olmedo J M 2017 Application of motion capture technology for sport performance analysis *Retos* **32** 241-247.
[3] Tian Y, Meng X, Tao D, Liu D and Feng C 2015 Upper limb motion tracking with the integration of IMU and Kinect *Neurocomputing* **159** 207-218.
[4] Tao W, Liu T, Zheng R and Feng H 2012 Gait analysis using wearable sensors *Sensors* **12** (2) 2255-2283.
[5] Zhang Z, Huang Z and Wu J 2009 Hierarchical Information Fusion for Human Upper Limb Motion Capture *12th International Conf. on Information Fusion Seattle WA USA* no. 978-0-9824438-0-4.
[6] Peppoloni L, Filippeschi A, Ruffaldi E and Avizzano C A 2013 A novel 7 degrees of freedom model for upper limb kinematic reconstruction based on wearable sensors *IEEE 11th International Symp. on Intelligent Systems and Informatics* no. 978-1-4799-0305-4/13.
[7] Verplaetse C 1996 Inertial proprioceptive devices: self-motion-sensing toys and tools *IBM Systems Journal* **35** 639 – 650.
[8] Lanka J 2000 Shot putting biomechanics in sport (eds. Zatsiorsky V.M.) 435-457.
[9] Marhold G 1973 Biomechanical analysis of the shot put *Biomechanics IV*, W R. C. Nelson and C. A. Morehouse (eds) Baltimore: University Park Press 502-509.
[10] Duran J 2002 Manual Básico de Atletismo Tomo II: Lanzamientos. R.F.E.A: Madrid.
[11] Karampatsos G, Terzis G and Georgiadis G 2011 Muscular strength, neuromuscular activation and performance in discus *Journal of Physical Education and Sport* **11** (4) 369 – 375.
[12] Badura M 2010 Biomechanical analysis of the discus at the 2009 IAAF World Championships in Athletics *New Studies in Athletics* no. 3/4/2010.
[13] Leigh S, Liu H, Hubbard M and Yu B 2010 Individualized optimal release angles in discus throwing *Journal of Biomechanics* **43** 540–545.
[14] Saratlija P, Zagorac N and Babić V 2013 Influence of kinematic parameters on result coll *Antropol* **37** Suppl. 2 31–36.
[15] Kelley J 2014 A camera calibration method for a hammer throw analysis tool *Procedia Engineering* **72** 74 – 79.
[16] Błażkiewicz M, Łysoń B, Chmielewski A and Wit A 2016 Transfer of mechanical energy during the shot put *Journal of Human Kinetics* **52** 139-146.
[17] Acharyya B and Mondal S K 2017 A study of kinematic release parameters of male and female shot putters *Asian Journal of Multidisciplinary Studies* **5** (6).
[18] Zhang Z Q, Wong W C and Wu J K 2011 Ubiquitous human upper-limb motion estimation using wearable sensors *IEEE Transactions On Information Technology In Biomedicine* **15** (4).
[19] Yang H, Shao L, Zheng F, Wang L and Song Z 2011 Recent advances and trends in visual tracking: A review *Neurocomputing* **74** 3823-3831.
[20] Frantz D D, Wiles A D, Leis S E and Kirsch S R 2003 Accuracy assessment protocols for electromagnetic tracking systems *Physics in Medicine & Biology* **48**.
[21] Höflinger F, Müller J, Zhang R, Reindl L M, Member IEEE, and Burgard W 2013 A wireless micro inertial measurement unit (IMU) *IEEE Transactions On Instrumentation And Measurement* **62**.
[22] Ahuja A and Ghosh A K 1985 Pre-Asiad 82 injuries in elite Indian athletes *Br J Sports Med* **19** 24-6.
[23] Brice S M, Hurley M and Phillips E J 2018 Use of inertial measurement units for measuring torso and pelvis orientation, and shoulder–pelvis separation angle in the discus throw *International Journal of Sports Science and Coaching* **13** (6) 985-992.
[24] Wang Y, Li H, Wan B, Zhang X and Shan G 2018 Obtaining vital distances using wearable inertial measurement unit for real-time, biomechanical feedback training in hammer-throw *Appl. Sci.* **8**.
[25] Wang Y, Li H and Shan G 2018 Acquiring the distance data with inertial measurement unit in a wearable device for the training of hammer throwers *Proceedings - 14th International Conference on Computational Intelligence and Security, CIS 2018*, 8564355 492-495.
[26] Saračević Z, Atikovic A, Štuhec S S and Čuk I 2018 How kinematics influences shot put results in track and field of international level athletes (a case study) *Journal of Physical Education and Sport* no. 18 Supplement 1 Art 73 512 – 516.
[27] Dinu D, Bidiugan R, Natta F and Houel N 2012 Preliminary study of accuracy and reliability of high-speed human-motion tracking using miniature inertial sensors *Procedia Engineering* **34** 790 – 794.

IOP Conf. Series: Materials Science and Engineering **670** (2019) 012052 doi:10.1088/1757-899X/670/1/012052

Tribological properties of DLC coating under lubricated and dry friction condition

M H Sulaiman [1, *], **R N Farahana** [1], **M N Mustaffa** [1] **and K Bienk** [2]

[1] School of Mechatronic Engineering, Universiti Malaysia Perlis, Pauh Putra Campus, 02600 Arau, Perlis, Malaysia
[2] CemeCon Scandinavia A/S, Navervej 28, DK-8382 Hinnerup, Denmark

*Corresponding author: hafissulaiman@unimap.edu.my

Abstract. Increasing demands on reducing abundant use of environmentally hazardous lubricants like chlorinated paraffin oil in production motivates extensive research on dry friction condition. However, the dry friction condition is not favourable, and it causes high friction and severe wear problems. This paper presents the tribological behaviour of a new hard coating, namely double-layer DLC/TiAlN, deposited onto the tool steel surface via a Physical Vapour Deposition (PVD) process. A pin-on-disk tribometer was used to study the long-term performance of DLC/TiAlN coated tool steel under lubricated and dry friction conditions. Surface damage after the experiment was observed by using a light optical microscope (LOM) and also, analysed by measuring wear scar profiles with a tactile roughness profilometer. The results were compared with uncoated tool steel. Based on the experimental result, applying the DLC/TiAlN coating to the tool steel surface has led to a significant reduction of friction and almost no wear scar was observed. The study proved that the DLC/TiAlN coating contributes to the better wear life of the tool steel, thereby improving the tool lifetime.

1. Introduction

In all mechanical systems, friction and wear happen when moving components in contact with each other. Poor lubrication has always been the reason to high friction, and leads to severe wear and surface damage. Often the industries do not have a specific procedure on how to treat lubricants, and they never check the lubricants or ask for certifications [1]. This is due to the lack of fundamental understanding on the issues which could be attributed to the lack of research and development activities among them. Highly skilled workers are the only one that can solve these issues. Large amount of efficient but hazardous lubricants like chlorinated paraffin oils are seen as the effective solution which produce an instant result. This is favourable by the industries that see the instant profit and high productivity as their priority.

Instead of utilizing lubricants, applying hard coating on tool substrate is one of the common methods to reduce friction and to prevent wear. A great number of hard coatings are available for protecting the tool steel surface, such as TiN, CrN, TiAlN [2], AlCrN [3], TiC and TiCN [4,5]. At present, Diamond-like Carbon (DLC) coating has gained acceptance in various industries, for example, automotive, medical, and tooling components, owing to its several attractive properties, including the low friction coefficient, high hardness, good wear resistance, and protection of the tool steel surface. These properties make DLC coatings suitable for many tribological applications in mechanical systems, such as valve train tappets, gears, and piston pins [6,7]. Unfortunately, depositing single DLC coating

layer onto the tool steel surface alone is insufficient, since the single-layer DLC coating can easily be delaminated and thereby impractical to protect the tool steel surface from severe wear and surface damage. A comparison between multi-layer coatings and a single layer deposited onto the substrate tool [8,9] has shown that the multilayer coating was able to add degree of freedom in tailoring the coating properties [10,11]. This is related to the combination of toughness and adhesion strength of the interlayer coating [12], which plays an important role in tribological system.
In the present study, the long-term tribological performance of a new hard coating, namely DLC/TiAlN, deposited on tool steel Vanadis 4 by PVD process, were evaluated and compared with uncoated tool steel. The present paper investigates the tribological behaviors of both uncoated and coated tool steel under lubricated and dry friction condition as one of the possible efforts in improving the tribological conditions that exist in almost all mechanical components of the two sliding surfaces.

2. Materials and method

2.1. Materials
The tool steel material is a powder metallurgical cold work tool steel (Vanadis 4) with high C and Cr contents that was alloyed with Mn, Mo, Si, and V. The tool steels were through-hardened, tempered to 62 HRC, and subsequently polished to a surface roughness *Ra* of 0.02 μm before the coating was applied. The ball indenter material fitted to the pin holder was a commercially available steel, 100 Cr6, with a ball diameter of 6 mm. Table 1 presents the mechanical properties and the compositions of the test materials used in the experiments.

Table 1: Compositions and mechanical properties of the test materials.

Components	Composition	Mechanical Properties		
		Density ρ (g/cm^3)	Poisson ratio ν	Elastic modulus E (GPa)
Tool steel (Vanadis 4)	1.4% C, 0.4% Si, 0.4% Mn, 4.7% Cr, 3.5% Mo, 3.7% V	7.56	0.3	200
Indenter ball (Steel 100 Cr6)	1.0% C, 0.25% Si, 0.35% Mn, 1.5% Cr	7.81	0.3	210

2.2. Coating
A double-layer a-C:H type DLC/TiAlN coating was studied, and its performance was compared with the uncoated tool steel as shown in Figure 1(a). The double-layer DLC/TiAlN coating film shown in Figure 1(b) was deposited onto the tool steel substrates (Vanadis 4, with a surface hardness of 62 HRC) via a Physical Vapour Deposition (PVD) process based on unbalanced magnetron sputtering. The tool steel roughness after coating was the same as before: *Ra* = 0.02 μm. The top-layer a-C:H type DLC coating was chosen because it has mixture of carbon bonds of sp^3, sp^2, and even sp^1, with the possible presence of hydrogen, in the DLC that promotes a good combination of high wear resistance and low friction for preventing wear [13,14]. Meanwhile, the inter-layer TiAlN was chosen owing to the high content of Al in the TiAlN coating, which results in high hardness, reducing the wear and extending the tool steel lifetime for improved productivity [15]. Figure 1(c) shows the a-C:H type DLC supported by two other coating films—CrN and CrCN—which acted as bonding layers that were deposited onto the TiAlN-coated die surface. Coating binders, CrN, and Cr interlayer coatings were added to the DLC coating in the form of a gradient coating structure. The mechanical properties and surface characteristics of the test coating are presented in Table 2, where the uncertainty values represent the observed variations in the investigated areas.

2.3. Tribological test

A pin-on-disk tribological test shown in Figure 2 was performed to analyze the long-term performance of the DLC coated and uncoated tool steel surfaces that reproduce the motions and forces that occur in mechanical systems engineering applications, i.e. bearings, internal combustion engines, metal cutting, and metal forming. The test parameters applied in the tribological experiment can be seen in Figure 2 (right). Figure 3 presents a tactile roughness profilometer and a light optical microscope (LOM) used to investigate wear scar on the DLC coated and uncoated tool steel surfaces after the experiments.

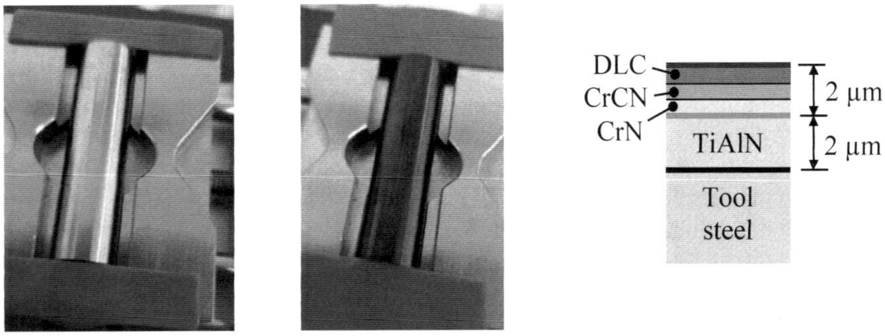

(a) (b) (c)

Figure 1: Test specimen made of tool steel Vanadis 4 – a) uncoated, b) double-layer DLC/TiAlN, and c) DLC/TiAlN coating structures.

Table 2: Mechanical properties and surface characteristics of the DLC/TiAlN coating.

Coating	Thickness (μm)	Hardness (HV)	Roughness *Ra* (μm)
DLC/TiAlN	4 ± 0.40	3,000	0.026

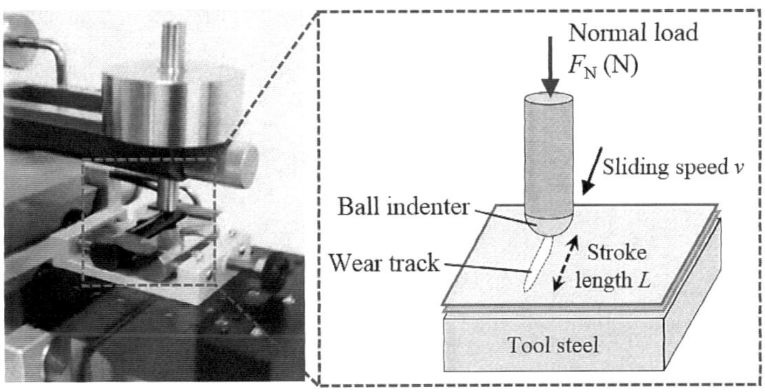

Test parameters	Values
Sliding speed v	100 mm/s
Normal load F_N	10 N
Stroke length L	16 mm
Max. strokes	500 laps
Temperature	32 °C

Figure 2: Schematics of the laboratory pin-on-disk tribological test (left) and test parameters applied in the experiment (right).

3. Results and discussion

The experiments were carried out in both lubricated and dry friction condition for a friction pair consisting of a 100Cr6 steel ball and a Vanadis 4 tool steel flat surface with and without a double-layer a-C:H DLC/TiAlN coating with a contact load of 10 N, at ambient temperature of 32+/- 1°C and a relative moisture of 68 +/- 2%, with the constant sliding rate and the total sliding distance being 100 mm/s and 8000 mm, respectively.

The diagram in Figure 3(a) shows the long-term tribological performance of the DLC coated and uncoated tool steel exposed to lubricated and dry friction condition. The lowest value of the friction coefficient in steady-state condition was observed for the specimens with a-C:H type DLC/TiAlN coatings in both lubricated and dry friction condition. The highest value of the friction coefficient was reported for the Vanadis 4 tool steel specimens without coatings tested under both lubricated and dry friction condition. Regardless of lubrication condition, it should be noted that the friction coefficient was much lower for the tool steel specimen with the DLC/TiAlN coating than for the specimen without coating. This could be attributed to the excellent wear-resistant properties of the DLC coating, in which supported by an improved adhesion strength from the TiAlN interlayer coating [16]. These combined characteristics are essential for improving the tool steel lifetime.

The influence of coating is significant. Larger friction coefficient can easily be observed in Figure 3(a) when no coating is applied to the tool steel during operation. Running the long-term experiment has revealed that the DLC/TiAlN coating is capable to reduce the frictional effects and reached a stable value after long-hour experiment. In order to measure this noticeable improvement for the DLC/TiAlN coated tool, roughness measurement on the wear scar profiles were employed with an optical profilometer for each sample. The wear scar images were captured using LOM. Figure 3(b) presents the wear scar images of the DLC coated and uncoated tool under dry friction condition by analyzing the depth and width of both worn surfaces after the pin-on-disk experiment. Based on these results, it is possible to see only minor scratches that is found on the wear track of the DLC/TiAlN coated tool. This can be related to the excellent toughness and good wear resistance of the DLC/TiAlN coating than can function with and without the presence of oil film. These characteristics are essential proof, in which the coating can contribute to the better wear life of the tool steel, thereby improving the tool lifetime.

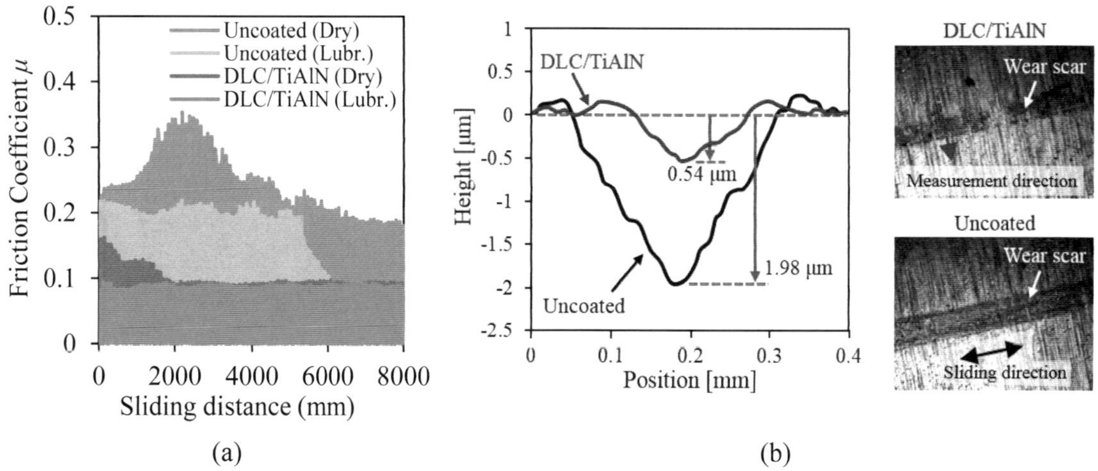

Figure 3: a) Friction coefficient and b) wear scar profiles of the uncoated and coated tool steels.

4. Conclusion

The tribological properties of the DLC/TiAlN coated and uncoated tool steels were evaluated successfully in both lubricated and dry friction condition. The present study adopted a pin-on-disk tribometer for emulating tribological conditions that exist in almost all mechanical components of the two sliding surfaces. The experimental results indicate that in the presence of no lubricant at all, the values of the friction coefficient were significantly lower for the specimen made of Vanadis 4 tool steel coated with a-C:H DLC/TiAlN than for the uncoated specimen. As shown by the LOM images and surface profiles of the worn surfaces, the DLC/TiAlN coating contributes to almost no surface damage either in lubricated condition or in dry friction condition, and this is directly related to a significant reduction in the friction coefficient. The use of DLC as the top layer coating and TiAlN as an interlayer coating in the present study revealed that their combined excellent tribological properties, including low-friction, anti-wear, high toughness and good adhesion strength, provide an environmentally-friendly alternative to outperform conventional approaches utilizing hazardous lubricants in production, which are toxic.

Acknowledgements

This work was supported by the Taiho Kogyo Tribology Research Foundation TTRF [grant number 9008-00015] and Short-term Grant Universiti Malaysia Perlis [grant number 9009-00059]. Special gratitude is dedicated to Tribology & Precision Forming (Tribo-Forming) research group for the fruitful discussions, equipment and technical assistance.

References

[1] Chamberlain S 2003 Combating tool wear with coating *Stamp. J.* **1**.
[2] Shoba Ch, Prasad D S 2015 Cathodic arc TiAlN PVD coatings for machining applications *Mater. Sci. An Indian J.* **12** (4) 142–6.
[3] Liu A, Deng J, Cui H, Chen Y and Zhao J 2012 Friction and wear properties of TiN, TiAlN, AlTiN and CrAlN PVD nitride coatings *Int. J. Refract. Met. Hard Mater.* **31** 82–8.
[4] Norrby N, Johansson-Jõesaar M P and Odén M 2014 Improved metal cutting performance with bias-modulated textured Ti0.50Al0.50N multilayers *Surf. Coatings Technol.* **257** 102–7.
[5] Mayrhofer P H, Hörling A, Karlsson L, Sjölén J, Larsson T, Mitterer C and Hultman L 2003 Self-organized nanostructures in the Ti-Al-N system *Appl. Phys. Lett.* **83** 2049–51.
[6] Cueva G, Sinatora A, Guesser W L and Tschiptschin A P 2003 Wear resistance of cast irons used in brake disc rotors *Wear* **255** 1256–60.
[7] Geißdörfer S, Rosochowski A, Olejnik L, Engel U and Richert M 2008 Micro-extrusion of ultrafine grained copper *Int. J. Mater. Form.* **1** 455–8.
[8] Bech J, Bay N and Eriksen M 1998 A study of mechanisms of liquid lubrication in metal forming *CIRP Ann. - Manuf. Technol.*
[9] Vetter J 2014 60years of DLC coatings: Historical highlights and technical review of cathodic arc processes to synthesize various DLC types, and their evolution for industrial applications *Surf. Coatings Technol.* **257** 213–40.
[10] Bobzin K, Brögelmann T, Kruppe N C, Arghavani M, Mayer J and Weirich T E 2017 Plastic deformation behavior of nanostructured CrN/AlN multilayer coatings deposited by hybrid dcMS/HPPMS *Surf. Coatings Technol.* **332** 253–61.
[11] Deng J and Braun M 1995 DLC multilayer coatings for wear protection *Diam. Relat. Mater.* **4** 936–43.
[12] Sulaiman M H, Christiansen P and Bay N 2017 A study of DLC coatings for ironing of stainless steel *36th International Deep Drawing Research Group (IDDRG)* 1–6.
[13] Podgornik B, Hogmark S and Sandberg O 2004 Influence of surface roughness and coating type on the galling properties of coated forming tool steel *Surf. Coatings Technol.* **184** 338–48.
[14] Carlsson P E R 2005 *Surface engineering in sheet metal forming* (Uppsala University).
[15] Neergaard A de 2017 *Personal communication with M.H. Sulaiman.*
[16] Sulaiman M H, Christiansen P and Bay N 2017 A study of anti-seizure tool coatings for ironing of stainless steel *J. Tribol. – Spec. issue WTC2017, Accept. Publ.* **17** 1–11.

The wettability characteristics of DLC coating for tribological engineering applications

M H Sulaiman [1, *], **R N Farahana** [1], **L Tajul** [1] **and H Jaafar** [2]

[1] School of Mechatronic Engineering, Universiti Malaysia Perlis, Pauh Putra Campus, 02600 Arau, Perlis, Malaysia
[2] School of Manufacturing Engineering, Universiti Malaysia Perlis, Pauh Putra Campus, 02600 Arau, Perlis, Malaysia

*Corresponding author: hafissulaiman@unimap.edu.my

Abstract. Hard coating deposited to tool steel surface can greatly improve wear resistance and reduce sticking. Since solid-liquid interactions are present in every lubricated tribological contact, the present study is aimed to understand the physical phenomena of contact interactions between the DLC coated surface and the liquid. In this study, double-layer DLC/TiAlN coating was prepared via Physical Vapour Deposition (PVD) process. The anti-sticking properties were assessed using contact angle measurements using two liquids with distinctly different viscosity – water and oil. No significant differences found in the contact angle values for both liquid properties. The results revealed that the DLC/TiAlN coated tool steel surfaces exhibit hydrophobic behaviour with high contact angle values. With a lower surface energy of the DLC/TiAlN coating in comparison to uncoated surface, this suggests that the DLC/TiAlN coating is a good hard coating candidate since it has a lower adhesion resistance and an improved release performance.

1. Introduction

A great number of hard coatings are available to protect the tool steel surface for a long-run production, such as TiN, CrN, TiAlN [1], AlCrN [2], TiC and TiCN [3,4]. Recently, Diamond-like Carbon (DLC) coating has gained acceptance in various industries, for example, automotive, medical, and tooling components, owing to several attractive properties, including their low friction coefficient, high hardness, good wear resistance, and protection of the tool steel surface. These properties make DLC coatings suitable for many tribological applications in mechanical systems, such as valve train tappets, gears, and piston pins [5,6]. Many studies have shown that the mixture of carbon bonds of sp^3, sp^2, and even sp^1, with the possible presence of hydrogen, in the DLC promotes significant effects on friction and wear to protect the tool steel surface from severe wear and surface damage [7–9]. Adding interlayer coating films, ie. TiAlN [10,11], to the DLC coated tool steel can furthermore enhance degree of freedom in tailoring the coating property [12,13]. Thereby, extending the tool steel lifetime for improved productivity. In addition to wear resistance, corrosive resistance and thermal stability, anti-sticking is one of the key important properties in many tribological engineering applications, including tribology and interface nanotechnology. The anti-sticking property of coating affects the release performance of steel surface [7], whereby an improved release performance requires low surface energy. The coating's anti-sticking property is related to the polar components of surface energy. Low polar components correspond with increased water-repellency as well as higher contact angle value. A surface with contact angle for water below 90° can be called a hydrophilic surface, while a surface with contact angle greater than 90° is hydrophobic. A higher contact angle value corresponds with improved release performance for the steel surface. In the present paper, the anti-sticking properties in relation to the hydrophobic and

IOP Conf. Series: Materials Science and Engineering **670** (2019) 012053 doi:10.1088/1757-899X/670/1/012053

hydrophilic characteristics of uncoated and double-layer a-C:H type DLC/TiAlN coated tool steels were investigated and the relationship between the surface roughness and their respective contact angle and surface energy were also discussed.

2. Materials and method

2.1. Materials and coating

The tool steel material was a powder metallurgical cold work tool steel (Vanadis 4) with high C and Cr contents that was alloyed with Mn, Mo, Si, and V. The tool steels were through-hardened, tempered to 62 HRC, and subsequently polished to a surface roughness *Ra* of 0.02 μm before the coating was applied. Table 1 presents the mechanical properties and the compositions of the test materials used in the experiments. The palm oil with kinematic viscosity of 95 cSt was selected to represent the oil. A double-layer a-C:H type DLC/TiAlN coating was studied, and its performance was compared with the uncoated tool steel as shown in Figure 1(a). The double-layer DLC/TiAlN coating film shown in Figure 1(b) was deposited onto the tool steel substrates (Vanadis 4, with a surface hardness of 62 HRC) via a Physical Vapour Deposition (PVD) process. The tool steel roughness after coating was the same as before: *Ra* = 0.02 μm. Figure 1(c) shows the a-C:H type DLC supported by two other coating films—CrN and CrCN—which acted as bonding layers that were deposited onto the TiAlN-coated die surface. Coating binders, CrN, and Cr interlayer coatings were added to the DLC coating in the form of a gradient coating structure. The mechanical properties and surface characteristics of the test coating are presented in Table 2, where the uncertainty values represent the observed variations in the investigated areas.

Table 1: Compositions and mechanical properties of the test materials.

Components	Composition	Mechanical Properties		
		Density ρ (g/cm^3)	Poisson ratio v	Elastic modulus E (GPa)
Tool steel (Vanadis 4)	1.4% C, 0.4% Si, 0.4% Mn, 4.7% Cr, 3.5% Mo, 3.7% V	7.56	0.3	200

Table 2: Mechanical properties and surface characteristics of the DLC/TiAlN coating.

Coating	Thickness (μm)	Hardness (HV)	Roughness *Ra* (μm)
DLC/TiAlN	4 ± 0.40	3,000	0.026

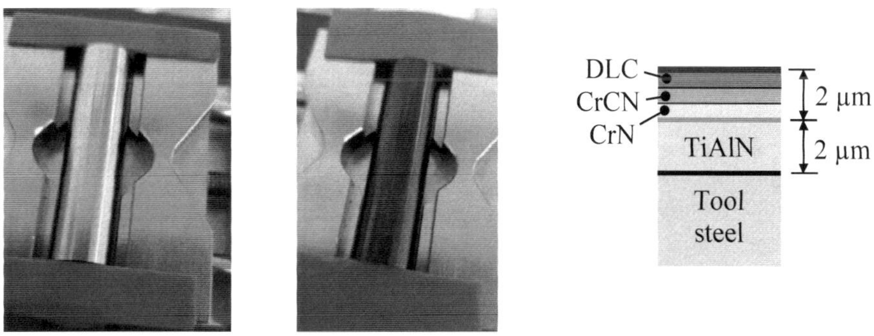

(a) (b) (c)

Figure 1: Test specimen made of tool steel Vanadis 4 – a) uncoated, b) double-layer DLC/TiAlN, and c) DLC/TiAlN coating structures.

2.2. Wetting analysis

The wetting properties of surfaces (uncoated and coated) were assessed in terms of the change in free energy and the work of adhesion that werecalculated from contact angle and interfacial tension values.

IOP Conf. Series: Materials Science and Engineering **670** (2019) 012053 doi:10.1088/1757-899X/670/1/012053

The contact angle measurement was performed using camera and ImageJ with drop analysis plug-in, see Figure 2. The drop analysis plug-in calculates the contact angle of a liquid drop on a flat uncoated and coated tool steel surface using the sphere approximation [θ=2tan^{-1} (2h/l)] and the ellipse approximation. The equilibrium of forces among the surface tensions at the 3-phase boundary were described by Young's Equation below [7].

$$\gamma_{SV} = \gamma_{SL} + \gamma_{LV}\cos\theta \tag{1}$$

where, γ_{SV}, γ_{SL} and γ_{LV} are respectively the surface energy of the solid–vapor, solid–liquid and liquid–vapor interfaces, and θ is the equilibrium contact angle.
The surface energy can be determined by measuring contact angles and using Fowkes method as described below [14].

$$\gamma_s = \gamma_s^d = \gamma_l^2(1+\cos\Theta)^2/(4\gamma_s^d) \tag{2}$$

If the measuring liquid is a dispersion one, i.e., the one that can be characterised by the dispersion interaction only, then $\gamma_l = \gamma_l^d$ and this Equation 2 simplifies to the following formula:

$$\gamma_s = 0.25\gamma_l(1+\cos\Theta)^2 \tag{3}$$

where, θ is contact angle, γ_s is surface energy, γ_l is surface tension of liquid. In this study, the surface tension $\gamma_{l\ oil}$ is 35 mN.m^{-1} and $\gamma_{l\ water}$ is 72.75 mN.m^{-1}.

Atomic Force Microscopy (AFM) analysis was used to evaluate surface morphology and roughness of the coated and uncoated tool steel surfaces. The uncoated and coated samples were cleaned in acetone for 10 minutes before contact angle measurements and AFM analysis.

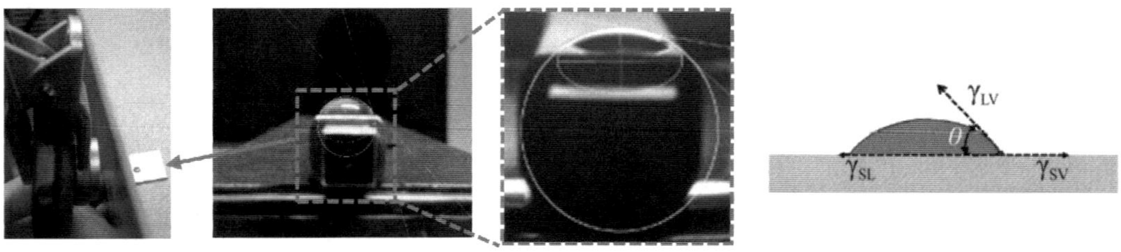

Figure 2: Contact angle measurement using camera and ImageJ with drop analysis plug-in (left) and schematic of liquid drop on solid surface with surface tension indicated as arrows originating at the 3-phase interface line (right).

3. Results and discussion

Contact angle of specimen surfaces was measured to estimate the characteristic of hydrophilic or hydrophobic by means of liquid drops. The contact angle measurements were carried out at an ambient temperature of 32+/- 1°C and a relative moisture of 68 +/- 2%. The contact angle measurements for both test specimens using water and lubricating oil were summarized in Table 3. It can be seen in Figure 3(a) and Figure 3(b) that no significant differences when using water and lubricating oil for this study. From Figure 3(a), the contact angle of the DLC/TiAIN coated tool steel is slightly higher than the uncoated ones. From this, it is clearly shown that both of the test specimens exhibit hydrophobic surface properties since the contact angle values greater than 90°.
The surface tension measurements for both of the test specimens were calculated using Equation 2 and Equation 3. As seen in Figure 3(b), the uncoated tool steel surface exhibit a relatively high surface energy, 69.73 mJ/m^2, in comparison to the DLC/TiAIN coated tool with a very low surface energy, 3.22 mJ/m^2, indicating weaker interactions with lubricating medium, and thus also poorer wetting. This explains that the DLC coating cannot form strong polar interactions with counter surfaces or liquids. The significant differences in the surface energy can be due to the coating properties such as oxidation, and even the crystal structure of the DLC itself can have significant influence on the surface energy as discovered by Y. Yang et al. [7].

Table 3: Contact angle measurements.

Liquid medium	Surface condition	Angle (°)	Theta C (°)	Theta Left (°)	Theta Right (°)	Theta E (°)
Water	Uncoated	113.39	99.9	-18	159.7	70.8
	DLC/TiAlN	121.57	170.9	-13.1	165.6	76.3
Oil	Uncoated	157.17	170.1	-12.3	162.2	75.0
	DLC/TiAlN	158.75	172.4	-12.1	139.5	63.7

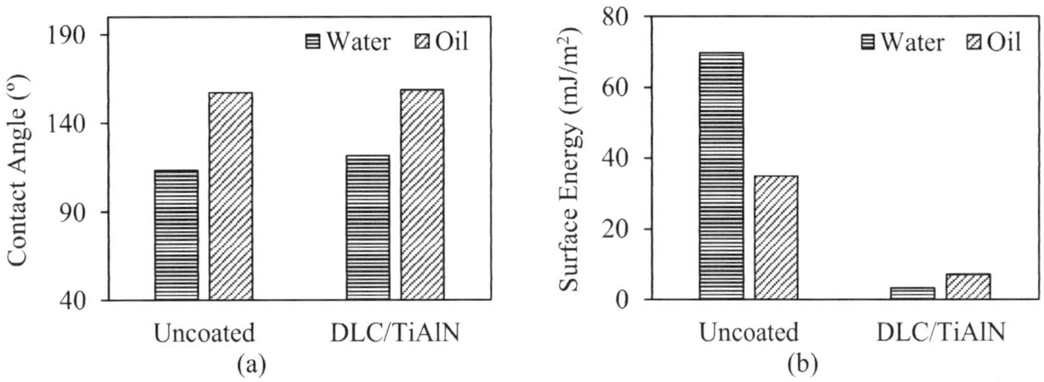

Figure 3: a) Contact angle and b) Surface energy measurements of the uncoated and DLC/TiAlN coated tool steels.

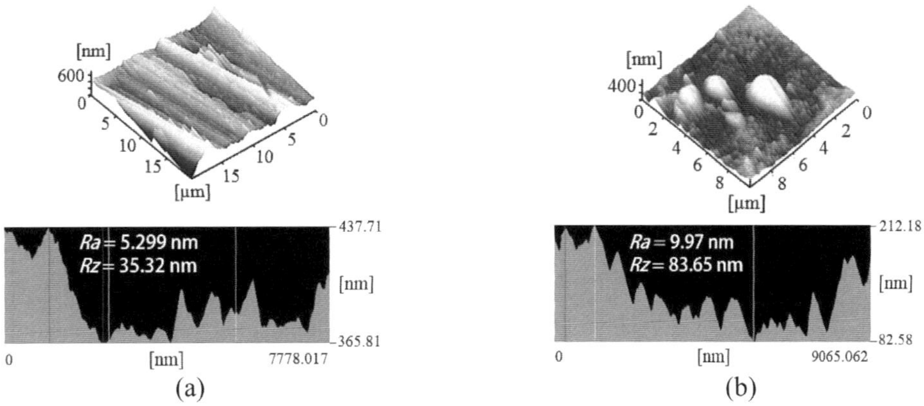

Figure 4: Surface morphology and roughness of a) the uncoated and b) DLC/TiAlN coated tool steels.

It is known that the changes in surface roughness and surface energy generated by the coating treatment could be the factor influencing the wettability characteristics. From Figure 4, the differences between the surface energy values of the samples can be related with the roughness, so the connection was investigated. Based on the AFM results, the average roughness (*Ra*) value for the uncoated (polished) tool steel surface was found to be ~ 5.3 nm, while the DLC/TiAlN coated surfaces had roughness values ranging from ~ 9 to 10 nm, indicating the measured three dimensional surface topographies of the DLC/TiAlN coated tool steel surface smoother than the uncoated tool steel surface. The difference in roughness supported the findings, in which smoother surface of the DLC/TiAlN coating caused low surface energy, indicating the surface energy is related to the roughness. The findings suggest that the DLC/TiAlN coated tool steel surface is a good hard coating candidate for the tool steel surface, which further improved release performance.

4. Conclusion

The wettability characteristics of uncoated and a-C:H type DLC/TiAlN coated tool steels were investigated. Physical vapour deposition (PVD) method was used to deposit the double-layer DLC/TiAlN coating on Vanadis 4 tool steel substrate. The experimental findings show that both uncoated and DLC/TiAlN coated samples have hydrophobic properties. The wetting analysis results show that the DLC/TiAlN coated tool steel has better wettability characteristics compared to the uncoated tool steel substrate. The contact angle measurements show that the surface energy values for the DLC/TiAlN coated tool steel is lower than the uncoated ones. The findings suggest the DLC/TiAlN coated tool steel has better adhesion ability compared to the uncoated tool steel. In other words, the DLC/TiAlN coated tool steel has a lower adhesion resistance compared to the uncoated tool steel. The lower surface energy (lower adhesion resistance) for the DLC/TiAlN coated tool steel was attributed mainly due to the rougher surfaces associated with the coated surfaces in comparison with the uncoated sample. The AFM analysis clearly show that the rougher the substrate surfaces, the lower the surface energy values.

Acknowledgements

This work was supported by the Taiho Kogyo Tribology Research Foundation TTRF [grant number 9008-00015] and Short-term Grant Universiti Malaysia Perlis [grant number 9009-00059]. Special gratitude is dedicated to Tribology & Precision Forming (Tribo-Forming) research group for the fruitful discussions, equipment and technical assistance.

References

[1] Shoba Ch, Prasad D S 2015 Cathodic arc TiAlN PVD coatings for machining applications *Mater. Sci. An Indian J.* **12** (4) 142–6.

[2] Liu A, Deng J, Cui H, Chen Y and Zhao J 2012 Friction and wear properties of TiN, TiAlN, AlTiN and CrAlN PVD nitride coatings *Int. J. Refract. Met. Hard Mater.* **31** 82–8.

[3] Norrby N, Johansson-Jõesaar M P and Odén M 2014 Improved metal cutting performance with bias-modulated textured Ti0.50Al0.50N multilayers *Surf. Coatings Technol.* **257** 102–7.

[4] Mayrhofer P H, Hörling A, Karlsson L, Sjölén J, Larsson T, Mitterer C and Hultman L 2003 Self-organized nanostructures in the Ti-Al-N system *Appl. Phys. Lett.* **83** 2049–51.

[5] Cueva G, Sinatora A, Guesser W L and Tschiptschin A P 2003 Wear resistance of cast irons used in brake disc rotors *Wear* **255** 1256–60.

[6] Geißdörfer S, Rosochowski A, Olejnik L, Engel U and Richert M 2008 Micro-extrusion of ultrafine grained copper *Int. J. Mater. Form.* **1** 455–8.

[7] Ahmed M H and Byrne J A 2012 Effect of surface structure and wettability of DLC and N-DLC thin films on adsorption of glycine *Appl. Surf. Sci.* **258** 5166–74.

[8] Podgornik B, Hogmark S and Sandberg O 2004 Influence of surface roughness and coating type on the galling properties of coated forming tool steel *Surf. Coatings Technol.* **184** 338–48.

[9] Carlsson P E R 2005 *Surface engineering in sheet metal forming* (Uppsala University).

[10] Sulaiman M H, Christiansen P and Bay N 2017 A study of DLC coatings for ironing of stainless steel *IOP Conf. Ser. J. Phys.* **896** 1–7.

[11] Sulaiman M H, Christiansen P and Bay N 2017 A study of anti-seizure tool coatings for ironing of stainless steel *J. Tribol. – Spec. issue WTC2017* **17** 1–11.

[12] Bobzin K, Brögelmann T, Kruppe N C, Arghavani M, Mayer J and Weirich T E 2017 Plastic deformation behavior of nanostructured CrN/AlN multilayer coatings deposited by hybrid dcMS/HPPMS *Surf. Coatings Technol.* **332** 253–61.

[13] Deng J and Braun M 1995 DLC multilayer coatings for wear protection *Diam. Relat. Mater.* **4** 936–43.

[14] Kalin M and Polajnar M 2014 The wetting of steel, DLC coatings, ceramics and polymers with oils and water: The importance and correlations of surface energy, surface tension, contact angle and spreading *Appl. S* **293** 97–108.

IOP Conf. Series: Materials Science and Engineering **670** (2019) 012054 doi:10.1088/1757-899X/670/1/012054

Corrosion resistance of PVD hard coatings for tribological engineering applications

M H Sulaiman [1, *], **A O C Sheng** [1], **R N Farahana** [1] **and K Bienk** [2]
[1] School of Mechatronic Engineering, Universiti Malaysia Perlis, Pauh Putra Campus, 02600 Arau, Perlis, Malaysia
[2] CemeCon Scandinavia A/S, Navervej 28, DK-8382 Hinnerup, Denmark

*Corresponding author: hafissulaiman@unimap.edu.my

Abstract. The present paper focuses on corrosion resistance of hard coatings deposited on tool steel via Physical Vapour Deposition (PVD) process. Three coating types were investigated – TiN, TiCN and DLC/TiAlN. An immersion corrosion test was performed on the PVD coated tool steels in accordance to ASTM G31 standard. With immersion corrosion test in aggressive media of 3.5% NaCl solution at ambient temperature of 33°C and exposure duration of up to 72 hours, the presence of the PVD coatings on the tool steel surface have improved the corrosion resistance of the tool steel surfaces. The results showed that the DLC/TiAlN coated tool steel has the highest corrosion resistance with corrosion rate of 36.58 mm/year, followed by TiCN and TiN coatings with corrosion rate of 42.10mm/year and 96.86 mm/year, respectively. The uncoated tool steel exhibits the lowest corrosion resistance of 140.61 mm/year. The present study suggests that the DLC coating deposited on SKD11 tool steel substrate has better stability than that of TiCN- and TiN-coated surfaces because the DLC can prevent the corrosion more efficiently than TiCN and TiN.

1. Introduction

Metal forming is an effective means of producing thousands of workpiece components. However, it does not come without the manufacturing challenge. A highly localized pressure of the long-term sliding surfaces between tool/workpiece contacts can cause mechanical stress, wear and corrosion problems on the tool surfaces, and thus calling for the need of coatings with high hardness and corrosion resistance. Physical Vapour Deposition (PVD) hard coating deposited onto tool steel substrate is one of the common methods to protect the tool steel substrate from surface damage, thereby prolong its lifetime in producing good product quality. A great number of hard coatings are available and they have been used in the industry for protecting the tool steel surface, such as TiN, CrN, TiAlN [1], AlCrN [2], TiC and TiCN [3,4].

Recently, Diamond-like Carbon (DLC) coating has gained acceptance in various industries, for example, automotive, medical, and tooling components, owing to its several attractive properties, including their low friction coefficient, high hardness, good corrosion resistance, good wear resistance, and protection of the tool steel surface. These properties make the DLC coating is suitable for many tribological applications in mechanical systems, such as valve train tappets, gears, and piston pins [5,6]. Unfortunately, depositing single DLC coating layer onto the tool steel surface alone is insufficient, since the single-layer DLC coating can easily be delaminated and shorten the tool lifetime [7,8]. Thereby, the single-layer DLC coating is impractical to protect the tool steel surface from severe wear and surface damage. As a result, this early coating failure could expose the tool steel substrate to corrosive environments, which expedite the corrosion behaviour of the tool steel substrate and reduce the lifetime of the tool. Introducing multi-layer coatings onto the substrate tool [9,10] had shown that the multilayer coatings were able to add the degree of freedom in tailoring the coatings' property [11,12], but little information is available on the corrosion behaviour issues of the coating.

Even if the tool surfaces are coated with excellent protective hard coatings, the surface damage causing a slightly deeper scratches and the peel-off of the coating during metal forming operation could be one of the main factors contributing to corrosion, that generates a pathway for corrosive materials to reach out and react with the tool steel substrate [13] [14]. Therefore, the present paper examines the corrosion resistance of the hard tool coatings in aggressive media of NaCl solution. The investigated DLC/TiAlN coating will be studied and compared with the two commercially industrial hard coating, TiCN and TiN, for metal forming applications.

2. Materials and method

2.1. Materials

The tool steel material is a cold work tool steel SKD11 with high C and Cr contents that was alloyed with Mn, Mo, P, S, Si, and V. The tool steels were through-hardened, tempered to 60 HRC, and subsequently polished to a smooth surface roughness before the coating was applied. Table 1 presents the mechanical properties and the compositions of the tool steel used in the experiments.

Table 1: Compositions and mechanical properties of the test materials.

Components	Composition	Mechanical Properties		
		Density ρ (g/cm^3)	Poisson ratio v	Elastic modulus E (GPa)
Tool steel (AISI D2)	1.4% C, 0.6% Mn, 11% Cr, 0.03% P, 0.03% S, 0.04% Si, 0.8% Mo, 0.2% V	7.58	0.3	210

2.2. Coating

Two commercially available hard coatings, TiCN and TiN, and a new coating, a double-layer a-C:H type DLC/TiAlN, were selected and deposited on the tool steel substrates, see Figure 1. Their corrosion resistance were compared with the uncoated tool steel. The investigated coating films shown in Figure 1 was deposited onto the tool steel substrate (SKD11, with a surface hardness of 60 HRC) via a Physical Vapour Deposition (PVD) process. The tool steel roughness after coating was the same as before ~ $Ra = 0.02 - 0.1$ µm. The mechanical properties and surface characteristics of the coatings are presented in Table 2, where the uncertainty values represent the observed variations in the investigated areas.

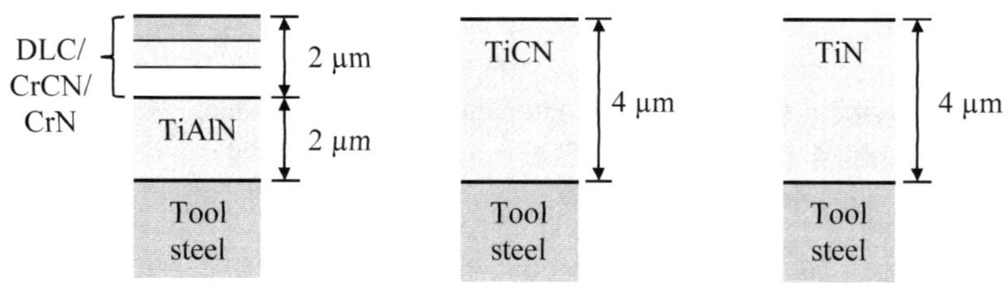

Figure 1: Structure of hard coatings deposited onto the tool steel SKD11 for the experiments.

Table 2: Mechanical properties and surface characteristics of the DLC/TiAlN coating.

Coating	Thickness (µm)	Hardness (HV)	Roughness Ra (µm)
DLC/TiAlN	4 ± 0.2	3,000	0.026
TiCN	4 ± 0.3	3,000	0.089
TiN	4 ± 0.2	2,000	0.092

2.3. Immersion corrosion test

Immersion corrosion test was performed to determine the corrosion resistance of the test specimens coated with the hard coatings according to ASTM G31 standard [15]. As seen in Figure 2, the scratches were made on the test specimen before the corrosion test to generate high porosity surface that serve as a pathway for corrosive materials to reach out to the SKD11 tool steel substrate. The materials and apparatus needed for the immersion corrosion test are shown in Figure 3. The test solution used in this immersion corrosion test was sodium chloride solution with concentration of 3.5% out of 100 mL distilled water. The experiments were carried out up to 72 hours in an oven with temperature of 45 °C [16][17]. The mass of each specimen was measured before and after the experiments using analytical balance meter.

With assumption that a localized or an internal corrosion is not present before the experiments, the average corrosion rate in accordance to ASTM G31 is given in the following formula:

$$\text{Corrosion rate} = \frac{KW}{ATD} \quad (1)$$

where, K is constant value (unit) to express corrosion rate 8.76 x 10^4 mm/year [15], T is time of exposure in hours, A is area in cm^2, W is mass loss in g and D is density in g/cm^3.

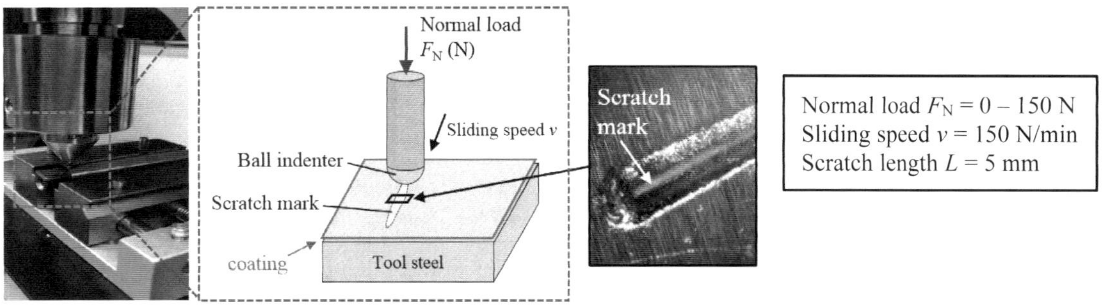

Figure 2: Scratches made on the test specimen by a 200 µm stylus tip made of diamond.

Figure 3: Test materials and apparatus needed for the corrosion test.

3. Results and discussion

After 72 hours of immersion corrosion experiments on all four test specimens, including the three tool steel SKD11 substrates that were coated with TiN, TiCN and DLC/TiAlN, it can be seen that the corrosion has took place in some of the test specimens. As expected, the uncoated tool steel exposed to the immersion solution containing 3.5% NaCl has the highest mass and corrosion rate in comparison to the coated test specimens. This is because of no protective barrier against the corrosion agents on the uncoated tool steel. Therefore, the corrosive agents and substances like NaCl can easily reacted to the uncoated tool steel surfaces, causing it to degrade within a short period of time and exhibits more surface damage in comparison to the coated tool steel specimens. In the case of the tool steels coated with different hard coatings, TiCN tends to be corroded more easily compared to the other coatings. Among all coatings, DLC/TiAlN has the highest corrosion resistance. The double-layer DLC/TiAlN enhanced the corrosion behaviour of the coating and acted as an excellent protective coating against NaCl even there is some scratches present on the test specimen. As seen in Figure 5, pitting corrosion appeared on the side of scratch track. The location or area that pitting corrosion is more likely to occur is somewhere around crack or pores, which in this case, caused by the scratches made before the immersion corrosion test. It can also be observed that the pits most likely to grow downward, deep inside the scratch surface area. The observations on each test specimen after the experiments suggests that the DLC/TiAlN coating deposited on SKD11 tool steel substrate has better stability than that of TiCN- and TiN-coated surfaces. This indicates that the DLC can prevent the corrosion more efficiently than TiCN and TiN.

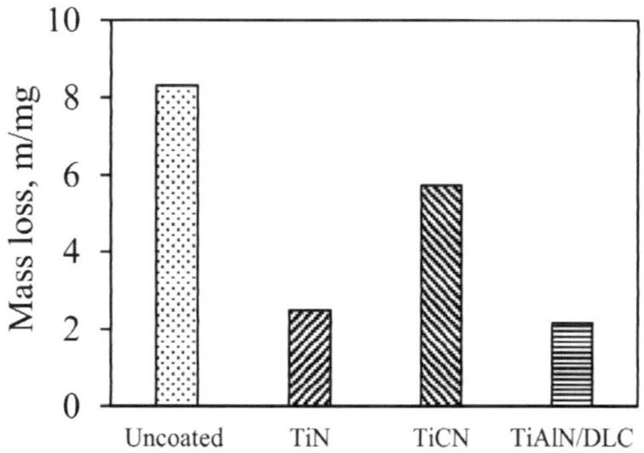

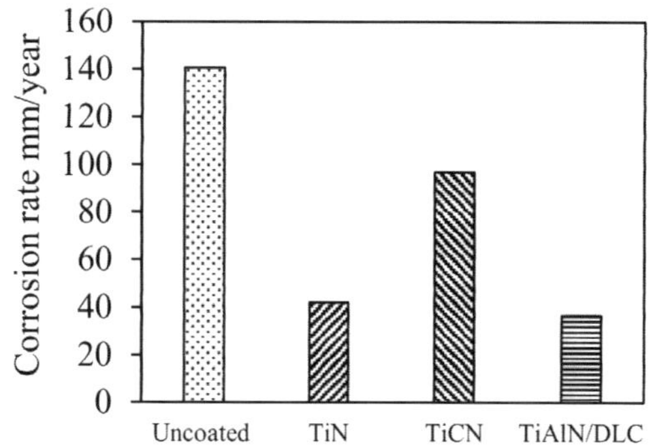

Figure 4: Measurements of mass loss (left) and corrosion rate (right) of each test specimen after the experiments.

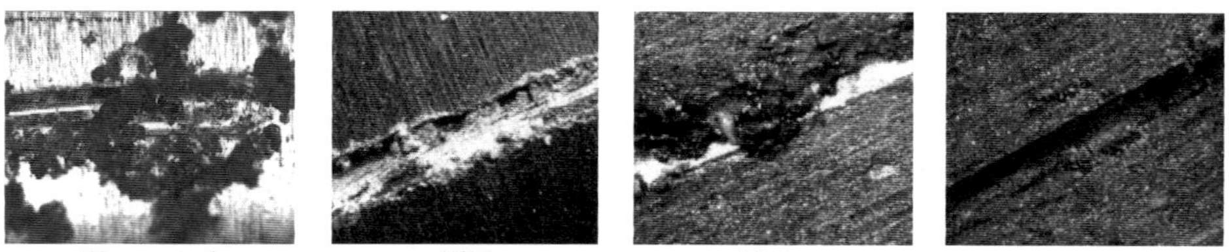

Figure 5: Corrosion occurred on each test specimen after the experiments.

4. Conclusion

PVD hard coating is becoming an emerging alternative for the metal forming industry to produce large quantity of high quality metal components with improved performance in a more eco-friendly and safe way. In the present work, three PVD hard coatings were characterized in terms of corrosion behaviour. The DLC/TiAlN PVD coating showed the best compromise between good corrosion resistance, making it a promising choice for specific applications in metal forming. It is worth noting that the DLC/TiAlN PVD coating studied in the present work was in the form of a double layer with a thickness comparable to penetration depth. Despite this, its performance was quite encouraging and could be further improved with increasing thickness. In applications where high-localized loads and creeps are expected, it is preferable, as in the case of forming of advanced high strength steels, to choose double layer over multilayer PVD coating to avoid delamination. The same situation is valid for corrosion resistance, where the fragility of the layer can lead to cracks and consequently to localized attacks with a worsening of the protective properties, especially in aggressive environments.

Acknowledgements
This work was supported by the Taiho Kogyo Tribology Research Foundation TTRF [grant number 9008-00015]. Special gratitude is dedicated to Tribology & Precision Forming (Tribo-Forming) research group for the fruitful discussions, equipment and technical assistance.

References
[1] Shoba Ch, Prasad D S 2015 Cathodic arc TiAlN PVD coatings for machining applications *Mater. Sci. An Indian J.* **12** (4) 142–6.
[2] Liu A, Deng J, Cui H, Chen Y and Zhao J 2012 Friction and wear properties of TiN, TiAlN, AlTiN and CrAlN PVD nitride coatings *Int. J. Refract. Met. Hard Mater.* **31** 82–8.
[3] Norrby N, Johansson-Jõesaar M P and Odén M 2014 Improved metal cutting performance with bias-modulated textured Ti0.50Al0.50N multilayers *Surf. Coatings Technol.* **257** 102–7.
[4] Mayrhofer P H, Hörling A, Karlsson L, Sjölén J, Larsson T, Mitterer C and Hultman L 2003 Self-organized nanostructures in the Ti-Al-N system *Appl. Phys. Lett.* **83** 2049–51.
[5] Cueva G, Sinatora A, Guesser W L and Tschiptschin A P 2003 Wear resistance of cast irons used in brake disc rotors *Wear* **255** 1256–60.
[6] Geißdörfer S, Rosochowski A, Olejnik L, Engel U and Richert M 2008 Micro-extrusion of ultrafine grained copper *Int. J. Mater. Form.* **1** 455–8.
[7] Sulaiman M H, Christiansen P and Bay N 2017 A study of DLC coatings for ironing of stainless steel *IOP Conf. Ser. J. Phys.* **896** 1–7.
[8] Sulaiman M H, Christiansen P and Bay N 2017 A study of anti-seizure tool coatings for ironing of stainless steel *J. Tribol. – Spec. issue WTC2017* **17** 1–11.
[9] Bech J, Bay N and Eriksen M 1998 A Study of Mechanisms of Liquid Lubrication in Metal Forming *CIRP Ann. - Manuf. Technol.*
[10] Vetter J 2014 60years of DLC coatings: Historical highlights and technical review of cathodic arc processes to synthesize various DLC types, and their evolution for industrial applications *Surf. Coatings Technol.* **257** 213–40.
[11] Bobzin K, Brögelmann T, Kruppe N C, Arghavani M, Mayer J and Weirich T E 2017 Plastic deformation behavior of nanostructured CrN/AlN multilayer coatings deposited by hybrid dcMS/HPPMS *Surf. Coatings Technol.* **332** 253–61.
[12] Deng J and Braun M 1995 DLC multilayer coatings for wear protection *Diam. Relat. Mater.* **4** 936–43.
[13] Davis C A, Knowles K M and Amaratunga G A J 1995 Cross-sectional structure of tetrahedral amorphous carbon thin films *Surf. Coatings Technol.* **76–77** 316–21.
[14] In C B, Kim S P and Chun J S 1994 Corrosion behaviour of TiN films obtained by plasma-assisted chemical vapour deposition *J. Mater. Sci.* **29** 1818–24.
[15] ASTM Standard G-31 1999 Standard Practice for Laboratory Immersion Corrosion Testing of Metals, Annual Book of ASTM Standards *ASTM Int. West Conshohocken, USA.* **3.02** 101.
[16] Jellesen M S, Christiansen T L, Hilbert L R and Møller P 2009 Erosion-corrosion and corrosion properties of DLC coated low temperature gas-nitrided austenitic stainless steel *Wear* **267** 1709–14.
[17] Cicek H 2018 Wear behaviors of TiN/TiCN/DLC composite coatings in different environments *Ceram. Int.* **44** 4853–8.

IOP Conf. Series: Materials Science and Engineering **670** (2019) 012055 doi:10.1088/1757-899X/670/1/012055

The physical properties observation of oil palm empty fruit bunch (OPEFB) chemical treated fibres

M K Faizi[1,2,a], Shahriman A B[1,2], M S Abdul Majid[1], M N M Abdul Rahman[1], Shamsul B M T[3], Ng Y G[3], Z M Razlan[1,2], W K Wan[1], M S M Hashim[1,2] and M A Rojan[1]

[1]School of Mechatronic Engineering, Universiti Malaysia Perlis, Pauh Putra Campus 02600 Arau, Perlis, Malaysia.

[2]Motorsport Technology Unit, Universiti Malaysia Perlis, Pauh Putra Campus 02600 Arau, Perlis, Malaysia.

[3]Faculty of Medicine and Health Science, Universiti Putra Malaysia, 43400 UPM Serdang, Selangor Darul Ehsan.

[a]mkhairulfaizi@unimap.edu.my

Abstract. The key objective of this work was to observe the effect of chemical treatments on the oil palm empty fruit bunch (OPEFB) fibres. The OPEFB fibres were treated with sodium hydroxide (NaOH) by 3% of concentration and 7 hours of soaking time. The treated OPEFB fibres were subjected to Fourier Transform Infrared (FTIR) spectroscopy and X-Ray Diffraction (XRD). The FTIR spectroscopy and XRD performed using Spectrum RXI and Shimadzu XRD 6000 respectively. The FTIR result shows that the physical properties of the treated OPEFB fibres changed due to amount of amorphous materials removed. While the XRD results show the major crystalline peak occurred at $2\theta = 29.8°$.

Keywords: Natural fibre reinforced composites; natural fibres; alkali treatment, silane treatment, cellulosic.

1. Introduction

Due to better weight proportion, natural fibre composite is a promising future composite material [1]. Presently in Malaysia, the potential natural fibre to be explored is oil palm empty fruit bunch (OPEFB). Malaysia is one of successful country in oil palm industry in the world that contributes to the major national financial development. Though, only 10% of the oil palm tree aggregate biomass delivered as useful oil, the remaining 90% was produced as wastes [2]. The wastes consist of OPEFBs, trunks, mesocarp fibres, kernel shells, and fronds [3]–[5]. Most of wastes completely recycled as secondary purposes such as trunks, both mesocarp fibres and kernel shells, and fronds used as plywood, boiler organic fuel, and food for livestock animals respectively [6]. However, no secondary cycle for OPEFB specifically. The oil palm millers tend to dispose the OPEFB by combustion and it always exposed to the pollution issue. Nevertheless, based on inclusive observations, the OPEFB fibre waste potentially to be converted into beneficial purposes especially as optional natural fibre composite [7], [8]. As for this purpose, chemical treatment on the fibre is a common approach for constituent modification to ensure the good fibre-matrix consolidation [9]–[13]. The sodium hydroxide (NaOH) was used for the treatment. In this paper, the physical properties of the OPEFB fibres are comprehensively discussed.

2. Materials and methods

To complete this study, the OPEFB fibres had supplied by United Oil Palm Industries, Nibong Tebal, Pulau Pinang. The fibres undergo the water retting process about a week to remove the dirt and the residual oil. Then the fibres naturally dried before extracted hair by hair. Prior the chemical treatment, the single fibres required to naturally dry to remove the moisture content. The fibres were then treated with 3% concentration of NaOH and soaked at room temperature about 7 hours. Next, the treated fibres observed with Fourier Transform Infrared (FTIR) spectroscopy, X-Ray Diffraction (XRD), and moisture content testing. These FTIR spectroscopy, XRD, and moisture content testing performed by using Spectrum RXI, Shimadzu XRD 6000, and FCD 3000 oven respectively.

3. Result and discussion

The FTIR spectra of 3% NaOH treated OPEFB fibre is presented in Figure 1. Based on Figure 1, the treated fibre exhibits well defined bands within its spectra with approximately 785.9, 835.82, 1019, 1322.03, 1447.19, 1555.9, 1650.44, 2022.02, 2167.69, 2352.32, 2921.21, 3270.62, and 3447.15 cm^{-1}. The treated fibre displays two main absorbance regions which are 2800 – 3600 cm^{-1} for high wave number and 700 – 2500 cm^{-1} for low wave number. This shows that the FTIR spectra obtained have similar functional groups that appear in most natural fibres study [14]. The higher and broad absorption band at 3447.15 cm^{-1} represents the stretching vibration of O-H groups due to the moisture content where hydroxyl groups are found in the hemicellulose, lignin, and cellulose [14]. The absorption at 2921.21 cm^{-1} is related to the asymmetric stretching vibration of the CH_2 group meanwhile the absorbance at 1650.44 cm^{-1} in the spectra indicates the absorption of water [15]. The absorption band at 1447.19 cm^{-1} is associated with the intermolecular hydrogen symmetric bending of CH_2 and designed as a crystalline band while the peak at 835.82 cm^{-1} which has the most significant peak intensity that continually increases with NaOH treatment and also the subsequent size reduction treatments associated with C-H rock vibration of cellulose observed in the fibre [15]. These features are characteristics of the β-glucosides that indicates the bond between anhydroglucose units of the cellulose units that appear in the fibres [16].

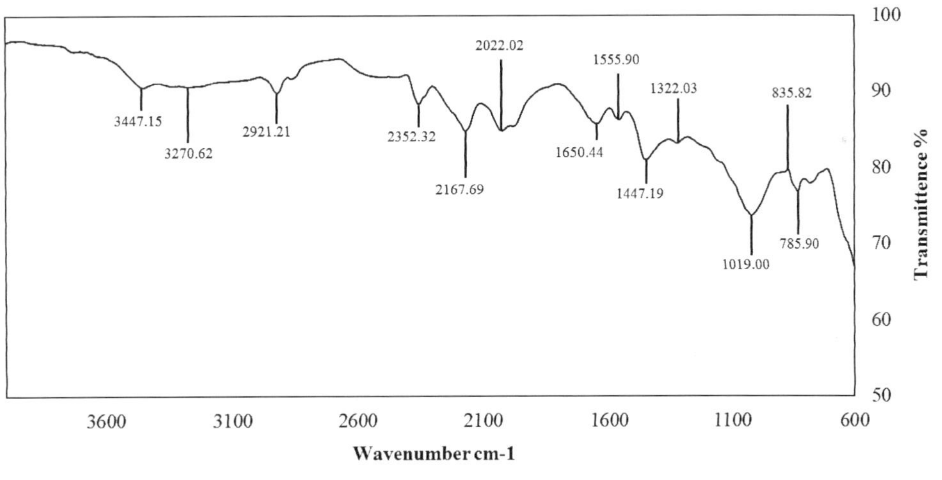

Figure 1. FTIR for 3% NaOH treated OPEFB fiber

Figure 2 exhibits the XRD diffractogram of the treated OPEFB fibre. It is common acknowledge that cellulose consists of both crystalline and amorphous regions. It can be seen from Figure 2 that the major crystalline peak (high intensity diffraction) of the treated OPEFB fibre occurred at $2\theta = 29.8°$ which represents the cellulose crystallographic plane and low intensity diffraction peak occurred at $2\theta = 28.5°$. The diffraction peak at $2\theta = 29.8°$ is sharp indicating the increase in crystallinity and thus related to the increases of the rigidity of the cellulose structure which can lead to a higher tensile strength of the fibre [17]. Other well defined diffraction peaks are occurred at 2θ = 21.46° and 2θ = 26.86°.

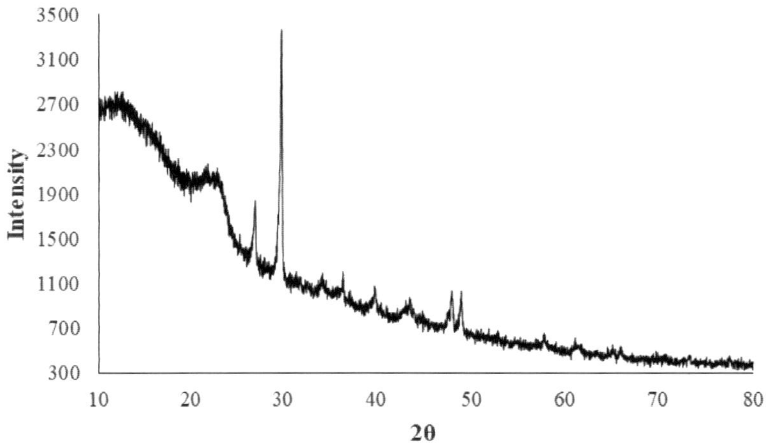

Figure 2. XRD plot of treated OPEFB fibre

4. Conclusions

The physical properties of the NaOH treated OPEFB fibres were investigated in this paper. The treated OPEFB fibres were subjected to Fourier Transform Infrared (FTIR) spectroscopy and X-Ray Diffraction (XRD). The FTIR spectroscopy displays two main absorbance regions which are 2800 – 3600 cm^{-1} for high wave number and 700 – 2500 cm^{-1} for low wave number. This shows that treated OPEFB fibres have similar functional groups that appear in most natural fibres study. The FTIR result also proves that the physical properties of the treated OPEFB fibres changed due to amount of amorphous materials removed. Though, the XRD results show the major crystalline peak occurred at 2θ = 29.8° while low intensity diffraction peak occurred at 2θ = 28.5°. Eventually, this knowledge will further support the feasibility of utilising OPEFB fibres as reinforcing materials as polymer composites.

Acknowledgement

The authors would like to be obliged to Universiti Malaysia Perlis (UniMAP) for providing laboratory facilities, Motorsport Technology Unit (MOTECH) UniMAP for providing financial assistance, and Universiti Putra Malaysia (UPM) as collaboration partnership.

References

[1] Faizi M K et al. 2017 An overview of the Oil Palm Empty Fruit Bunch (OPEFB) potential as reinforcing fibre in polymer composite for energy absorption applications *MATEC Web Conf.* **90** 01064.

[2] Awalludin M F, Sulaiman O, Hashim R and Nadhari W N A W 2015 An overview of the oil palm industry in Malaysia and its waste utilization through thermochemical conversion, specifically via liquefaction *Renew. Sustain. Energy Rev.* **50** 1469–1484.

[3] Abdullah N and Sulaim F 2013 The oil palm wastes in malaysia *Biomass Now - Sustain. Growth Use..*

[4] Aljuboori A H R 2013 Oil Palm Biomass Residue in Malaysia: Availability and Sustainability *Int. J. biomass renewables OIL* **2** (1) 13–18.

[5] Sulaiman F, Abdullah N, Gerhauser H and Shariff A 2011 An outlook of Malaysian energy, oil palm industry and its utilization of wastes as useful resources *Biomass and Bioenergy* **35** (9)

IOP Conf. Series: Materials Science and Engineering **670** (2019) 012055 doi:10.1088/1757-899X/670/1/012055

3775–3786.

[6] Faizi M K, Shahriman A B, Abdul Majid M S, Ahmad Z A, Shamsul B M T and Ng Y G 2018 The Effect of Multiple Surface Treatments on Oil Palm Empty Fruit Bunch (OPEFB) Fibre Structure The Effect of Multiple Surface Treatments on Oil Palm Empty Fruit Bunch (OPEFB) Fibre Structure *J. Phys. Conf. Ser.* 0–5.

[7] Faizi M K, Shahriman A B, Abdul Majid M S, Ahmad Z A, Shamsul B M T and Ng Y G 2018 The Effect of Alkaline Treatments with Various Concentrations on Oil The Effect of Alkaline Treatments with Various Concentrations on Oil Palm Empty Fruit Bunch (OPEFB) Fibre Structure *J. Phys. Conf. Ser.* 3–8.

[8] Faizi M K, Shahriman A B, Abdul Majid M S, Ahmad Z A, Shamsul B M T and Ng Y G 2017 The effect of alkaline treatments soaking time on oil palm empty fruit bunch (OPEFB) fibre structure *J. Phys. Conf. Ser.* **908** (1) 1–5.

[9] Mohammad Asim M R I, Mohammad Jawaid and Khalina Abdan 2016 Effect of alkali and silane treatments on mechanical and fibre-matrix bond strength of kenaf and pineapple leaf fibres *J. Bionic Eng.* **13** (3) 426–435.

[10] Orue A, Jauregi A, Unsuain U, Labidi J, Eceiza A and Arbelaiz A 2016 The effect of alkaline and silane treatments on mechanical properties and breakage of sisal fibers and poly(lactic acid)/sisal fiber composites *Compos. Part A Appl. Sci. Manuf.* **84** 186–195.

[11] Cai M et al. 2015 Influence of alkali treatment on internal microstructure and tensile properties of abaca fibers *Ind. Crops Prod.* **65** 27–35.

[12] Ridzuan M J M, Majid M S A, Afendi M, Azduwin K, Kanafiah S N A and Dan-mallam Y 2015 The effects of the alkaline treatment's soaking exposure on the tensile strength of napier fibre *Procedia Manuf.* **2** 353–358.

[13] Kabir M M, Wang H, Lau K T and Cardona F 2013 Effects of chemical treatments on hemp fibre structure *Appl. Surf. Sci.* **276** 13–23.

[14] Khalil H P S A, Yusra A F I, Bhat A H and Jawaid M 2010 Cell wall ultrastructure, anatomy, lignin distribution, and chemical composition of Malaysian cultivated kenaf fiber *Ind. Crop. Prod.* **31** 113–121.

[15] Johar N, Ahmad I and Dufresne A 2012 Extraction, preparation and characterization of cellulose fibres and nanocrystals from rice husk *Ind. Crop. Prod.* **37** (1) 93–99.

[16] Ridzuan M and Mohd B I N 2016 Development and characterisation of hybrid napier / glass fibre reinforced epoxy composites 1441411504.

[17] Chauhan A and Chauhan P 2017 XRD: A Pioneer Technique for Characterizing the Polymer and Fiber *J. Text. Sci. Eng.* **07** (03) 1000303.

Design and characterization of textured surfaces for metal forming applications

M H Sulaiman [1, *], M U Kamel [1], R N Farahana [1] and J A Wahab [2]

[1] School of Mechatronic Engineering, Universiti Malaysia Perlis, Pauh Putra Campus, 02600 Arau, Perlis, Malaysia
[2] School of Material Engineering, Universiti Malaysia Perlis, Pauh Putra Campus, 02600 Arau, Perlis, Malaysia

*Corresponding author: hafissulaiman@unimap.edu.my

Abstract. Surface texture has been the state-of-the-art in many tribological engineering applications to enhance lubrication as well as to improve wear resistivity. The present work focuses on textured surfaces manufactured by micro-milling and excimer laser texturing techniques on flat, hard tool steel surfaces. In order to promote lubrication entrapment and possible escape by micro-hydrodynamic lubrication mechanism, shallow, longitudinal grooves oriented perpendicular to sliding direction were designed and manufactured on the tool steel surfaces. The morphological characterization of the textured surfaces was performed using a tactile roughness profilometer, Light Optical Microscope (LOM), Scanning Electron Microscopy (SEM) and Atomic Force Microscopy (AFM) measurements. It is found out that both surface texturing techniques are feasible for the surface texturing purposes. Assessment of resulting textured surface topographies by means of LOM, SEM and AFM analysis has revealed that the positive result regarding the desired textured surface profile by the excimer laser technique was obtained, owing to its ability to produce shallow groove depth and smaller groove angle than the micro-milled textured surface. This has been confirmed by observation on the laser textured surface topographies by using LOM, SEM and AFM measurements.

1. Introduction

In many mechanical engineering applications, improper lubrication is not desirable that would results in poor tribological performance and thereby, shortens the lifetime of two metal sliding surfaces. Often, large amount of efficient but hazardous lubricants like chlorinated paraffin oils are seen as the effective solution and able to produce instant result. This is favourable by industry that sees instant profit and high productivity as their priority. The good boundary oils alone cannot solve tool wear problem. Correct tribological contact design and manufacture are critically important in reducing wear. Besides, combating the wear with thin-film coatings, surface treatments and anti-seizure steel materials are common practices applied in industry to prolong the life of the sliding surfaces. These are, however, depends on the type and cause of the wear [1]. These days, the application of surface texturing to facilitate lubrication in mechanical engineering applications such as in metal forming, bearings and internal combustion engines is well known.

Texturing of workpiece surfaces to promote lubrication in metal forming has been applied for several decades, while the tool surface texturing is rather new. Texturing workpiece surfaces by using large rolls roughened by Shot Blast Texturing (SBT) or Electro Discharge Texturing (EDT) [2] has been applied to promote mechanical lubricant entrainment, pressurization and possible escape by micro-hydrodynamic lubrication [3,4]. However, it is less feasible in large-scale, multi-stage sheet metal forming production, since the textured workpiece surfaces will flattened out after first forming operation

[5]. A few tests of surface engineered deep drawing tools [6] have shown very promising results, indicating that textured tool surfaces may provide mechanical lubrication systems which can substitute the chemical ones, and thereby replacing the environmentally hazardous lubricants with environmentally benign ones. This positive result has shown that creating micro texturing patterns on tool surfaces would be economically feasible, since the textured tool surfaces can be utilized to produce thousands of workpiece components.
The present paper studies textured tool surfaces for metal forming applications with a focus on a small groove angle, shallow groove depth, and oblong grooves oriented perpendicular to the sliding direction, with varying distances between the grooves. The manufacturing of the textured tool surfaces by using laser micro-machining and conventional micro-milling were studied comprehensively. The resulting textured tool surface topographies were evaluated by means of LOM, SEM and AFM analysis.

2. Materials and method

2.1. Tool material

The tool steel material is a cold work tool steel SKD11 with high C and Cr contents that was alloyed with Mn, Mo, P, S, Si, and V. The tool steels were through-hardened, tempered to 62 HRC, and subsequently polished to a surface roughness *Ra* of 0.1 μm before the surface texturing was applied. Table 1 presents the mechanical properties and the compositions of the tool steel used in the experiments.

Table 1: Compositions and mechanical properties of the test material.

Components	Composition	Mechanical Properties		
		Density ρ (g/cm^3)	Poisson ratio ν	Elastic modulus E (GPa)
Tool steel (AISI D2)	1.4% C, 0.6% Mn, 11% Cr, 0.03% P, 0.03% S, 0.04% Si, 0.8% Mo, 0.2% V	7.58	0.3	210

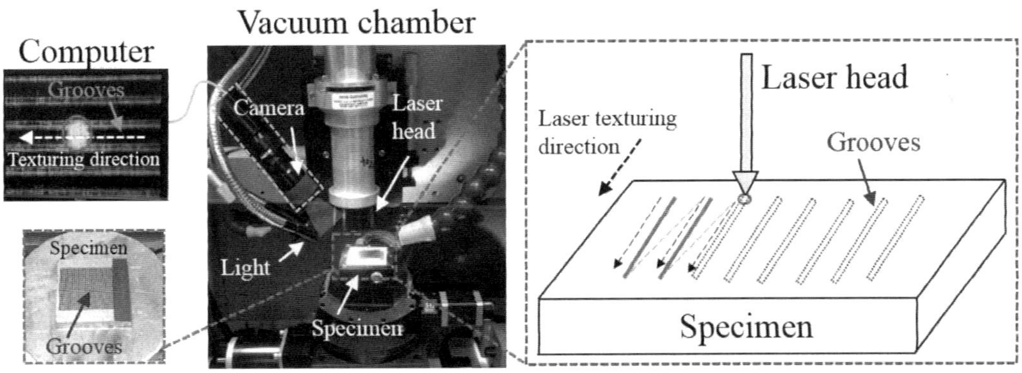

Figure 1: Schematics of laser texturing of oblong shaped grooves on tool surfaces.

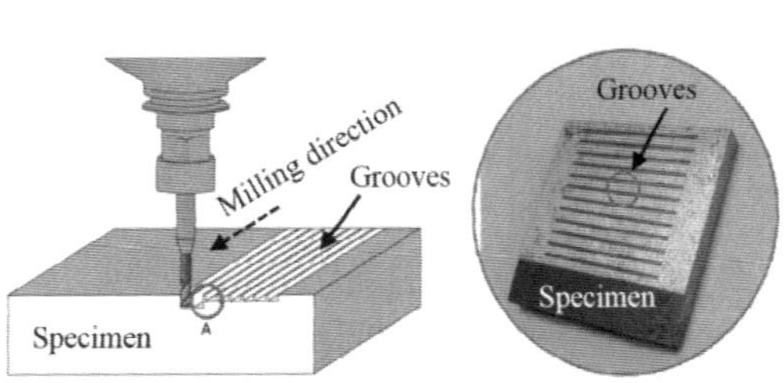

Figure 2: Mechanical micro-milling to manufacture shallow, oblong shaped grooves on tool surfaces.

2.2. Preparation of textured surfaces by excimer laser and mechanical milling

RapidX-250 KrF excimer laser micromachining and manual-operated conventional milling machine were used to manufacture textured surfaces on tool steels SKD11 with a roughness *Ra* of approximately ~ 0.1 µm, see Figure 1 for laser surface texturing technique and Figure 2 shows conventional manual-control micro-milling texturing technique. In the case of laser surface texturing, semi-automatic laser machine was controlled by a highly flexible Computer Numerical Control (CNC) A3200 program. AutoCAD technical details of the desired textured surfaces were first drawn and transferred to BobCAD software before the A3200 program started the laser texturing on the hard steel SKD11. In this study, three surface texture parameters are emphasized to promote micro-hydrodynamic lubrication mechanism [7] and to avoid mechanical interlocking of the workpiece sliding into and out of the groove valleys of the textured tool surfaces [8], which are 1) oblong grooves oriented perpendicular to the drawing direction, 2) small groove angle γ and 3) shallow groove depth d. Four different texture designs were chosen with a width of flat plateaus between the lubricant grooves $x = 0.23, 0.46, 0.92$ mm and 1.61 mm, respectively. Figure 3 shows the groove angle γ and the groove depth d were chosen to be 5° and 0.01 mm respectively. The groove angle was chosen according to recommendations by Popp and Engel [9], the depth was determined by the width, which was chosen small enough to ensure a sufficient number of grooves along the tool/workpiece interface. The length of the grooves were decided to be smaller than the workpiece width in order to ensure pressure build-up of the trapped lubricant. Table 2 lists the surface texture parameters as calculated by Equation 1 and 2.

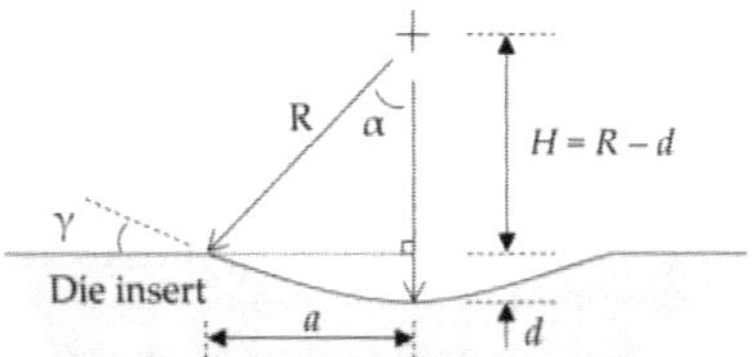

Figure 3: Design of groove geometry.

Table 2. Surface texture parameters.

Parameters	Value
Groove angle γ (°)	5
Groove width $w = 2a$ (mm)	0.23
Groove depth d (mm)	0.01
Groove ratio d/w	0.05

$$\tan\gamma = \frac{d}{a} \tag{1}$$

$$R^2 = a^2 + H^2 = a^2 + (R - d)^2 \tag{2}$$

2.3. Surface topography measurements

Roughness profiles of the textured surfaces were evaluated using Mitutoyo Formtracer CS-3100 profilometer, see Figure 4. To further evaluate resulting grooves of the textured surfaces, Light Optical Microscope (LOM), Scanning Electron Microscope (SEM) and Atomic Force Microscopy (AFM) were used to evaluate surface topography of the textured surfaces. The samples were cleaned in acetone for 10 minutes before analysis.

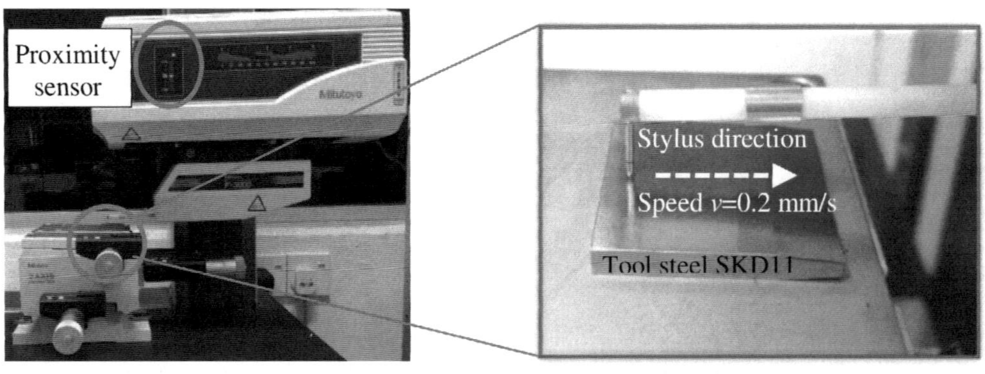

Figure 4: Measured roughness profiles of the resulting textured surface topographies.

3. Results and discussion

Figure 5 represents the resulting, measured grooves of nominal dimensions: length y = 20 mm, angle γ = 3° ± 0.5°, width w = 0.23 mm ± 0.1 mm, depth d = 7 µm ± 1 µm and distance between grooves of x = 0.23, 0.46, 0.92 and 1.61 mm. It is noticed that the groove depths were reached within the tolerance gap, whereas the groove angles turned out to be somewhat smaller than aimed at. This is, however, only promoting the micro-hydrodynamic lubrication mechanism and preventing mechanical interlocking.

As seen in Figure 6, roughness profile measurements showed that the mechanical micro-milling texturing technique is feasible for creating desired depth and width of the grooves, however, shallower grooves is best to be manufactured by excimer laser. Interestingly, smooth bottom groove valleys were observed when manufactures of grooves using excimer laser technique, see AFM result shown in Figure 6 to the bottom right image. This smooth groove valleys would be beneficial in reducing stress concentration that leading to localized crack initiation. Additionally, the resulting textured surfaces made by the laser will be cost savings since this would eliminate the problems of polishing the textured surfaces after the texturing operation due to the edge burr present on both groove ends near to the plateau when manufacturing the grooves using micro-milling, see SEM images shown in Figure 6 to the bottom left. It has been realized that, in case of time consumption when it comes to manufacturing the grooves, the excimer laser technique required more times in comparison to the manual operated micro-milling technique. This would probably adding to the manufacturing costs.

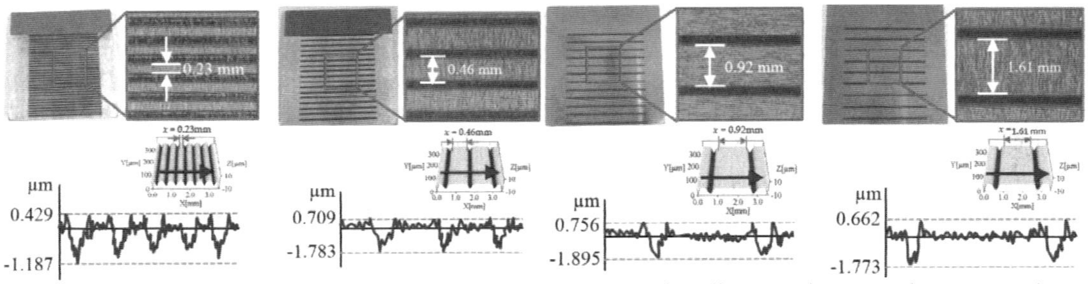

Figure 5: Laser textured tool surfaces concentrating on varying distances between the grooves (x = 0.23, 0.46, 0.92, and 1.61 mm) of the similar sizes of groove depth d and groove angle γ.

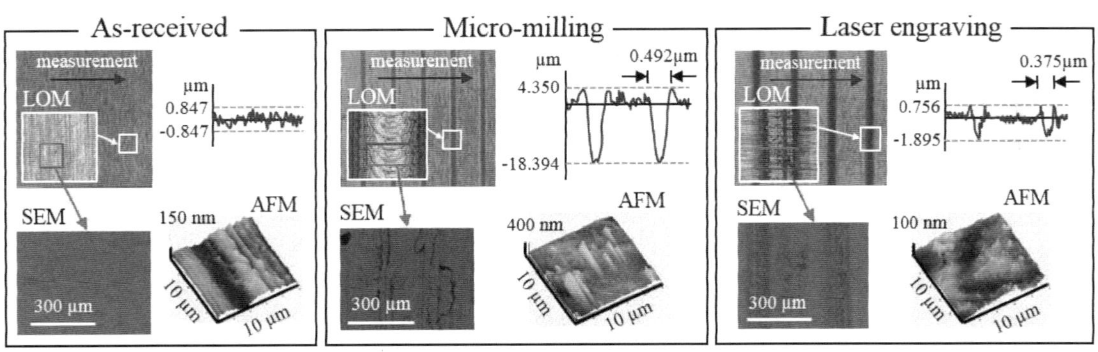

Figure 6: Resulting textured surfaces by means of LOM, SEM and AFM analysis, for as-received tool steel SKD11 condition (left), and the tool steels manufactured by mechanical micro-milling (middle) and excimer laser techniques (right).

4. Conclusion

Applying textured surface on tool steel surfaces is feasible to improve the tribological performance. Implementing laser texturing technique as a method to generate micro-texturing on a hard tool material with desired groove angle, groove depth, groove width and distance between grooves was successfully performed. With the aim to evaluate the resulting textured surface made by the excimer laser technique, the evaluation on surface roughness profiles of each textured tool surface was carried out and comparison was made to the conventional micro-milling technique. Positive result regarding the desired textured surface profile by the excimer laser technique was obtained, owing to its ability to produce shallow groove depth and smaller groove angle than the micro-milled textured surface. This has been confirmed by observation on the laser textured surface topographies by using LOM, SEM and AFM measurements. Further assessments of resulting textured surfaces by advanced surface profilometer is deemed necessary to address the contribution of constituent spatial components with varied amplitudes and wavelengths. This could give more meaningful measured textured surface parameters in developing a functional correlation between a manufacturing process and its corresponding surface profile.

Acknowledgements
This work was supported by the Taiho Kogyo Tribology Research Foundation TTRF [grant number 9008-00015] and Short-term Grant Universiti Malaysia Perlis [grant number 9009-00059]. Special gratitude is dedicated to Tribology & Precision Forming (Tribo-Forming) research group and AMBIENCE UniMAP for the fruitful discussions, equipment and technical assistance.

References
[1] Klocke F, Maßmann T and Gerschwiler K 2005 Combination of PVD tool coatings and biodegradable lubricants in metal forming and machining *Wear* **259** 1197–206.
[2] Kijima H and Bay N 2007 Contact conditions in skin-pass rolling *CIRP Ann. - Manuf. Technol.* **56** 301–6.
[3] Bay N, Bech J I, Andreasen J L and Shimizu I 2002 Studies on micro plasto hydrodynamic lubrication in metal forming *Metal Forming Science and Practice: A state-of-the-Art Volume in Honorary of Prof. J.A. Schey.* (Woodhead Publishing Limited) 115–34.
[4] Sulaiman M H, Farahana R N, Ridzuan M J M, Jaafar H, Tajul L and Wahab J A 2019 Manufacturing strategies and tribology of tailored tool surface topographies for enhanced lubrication in metal forming *Proceedings of 3rd Mytribos Symposium* 47–51.
[5] Ceron E 2013 *New tribo-systems for sheet metal forming of advanced high strength steels and stainless steels.*
[6] Lindvall F, Bergstrom J, Krakhmalev P and Bay N 2010 The effect of grinding and polishing procedure of tool steels in sheet metal forming *Proceedings of the 4th International Conference on Tribology in Manufacturing Processes (ICTMP)* 603–12.
[7] Sulaiman M H, Christiansen P and Bay N 2017 Influence of tool texture on friction and lubrication in strip reduction *Procedia Eng.* **207** 2263–8.
[8] Sulaiman M H, Christiansen P and Bay N 2017 The influence of tool texture on friction and lubrication in strip reduction testing *Lubricants* **5** 1–11.
[9] Sulaiman M H 2017 *Development and testing of tailored tool surfaces for sheet metal forming* (Technical University of Denmark (DTU)).

Mechanical performance of wood poly-lactic acid 3d part under different printing parameter in fused filament fabrication

S F Khan[1,2], H Zakaria[1], M A Asyraf, M A M Saad[1,2]

[1]School of Mechatronic Engineering, Universiti Malaysia Perlis, Pauh Putra Main Campus, 02600 Arau, Perlis, Malaysia.
[2]Additive Manufacturing and Design Lab, School of Mechatronic Engineering, Universiti Malaysia Perlis, Malaysia.

*corresponding author: sfkhan@unimap.edu.my

Abstract. Mechanical properties of parts from fused filament fabrication has been made known to be directionally dependent or in other word anisotropy. Several common process parameters were varied to study their impact on the mechanical properties, in particular compression and bending strength of the end product in the material of wood PLA. The Taguchi method was used to design and simplify the experiment without compromising the experiment's efficiency. S/N ratio from Taguchi analysis was interpreted for determining the optimum combination of process parameter for the highest compressive and bending strength. The build orientation angle is the major factor contributing to higher compression and bending strength recorded followed by infill density and lastly layer thickness. The maximum value for compressive stress is 98.9 MPa with 0° built orientation, 0.15mm layer thickness and 60% infill, while for bending stress it is 73.6 MPa with 0° built orientation, 0.20mm layer thickness and 100% infill.

1. Introduction

Fused filament fabrication (FFF) is an additive manufacturing (AM) technology used in the construction of 3D products [1]. In detail, it is a rapid prototyping and production technique that adds molten plastic layer by layer to form a 3D product. FFF also known as fused deposition modeling (FDM). Poly-lactide-acid (PLA) filament is known to be easy to print [2] and because of that, a substantial amount of research work on PLA had been performed. Sood et al. [3], stated the mechanical properties are influenced by layer thickness, orientation, raster angle, raster width and air gap. However, there is a small number of researched involving PLA that is combined with other materials as to alter the mechanical properties. Hence, the research work presented is in this paper is of the mechanical properties for filaments from PLA fused with wood. It presents the effect of print orientation, layer thickness and in-fill density on compressive and flexural strength.

2. Design of Experiment

This research used the Taguchi method of orthogonal array design to study the whole spectrum of the parameter used with reduced number of experiments [4].and was conducted for L_9 orthogonal array. Calculation on the degree of freedom (DOF) is necessary before any orthogonal array is chosen to be implemented. The chosen array must have the same or greater number of runs compared to the total DOF calculated. Table 1 show the orthogonal array of L_9 used in this research, with 9 experimental runs in this research. The parameter tested were print orientation, layer thickness and in-fill density with others parameter were set to Ender 3 machine default setting. The experiment was conducted in open environment of the Additive Manufacturing Laboratory. The Wood PLA filaments were obtained from Magma with 20% by weight of real wood powder [5].

Table 1. Taguchi Model for Experiment Run

Run	Orientation (angle°)	Layer Thickness (mm)	In-fill Density (%)
1	0	0.10	60
2	0	0.15	80
3	0	0.20	100
4	45	0.10	80
5	45	0.15	100
6	45	0.20	60
7	60	0.10	100
8	60	0.15	60
9	60	0.20	80

3. Testing

The compression test was conducted with reference to standard ASTM D695. Table 2 show the result obtained for compression test while Table 3 is the result for bending test. As shown in Table 2, the maximum compression value is 98.9MPa in Run 1, while the intermediate value is 21.1MPa in Run 5 and the minimum compressive stress recorded is 9.6MPa in Run 8. While in Table 3, the maximum, intermediate and minimum value for bending stress is 73.6MPa in Run 3, 49.6MPa in Run 4 and 30.0 MPa in Run 8; respectively.

Table 2. Tabulated Raw Data of Wood PLA for Compression Test

Run	Result Compressive Stress (MPa)					Average (MPa)
	n1	n2	n3	n4	n5	
1	95.4	94.7	***98.9***	98.1	95.9	96.6
2	94.7	93.8	95.0	95.1	95.3	94.8
3	95.0	93.9	95.7	94.9	80.4	92.0
4	19.5	17.3	17.6	18.2	17.8	18.1
5	22.2	18.9	***21.1***	19.9	22.9	21.0
6	11.6	10.2	10.2	11.0	10.3	10.7
7	26.5	26.5	25.7	24.0	24.8	25.5
8	9.9	***9.6***	10.3	10.5	10.6	10.2
9	12.6	11.1	15.7	17.3	13.9	14.1

Table 3. Tabulated Raw Data of Wood PLA for Bending Test

Run	Result Bending Stress (MPa)					Average (MPa)
	n1	n2	n3	n4	n5	
1	45.1	-	50.5	49.1	47.6	48.1
2	54.6	49.0	-	56.8	44.6	51.2
3	***73.6***	73.1	-	70.4	61.7	69.7
4	45.9	***49.6***	50.8	52.1	40.5	47.8
5	42.7	42.2	41.2	39.8	-	41.5
6	50.3	-	50.5	48.1	45.2	48.5
7	41.0	41.3	-	40.8	44.7	42.0
8	30.3	30.4	***30.0***	30.4	32.1	30.7
9	49.8	51.6	36.0	48.7	50.7	47.4

4. Analysis of Results

Taguchi method uses a statistical measure of performance, called as signal-to-noise (S/N) ratio. The S/N ratio is a logarithmic function of desired output serves as objective functions for optimization. The S/N ratio is the ratio of the mean (signal) to the standard deviation (noise). This ratio is a measure of robustness used to identify control factors that reduce variability in a product or process by minimizing the effect of uncontrollable as Nominal the best (NB), Lower the better (LB) and Higher the better (HB). The S/N ratio for each quality characteristic can be computed independently and regardless of the category of the performance characteristics. Figure 1 show the result for compression test and bending test.

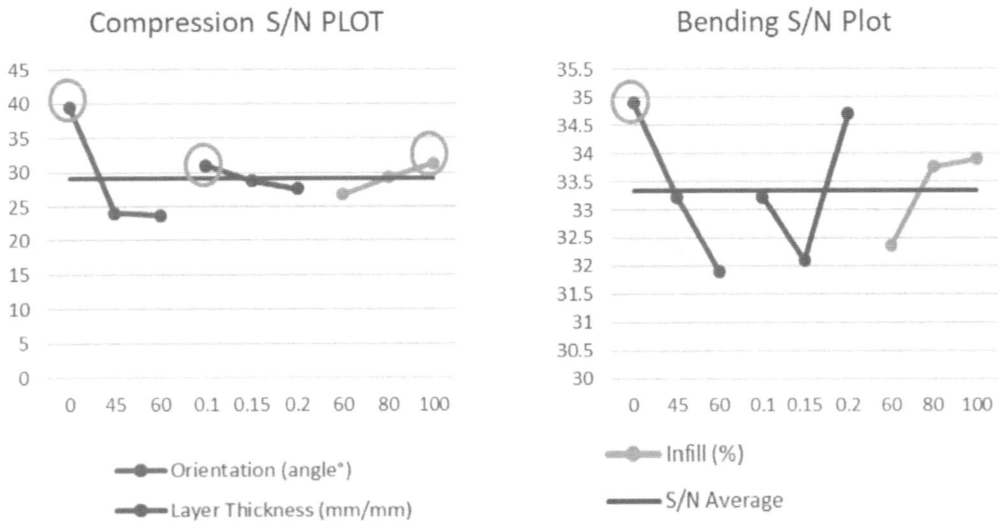

Figure 1. Compression S/N Plot and Bending S/N Plot.

4.1 ANOVA Calculation for Determining Ultimate Strength

ANOVA is a short for analysis of variances is a statistical technique for testing if three or more population means are equal. In this research, to support the S/N plot, ANOVA tables are developed and show the percentage influences of the factor for the compression (Table 4) and bending test (Table 5).

Table 4. ANOVA Table for Compression Test

Parameter	Factor	Degree of Freedom (DOF) (F_n)	Sum of Square (S_n)	Variance (V_n)	F-ratio	Contribution (%)
A	Orientation	2	486.65	243.32	125.08	89.64
B	Layer Thickness	2	17.86	8.93	4.59	2.59
C	Infill Density	2	30.15	15.07	7.75	4.88
Error	-	2	3.89	1.94	-	2.89
Total	-	8	538.56	-	-	100

Table 5. ANOVA Table for Bending Test

Parameter	Factor	Degree of Freedom (DOF) (F_n))	Sum of Square (S_n)	Variance (V_n)	F-ratio	Contribution (%)
A	Orientation	2	13.54	6.77	115.29	47.61
B	Layer Thickness	2	10.22	5.11	87.01	35.83
C	Infill Density	2	4.31	2.15	36.76	14.90
Error	-	2	0.11	0.058	-	1.66
Total	-	8	28.20	-	-	100

IOP Conf. Series: Materials Science and Engineering **670** (2019) 012057 doi:10.1088/1757-899X/670/1/012057

4.2 Scanning Electron Microscope (SEM) Analysis

SEM was used in this research to analyse the surface morphology of failure. Both Torrado et al. and Sood et al. utilized SEM to analyse fracture surface of the 3D printed specimens to provide a vision to the mode of failure [6] and to have a clear image of air gap and void between raster extruded for the study in the dimensional accuracy of FDM parts [4]. The SEM results in Figure 4 to Figure 21 showed the microstructure at the surface of failure for compression and bending test. The specimens with highest, intermediate and lowest charted compression and bending strength are inspected. As aforementioned in Table 2, the specimen for compression test with the lowest properties is specimen number 8.2 (Figure 2, 3, & 4) followed by 5.3(Figure 5, 6, & 7) for the intermediate and 1.3(Figure 8, 9, & 10) for the highest respectively. Whereas in Table 3, the lowest bending or flexural strength scored is specimen number 8.3, while 4.2 for the intermediate and 3.1 for the highest. In compression test, the specimen fractured at an angle of 60° and 45° (Figure 4 and 5) which are similar to the build orientation used to print the specimen. Due to a lower bond between raster and with significantly larger and more air gap formed (Figure 3 & 4), specimen 8.4 has the lowest compressive strength as compared to specimen 5.3 (Figure 6 & 7) and 1.3 (Figure 9 & 10).

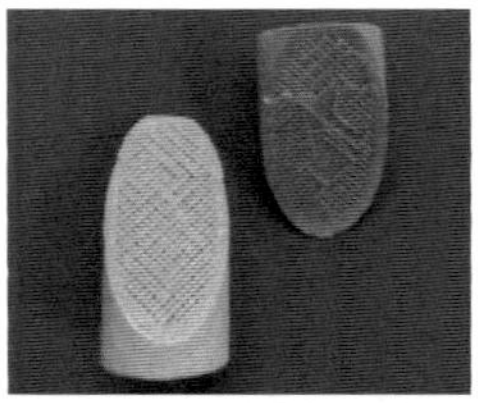

Figure 2. Fracture of Specimen 8.2

Figure 3. x50 Zoom

Figure 4. x100 Zoom

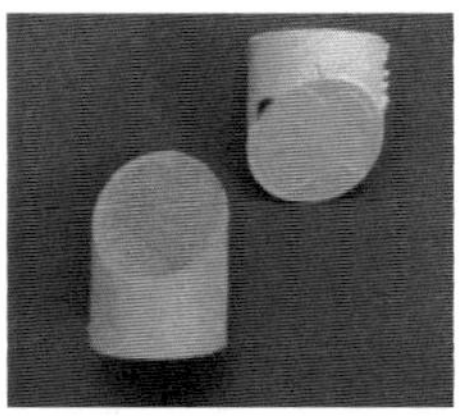

Figure 5. Fracture of Specimen 5.3

Figure 6. x50 Zoom

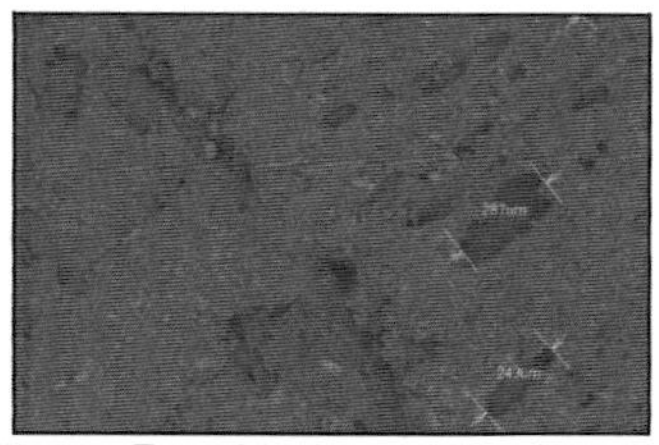

Figure 7. x100 Zoom

Figure 8. Fracture of Specimen 1.3

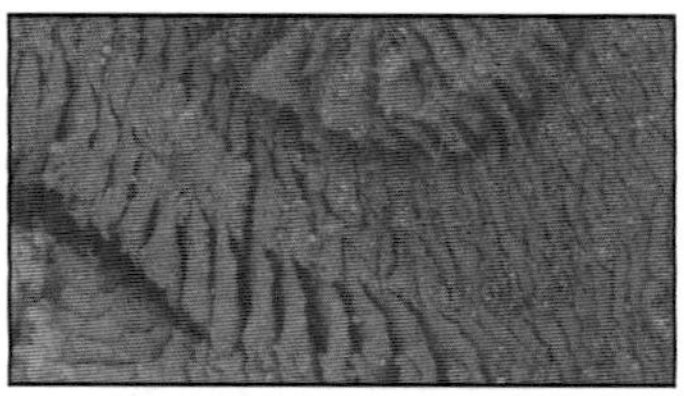

Figure 9. x50 Zoom

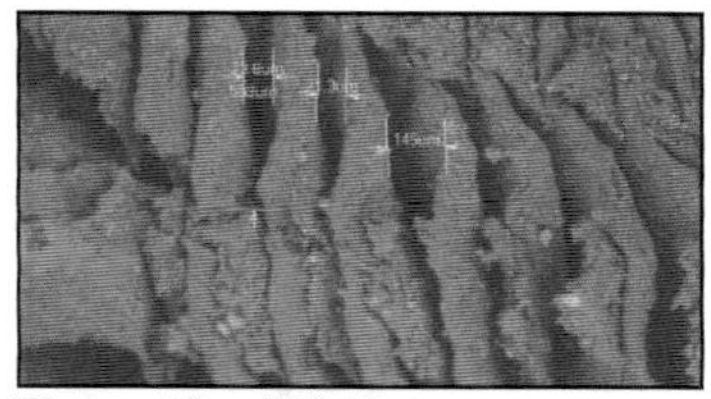

Figure 10. x100 Zoom

Figure 11. Fracture of Specimen 8.3

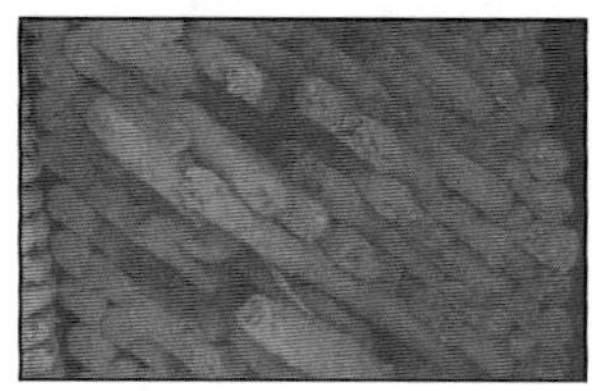

Figure 12. x50 Zoom

Figure 13. x100 Zoom

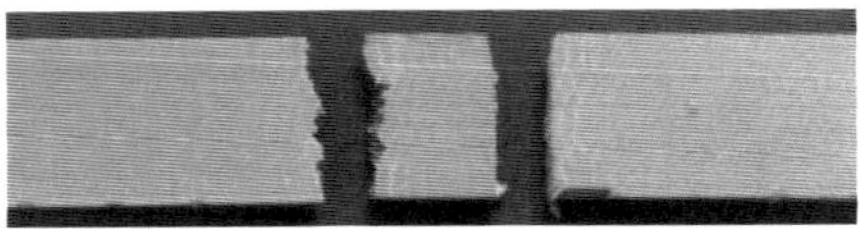

Figure 14. Fracture of Specimen 4.2

Figure 15. x50 Zoom

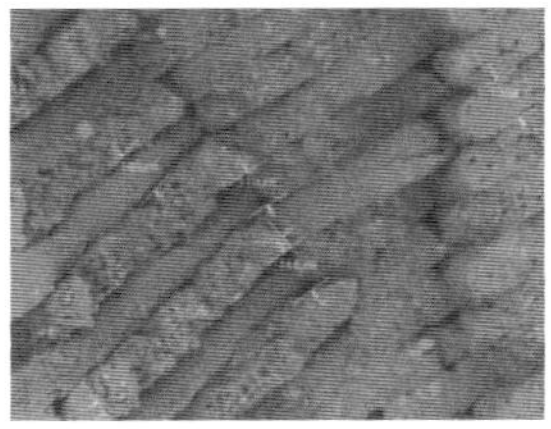

Figure 16. x100 Zoom

Figure 17. Fracture of Specimen 3.1

Figure 18. x50 Zoom

Figure 19. x100 Zoom

5. Conclusion

In regards to the highest compression and bending strength for wood PLA, it is apparent through to the S/N ratio plot, the build orientation angle is the major factor contributing to higher compression and bending strength recorded followed by infill density and lastly layer thickness. A drastic increase of compression and flexural strength is being observed when the specimen was printed at 0° angle of build orientation with an in-fill density is being increased from 60% to 100%, due to the decrease in void formation and air gap in between raster which contribute to the higher bonding surface of the adjacent raster, thus increase load-bearing area. As for the layer thickness on compression and bending strength, there are not much different trend between both experimental. Despite of the significance, wood PLA charted highest in both compression and bending strength when printed in a 0° manner. On the subject of build orientation, wood PLA had a same optimum printing condition in both experimental. Both compression and strength test set, the strength is decreasing by increasing the build orientation angle from 0° angle to 60°.

Acknowledgement

The authors would like to express profound gratitude and deep appreciation to the School of Mechatronics Engineering, School of Manufacturing Engineering and Universiti Malaysia Perlis in providing researching platform well as providing the facilities and equipment.

References

[1] Techopedia, Fused Filament Fabrication (FFF), Polylactic Acid (PLA), Acrylonitrile Butadiene Styrene (ABS), 2018. [Online]. Available: https://www.techopedia.com/definition/29379/polylactic-acid-pla [Accessed: 4-Oct-18].

[2] Tyler Koslow, All3dp.com, 2018 PLA filament guide – explained, compared & all blends. 2018. [Online]. [Accessed: 6-Oct-18].

[3] Krishnaiah K and Shahabudeen P 2013 Applied design of experiments and taguchi methods, PHI Learning Private Limited, New Delhi.

[4] Sood A K, Ohdar R and Mahapatra S 2010 Parametric appraisal of mechanical property of fused deposition modelling processed parts Mater Des, **31** (1) 287-295.

[5] magma3dp.com, Magma PLA wood composite wood filament 1.75mm 1.00 kg. [Online]. Available: http://magma3dp.com/magma-exotic [Accessed: 4-Oct-18].

[6] Torrado, Angel & Shemelya, Corey & D. English, Joel & Lin, Yirong & B. Wicker, Ryan & Roberson, David 2015 Characterizing the effect of additives to abs on the mechanical property anisotropy of specimens fabricated by material extrusion 3d printing. additive manufacturing **6** 16-29. 6.10.1016/j.addma.2015.02.001.

IOP Conf. Series: Materials Science and Engineering **670** (2019) 012058 doi:10.1088/1757-899X/670/1/012058

Overview of cellulose nanocrystals: extraction, physicochemical properties and applications

W L Lim[1], A A N Gunny[2,3] and F H Kasim[1,2]

[1] School of Bioprocess Engineering, Universiti Malaysia Perlis, Kompleks Pusat Pengajian Jejawi 3, 02600, Arau, Perlis, Malaysia
[2] Centre of Excellence for Biomass Utilization, School of Bioprocess Engineering, Universiti Malaysia Perlis, 02600, Arau, Perlis, Malaysia
[3] Department of Chemical Engineering Technology, Faculty of Engineering, Universiti Malaysia Perlis, Kampus UniCITI Alam, Sungai Chuchuh, 02100, Padang Besar, Perlis, Malaysia

Abstract. The development of cellulose nanocrystal (CNC) has attracted significant attention over the past few decades owing to its unique and superior properties. To date, acid hydrolysis process is still the most frequently used method to treat cellulosic material for the production of CNC. The main reason being that acid hydrolysis method is highly effective in removing the amorphous regions within the cellulose microfibrils while retaining the crystalline domains, thereby yielding CNC. CNC exhibits various outstanding characteristics such as large specific surface area, high aspect ratio, good thermal and mechanical properties, lightweight, high stiffness and high tensile strength. Owing to this, CNC has been subjected into a wide array of research, including implementation as reinforcing agent in nanocomposites and sustainable bio-materials to be utilized in various potential areas of industrial application. Overall, this review summarizes the current knowledge on the isolation, physicochemical properties and applications of CNC.

1. Nanocellulose

Nanocellulose refers to cellulose in the nanometer scale, also known as cellulose nanomaterial which commonly derived from disintegration of the naturally occurring biopolymer or produced by bacteria [1]. Nanocellulose can be divided into three main types; cellulose nanocrystal (CNC), cellulose nanofibril (CNF) and bacterial cellulose (BC). Although all these three types are relatively similar in chemical composition, but somehow different in morphological structure, particle size, crystallinity and physicochemical properties due to the varied sources and extraction methods [2]. The elementary fibrils in cellulosic fiber generally consist of highly ordered regions (crystalline structure) and disordered regions (amorphous structure) which can be segregated through mechanical, chemical and enzymatic methods in order to produce either CNF or CNC. In general, Nanocellulose exhibits excellent physical and mechanical properties due to the presence of hydroxyl groups and strong hydrogen bonding network [3].

1.1. Cellulose Nanocrystal (CNC)

Cellulose nanocrystal with other terminologies also known as nanocrystalline cellulose (NCC), cellulose nanowhisker (CNW) or rod-like cellulose microcrystals. Individual cellulose molecules are linked together to form elementary fibrils. Each elementary fibril contains nano-sized and highly crystalline rod shaped fragments, commonly referred as CNC [2]. Each CNC rod generally has a width about 2-30 nm and length of few hundred nanometers. However, the physical dimensions of CNC rod are vary depending on several factors, including the source of the cellulose, hydrolysis conditions and the ionic strength of acid. CNCs have high distribution in length due to the diffusion-controlled nature of the acid hydrolysis process [4]. CNCs can be extracted from many different cellulosic resources such as cellulose enriched pulp, microcrystalline cellulose and cotton via acid hydrolysis method [5].

2. Extraction of CNC

The initial concept of isolating CNC using acid hydrolysis method was introduced by Nickerson and Habrle [6]. In the 1950s, a stable colloidal suspension of CNC was first obtained by Rånby through a controlled sulfuric acid-catalyzed degradation of cellulose fibers [7]. During the controlled acid hydrolysis process, amorphous regions of elementary nanofibril are preferentially hydrolyzed when the hydronium ions infiltrate the cellulose chains in amorphous domains, triggering hydrolytic cleavage of the glycosidic bonds by releasing individual crystallites while leaving the crystalline domains intact [4].

There are many different types of strong inorganic acids such as sulphuric acid (H_2SO_4), hydrochloric acid (HCl), phosphoric acid (H_3PO_4), hydrobromic acid (HBr), phosphotungstic acid (PTA) and nitric acid (HNO_3) have been successfully used for the production of CNCs [8]. All these strong inorganic acids are diluted to the desired concentration before treating with cellulosic fiber under strictly controlled conditions of temperature, agitation speed and reaction time. Among all the inorganic acids, H_2SO_4 and HCl are the most common inorganic acids used for CNC preparation. Many studies have reported that 55-65 wt. % of H_2SO_4 or 25-30 wt. % of HCl at moderate temperature (40-60 °C) were applied in acid hydrolysis along with subsequent mechanical and ultrasound disintegration of the hydrolyzed cellulose in aqueous suspension [9].

However, there are many alternative methods in extracting and improving the yield of CNC have been introduced since the beginning of twenty-first century such as cation-exchange resin-catalyzed hydrolysis [10], ultrasonic-assisted sulphuric acid hydrolysis [11], two-stage mildly acidic ionic liquid-mediated hydrolysis [12] and acidic deep eutectic solvents (DESs) treatment [13]. Owing to the emergence of DES, choline chloride (ChCl) has been successfully combined with organic acids such as oxalic acid, p-toluenesulfonic acid monohydrate and levulinic acid to produce acidic DES which used in acid hydrolysis of cellulosic fibers. In addition, Sirviö reported that these acidic DESs could provide a green alternative pathway for CNC production [13].

3. Physicochemical properties of CNC

3.1. Specific surface area

Specific surface area (SSA) can be defined as the total surface area of a material per unit of mass. The SSA of a non-porous material increases as the size of a particle decreases. For nanocellulose, the SSA can be measured from the average geometrical dimensions of the particles, assuming that each CNC rod-like geometry with a density of 1.5 or 1.6 g/cm^3 [4]. Moreover, the SSA can be also determined by Brunauer-Emmett-Teller (BET) surface area analysis from nitrogen gas adsorption isotherms at 77k using a surface area or porosity analyzer [14]. However, the determination of SSA using nitrogen gas adsorption isotherm is often inaccurate due to the irreversible aggregation of the CNC upon oven drying [4]. Spray dried CNC, freeze dried CNC and spray-freeze dried CNC extracted from bleached kraft pulp with SSA of 0.5 m^2/g, 17.2 m^2/g and 56.4 m^2/g, respectively were reported by Abdallah and Kamal [15]. Therefore, the CNC with high specific surface area is potentially useful in various nano-applications, especially applied as porous template for the preparation of aerogel [4].

3.2. Thermal stability

Thermal stability is one of the important physical property which plays an essential role in preparation of the CNC reinforced nanocomposites. The thermal stability of CNC can be determined using thermogravimetric analysis (TGA). For CNC, the thermal stability is varied depending on the types of surface chemistry [16]. For instance, the introduction of negatively charged sulphate half ester groups on the CNC surface obtained via dilute acid hydrolysis using sulphuric acid significantly affect the thermal stability of CNC. During the acid hydrolysis, partial esterification of hydroxyl groups of cellulose along with substitution by sulphonic acid groups ($-SO_3H$) occur, significantly reduces the thermal stability of CNC. Morais reported that the pure cellulosic fiber obtained from cotton linter decomposes with a maximum rate at 340 °C while the isolated CNC degrades at 288 °C [17]. Moreover, the increasing of sulfonic acid groups content in CNC also leads to lower thermal stability [18]. However, the thermal stability of CNCs isolated via HCl and H_3PO_4 acid hydrolysis tend to be higher in which decompose at 325 °C and 350 °C respectively. Therefore, CNC with lower thermal stability may limit its use as reinforcing filler to produce bio-nanocomposites at high temperature in manufacturing process [16].

3.3. Aspect ratio

Aspect ratio is defined as the ratio of the length to the width. It is an important parameter to determines the anisotropic phase formation and reinforcing properties. In general, the aspect ratio of CNCs which isolated from cellulosic materials such as cotton, sisal, flax and jute via acid hydrolysis ranging between 16 and 50 [2]. Some literature reported that the CNC prepared from the rare marine creatures, also known as tunicates has higher aspect ratio (70-100) [19]. In fact, CNC with higher aspect ratio has greater reinforcement capacity when added into the nanocomposite materials. Consequently, this can improve the mechanical properties of the nanocomposite such as tensile modulus due to the alignment of high aspect ratio filler particles which increases the surface area of the fillers [20]. Owing to this, CNC with higher aspect ratio has great potential in filter and composite membrane applications for air and water purification [4].

4. Application of CNC

CNC has been used in various applications owing to its availability and remarkable physical properties. For instance, CNC exhibits high thermal conductivity and low coefficient of thermal expansion which may suitable for the use in applications such as electronic devices and packaging. Moreover, CNC often chosen as the nano-sized filler that act as reinforcement to improve the mechanical and barrier properties of the matrix in bio-nanocomposite material. This is due to the CNC generally has a high aspect ratio and large specific surface area that could provide better reinforcing effects [21].

CNC has also been applied in the production of hydrogels and aerogels owing to its light weight and toughness. The hydrophilic surfaces of CNC serve as building blocks to form the gels, which enables crosslinking with other different metals, organic materials and polymers via hydrogen bonding, covalent bond or ionic interactions [22]. In coating industry, the durability of water soluble paints can be improved by incorporating CNC. The reason being that CNC acts as the thickening and stabilizing agent that enhances the mechanical properties, scratch resistance and UV-shielding property [23]. CNC is also serving as an important cosmetic ingredient that improves the sensation and nourishment in facial and body applications. This is owing to the CNC having high dispersion stability in water and water holding capability for a long period of time [4]. Furthermore, CNC can form a very dense and resistant mats because of its strength to weight ratio which is eight times stronger than stainless steel. This feature allows the CNC to be applied in manufacturing of bullet-proof vests and body armor.

In medical application, CNC is widely utilized due to its excellent physical properties such as extended surface area, surface functionality, biocompatibility, biodegradability and low toxicity. This allow the CNC to be used in drug delivery, enzyme or protein immobilization, tissue skeleton for cell culture and macroscopic biomaterials [24]. Moreover, CNC has been added into composite membrane and adsorbent materials for the removal of contaminants. This is attributed to its high aspect ratio and specific surface area, large retention capacity and environmental inertness. Consequently, the pollutants such as heavy metal ions are binding to the active sites, thereby enhancing the filtration efficiency [25].

5. Conclusion

In conclusion, CNC is a rod-like nanoparticle that can be isolated from cellulosic sources through acid hydrolysis process. CNC generally has many outstanding physical properties, including low density, high aspect ratio and specific surface area, good thermal and mechanical properties that plays a vital role as an intermediate in sustainable biomaterials and nanocomposites. Therefore, the transformation of cellulose into CNC is perceived as a milestone in the field of materials science and engineering. This is proven by the global attention in which CNC have been utilized in various applications such as electronic devices, packaging, bio-nanocomposites, medical, filter, membranes and etc.

Acknowledgments

The authors thank to Ministry of Education for financing the research under Fundamental Research Grant Scheme (FRGS 9003-00680)

References

[1] Shak K P Y, Pang Y L and Mah S K 2018 *Beilstein J Nanotechnol* **9** 2479-2498.
[2] Moon R J, Martini A, Nairn J, Simonsen J and Youngblood J 2011 *Chem Soc Rev* **40** 3941-3994.
[3] Tayeb A, Amini E, Ghasemi S and Tajvidi M 2018 *Molecules* **23** 2684.
[4] Dufresne A 2019 *Curr Forest Report* **5** 76-89.
[5] Beck-Candanedo S, Roman M and Gray D G 2005 *Biomacromolecules* **6** 1048-1054.
[6] Nickerson R F and Habrle J A 1947 *Ind Eng Chem* **39** 1507-1512.
[7] Rånby B G 1951 *Discuss Faraday Soc* **11** 158-164.
[8] Habibi Y, Lucia L A and Rojas O J 2010 *Chem Rev* **110** 3479-3500.
[9] Hubbe M A, Rojas O J, Lucia L A and Sain M 2008 *Bioresources* **3** 929-980.
[10] Tang L, Huang B, Ou W, Chen X and Chen Y 2011 *Bioresource Technol* **102** 10973-10977.
[11] Tang Y, Yang S, Zhang N and Zhang J 2014 *Cellulose* **21** 335-346.
[12] Mao J, Heck B, Reiter G and Laborie M P 2015 *Carbohyd Polym* **117** 443-451.
[13] Sirviö J A, Visanko M and Liimatainen H 2016 *Biomacromolecules* **17** 3025-3032.
[14] Rouquerol J, Avnir D, Fairbridge C W, Everett D H, Haynes J M, Pernicone N, Ramsay J D F, Sing K S W and Unger K K 1994 *Pure Appl Chem* **66** 1739-1758.
[15] Abdallah W and Kamal M R 2018 *Cellulose* **25** 5711-5730.
[16] Camarero Espinosa S, Kuhnt T, Foster E J and Weder C 2013 *Biomacromolecules* **14** 1223-1230.
[17] Morais J P S, Rosa M de F, de Souza Filho M de sá M, Nascimento L D, do Nascimento D M and Cassales A R 2013 *Carbohyd Polym* **91** 229-235.
[18] Lin N and Dufresne A 2014 *Nanoscale* **6** 5384-5393.
[19] Amiralian N, Annamalai P K, Garvey C J, Jiang E, Memmott P and Martin D J 2017 *Cellulose* **24** 3753-3766.
[20] Landry V, Alemdar A and Blanchet P 2011 *Forest Prod J* **61** 104-112.
[21] Azizi Samir M A S, Alloin F and Dufresne A 2005 *Biomacromolecules* **6** 612-626.
[22] Lavoine N and Bergström L 2017 *J Mater Chem A* **5** 16105-16117.
[23] Mascheroni E, Rampazzo R, Ortenzi M A, Piva G, Bonetti S and Piergiovanni L 2016 *Cellulose* **23** 779-793.
[24] Du H, Liu W, Zhang M, Si C, Zhang X and Li B 2019 *Carbohyd Polym* **209** 130-144.
[25] Abouzeid R E, Khiari R, El-Wakil N and Dufresne A 2019 *Biomacromolecules* **20** 573-597.

IOP Conf. Series: Materials Science and Engineering **670** (2019) 012059 doi:10.1088/1757-899X/670/1/012059

Laboratory development of low speed wind tunnel for educational purposes

***R A Siregar and K Umurani**

University of Muhammadiyah Sumatera Utara, Jalan Kapten Muchtar Basri No. 3 Medan, Indonesia

*corresponding author: rakhmadarief@umsu.ac.id

Abstract. In this paper presents the design and development of wind tunnel for education purposes. The wind tunnel is designed as an open-type, low speed, able to provide wind speeds up to 20 m/s and maximum turbulent intensity is 5%. A 70 cm tube axial fan is driven by a 3 hp motor using a variable frequency drive to allow continuously variable wind speeds. The size of testing chamber is about 40 cm by 40 cm, large enough to conduct experiments such as drag coefficient, airfoil and wing. Preliminary test of air flow in testing chamber shows that wind tunnel is particularly suitable not only for education but also for aerodynamic researches.

1. Introduction

Research with the application of wind tunnels as a very intense tool used in aeronautics problems and has been so solid in terms of reliability in that field over the last few decades. Parameters such as pressure variation, number of Mach, density and temperature distribution, noise and acoustics have been the concern of many researchers. The most important performance measure in a wind tunnel depends on its turbulence, the rate of fluctuation in the unstable velocity of the average flow rate [1]. Wind tunnel designers seek to improve the flow quality of wind tunnels by using the best design rules and some optimization to improve the role of wind tunnels in industrial design. Currently wind tunnel testing is in the first stage in design for all aircraft [2]. This acceptance is so widespread that people are tempted to use the tunnel to study other airborne phenomena not encountered in the field of aeronautics. The analysis of contraction wall shapes has been published in designing of the open-type and low-speed wind tunnel using a variety of analytical and numerical techniques [3-5]. By using CFD software, flexibility in designing the expanded generally to the contractions of arbitrary cross-section and wall profile [6-7]. There are many types of different wind tunnel designs available for variety applications. Based on wind tunnel wind velocity there are wind tunnel for subsonic, transonic, supersonic and hypersonic speed. An advanced commercial wind tunnel sold commercially is considered very expensive. Since the knowledge of aerodynamics has become the basis for undergraduate students, some universities have built their own wind tunnels [9-10]. This paper aims to design and develop open-type wind tunnel at affordable prices capable of conducting experiments for educational purpose as well as aerodynamic researches.

2. Design Concept

2.1 Contraction ratio

The wind tunnels is designed for laboratory purposes with the primary consideration of having an inexpensive middle size, capable of performing aerodynamic force and moment testing as well as drag coefficients. The main objective is for teaching as an effort to increase theoretical knowledge. To accommodate these requirements, the wind tunnel is designed with modular, open-type construction and has components that can be easily modified for experimental work. The main modules of open-type of wind tunnel are contraction cone, testing chamber and diffuser. The dimensions of the wind tunnel are referred to literature data. The work of Mehta et.al [3-5] in analyzing the construction ratio inspired the current design. As mention in their paper, the adequate contraction ratios of wind tunnel is about 6 to 10 for small and low-speed wind tunnels that suitable to the current requirement. For the present design the 7.5 ratio contractions is selected and this dimension is very close to the dimensions of C-wind tunnel as shown in Table 1. The wall contour contraction of the current design, E-wind tunnel, are inserted into J.H Bell and Mehta [4] result and re-produced as presented in Fig. 1.The complete dimension of the current design can be seen in Fig. 2.

2.2 Wind Tunnel Development

A steel plate of 5 mm thick is used to form the contraction cone and diffuser modules. The plate is carefully bent to develop the appropriate shape. To form contraction contour is the most challenging job, starting forming the mold up to welded the plate (see Fig. 3). The diffuser module utilized 3-blade axial fan with a diameter of 70 cm to suck air that belt driven by a 3 hp motor as shown in Fig. 4(a). The speed of the motor is controlled by an inverter that can be adjusted ranging from 0 to 50 Hz. The current setting provides the airspeed range from 4 to 20 m/s. The test chamber module (40 cm x 40 cm x 60 cm) is designed with transparent windows. The completed wind tunnel is presented in Fig. 4(c).

The total production cost of the machine is about Rp. 10 million (US$750). The highest cost contributors are development of contraction cone, diffuser, fan, controllers and motor while frame and machining is relatively inexpensive due to the precision works is not so high.

3. Preliminary test result and discussion

In order to test the performance of wind tunnel, a simple test was performed with anemometer as the sole instrument. The constant motor speed is set to the fan so it is expected to suck air in the laminar stage in the testing chamber. The first test is to measure the turbulent intensity by using equation,

Table 1. Contractions dimension of wind tunnels for comparison.

No	Wind tunnel	Contraction ratio	Inlet height x width (cm)	Exit height x width (cm)	Contraction length (cm)	Length to height ratio	Wall contour shape
A	Shear layer tunnel (NASA Ames)	10	38 x 76	7.6 x 38	91	1.2	"Eye" design
B	Mixing layer tunnel (NASA Ames)	7.7	91 x 137	91 x 18	244	0.89	Fifth-order polynomial
C	Boundary layer tunnel (NASA Ames)	7.5	120 x 100	20 x 80	120	1.0	Fifth-order polynomial
D	Boundary layer tunnel (Imperial College)	9	114 x 76	13 x 76	122	1.07	"Eye" design
E	Boundary layer tunnel (UMSU)	7.5	120 x 100	40 x 40	120	1.0	"Eye" design

$$T.I = \frac{u'}{U} \tag{1}$$

where u' is the root-mean-square of the turbulent velocity fluctuations and U is the mean velocity at the same location over same time period. Anemometer is placed and data is collected at the centre line for same time period. Figure 5 shows the results of wind velocity for time about 100 seconds with the mean velocity of about 10.52 m/s. By using Eq. 1, it was found that the turbulent intensity was 3.69 %. Based on this preliminary test data, the maximum turbulent intensity limit below 5% has been satisfied.

For the second tests, above the testing chamber, vertical and horizontal rail pits are made to observe the distribution of wind velocity. Data for each position in Fig. 6 are an average data of 10-15 seconds. It can be seen laminar stage occurred at the testing chamber.

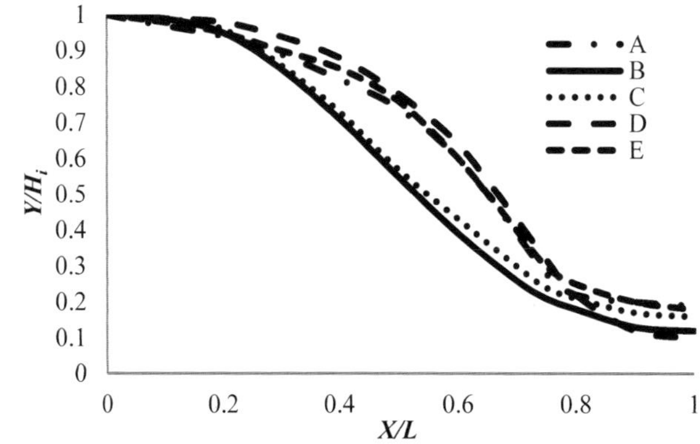

Figure 1. Comparison of wall contour of six contractions.
(L is the contraction length and H_i is the inlet height).

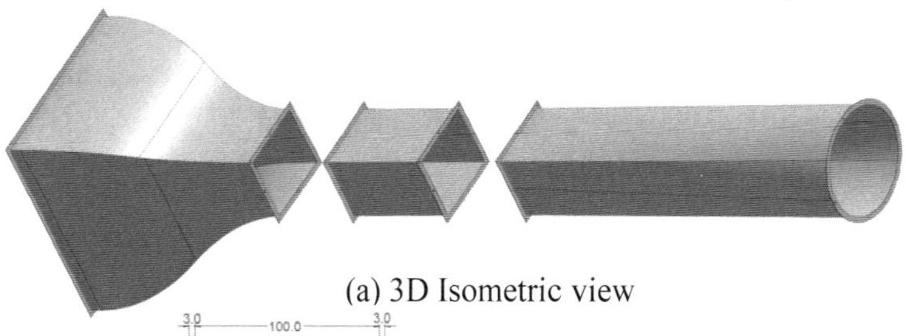

(a) 3D Isometric view

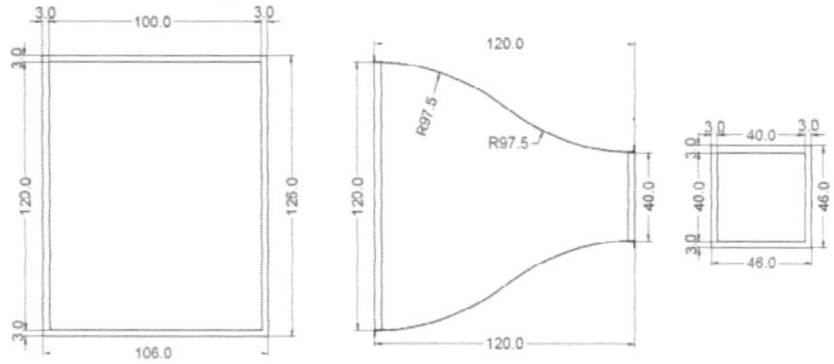

(b) Contraction Cone

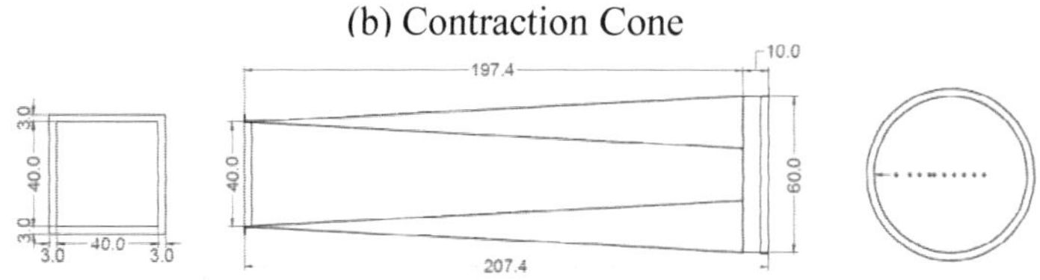

(c) Diffuser

Figure 2. CAD drawing of wind tunnel.

(a) Contraction cone

(b) Diffuser

Figure 3. Construction of wind tunnel.

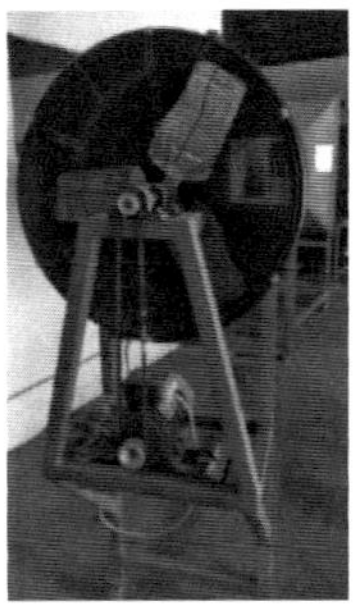

(a) Three-blade fan and motor

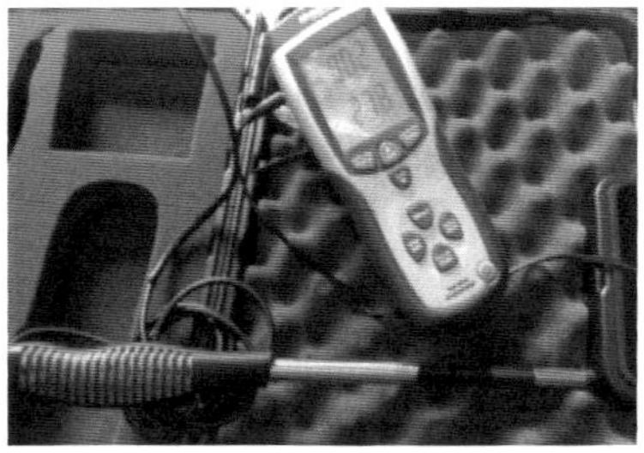

(b) Anemometer

(c) Completed wind tunnel

Figure 4. Completed development of UMSU wind tunnel.

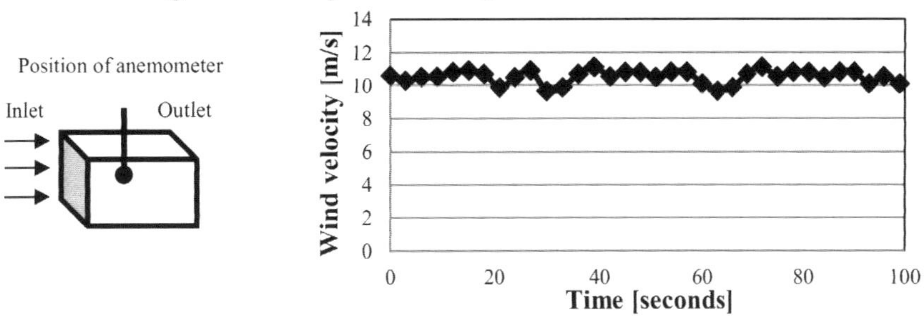

Figure 5. Wind velocity vs. time on centre line of test section.

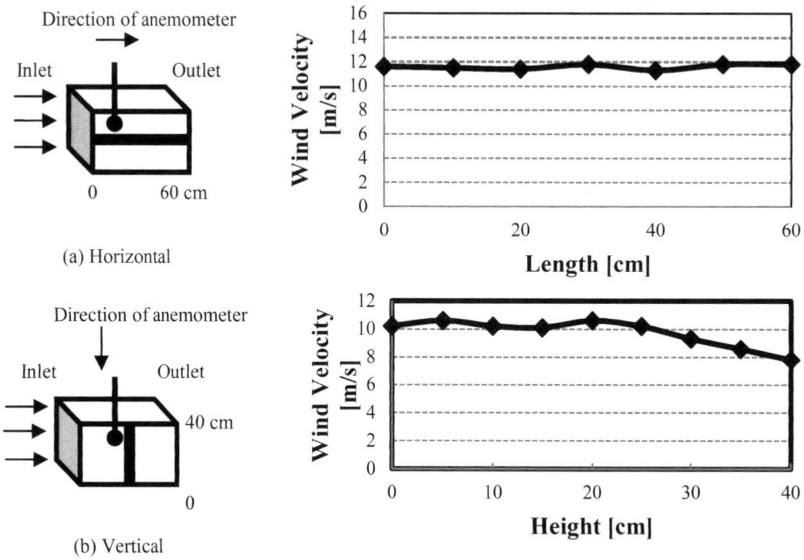

Figure 6. Distribution of wind velocity on centre line of test section.

4. Conclusion

In this paper, the UMSU wind tunnel has been successfully designed and built as a modular, small scale, low-speed wind tunnel and has been evaluated for educational purposes. It will allow mechanical engineering students and faculty staff to have access to a local wind tunnel where a variety of testing and experimentation on aerodynamics field. With wind speeds of up to 20 m/s and a cross sectional area test chamber as large as 40 cm by 40 cm, it possible to use it for a variety experimental works.

References

[1] Frank M W 2006 *Viscous Fluid Flow* Third Ed. McGraw Hill.
[2] Douglas A J 2014 *Flying beyond the stall: the X-31 and the advent of supermaneuverability* National Aeronautics and Space Administration.
[3] James H B and Rabindra D M 1988 *Joint Institute for Aeronautics and Acoustics*, Stanford University.
[4] Bell J H and Mehta R D 1989 *American Institute of Aeronautics and Astronautics Journal* **27-3** 372-374.
[5] Mehta R D and Bradshaw P 1979 *The aeronautical journal of the royal aeronautical society* 443-449.
[6] Sargison J E, Walker G J and Rossi R 2004 *Proc. 15th Australasian Fluid Mechanics Conference.*
[7] Ahmed D E and Eljack E M 2014 *Proc.10th International Conference on Heat Transfer, Fluid Mechanics and Thermodynamics*.
[8] Mohod S W and Aware M V 2011 *Int. Journal of Engineering, Science and Technology* **3** (5) 73-82.
[9] Bunlung N, Somporn S, Somchai C 2007, *Int. Energy Journal* **8** 21-28.
[10] Yong T H and Dol S S 2015 *IOP Conf. Series: Materials Science and Engineering* **78** 012039.

IOP Conf. Series: Materials Science and Engineering **670** (2019) 012060 doi:10.1088/1757-899X/670/1/012060

Optimization of process parameters in fused filament fabrication (FFF) utilizing poly lactic acid (PLA)

***H Zakaria[1], S F Khan[1,2] and M I Zulkafli[1]**

[1] School of Mechatronic Engineering, Universiti Malaysia Perlis, Malaysia.
[2]Additive Manufacturing and Design Lab, School of Mechatronic Engineering, Universiti Malaysia Perlis, Malaysia

***corresponding author: hafizawati@unimap.edu.my**

Abstract. This research was to find an optimum parameter in Fused Deposition Modelling, FDM. They are various type of printing parameter in 3D printing. The parameter that was focused on in this research are orientation, print head and layer thickness while other parameters will be fixed. This parameter was evaluated the mechanical properties of the printed parts. The mechanical properties will be performed on the specimen are tensile test and 3 point-bending tests. The data were analyzed using Taguchi's Method with orthogonal array of 9 level and Analysis of Variance, ANOVA. The highest of ultimate tensile strength was 34 MPa where the parameter were 34.592 MPa. The highest of flexural strength was 95.542 MPa. The optimum parameter for the tensile test was 25-degree of orientation, 0.4 mm of print head and 0.2 for layer thickness. The optimum parameter for 3-point-bending test was 0-degree of orientation, 0.4 mm of print head and 0.15 mm of layer thickness. This parameter did the mechanical test again for the validation test. The average ultimate tensile strength value for 25-degree of orientation, 0.4 mm of print head and 0.2 for layer thickness was 30.558 MPa. While the average of flexural strength for 0-degree of orientation, 0.4 mm of print head and 0.15 mm of layer thickness was 95.557 MPa.

1. Introduction

Additive manufacturing is the technology that produce 3 dimensional part via printing. This technology had already seen vast development and research done for the past few years [1]. These technologies are known to be independence from geometric constraints [2]. Fused filament fabrication (FFF) is one of the additive manufacturing systems that show an excellence work with reliable cost alternative in additive manufacturing [3]. There is various type of material used in FFF these days in 3- dimensional printing such as acrylonitrile butadiene styrene (ABS) and poly lactic acid (PLA). This material can be used as a filament to be extruded from the nozzle that had been heated and produce 3D objects layer by layer from bottom to the top. Poly lactic acid is the preferable material for prototyping or as master pattern in casting since it is easy to print and have excellent mechanical properties as compare to other plastic except composite material [4].

Additive manufacturing (AM) creates three-dimensional (3D) physical objects by the aid of computer- aided design (CAD) software [5]. The process of 3-dimensional printing is by melting and extruding material which creating an object in a layer by layer manner to form 3-dimensional objects [6]. On the other hand, the additive manufacturing provides the technique in an efficient way to make the complicated geometries at a shorter produce of time cycle and in lower of cost [7]. Furthermore, this technology is broadly used in engineering to custom the prototype, pre-surgical models, functional models and conceptual models [8].

2. Methodology

2.1. Differential Scanning Calorimetry

Differential scanning calorimetry analysis of the filament was conducted under 0.833 mL/s of nitrogen gas flow by using DSC Q10 from instrument of Thermal Analysis. The poly-lactic acid sample was weighted about 5 mg. Then, the sample was put into the aluminium pan before put into the thermal analysis machine.

2.2. Printing setup

Ideamaker was used to set-up the varied parameters of print head, orientation and layer thickness prior to the printing process. The others parameter which are not considered their effect to the mechanical properties in this study are set as default value. The varied values for print head, orientation and layer thickness are identified by bench marking on finding of previous researchers. The values were then being used to set on the Ideamaker software [5]. The geometrical solid CAD were saved in Standard Tessellation Language (STL) file format where 3D printer can read the format. Some common adjustment are needed by the STL file to fix the size, orientation and position to test the specimen after printing [9]. Ideamaker is the software had an adjustment and decoded the G- Code. 45 samples of dog-bone shape samples were printed and 45 samples of rectangular bar were printed respectively.

2.3. Tensile test and 3-point-bending test

Two mechanical tests were carried out to determine the structural behavior of the specimen which are tensile test and 3-point-bending test. The mechanical test were conducted using AG-IS Shimadzu Universal Testing Machine. Five samples were tested and the tensile strength or flexural strength values were averaged for each level. For each mechanical test, nine level are to be set based on Taguchi's orthogonal array of 9 Level. Based on the bending test standard ASTM D790, the thermoplastic (Poly-Lactic Acid, PLA) is set at support-span-to-depth ratio of 16:1. The speed of the tensile test at constant rate of 83.3 μm/s. Meanwhile the speed of 3-Point Bending test was taking at 33.3 μm/s [10] [11].

2.4. Analysis of Variance (ANOVA) and Pooling Technique

By taking the data from Table 3 and Table 4, the ANOVA analysis of the important of all major and their relations by comparing the mean square against the evaluated specific confident levels of an estimate of the experimental error. Pooling is usual practice of disregarding of the influence of the column where the consideration are not significant. When the error of Degree of Freedom is non-zero, the test for performance could be performed. However, pooling method should start with the weakest factor.

3. Results and discussion

3.1 Differential Scanning Calorimetry

By referring the Figure 1, it is observed that at the first heat, a glass transition temperature are around 56 ^{O}C. From this point, that the properties of the mechanical test of the Poly-Lactic acid is changed from elastic material to brittle material. This is due to when chain movement is changed. Then, the cold crystallization is started around 83.75 ^{O}C and the melting temperature is about 129.43^{O}C. However, for second heat for this Poly-Lactic Acid sample with a rate of 10 ^{O}C/min, shows that there is no evidence of the glass transition temperature, cold crystallization temperature and the peak of the melting temperature.

3.2 Tensile test

It is observed the high of ultimate tensile stress is 34.592 MPa. The combination of printing parameter are 25 degree of orientation, 0.2 millimetre of print head and 0.15 millimetre of layer thickness. Table 1 shows that parameter ranking for mean of Signal-to-Noise ratio for tensile test. It can be observed that the combination of optimum parameter are level 2 in orientation, level 3 in print head and level 3 for layer thickness. Figure 2 until figure 4 show the level of orientation, the level of print head and the level of layer thickness.

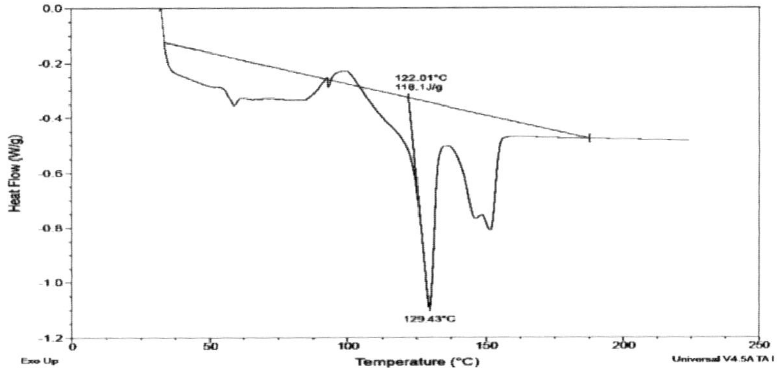

Figure 1: DSC graph

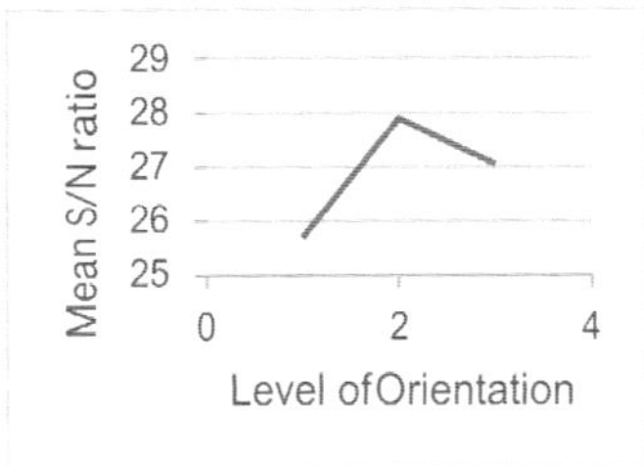

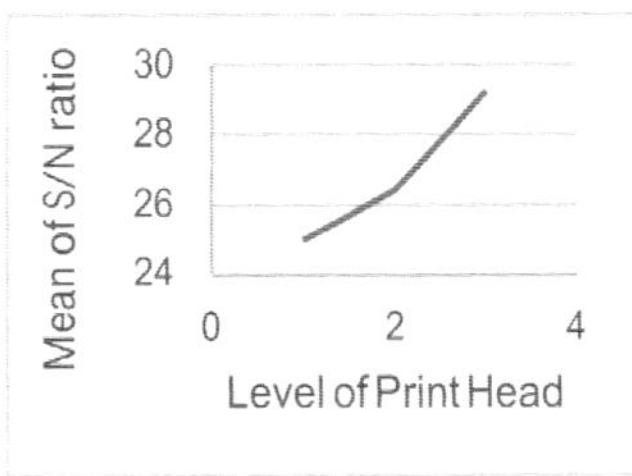

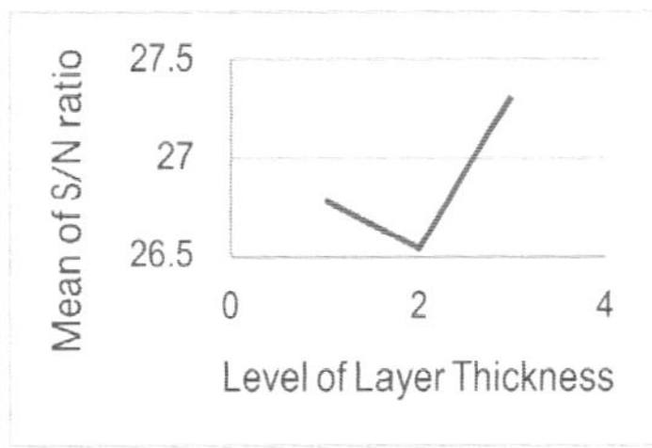

Figure 2: Level of Orientation **Figure 3:**Level of Print Head **Figure 4:**Level of Layer Thickness

Table 1: Parameter ranking for mean of Signal-to-Noise ratio for tensile test

Parameters	Level 1 (dB)	Level 2 (dB)	Level 3 (dB)	Main effect	Rank
Orientation	25.686	27.898	26.790	2.212	2
Print Head	24.965	26.427	26.540	4.279	1
Layer Thickness	26.790	26.540	27.310	0.770	3

3.3 3-point-bending test

The highest value of the flexural strength produce was 94.577 MPa. The combination of the parameter was 45-degree in orientation, 0.4 millimeter of the print head and 0.1 millimeter of the layer thickness. Table 2 shows that parameter ranking for a mean of Signal-to-Noise ratio for 3-point-bending test. Figure 5 until figure 7 shows that the level of orientation, the level of print head and the level of layerthickness.

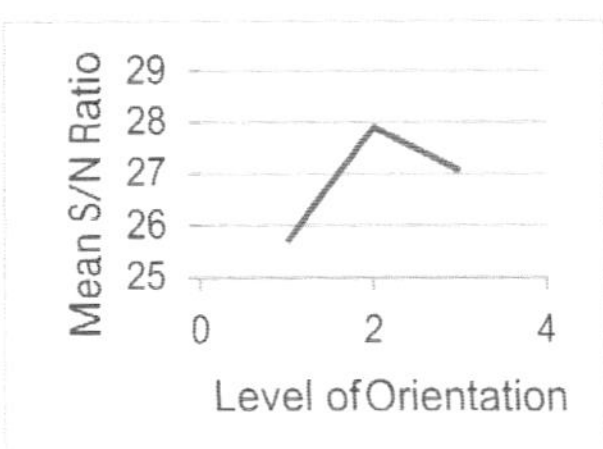

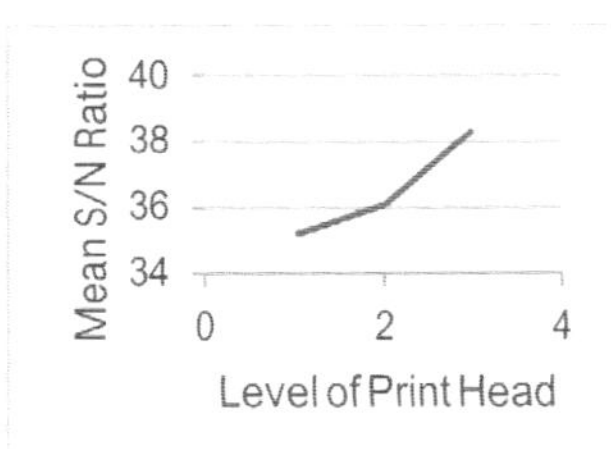

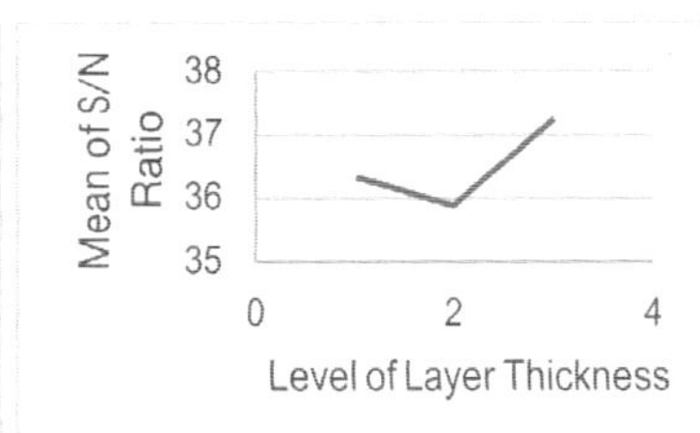

Figure 5: Level of Orientation **Figure 6:**Level of Print Head **Figure 7**:Level of Layer Thickness

Table 2: Parameter ranking for a mean of Signal-to-Noise ratio for 3-point-bending test

Parameter	Level 1 (dB)	Level 2 (dB)	Level 3 (dB)	Main Effect	Rank
Orientation	37.6659	35.3741	36.5409	2.2920	2
Print Head	35.1782	36.0703	38.3324	3.1542	1
Layer Thickness	36.9900	37.5442	37.2516	1.7074	3

3.4 Data analysis using Analysis of Variance

Table 3: ANOVA for Tensile test

Source	DF	Adj SS	Adj MS	F-Value	P-Value	Contribution (%)
Orientation	2	7.4757	3.738	0.51	0.663	14.52
Print Head	2	28.381	14.190	1.93	0.341	55.14
Layer Thickness	2	0.923	0.461	0.06	0.941	1.79
Error	2	14.6866	7.343			
Total	8	51.467				

It can be observed that from the Table 3, the print head contributes the most toward the effect of tensile strength at 55.14% compares to orientation where the contribution value is 14.52%. Layer thickness gives the lowest contribution value of 1.79%. As the layer thickness pooled, confidence interval of main effect cannot be calculated. Layer thickness in this case is considered as not significant, the performance at optimum condition cannot be estimated. Therefore, the printing process parameters which are orientation, print head and layer thickness were not optimized.

Table 4: ANOVA for 3-Point-Bending test

Source	DF	Adj SS	Adj MS	F-Value	P-Value	Contribution (%)
Orientation	2	14.044	7.022	4.91	0.169	34.55
Print Head	2	15.861	7.931	5.55	0.153	39.02
Layer Thickness	2	7.879	3.939	2.76	0.266	19.38
Error	2	2.858	1.429			
Total	8	40.642				

It can be observed from Table 4, print head gives contributes the most toward the effect of flexural strength at value of 39.02% compares to orientation where the contribution value is 34.55%. Layer thickness gives the lowest contribution value of 19.38%. Since layer thickness pooled, the confidence interval of the main effect cannot be calculated. Layer thickness in thos case is considered as not significant, the performance at optimum condition was not estimated. Therefore, the processing parameter which are orientation, print head and layer thickness were not optimized.

To solve these issues, first, the machine needs to be calibrated in term of alignment of the print bed. This is because the alignment of the print bed was not stable. Dwell small variation of print head and layer thickness, the post-print bed platform is sensitive to the outcome of printed sample. Next, the value of the varied parameter which are orientation, print head, and layer thickness must be reidentified. These value which had been set in table 3 based on previous researchers are not suitable to be setting in the experiment where it leads unable to pool the weakest factor that is layer thickness. Table 5 shows the suggestion parameter which helps to get the significant optimum parameter.

Table 5: Printing parameters with level

Parameters	Levels		
	1	2	3
Orientation	0°	25°	45°
Print Head	0.2 mm	0.3 mm	0.4 mm
Layer thickness (mm)	0.1 mm	0.15 mm	0.2 mm

After that, the software which been used must be changed from Ideamaker to Cura. This is because the g-code generated by different software might produce different layer-by-layer of the printed nozzle. Then, the temperature of the nozzle must be changed from 200° C to 201° C. The temperature difference was 1^{O} C. However, from this different of values, it could change the error of the printing process where it leads to getting to significant optimum process parameters. Furthermore, 5 samples taken on each level must be changed to 6. This is because the more sample taken to do the mechanical test will produce accurately the data.

4. Conclusion

In conclusion, the objectives of this research had achieved which are to evaluate the mechanical properties of polylactic acid for Fused Filament Fabrication, FFF printed parts, to investigate the significant processing parameter which are orientation, print head and layer thickness and to determine the optimum printing parameter.

Acknowledgment

The authors would like to express profound gratitude and deep appreciation to the School of Mechatronics Engineering, School of Manufacturing Engineering and Universiti Malaysia Perlis in providing researching platform well as providing the facilities and equipment.

Reference

[1] Claver J and Mar A 2018 The influence of manufacturing parameters on the mechanical behaviour of PLA and ABS pieces manufactured by FDM: a comparative analysis.
[2] Michopoulos J G 2018 Multiscale topology optimization for additively manufactured objects 1–9.
[3] Godec D, Spoerk M, Kukla C and Holzer C 2018 3D printing conditions determination for feedstock used in fused filament fabrication (FFF) of 17-4PH stainless steel parts **57** 117–120.
[4] Khan S F, Zakaria H, Chong Y L, Saad M A M and Basaruddin K 2018 Effect of infill on tensile and flexural strength of 3D printed PLA parts *IOP Conference Series: Materials Science and Engineering* **429**.
[5] Khan S F, Zakaria H, Chong Y L, Saad M A M and Basaruddin K 2018 Effect of infill on tensile and flexural strength of 3D printed PLA parts *IOP Conference Series: Materials Science and Engineering* **429**.
[6] Saxena A 2016 Comprehensive study on 3D printing technology.
[7] Ezoji M, Razavi-Nouri M and Rezadoust A M 2017 Effect of raster angle on the mechanical properties of PLA 3D printed articles 12 *The International Seminar on Polymer Science and Technology Effect of Raster Angle on the Mechanical Properties of PLA 3D Printed Articles* 3–5.
[8] Othman F M, Fadhil T and Hind A 2018 Influence of process parameters on mechanical properties and printing time of FDM PLA printed parts using design of experiment **8** (7) 65–69.
[9] Salleh A F et al. 2007 Preliminary study on the influence of slice thickness and support parameters on part defects of FDM prototype 25–26.
[10] Materials P and Materials E I 2015 Standard test method for tensile properties of plastics 1.
[11] Claver J and Mar A 2018 The influence of manufacturing parameters on the mechanical behaviour of PLA and ABS pieces manufactured by FDM: a comparative analysis.

IOP Conf. Series: Materials Science and Engineering **670** (2019) 012061 doi:10.1088/1757-899X/670/1/012061

CFD analysis of hydrodynamic lubrication effects of micro textured surface

M H Sulaiman [1,*], M F S Zaid [1], R N Farahana [1] and M J M Ridzuan [1]

[1] School of Mechatronic Engineering, Universiti Malaysia Perlis, Pauh Putra Campus, 02600 Arau, Perlis, Malaysia

*corresponding author: hafissulaiman@unimap.edu.my

Abstract. A great number of lubricated mechanical components everywhere in life. Improvement in efficiency would lead to greater cost savings in many engineering applications. In the present paper, an approach to increase the efficiency in hydrodynamic lubrication for applications in metal forming processes by surface texturing was studied using Computational Fluid Dynamics (CFD) simulation. A two-dimensional fluid domain containing a flat upper wall with a tangential velocity and a static lower wall containing an oblong groove of the same sizes was studied. This includes the effect of varying distances between grooves on lubrication. With an isothermal lubrication flow condition and parallel walls, the pressure in the lubricant domain produces largest load carrying capacity and wall shear force, indicating large contact area contributes to an increase of frictional force. The present study revealed that the positive influence on hydrodynamic lubrication performance can be achieved by introducing a textured surface. Larger distance between the grooves has effectively produced a pressure build-up in the lubricant domain introducing a larger load carrying capacity. It is seen that a load carrying capacity exists for a small distance between the grooves.

1. Introduction

Lubrication is important to reduce friction and to prevent wear of the sliding surfaces in many engineering applications. Poor lubrication has always been the reason to inferior surface finish and production stops in metal forming operations. The good boundary oils alone cannot solve tool wear problem. Combating tool wear with thin-film coatings, surface treatments and anti-seizure tool materials are common practices applied in industry to prolong the life of tooling. These are, however, depends on the type and cause of tool wear [1]. These days, the application of surface texturing to facilitate lubrication in engineering applications such as in stamping and forging is well known. While texturing of workpiece surfaces to promote lubrication in metal forming has been applied for several decades, tool surface texturing is rather new. Texturing workpiece surfaces by using large rolls roughened by Shot Blast Texturing (SBT) or Electro Discharge Texturing (EDT) [2] has become a state-of-the-art for several decades to promote mechanical lubricant entrainment, pressurization and possible escape by micro-hydrodynamic lubrication [3]. However, it is less feasible in large-scale, multi-stage sheet metal forming production, since the textured workpiece surface will flattened out after first forming operation [4]. A few tests of surface engineered deep drawing tools [5] have shown very promising results, indicating that textured tool surfaces may provide mechanical lubrication systems which can therefore substitute the chemical ones, and thereby replacing the environmentally hazardous lubricants with environmentally benign lubricants [6]. This positive result has shown that creating micro texturing patterns on tool surface would be economically feasible, since the textured tool surface can be utilized to produce thousands of workpiece components [7]. In the present paper, the hydrodynamic performance of micro-textured surface with varying distances between grooves applied on metal forming tool surfaces, such as pressure distribution streamlines and wall forces, were studied.

2. Materials and method

2.1. Micro-textured surface

Three surface texture parameters were emphasized to promote micro-hydrodynamic lubrication mechanism [8] and to avoid mechanical interlocking in the groove valleys [9];1) oblong grooves oriented perpendicular to the drawing direction, 2) small groove angle γ and 3) shallow groove depth d. Three different texture designs were chosen with a width of flat plateaus between the lubricant grooves $x = 0.23$, 0.46 and 0.92 mm, respectively. Figure 1 shows the groove angle γ and the groove depth d were chosen to be 5° and 0.01 mm respectively. The groove angle was chosen according to recommendations by Sulaiman et al. [7], the depth was determined by the width, which was chosen small enough to ensure a sufficient number of grooves along the tool/workpiece interface. The length of the grooves were decided to be smaller than the workpiece width in order to ensure pressure build-up of the trapped lubricant. Table 1 lists the surface texture parameters as calculated by Equation 1 and 2. Figure 2 represents the resulting, measured grooves of nominal dimensions: length $y = 16$ mm, angle $\gamma = 3° \pm 0.5°$, width $w = 0.23$ mm $\pm$ 0.1 mm, depth $d = 7$ µm $\pm$ 1 µm and distance between grooves of $x = 0.23$, 0.46 and 0.92 mm. It is noticed that the groove depths reached within the tolerance gap, whereas the groove angles turned out to be somewhat smaller than the aimed at. This is, however, only promoting the micro-hydrodynamic lubrication mechanism and preventing mechanical interlocking [9].

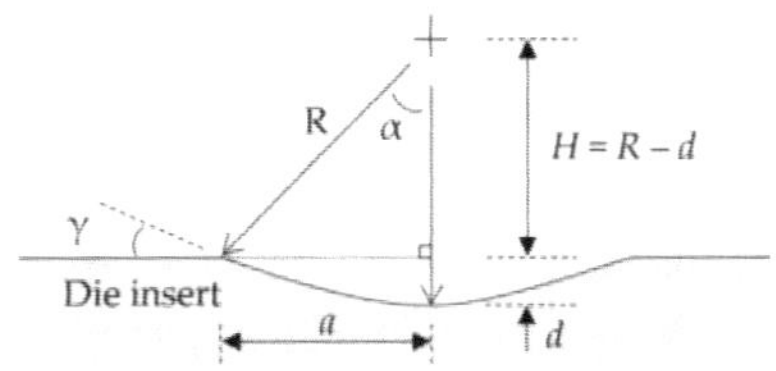

Figure 1: Design of groove geometry.

Table 1. Surface texture parameters.

Parameters	Value
Groove angle γ (°)	5
Groove width $w = 2a$ (mm)	0.23
Groove depth d (mm)	0.01
Groove ratio d/w	0.05

$$\tan\gamma = \frac{d}{a} \tag{1}$$

$$R^2 = a^2 + H^2 = a^2 + (R - d)^2 \tag{2}$$

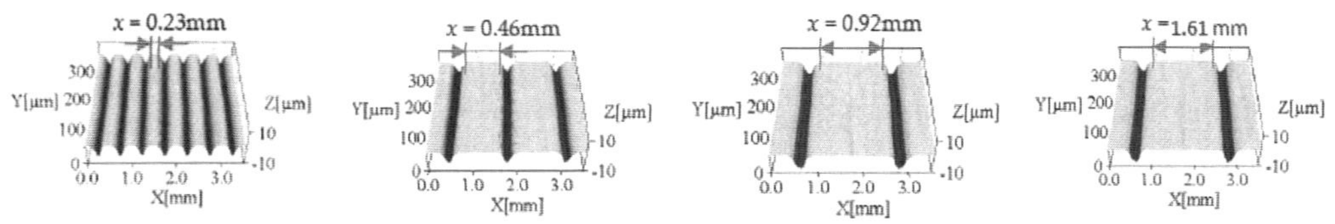

Figure 2: Micro-textured of tool surfaces concentrating on varying distances between the grooves (x = 0.23, 0.46, 0.92, and 1.61 mm) of the same sizes of groove depth d and groove angle γ.

2.2. CFD analysis of textured surfaces

The present work was carried out using Computational Fluid Dynamics (CFD) FLUENT v6.3. Simulation of a fully wetted smooth upper wall with two different tangential velocities over a static lower wall containing an oblong groove of the same sizes separated by a 40 μm of lubricating film was performed for 2D steady-state flow of viscous incompressible fluid (density of 940 kg/m^3 and dynamic viscosity of 1.06 kg/ms). The two different drawing speeds $\upsilon = 240$ mm/s and 65 mm/s were

IOP Conf. Series: Materials Science and Engineering **670** (2019) 012061 doi:10.1088/1757-899X/670/1/012061

applied with micro-textured tool features. The high and low speeds were intended to identify possible influence of micro-hydrodynamic lubrication mechanism. The convergence criteria was set to ε 10^{-4}.

3. Results and discussion

3.1. Effects of distance between grooves

Figure 3 shows comparisons of pressure distributions as a function of the sliding distance along the walls in the *y*-direction for the two flat parallel walls separated by a thin lubricating film. The smooth, non-textured lower static surface produces the highest pressure and the highest shear force between all surfaces investigated. This is in fact due to the greater contact area, which in turn gives rise to a larger frictional force. As seen in Figure 3, grooves can be beneficial in reducing the large contact area, thereby reducing the pressure distribution in order to reduce the friction between two sliding surfaces, meanwhile, the lubricant pressures reached its high values on smooth, flat parallel surfaces when no grooves are present, indicating that the grooves influencing the hydrodynamic performance.

The benefits of applying grooves is especially pronounced when the distance between the grooves are larger, in which allows lubricant entrapment in the groove and subsequently the lubricant escape out of the groove, as experimentally and analytically proved in ref. [8]. This implies that the larger the distance between the grooves, the lower the lubricant pressure, see Figure 4(a). Same effects can also be observed on the shear force as shown in Figure 4(b). The results explained that the hydrodynamic lubrication efficiency depends on pressure (representing the load carrying capacity) and shear force in the *y*- and *x*-direction, respectively.

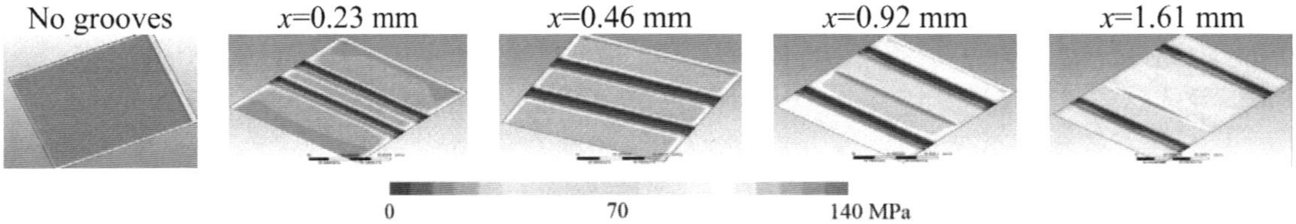

Figure 3: Pressure distributions in the fluid domain of the smooth, non-textured and textured surfaces.

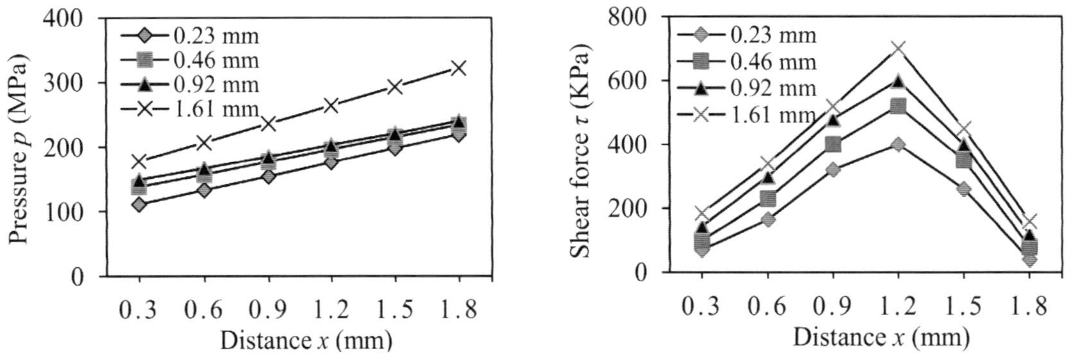

(a) (b)

Figure 4: Distributions of a) pressure and b) shear force in varying distances between the grooves *x*.

3.2. Effects of sliding speed

The influence of high and slow sliding speeds on hydrodynamic performance in the case of small and large difference between the grooves were studied. Figure 5 shows the load carrying capacity and shear force increases tremendously when the high sliding speed is applied. This is may be due to the relationship between viscosity and micro-hydrodynamic lubrication mechanism [9]. The positive influence of high sliding speed on hydrodynamic performance is as expected, owing to the frictional force produced by the lubricating film. This simulation is in agreement with the experimental results

[10] and the CFD simulation results [11] from the previous studies on textured surface and their influence on frictional behaviour.

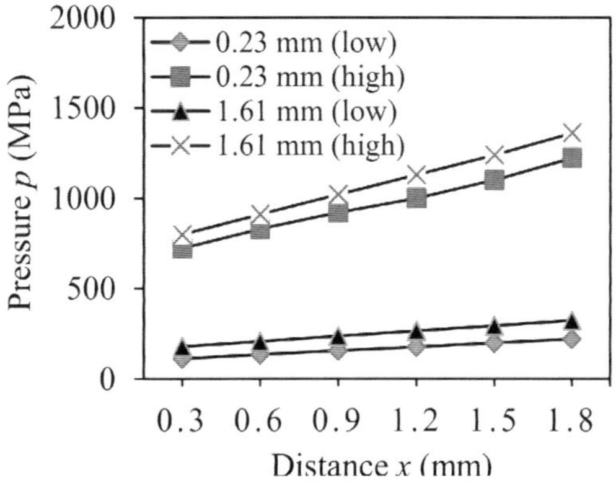

(a)

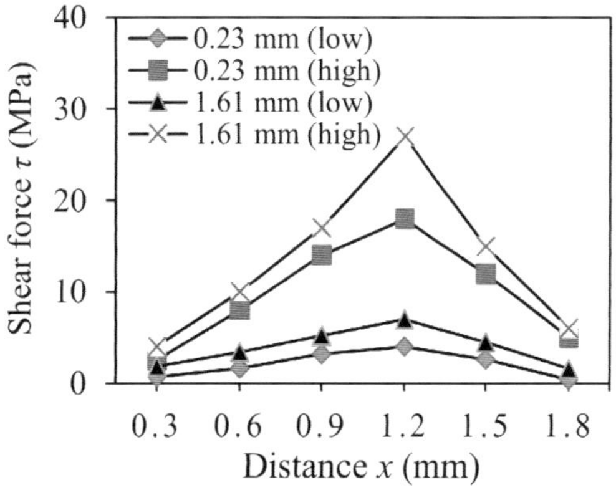

(b)

Figure 5: Effects of high and slow sliding speeds on distributions of a) pressure and b) shear force along the sliding surfaces for small and large distances between the grooves *x*.

4. Conclusion

CFD analysis on the effects of micro-textured surfaces of hydrodynamic lubrication performance was performed. The CFD simulation discussed on the effects of the varying distances between the grooves on the load carrying capacity and the shear force. Introduction of micro-textured grooves to the lower static surface affects the lubricant flow, wall shear force and pressure distributions. The smaller the distance between the grooves, the lower the load carrying capacity as well as the shear force, indicating the grooves influencing the hydrodynamic performance. Larger sliding speed contributes to a larger load carrying capacity and shear force in the lubricant domain. It was found out that an optimum amount of textured tool surface area exists which reduces frictional wall force and pressure between the lubricant and the two sliding surfaces. The textured surfaces were advantageous at greater sliding speeds which facilitate the possible escape of trapped lubricants by micro-hydrodynamic lubrication mechanism.

Acknowledgements

This work was supported by the Taiho Kogyo Tribology Research Foundation TTRF [grant number 9008-00015]. Special gratitude is dedicated to Tribology & Precision Forming (Tribo-Forming) research group for the fruitful discussions, equipment and technical assistance.

References

[1] Klocke F, Maßmann T and Gerschwiler K 2005 Combination of PVD tool coatings and biodegradable lubricants in metal forming and machining *Wear* **259** 1197–206.
[2] Kijima H and Bay N 2007 Contact conditions in skin-pass rolling *CIRP Ann. - Manuf. Technol.* **56** 301–6.
[3] Bay N, Bech J I, Andreasen J L and Shimizu I 2002 Studies on micro plasto hydrodynamic lubrication in metal forming *Metal Forming Science and Practice: A state-of-the-Art Volume in Honorary of Prof. J.A. Schey.* (Woodhead Publishing Limited) 115–34.
[4] Ceron E 2013 *New tribo-systems for sheet metal forming of advanced high strength steels and stainless steels.*
[5] Lindvall F, Bergstrom J, Krakhmalev P and Bay N 2010 The effect of grinding and polishing procedure of tool steels in sheet metal forming *Proceedings of the 4th International Conference on Tribology in Manufacturing Processes (ICTMP)* 603–12.
[6] Sulaiman M H 2017 *Development and testing of tailored tool surfaces for sheet metal forming* (Technical University of Denmark (DTU)).

[7] Sulaiman M H, Farahana R N, Ridzuan M J M, Jaafar H, Tajul L and Wahab J A 2019 Manufacturing strategies and tribology of tailored tool surface topographies for enhanced lubrication in metal forming *Proceedings of 3rd Mytribos Symposium* 47–51.
[8] Sulaiman M H, Christiansen P and Bay N 2017 Influence of tool texture on friction and lubrication in strip reduction *Procedia Eng.* **207** 2263–8.
[9] Sulaiman M H, Christiansen P and Bay N 2017 The influence of tool texture on friction and lubrication in strip reduction testing *Lubricants* **5** 1–11.
[10] B. Podgornik, L. M. Vilhena, M. Sedlaček, Z. Rek I Ž 2012 Effectiveness and design of surface texturing for different lubrication regimes *Meccanica* **47** 1613–1622.
[11] Vilhena L, Sedlaček M, Podgornik B, Rek Z and Žun I 2018 CFD modeling of the effect of different surface texturing geometries on the frictional behavior *Lubricants* **6** 15.

IOP Conf. Series: Materials Science and Engineering **670** (2019) 012062 doi:10.1088/1757-899X/670/1/012062

An experimental investigation of wind induced noise inside the small cavity in low wind speed

W M Hairudin[1], **N M Kamarudin**[1], **O L Ean**[2] **and N A Ismail**[1*]

[1]School of Aerospace Engineering, Universiti Sains Malaysia, 14300, Nibong Tebal, Malaysia.

[2]School of Mechanical Engineering, Universiti Sains Malaysia, 14300, Nibong Tebal, Malaysia.

*corresponding author: aenorilmi@usm.my

Abstract. An experimental study was conducted to investigate the wind noise produced inside the cavity model that replicate the two-way radio's cavity. The experiment conducted with varies of cavity distance from the wind source of P1 (0.01m), P3 (0.078m), and P5 (0.146m) at the wind speed of 3.75m/s. The cavity has a length to depth ratios of 1.4. This preliminary measurement was to highlight the relationship between cavity positions and low speed wind flow that affect the sound pressure level occurred inside the cavity. The study also look on the impact this factor on the performance of two-way radio that causes unclear signal interference during windy condition. The results show that by increasing the distance of cavity position, the sound pressure level decreased. It shows that the farthest distance from the wind source, the noise level reduces.

1. Introduction

A two-way radio is a radio that can receive and transmit radio signals. It has been used at large sites business such as hospitals, security, military, oilrig plant, and others to eliminate the obscure communications [1]. However, they faced a problem with unclear audio with the environmental background and noise since it is expose to the windy condition. Wind flow affects the quality of the signal receiver due to the presence of turbulent flow over the microphone's surface from the flow trapped inside the cavity that become the source of the noise. This noise can severely downgrade the quality of the signal for outdoor activities.

The phenomenon of wind-induced noise (WIN) in the cavities were studied by numerous applications in the past, such as in landing gear [2], car sunroof [3], vehicles [4-5], high-speed train (pantograph) [6], building [7], wind turbines [8] and musical instruments [9]. Wind induced noise is also under fluid resonant oscillation, namely called as aero-acoustic. The unsteady oscillations of the air often caused due to disturbance, such as cavity causing turbulence or vortices. The turbulence and vortices produced when the large pressure fluctuates by the impinging a shear layer on the leading edge and the disturbance impinge on the trailing edge and excite instability waves (acoustic). The acoustic wave then propagate downstream and grow recirculation flow inside the cavity. This cavity mechanism depicted in Figure 1.

There are limited literature study of wind induce noise that focuses on an electronic and mobile communication device such as two-way radio. Wind noise in microphones has been studied by [10, 11] to minimize the effect of wind noise at outdoor. They used the passive wind noise reduction method such as windscreens to reduce the outdoor noise. Alamshah and Zhao [12, 13] used windshields of 90 diameter with different from 20-60per inch (PPI)) porosity to minimize the wind-induced noise microphones. The result found that the 40 PPI windshields has an effective performance compare to other porosity.

The physical mechanism of the wind noise in the cavity of two-way radio still new and has not been fully understood. Study on the two-way radio is recently conducted in [14-16]. Fisol et al [14] studied the relationship between the flow speed and angle of attack with the noise level in two-way radio. They concluded that the higher flow speed had increased the noise level. It also shows that a significant increase in noise level when the radio has been oriented to the critical angle of attack at 15 degrees. Another work by [15] performed an investigation of the sound quality of two-way radio by measuring the effect of angles orientation, wind speed and threshold hearing level intelligibility of the received speech signal. The results have shown that the angle leads to worst wind-induced noise effect is at 90° to the leading edge, which is showing the highest sound pressure level. Another study of small cavity was conducted by Yow et al [16] who investigated the small vortices existence inside the small cavity. Due to less attention into the flow and acoustic analysis is taken into the cavity in two way radio, it is important to determine the noise level inside the cavity and clarify the method to suppress the cavity tones.

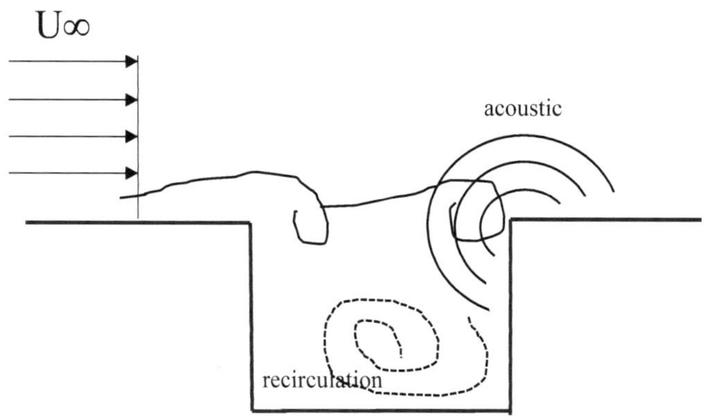

Figure 1: The cavity noise mechanism

In this paper, the effect of sound pressure level at different cavity distance was investigated. This is the preliminary part to choose the right distance cavity which produces pressure fluctuation as well as the noise inside the cavity. This study also contributes to cavity noise suppression research.

2. Experimental Method

A schematic view of a rectangular cavity model and cavity position show in Figure 2. Cavities can be classified based on the length to depth ratio (L/D), where L is the length and D is the depth of the cavity. There are three types of cavities; 'open' if $L/D < 10$, 'closed' if $L/D > 13$ and transitional for $10 < L/D < 13$ [17]. The dimensions of the cavity are L (14mm) and D (10mm). The length to depth ratio L/D is 1.4.

The experiments were carried out using closed loop low-speed wind tunnel facility with a rectangular test section having a cross-section with dimensions of 1.0m x 0.8m x 1.8m at the Science and Engineering Research Centre (SERC), University Sains Malaysia (USM). The intensity of free stream turbulence was 0.1%, at the maximum achievable speed of 80 m/s. Sound pressure level measurement (SPL) and spectrum acquisition are obtained by IMC device dynamic measurement system. Figure 3a shows the two microphones were used; the first is microphone M1 that located outside the wind tunnel near the fan motor to measure the ambient noise. Second, microphones M2, which are placed inside each of the cavity to measure the noise received. The second microphone was positioned at 76.5cm from wind source and placed in the same height at the centre of each cavity. Microphones are calibrated using sound calibrator measurement before measurement are made. An accelerometer is positioned in the upper of the test section to measure the mechanical vibration of the wind tunnel base and the source of the mechanical noise during the experiment, which could either come from the motor or the wind tunnel structure. The SPL measurement point is located at the bottom of the cavity, as shown in Figure 3(b). In this experiment, the sound level is recorded first when there is no wind in the wind tunnel, then followed with the wind tunnel is running with different cavities.

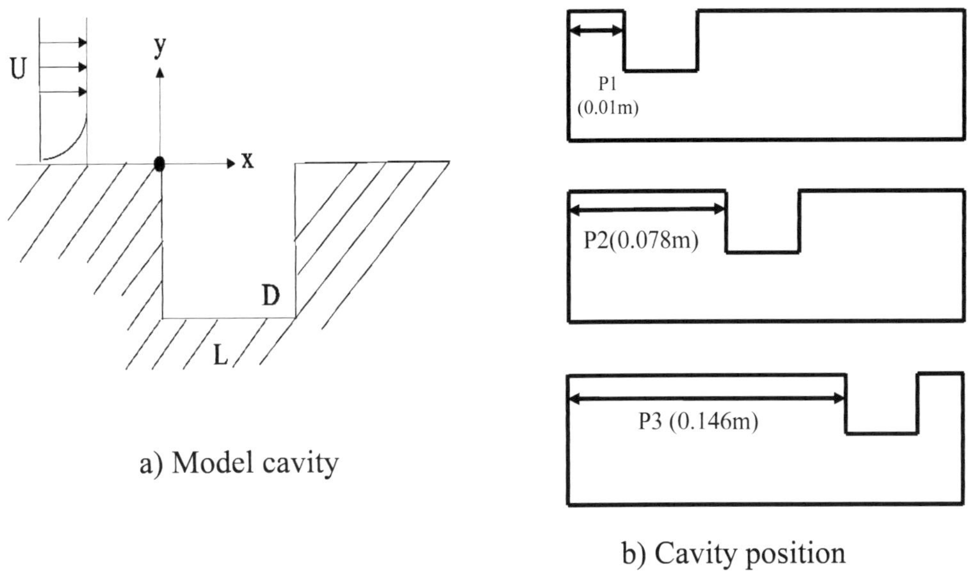

Figure 2: a) Schematic view of the cavity b) Cavities position

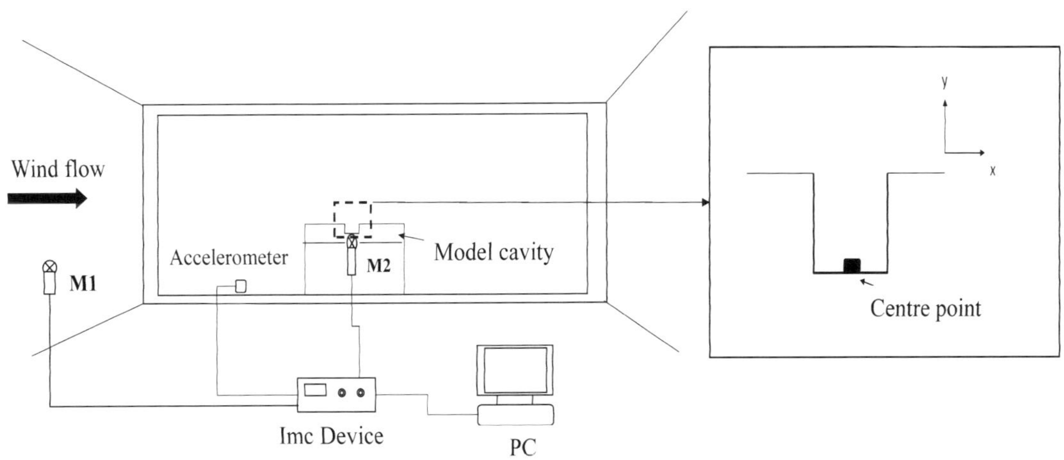

Figure 3: a) Schematic diagram for the experimental setup (b) Measurement point

3. Result and Discussion

Figure 4(a) shows the pressure fluctuation that occurred in cavity position (P3) at wind speed of 3.75m/s. In this figure, there is two significant vibration peaks mode; firstly is 48Hz and secondly at 100 Hz. It shows that the open cavity of two-way radio produced a very low frequency. Figure 4(b) shows the highest SPL of 65 dB at frequency 48 Hz under cavity position (P1) compared with the P2 and P3. The lowest SPL of 61 dB is the cavity position (P3) under the same frequency. It shows that the SPL of the cavity model (P3) is reducing about 4 dB from cavity model (P1). The farthest position is a suitable position to avoid any necessary noise induced by wind. This result is agreed with Fisol's finding [18].

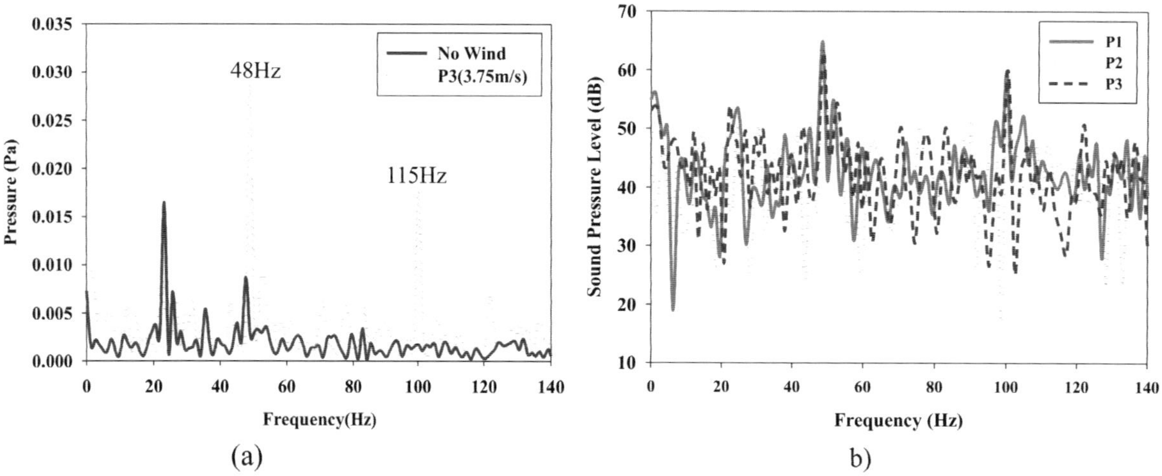

Figure 4: a) The pressure fluctuation (b) The sound pressure level (SPL)

4. Conclusion

In this study, the sound level measurement of different cavity position under low speed was presented. The basic conclusions are as followings; when cavity location is increased at a certain distance, the sound pressure level also increases, however, when cavity distance is far, the decibel level decreased due to the minimum pressure fluctuated produced. The cavity distance was observed to influence the noise level inside cavity of two-way radio.

5.Acknowledgement

The authors acknowledge the research work was financially supported by the Research Universiti Sains Malaysia Grant (1001/PAERO/8014005). Thanks to technical supports, Mr Wan Amri from Vibration Lab, School of Mechanical Engineering, Mr Mahmud Isa and Mr Mohamad Najhan Awang from School of Aerospace Engineering for the help.

References

[1] Rogers A L, Manwell J F and Wright S 2006 Wind turbine acoustic noise *Renewable Energy Research Laboratory, Amherst:* University of Massachusetts.

[2] Neittaanmäki P, Rossi T, Korotov S, Oñate E, Périaux J and Knörzer D. Landing gears aerodynamic interaction noise *Eur. Congr. Comput. Methods Appl. Sci. Eng.* 115–35.

[3] Wang Y, Deng Y, Yang Z, Jiang T, Su C, Yang X 2013 Numerical study of the flow-induced sunroof buffeting noise of a simplified cavity model based on the slightly compressible model *Proc. Inst. Mech. Eng. Part D J. Automob. Eng* **227** (8) 1187-99.

[4] Chen K H, Johnson J, Dietschi U and Khalighi B 2010 Wind noise measurements for automotive mirrors *SAE Int. J. Passeng. Cars - Mech. Syst.* **2** (1) 419–33.
[5] Alam F 2000 The effects of car a-pillar and and windshield geometry on local flow and noise 265.
[6] Thompson D J, Latorre Iglesias E, Liu X, Zhu J and Hu Z 2015 Recent developments in the prediction and control of aerodynamic noise from high-speed trains *International Journal of Rail Transportation* **3** (3) 119-50.
[7] Čeheľová D, Urbán D, Bielek B, Rychtáriková M and Roozen N B 2017 Literature review on wind induced sound on buildings *Akustika* **27** (1) 42-7.
[8] Gallo P, Fredianelli L, Palazzuoli D, Licitra G and Fidecaro F 2016 A procedure for the assessment of wind turbine noise *Applied Acoustics* **114** 213-17.
[9] Fabre B, Gilbert J, Hirschberg A and Pelorson X 2012 Aeroacoustics of musical instruments *Annual review of fluid mechanics* **44** 1-25.
[10] Shiau B S 1998. Measurement of turbulence characteristics for flow past porous windscreen *J Wind Engineering and Industrial Aerodynamics* **74** *521-30.*
[11] Shams Q A, Zuckerwar A J and Sealey B S 2005 Compact nonporous windscreen for infrasonic measurements *J Acoustical Society of America* **118** (3) 1335-40.
[12] Wang L, Zander A C and Lenchine V V 2012 Measurement of the self-noise of microphone wind shields *Measurement* **3** (7).
[13] Xu Y, Zheng Z C, Wilson D K 2011 A computational study of the effect of windscreen shape and flow resistivity on turbulent wind noise reduction *J Acoustical Society of America* **129** (4) 1740-7.
[14] Fisol U M M, Ripin Z M, Ismail N A, Saad A A, Hashim M K R, Chan P Y, Yow H T, Tan C H, Hii U H and Saw T S 2013 Wind noise analysis of a two-way radio *2013 IEEE International Conference on Smart Instrumentation, Measurement and Applications (ICSIMA)* 1-5.
[15] Hashim M K R, Ismail N A, Saad A A, Chan P Y, Fisol U M M, Ripin Z M, Yow H T, Tan C H, Hii U H and Saw T S 2013 Sound quality analysis for two-way radio under wind noise *2013 IEEE International Conference on Smart Instrumentation, Measurement and Applications (ICSIMA)* 1-5.
[16] Yow H T, Hii D, Saw A, Tan C H, Fisol U M, Ripin Z M, Ismail N A, Saad A A, Hashim M K and Yi C P 2013 Prediction of wind induced noise over bodies and small cavity *The Journal of the Acoustical Society of America* **134** (5) 4222.
[17] Tracy, Maureen B, and Plentovich E B 1997 Cavity unsteady-pressure measurements at subsonic and transonic speeds *NASA Technical Paper* 3669.
[18] Fisol M U 2016 *Experimental and computational fluid dynamics (CFD) analysis of wind-induced noise in two way radio* MSc. Thesis, University of Sains Malaysia, Pulau Pinang, Malaysia.

IOP Conf. Series: Materials Science and Engineering **670** (2019) 012063 doi:10.1088/1757-899X/670/1/012063

Enhancement of MODENAS CT115'S motorcycle muffler performance by optimization of backpressure

M A Rojan, Z Alias, S A Bakar and S Hashim

School of Mechatronic Engineering, Universiti Malaysia Perlis, Pauh Putra Main Campus, 02600 Arau, Perlis, Malaysia.

Abstract. A research study on the enhancement of motorcycle muffler to improve MODENAS CT115's engine performance was conducted. To obtain the optimum value of backpressure, velocity, and temperature, the hole diameter of perforated tube has been modified. The modifications made for each design by increasing the hole diameter to the front and back of the perforated tube. Both the original muffler and modified designs were tested to obtain optimum pressure values through the ANSYS Fluent simulation. According to the theory, the value of the hole diameter will affect the backpressure value in the muffler, hence enable design optimization for achieving the best engine performance. The largest hole diameter for perforated tube on the front shows the lowest backpressure from the results obtained. The percentage improvement for design 4 is 10.3% higher than design 1, design 2 and design 3.

1. Introduction

Exhaust gas is emitted by combustion of fuel such as petroleum, diesel or fuel oil. The exhaust is responsible to transporting the burnt exhaust, or combustion gas from the engine and out through the exhaust to the atmosphere. According to Suganeswaran et al., the flow resistance of exhaust gases will normally occur at this stage because of backpressure and give negative effect on engine efficiency [1]. The component of exhaust consists of the exhaust manifold, catalytic converter, muffler and connecting pipes. Muffler is one of the major components of exhaust system to reduce the intake and exhaust noise [2]. In the literature, there are various studies conducted on the effect of change in a diameter of perforated pipe by using computational fluid dynamics (CFD) analysis. The exhaust system is design to reduce the resistance gas flow to provide maximum efficiency. For example, Pangavhane et al., changed the dimensional of the perforated internal tube to evaluate its effect on the value of backpressure [3]. The backpressure rapidly falls down with the increase in the diameter of the hole. So, that the backpressure is increased at smaller hole diameter as compared to larger hole diameter. In this present work, an investigation was carried out via CFD analysis to investigate the effect of backpressure of exhaust gases on engine performance by varying the hole diameter of perforated tube. The geometric design improvement focuses on the front and back perforated tube of the muffler body.

2. Methodology

2.1. Data Collection

The existing Modenas CT115's motorcycle muffler was provided by MODENAS Research and Development Unit. Flow bench machine (Super Flow Bench 110 E) was used to measure volume flow rate, pressure, temperature and velocity of MODENAS CT115's motorcycle muffler. All measurements were done and recorded at MODENAS's facility. Table 1 tabulates the data obtained by using Super Flow Bench 110 E.

Table 1: Data obtained by using Super Flow Bench 110 E

Volume Flow Rate(m^3/s)	Inlet Pressure (inH_2O)	Inlet Temperature (°C)	Outlet Velocity (m/s)
2.3	10.00	34.30	11.8
5.4	10.22	32.79	12.2
12.3	10.14	33.44	15.8
14.4	10.12	34.00	16.5
17.2	10.11	33.71	18.1
19.5	10.13	34.59	18.2
20.6	10.22	34.79	17.9

2.2. Muffler model using CATIA software

The MODENAS CT115's motorcycle muffler was designed by using CATIA V5 software. For design improvement, four model of muffler were created by changing the hole diameter of perforated tube at the back and front perforated tube according to the available nearest drill bit size. For Design 1 and Design 2, the existing 4.92mm diameter at the back perforated tube was changed to 5.0mm and 5.1mm, respectively. Meanwhile , for Design 3 and Design 4, the existing 4.03mm diameter at the front perforated tube was changed to 4.1mm and 4.2mm, respectively. **Figure 1** shows the different hole dimension of perforated tube in the proposed muffler models.

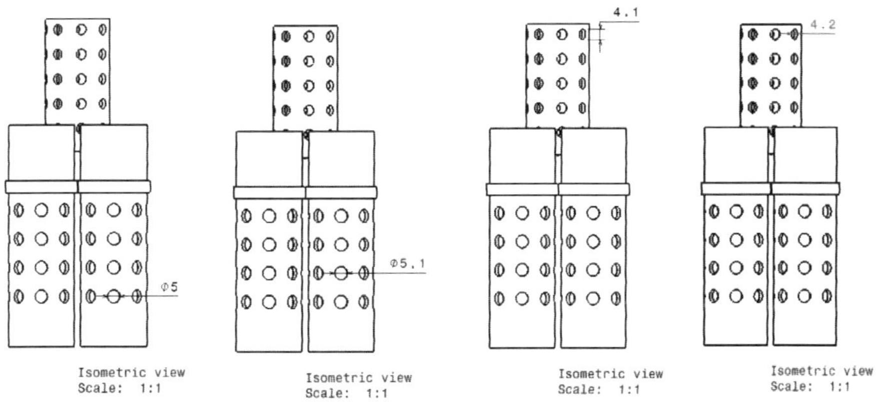

Design 1 Design 2 Design 3 Design 4

Figure 1: Different Muffler Model

2.3. Muffler model used for simulation

The model of the CATIA muffler was transferred to the ANSYS Workbench. The actual design of the air flow model was created based on the exact dimensions of the CT115 engine exhaust itself and then the model was imported into ANSYS Fluent software. The meshing involves surface meshing, volume meshing and setting up of boundary condition for the study of backpressure. The meshing was carried out carefully to avoid any failure in the analysis. For a boundary condition for this project, the pressure inlet was set at 2490.89 Pa and the inlet velocity at 14.55 m / s. To set the part of the design muffler as a full fluid for the simulation, the muffler drawing using CATIA was in the inner cavity to ensure that

IOP Conf. Series: Materials Science and Engineering **670** (2019) 012063 doi:10.1088/1757-899X/670/1/012063

the part body was assumed to be fluid when transferred to the ANSYS Fluent software. For existing design, the hole diameter at the back perforate tube is 4.92mm.

The simulation of the existing parameter was conducted and the result was compared to the experimental flow bench test result. After the comparison, one-dimensional mapping was validated. CFD simulation results for existing design were controlled to result in less than 5% error from the results obtained from flow bench test in order that the set boundary condition can be used for improvement design.

Result and Discussion

Figure 2 shows the streamline pressure and velocity for existing MODENAS CT115's muffler. The results of the simulation clearly show the occurrence of backward pressure inside the body of exhaust muffler. **Table 2** tabulates the the results of CFD analysis for inlet and outlet of existing MODENAS CT115's muffler.

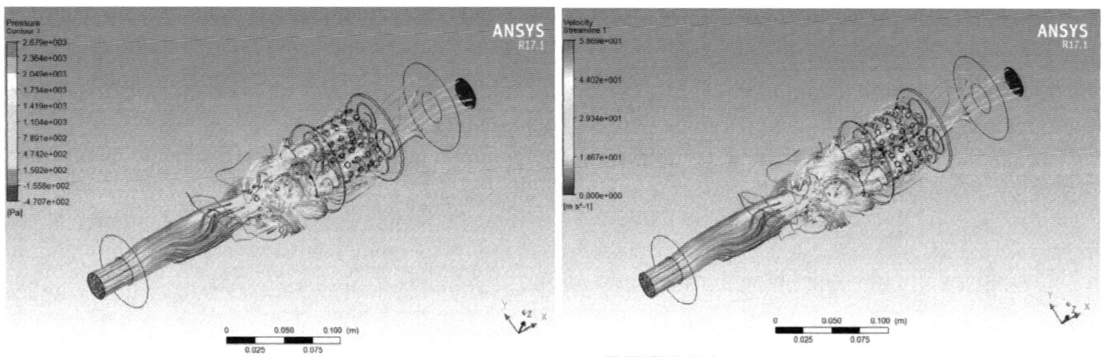

Figure 2: Pressure and velocity streamline for existing MODENAS CT115s

Table 2: CFD Analysis for Existing MODENAS CT115's Muffler

MODENAS CT115's Muffler	Inlet	Outlet
Pressure (Pa)	2490.89	4.354
Velocity (m/s)	14.55	11.24
Temperature (°C)	34.55	32.85

Figure 3 and **Figure 4** show the comparison of backpressure between existing MODENAS CT115's motorcycle muffler with proposed designs. The backpressure for Design 1 and Design 2 is lower than the backpressure for existing design. The decrease in backpressure is corresponded to the increase of hole diameter of the perforated tube. The same results can be notices in Design 3 and Design 4 where Design 4 achieved the lowest backpressure at 2229.9 Pa. It was found that the backpressure decreases when the hole diameter of the perforated tube increases. From the calculation, the optimization value of backpressure is 5.5% for Design 1, 2.6% for Design 2, 8% for Design 3 while for Design 4 is 10.3%.

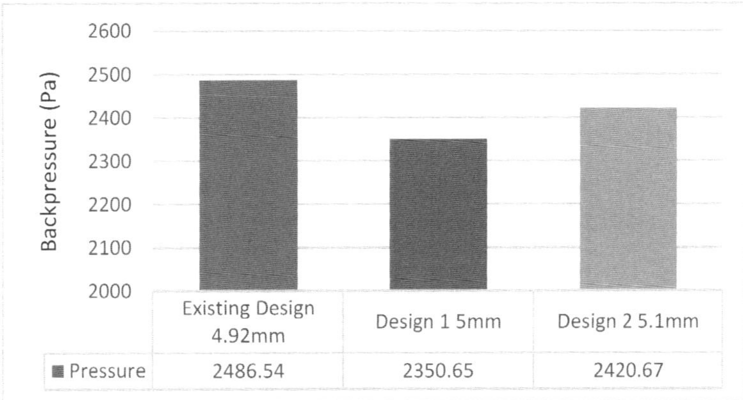

Figure 3: Comparison of Backpressure between MODENAS CT115's Muffler with Design 1 and Design 2

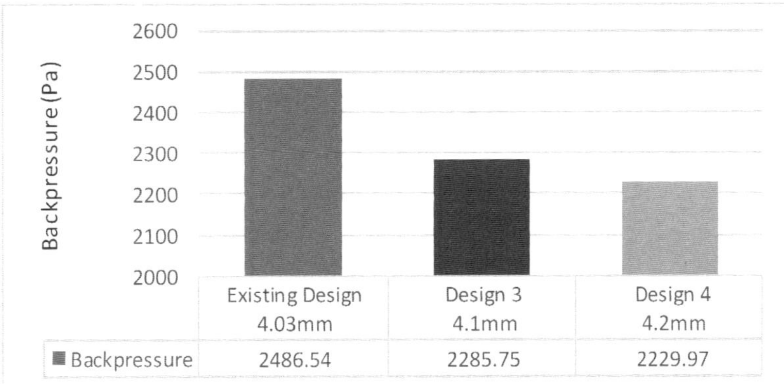

Figure 4: Comparison of Backpressure between MODENAS CT115's Muffler with Design 3 and Design 4

Figure 5 and Figure 6 show the outlet velocity for all proposed designs is greater than the existing design. From the results, it can be seen that, when the velocity increases, the backpressure also decreases. Overall, it is noted that Design 4 is the best design to increase engine performance for MODENAS CT115 since it gives lower backpressure with higher outlet velocity.

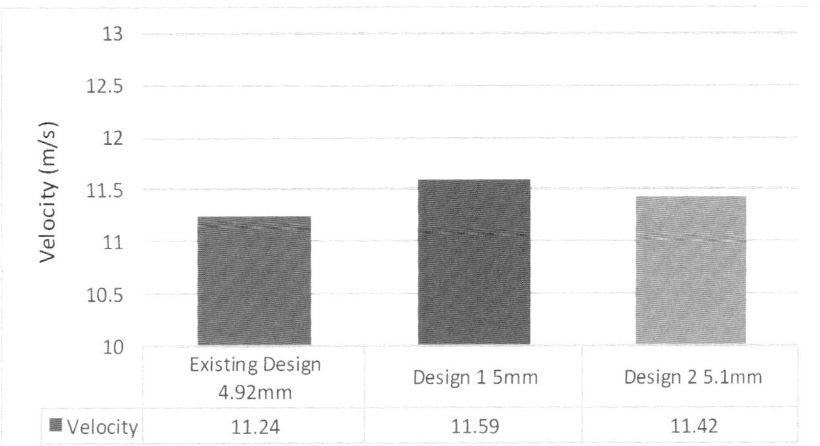

Figure 5: Comparison of outlet velocity between MODENAS CT115's with Design 1 and Design 2

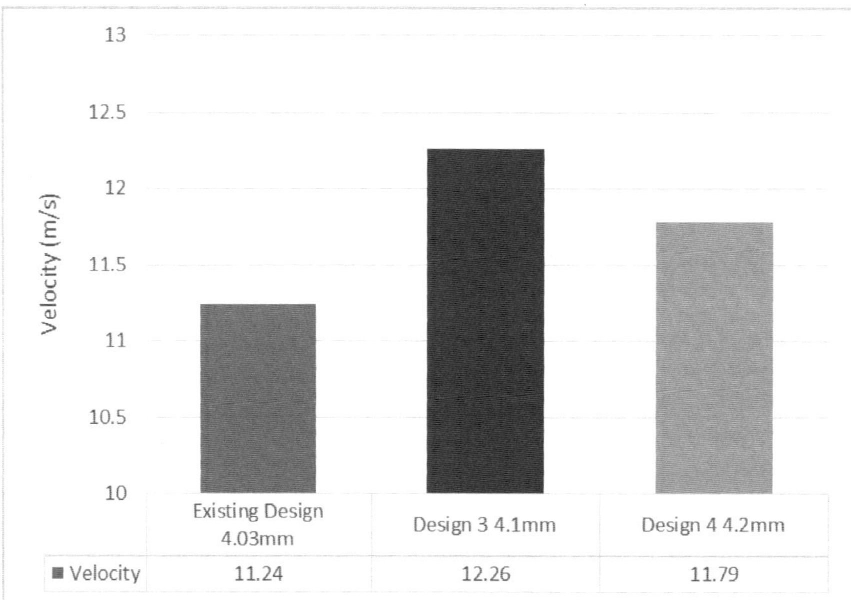

Figure 6: Comparison outlet velocity between MODENAS CT115's Muffler with Design 3 and Design 4

3. Conclusion

In this study, the authors investigated new muffler designs to enhance the engine performance of MODENAS CT115'S motorcycle by optimization of backpressure. As general finding, when the geometric design of the front and back perforated tube is changed the backpressure level decreases. It was found that the backpressure for Design 4 was 10.3% improved than the existing muffler design result. It can be concluded that the increase of the hole diameter will decrease the backpressure thus improves the performance of engine.

References

[1] Suganeswaran K, Parameshwaran R, Amirthamani S and Palanimohan D 2014 Engineering and technology design and optimization of muffler for manufacturing *International Journal of Innovative Research in Science* **3** (4) 282–290.

[2] Xue F and Sun B 2018 Experimental study on the comprehensive performance of the application of U-shaped corrugated pipes into reactive mufflers *Appl. Acoust.* **141** 362–370.

[3] Pangavhane S D, Ubale A B, Tandon V A and Pangavhane D R 2013 Experimental and CFD analysis of a perforated inner pipe muffler for the prediction of Backpressure *Int. J. Eng. Technol.* **5** (5) 3940–3950.

IOP Conf. Series: Materials Science and Engineering **670** (2019) 012064 doi:10.1088/1757-899X/670/1/012064

Study on deformation characteristics of tailored blanks having thickness distribution by successive forging process

N H Romzi[1], L Tajul[1], H Jaafar[2], M H Sulaiman[1] and M J M Ridzuan[1]

[1]School of Mechatronic Engineering, Universiti Malaysia Perlis, Pauh Putra Campus, 02600, Arau, Perlis, Malaysia.
[2]School of Manufacturing Engineering, Universiti Malaysia Perlis, Pauh Putra Campus, 02600, Arau, Perlis, Malaysia.

Abstract. Tailored blanks are used in car body-in-white parts for the purpose of reducing weight and improving safety. The reduction of weight in a car decreases the fuel consumption and global emissions. In this study, the deformation characteristics of tailored blanks having thicknesses distribution produced by successive forging were analysed using finite element method. The successive forging produced tailored blanks without joining similar to tailor rolling by means of presses. The blank was locally and repeatedly compressed using upper and lower punches hence compression load is small. The punch geometry was varied to understand the deformation behaviour of the blanks for feeding interval of 10 mm. Waving was observed on the sides of the tailored blanks. The width of the blanks after successive forging increases as the punch width increases and when using an inclined punch.

1. Introduction

Nowadays the demand of lightweight vehicle is increasing due to environmental concern as lightweight vehicle leads to reduced fuel consumption and global emission. Tailored blanks are semi-finished parts that are found in the car body-in-white for the purpose of weight reduction. Tailored welded blanks are generally produced by joining two or more sheets of different material, thickness or coating together by welding process. Welding and thickness differences of tailor welded blanks lead to reduced formability [1]. The sudden change in thickness increases stress concentration and decreases the formability of the tailor welded blank. The non-uniform material deformation of thin and thick blanks and the weld line movement of tailor welded blanks lead to tearing during forming [2].

Tailored rolled blanks on the other hand are tailored blanks having thickness distribution produced by rolling process without the need of joining. These blanks are produced by adjusting the roll gap between rollers to achieve the desired thickness distribution [3-4]. The continuous thickness variation at the transition zones reduces the stress concentration hence improves formability. Han et al. [5] investigated the manufacturing process of tailor rolled blanks having thickness distribution in longitudinal and latitudinal directions. Tailor rolled blanks however, are costly and the supply is limited.

Tailored blanks having thickness distribution in the longitudinal direction was also produced by successive forging [6]. The thickness was controlled by varying the feeding interval and also utilizing the elastic deformation of the tools and press. T. Kuboki et al. [7] produced blanks for micro parts by controlling the feeding interval using successive compression. S Jirathearanat et al. [8] analysed the successive forging process in producing tailored blanks having thickness distribution using wedge-shaped tool. This study analysed the deformation characteristics of tailored blanks having thickness

IOP Conf. Series: Materials Science and Engineering **670** (2019) 012064 doi:10.1088/1757-899X/670/1/012064

distribution under different punch geometry by finite element method. The effect of successive forging on the compression load, thickness change and surface quality were observed.

2. Approach of successive forging of tailored blanks having thickness distribution

Figure 1 shows the steps in successive forging process. It is a repetitive process where a small area of sheet metal is compressed at a time. After compression (see Figure 1(a)), the upper punch is retracted to allow for the feeding of the sheet (see Figure 1(b)) for the next compression (see Figure 1(c)). These steps are repeated until the desired length of compression is achieved. The feeding interval was 10 mm.

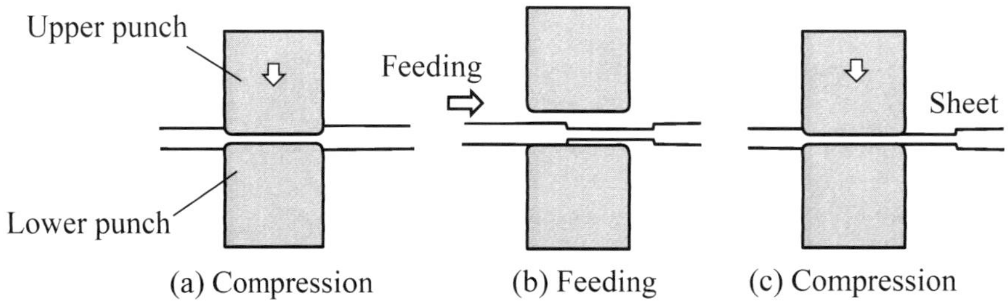

Figure 1: Steps in successive forging process

Finite element analysis was conducted using ABAQUS to understand the successive forging process and the deformation behaviour of the tailored blanks. To model the successive forging process, for the first compression, the upper punch compressed the blank as shown in Figure 2. Then both of the punches were retracted while the next set of punches move towards the sheet for the second compression. The second set of punches was located 10 mm apart from the first punches based on the feeding interval. The total number of compression was five. The stroke was set to 0.8 mm. The tools were modelled to be rigid while the sheet was elastic-plastic. The sheet was divided into solid hexahedral elements. The material for the blank is Stainless Steel 304 and the properties are as listed in Table 1. The initial length of the blank was 100 mm while the width and thickness of the blank were 20 mm and 3 mm, respectively.

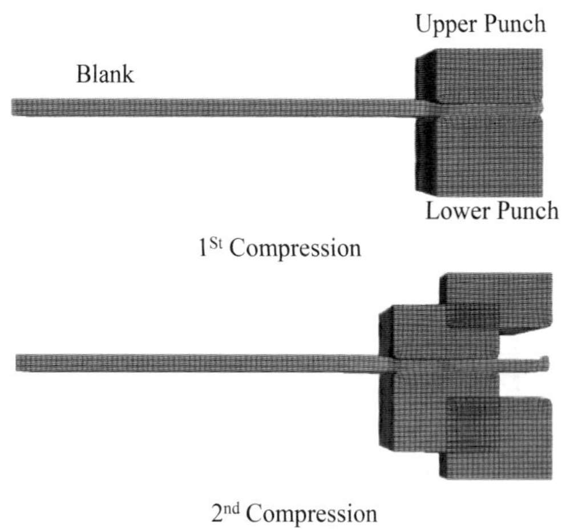

Figure 2: Successive forging process model

Table 1: Material properties

Properties	Value
Density (kg/m^3)	7900
Young's modulus (GPa)	193
Yield Strength (MPa)	215
Poisson's ratio	0.29

The simulation was conducted for three different conditions as illustrated in Figure 3. For condition 1, the punch width was 30 mm and the punch will compress the entire sheet in the width direction. Both conditions 1 and 2 used flat punches while condition 3 utilized inclined punch. The flat punch width was varied between 5 to 15 mm for condition 2 while for condition 3 the inclined punch angle were varied for $\alpha = 30^{\circ}$ and 45°.

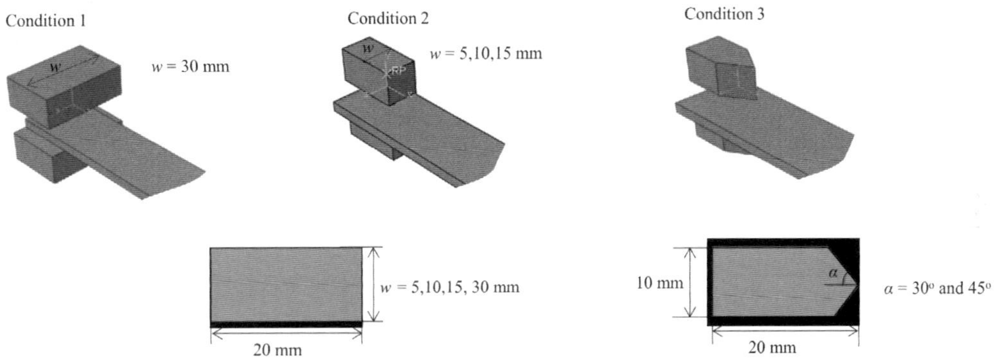

Figure 3: Successive forging conditions for feeding interval 10 mm

3. Result and discussion

The simulation result for blank after successive forging process for condition 1 is shown in Figure 4. The blank was compressed five times with constant punch stroke. The thickness after compression was 2.2 mm. Waving was observed on both sides of blank causing uneven width distribution due to increase in number of compression and large deformation in the thickness direction. W_1 is larger than W_2 and W_3 because the first compression involves large initial contact area as compared to the following compressions. Hence the compression load is high for the first compression while the load is reduced for the next compression as shown in Figure 5.

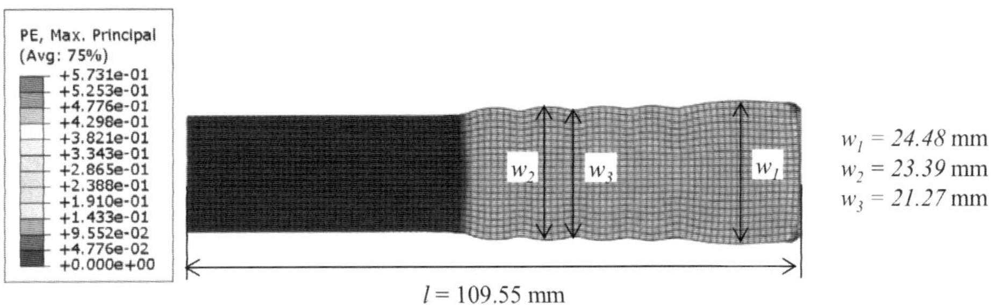

Figure 4: Simulation result of blank after successive forging process (condition 1)

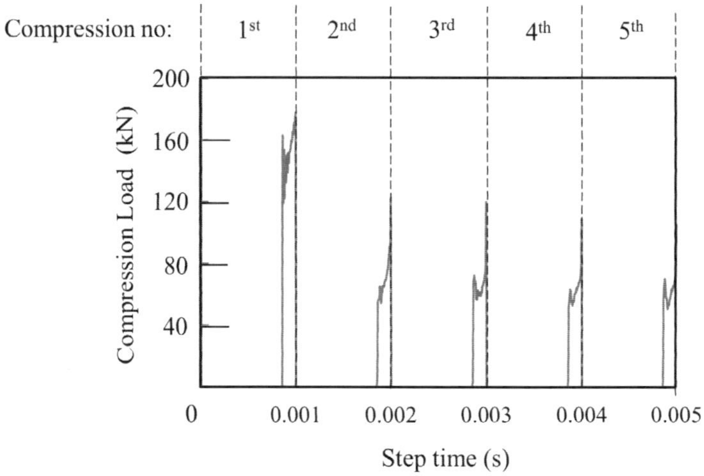

Figure 5: Compression load of blanks after successive forging process (condition 1)

Figure 6 shows the simulation result for blanks after successive forging process for condition 2 (flat punch with width 10 mm) and condition 3 (inclined punch with 45 ° angle and width 10 mm). The reduction in thickness occurred at the centre of the blanks and waving on both side of the blanks are minimum as compared to condition 1. The inclined punch produced more spreading with less elongation. The inclined punch allows more material to flow towards the side of the blank and hence increases the width.

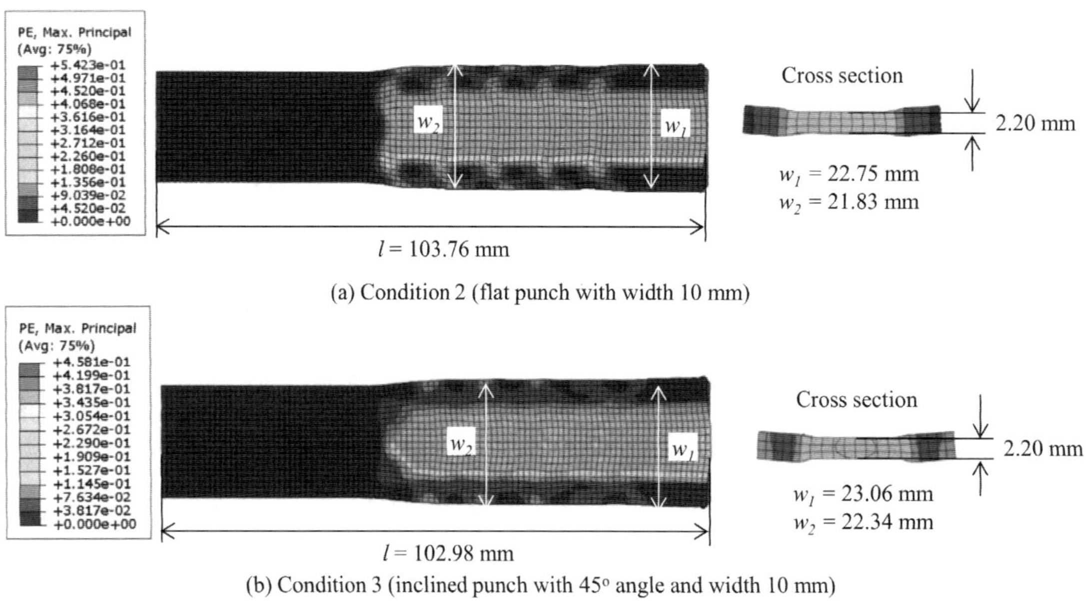

(a) Condition 2 (flat punch with width 10 mm)

(b) Condition 3 (inclined punch with 45° angle and width 10 mm)

Figure 6: Simulation result of blanks after successive forging process for conditions 2 and 3

Figure 7 shows the width of the blanks after successive forging for different punch width and angle. As the punch width increases the spreading of the blank increases. The inclined punch shows larger width distribution after five compressions as compared to the flat punch.

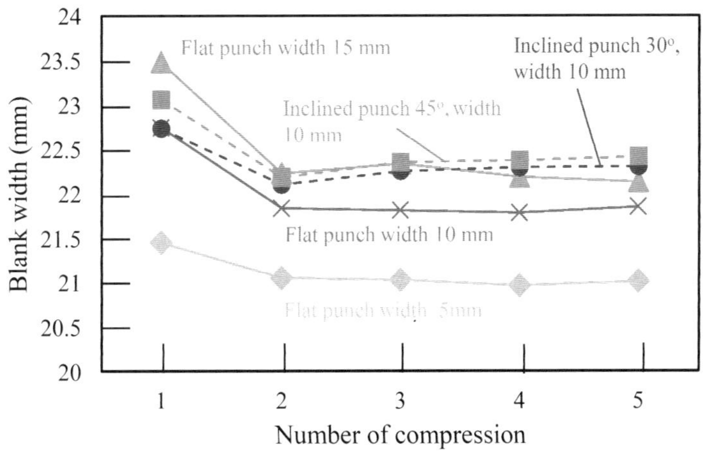

Figure 7: Width of blank after successive forging for conditions 2 and 3

4. Conclusion

Simulations on successive forging of tailored blanks having thickness distribution by finite element analysis have been conducted. Three forging conditions with different punch geometry were studied to understand the deformation characteristics of the tailored blanks. During successive forging, the compression load is small due to small contact area between punch and blank. However, waving was observed on both sides of the blanks. Waving was reduced for conditions 2 and 3 due to smaller punch width. This uneven width distribution will be further investigated in the future study. The blank spreads more as the punch width increases and when using an inclined punch.

References

[1] Habibi M, Hashemi R, Tafti M F and Assempour A 2018 Experimental investigation of mechanical properties, formability and forming limit diagrams for tailor-welded blanks produced by friction stir welding *Journal of Manufacturing Processes* **31** 310-323.

[2] Merklein M, Johannes M, Lechner M and Kuppert A 2014 A review on tailored blanks—Production, applications and evaluation *Journal of Materials Processing Technology* **214** 151-164.

[3] Kopp R, Wiedner C and Meyer A, 2005 Flexibly rolled sheet metal and its use in sheet metal forming *Advanced Materials Research* **6-8** 81-92.

[4] Ryabkov N, Jackel F, Putten K van and Hirt G 2008 Production of blanks with thickness transitions in longitudinal and lateral direction through 3D-Strip Profile Rolling *International Journal of Material Forming* **1** (Supplement 1) 391-394.

[5] Han S, Hwang T, Oh I, Choi M and Moon Y H 2018 Manufacturing of tailor-rolled blanks with thickness variations in both the longitudinal and latitudinal directions *Journal of Materials Processing Technology* **256** 172-182.

[6] Tajul L, Maeno T, Kinoshita T and Mori K 2017 Successive forging of tailored blank having thickness distribution for hot stamping *The International Journal of Advanced Manufacturing Technology* **89** 3731-3739.

[7] Kuboki T, Azrie A and Jin Y 2014 A new incremental in-plane bending of thin sheet metals for micro machine components by using a tiltable punch *CIRP Annals - Manufacturing Technology* **63** 249-252.

[8] Jirathearanat S, Taureza M and Atsushi D 2017 Successive Forming of Multi-thickness Plates by Using Wedge-shaped Tool *Procedia Engineering* **207** 1147–1152.

IOP Conf. Series: Materials Science and Engineering **670** (2019) 012065 doi:10.1088/1757-899X/670/1/012065

Fracture toughness of railway for higher speed rail corridors in Malaysia

M A Rojan, M S M Azmi, M S Abdul Majid, R Daud and K S Basaruddin

School of Mechatronic Engineering, Universiti Malaysia Perlis, Pauh Putra Main Campus, 02600 Arau, Perlis, Malaysia.

Abstract. When Keretapi Tanah Melayu Berhad (KTMB) introduced the High Speed Electrical Train System (ETS) in 2013, the fracture toughness of the railway rail remains relatively unknown. The aim of this study was to evaluate the fracture toughness of rail steel. The rail steel specimen was prepared according to ASTM E1820 and cut using electrical discharge machining (EDM) wire cut process. The specimen was pre-cracked using Biss Fatigue Machine to obtain a relatively accurate results followed by three point bend testing under Universal Testing Machine (UTM), in accordance to the standard of ASTM E1820. The fractured crack tip was then observed using an optical microscope. From the three point bend test, the fracture toughness was calculated and the behaviour of the steel failing was then studied and compared to previous studies. The fracture toughness of the steel was found to be 44.009 MPa√m, which is within the range of the various rail steels used around the world..

1. Introduction

In this study, the fracture toughness of the railway used by the ETS service is to be determined. Specifically, the subject of interest is the railway steel grade 900A International Union of Railways (UIC 54), which is used in the ETS railway. Fracture toughness is a material constant that is independent of the size and geometry of the cracked body. Fracture poses a danger in railways as fracture of the railway rail will lead to derailment. In general, fatigue and fracture are caused by loadings. In railway applications, rails are subjected to primary and secondary loading components. From the loading applied to the rail by the wheel, the rail is subjected to bending stress, axial stresses and Hertzian pressure during rolling contact. The failure of the rail steel has caused various incidents that was followed by detrimental outcomes, namely serious injuries and deaths aside from impact towards the economy and the various. Although, derailments due to rail failure are rare in Malaysia, low maintenance of the rail will eventually results in derailments. Hence, the purpose of this study is to evaluate the fracture toughness of UIC 54 rail profile according to ASTM E1820 standard as well as to study the crack surface of the fractured specimen of UIC 54 profile obtained from KTMB. The fracture toughness of UIC 54 rail steel will be compared with the other rail steels used in other countries.

2. Methodology

2.1 Material and Specimen Preparation

In this study, the material of interest is the rail steel used in Malaysia railways, which is of grade 900A, meeting the requirements of Codex UIC 54 standard. The weight of the 900A UIC 54 is noted to be 54 kg/m. The chemical composition of the rail profile is provided in Table 1.

The Single Edge Notched Bend (SENB) specimen was used in the evaluation of the fracture toughness of the UIC 54 rail profile. The specimens were prepared in accordance to the ASTM E1820 standard [2]. From the UIC 54 rail profile, the web region of the rail was used to fabricate the SENB specimens. From the rail profile obtained the desirable thickness of the specimen was 13 mm as the thickness of the web of the rail profile was 16 mm. Hence, the specimen thickness cannot exceed the thickness of the rail profile. The 13 mm was used as the base parameter in preparing the specimen. In accordance to ASTM 1820, the height of the specimen was 26 mm with a width of 117 mm. A notch was designed onto the specimen and below the minimum crack length so that the specimen can be pre-cracked to obtain a pre-cracked notch, which will enable the acquisition of more accurate results. Figure 1 shows a technical drawing of said specimen.

Table 1: Chemical composition of grade 900A (UIC 54) rail steel [1].

Quality	**Chemical Composition**							**Tensile Strength (MPa)**	**Elongation after Fracture**
	C	**Si**	**Mn**	**P** (max)	**S** (max)	**Cr**	**Ni**		
Grade 900 A (UIC 54)	0.6 – 0.8	0.1 – 0.5	0.8 – 1.3	0.04	0.04	-	-	880 – 1030	$\geq$ 10

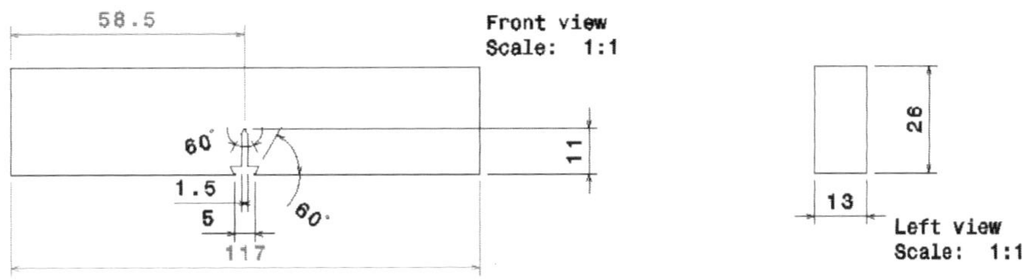

Figure 1: Full dimension of the SENB Specimen.

2.2 Mechanical Testing

A pre-cracking procedure was carried out using the fatigue testing machine (BISS) using a three point bend jig as shown in Figure 2. The pre-cracking was done using a determined cyclic load. The pre-cracking length of the SENB specimen was carried out in accordance to the ASTM E1820 [20]. For the load used during pre-cracking, the maximum calculated load was 5.6 kN. Therefore, the load used should not be more than 5.6 kN. The pre-cracking procedure was carried out with a maximum load of 4.5 kN and stress ratio, $R = 0.1$, where the minimum load was 0.45 kN. The frequency of load was set to 20 Hz.

The specimen was then tested for fracture toughness using UTM (Shimadzu AG-IS MS) which has a maximum load capacity of 250 kN. A three point bend jig was used to carry out the test as shown in Figure 3. For the load, a load rate of 5 mm/min was used to pre-crack the experiment. After the specimen was successfully cracked, the raw data were recorded and saved in .csv format.

3. Results and Discussion

3.1 Findings from Pre—cracking Process

As required, an additional 2 mm crack length was created through pre-cracking, with increasing the initial crack length from 11 mm to 13 mm. From the pre-cracking process, various findings were obtained. Figure 4 shows a graph of the crack length against the number of cycles. From the graph, it can be observed that significant crack propagation starts at the 1000th number of cycle. The trend of the propagation can be observed as increasing exponentially, where a sharp increase of the crack length can be observed near the end of the process.

Figure 2: Three point bend fixture for pre-cracking.

Figure 3: A specimen during the test.

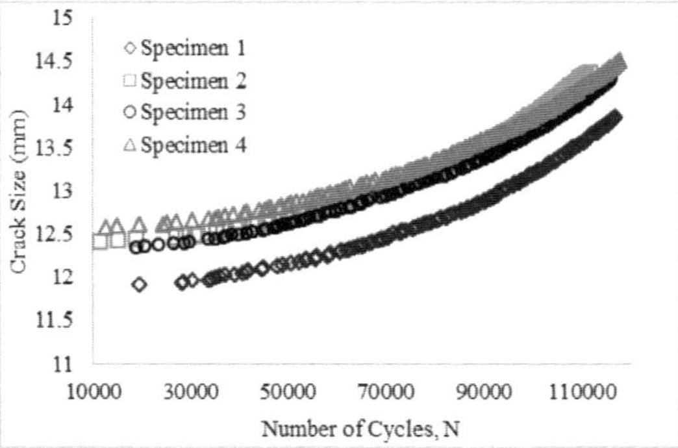

Figure 4: Crack Length against Number of Cycles.

Similarly, Figure 5 shows a graph of compliance against the number of cycles. Compliance is required to calculate the correction factor used in calculating the fracture toughness value of the specimen and is given by:

$$C_i = \frac{\Delta v_m}{\Delta P} \tag{1}$$

where, Δv_m is crack opening displacement at notched edge and ΔP is the force acting on the specimen.

However, as observed from Figure 5, the values differ as the number of cycle increases. The compliance increases exponentially with the number of cycles. In other words, the crack size increases as the value of the compliance increases significantly, resulting in the unreliable application of the compliance value. Hence, the initial value of the compliance is used in the calculation of the fracture toughness as it retains the original property of the specimen before external forces are applied to it.

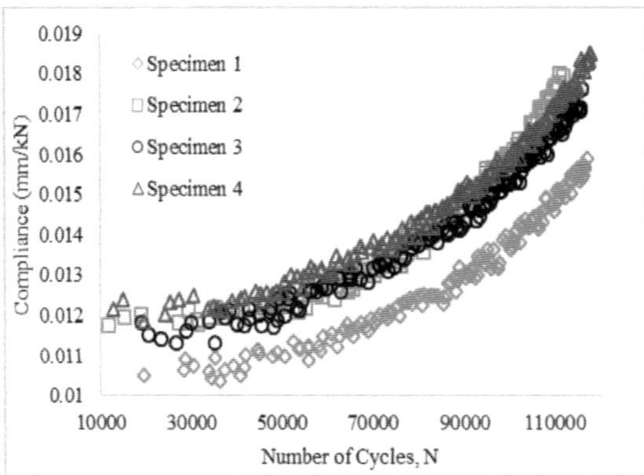

Figure 5: Compliance against Number of Cycles.

Log-log plots of *da/dN* against ΔK was also obtained from the process for each specimen, as shown in Figure 6 for specimen 1. From the graph, it can be observed that the pre-crack process happened at stage 1 of the fatigue crack growth behavior. The tangent line shows the beginning of the stage 2 of a metal's crack propagation behavior. By using the points on the line, the constant m can be calculated as the log-log gradient of the graphs, and thus later can be used to calculate C. Table 2 shows the comparison between the current study and a previous study done by Gurubaran [3].

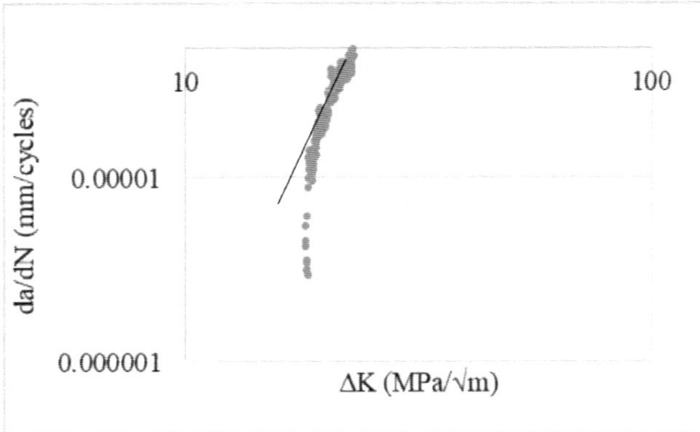

Figure 6: Log-log plots of *da/dN* against ΔK for specimen 1.

Table 2: Material constant value comparison for UIC 54 profile.

	Specimen	P_{max} (kN)	R	C	m
Gurubaran [3]	1	16	0.1	2.567E-13	3.455
	2	13	0.3	4.482E-13	3.270
	3	13	0.1	5.928E-13	3.223
Current Study	1	4.5	0.1	1.2894E-12	3.344
	2			1.8172E-12	3.180
	3			1.432E-12	3.205
	4			7.388E-13	3.401

3.2 Fracture Toughness Test of UIC54 Specimen

After the pre-cracking of the specimen was done, the specimen is then tested using the UTM with a three point bend jig. The test was done with a stroke speed of 5 mm/min. From the data obtained from the test as well as the pre-cracking process, the fracture toughness, K_{IC}, can be calculated using Eq. (2). Table 3 shows the results of the calculations.

$$K_{(i)} = \left[\frac{P_i S}{(BB_N)^{\frac{1}{2}} W^{\frac{3}{2}}}\right] f\left(\frac{a_i}{W}\right) \tag{2}$$

where,

$$f\left(\frac{a_i}{W}\right) = \frac{3\left(\frac{a_i}{W}\right)^{\frac{1}{2}}\left[1.99 - \left(\frac{a_i}{W}\right)\left(1 - \frac{a_i}{W}\right) * \left(2.15 - 3.93\left(\frac{a_i}{W}\right) + 2.7\left(\frac{a_i}{W}\right)^2\right)\right]}{2\left(1 + 2\frac{a_i}{W}\right)\left(1 - \frac{a_i}{W}\right)^{\frac{3}{2}}} \tag{3}$$

3.3 Comparison between Experimental and Simulation Results

Results from other researches are then collected from the literature to compare with the current study. It should be noted that the compared results are of different rail steel materials. Table 5 shows the results of the different rail steels where the fracture toughness study was conducted.

Table 3 : Comparison between Rail Steel Studies.

Steel Grade	Tensile Strength (MPa)	Fracture Toughness, K_{IC} (MPa√m)
R260 [4]	951	38.56 (Rail Head to Web)
		39.40 (Rail Flange)
R370CrHT [4]	1373	40.44 (Rail Head to Web)
		39.60 (Rail Flange)
R350 HT [5]	1194 (Head)	42.2 (Head)
	999 (Web)	68.1 (Web)
R260 (UIC 60) [5]	940 (Head)	34.1 (Head)
	907 (Web)	46.6 (Web)
R900a (UIC 860) [6]	924 (Head)	32.49 (Head)
R900a (UIC 54)	**880-1030**	**44.01 (Web)**

4. Conclusion

From this study, the behaviour of the crack size can be observed as increasing exponentially against the number of cycles. The compliance was also found to differ at different stage of the crack propagation as the compliance increases exponentially as well. The constant *m* was found to be similar to the study done by Gurubaran [3]. From the three point bend test done using UTM, the behaviour of the specimen was observed. The trend of the graph shows a brittle behaviour of the steel, showing minimal elastic behaviour. By using data obtained from the graph, the fracture toughness of the steel was found to be 44.009 MPa√m, which is acceptable as it is within the range of the various rail steels used around the world.

Acknowledgement

The author would like to acknowledge the support from the Fundamental Research Grant Scheme (FRGS) under a grant number of FRGS/1/2018/TK03/UNIMAP/03/6 from the Ministry of Education Malaysia.

References

[1] Vitez I, Vitez B and KrumesD 2005 UIC-recommendations for the use of rail steel grades *Metalurgija* **44** (2) 137-140.

[2] ASTM E1820-16 2016 *Standard Test Method for Measurement of Fracture Toughness* ASTM International, West Conshohocken, PA.

[3] Gurubaran P 2017 *Fatigue crack behavior on the wheel and railway for high speed rail corridors in Malaysia* Master, Universiti Malaysia Perlis.

[4] Maya-Johnson S, Ramirez A and Toro A 2015 Fatigue crack growth rate of two pearlitic rail steels *Engineering Fracture Mechanics* **138** 63-72.

[5] Tabatabaei A 2014 *Fracture and fatigue crack growth chracterization of conventional and head hardened railway rail steels* Masters of Science, Middle East Technical University.

[6] Christodoulou P, Kermanidis A and Haidemenopoulos G 2016 Fatigue and fracture behavior of pearlitic Grade 900A steel used in railway applications *Theoretical and Applied Fracture Mechanics* **83** 51-59.

IOP Conf. Series: Materials Science and Engineering **670** (2019) 012066 doi:10.1088/1757-899X/670/1/012066

Printing temperature, printing speed and raster angle variation effect in fused filament fabrication

H Zakaria[1], S F Khan[1,2], M F C Fee[1] and M Ibrahim[3]

[1] School of Mechatronic Engineering, Universiti Malaysia Perlis, Malaysia.
[2]Additive Manufacturing and Design Lab, School of Mechatronic Engineering, Universiti Malaysia Perlis, Malaysia.
[3] School of Manufacturing Engineering, Universiti Malaysia Perlis, Malaysia.

***corresponding author: hafizawati@unimap.edu.my**

Abstract. Additive Manufacturing becomes most widely uses in prototyping and production applications since 3D printer becomes as open-source. However, there are not provide standard value of printing process parameters to give best performance of mechanical properties of the printed parts. In additional, printed parts produced can be affected by the various parameters used, this nature of phenomena very complex. The purpose of this research is to evaluate the effects printing process parameters respond to its mechanical properties in tensile strength and flexural strength from Fused Filament Fabrication by using Polylactic Acid (PLA) with printing parameters, namely printing temperature, printing speed and raster angle. The specimens were printed by using Monoprice Maker Select 3D printer. Tensile test and three-point bending test were carried out. Taguchi Method involves in identification of proper control factors to obtain the optimum results of the process. Orthogonal Arrays (OA) are used to conduct a set of experiments. Results of these experiments analyze by using Analysis of Variance (ANOVA) to predict the quality of components produced and find influence of printing temperature, printing speed and raster angle on mechanical properties. The results show that raster angle have most influence on both mechanical properties with 68.72% contribution, printing speed 18.88% and temperature 1.49% contribution in tensile test while for bending test raster angle obtain 94.02% contribution ,printing speed give 2.28% and temperature only 1.67% contribution. Optimum value obtain for tensile test in this study shows temperature = 180°C, printing speed = 40 mm/s, and raster angle = 60°. In bending test gives temperature = 185°C, printing speed = 30 mm/s, and raster angle = 60°.

1. Introduction

The manufacture of part using FFF processes is subjected to numerous variables dependent upon thermophysical and/or chemical phenomena that are going to result in parts with different characteristics depending on the method used along with the parameters of the process. The nature of these phenomena is very complex, as they are also affected by the influence of related environmental and process conditions; for example, if the process is performed within a controlled atmosphere, the existence (or the lack thereof) of a heated bed and/or the heat transmission process affects the thermal gradients of the workpiece, particularly between layers. This research presents a study influence of the tensile mechanical behaviour of pieces produced using the Fused Filament Fabrication (FFF) additive manufacturing technique with respect to the parameter of printing process with most widely used material polylactide (PLA). The aim of this study is to the evaluate best parameters of printing temperature, printing speed and raster angle by optimizing on the mechanical performance of PLA test specimens.

2. Methodology

Material used is Polylactic Acid (PLA). 3D printing machine used is IIIP Monoprice 3D Printing. STL generated from Catia, and change into G code in the IdeaMaker slicer. Table 1 show varied parameter used for set of experiment design for Taguchi optimization method. This set experiment that conducted using Taguchi is appropriate because of their effective method of experiment and more precise. Set of experiment for both tensile and fatigue is show in Table 2. The microscopic structure and porosity of the filament will be analyse using metallurgy microscope. Next, the relationship between the PLA materials are analysed especially the aspect of microscopic structure, mechanical properties and others. The best PLA (specimen filament) parameter will be chosen, in the next step for future work.

Table 1. Parameter and level

Factor	Level			Unit
	1	2	3	
Printing Speed	30	35	40	mm/s
Printing Temperature	180	185	195	Degree
Raster Angle	-45/45	0/90	30/60	degree

Table 2. Taguchi table for both tensile and fatigue test

Raster Angle	Printing Speed	Printing Temperature
0/90	30	180
0/90	35	185
0/90	40	190
-45/45	30	190
-45/45	35	180
-45/45	40	185
30/60	30	185
30/60	35	190
30/60	40	180

90 samples were fabricated for tensile and flexural test. ASTM D638 and ASTM D790 [17][18] used to decide the geometry, velocity, load and procedure are followed. Tensile test is conducted by utilizing Shimadzu Universal Testing machine with minimum speed applied is 5 mm/mm and 5kN load. While flexural test was conducted at 2mm/mm speed with 5kN load.

3. Result and Discussion

3.1 Result for Tensile and Flexural Test

After the tensile and flexural tests are conducted the data obtained are recorded in Table 3 and Table 4. The highest tensile yield strength is 23.16MPa from level 3 with -45/45 raster angle, 40mm/s printing speed and 190-degree printing temperature. While the lowest (level 8) with printing parameter are set at 30/60 raster angle, 35mm/s printing speed and 190-degree printing temperature. The highest flexural yield strength is level 5 with 0/90 raster angle, 35mm/s printing diameter and 185-degree printing temperature with 63.48MPa strength. While the lowest is at level 8 similar to the tensile lowest level.

Table 3. Data of Tensile Strength

Exp No	Ultimate Tensile Strength (MPa)					
	1	2	3	4	5	Average (MPa)
1	12.38	16.30	15.93	16.10	13.64	14.87
2	16.79	15.44	16.42	15.20	17.25	16.22
3	23.78	23.03	22.67	22.80	23.52	**23.16***
4	17.10	15.40	15.20	16.00	17.80	16.30
5	19.85	20.47	19.20	21.11	20.17	20.16
6	14.09	13.73	14.83	14.51	13.94	14.22
7	16.67	20.10	19.33	17.82	18.03	18.39
8	13.86	12.23	10.91	9.73	12.07	11.76
9	21.16	21.20	23.16	22.41	22.62	22.51

Table 4. Data of Flexural Strength

Exp No	Maximum Force (kN)						Flexural Strength (MPa)
	1	2	3	4	5	Average (kN)	
1	0.034	0.025	0.025	0.030	0.026	0.028	20.63
2	0.050	0.056	0.044	0.061	0.039	0.050	36.91
3	0.087	0.069	0.072	0.065	0.085	0.076	56.10
4	0.066	0.047	0.051	0.049	0.067	0.056	41.34
5	0.094	0.081	0.084	0.090	0.081	0.086	**63.48***
6	0.038	0.031	0.013	0.032	0.021	0.027	19.93
7	0.086	0.091	0.080	0.082	0.091	0.086	63.40
8	0.023	0.023	0.013	0.034	0.035	0.019	14.03
9	0.038	0.035	0.044	0.040	0.035	0.039	38.78

Based on the data average value from Table 3 and Table 4, the morphological characteristic shows good interlayer bonding and uniformly distributed struts is the main factor that contributed toward highest tensile and flexural strength. Meanwhile, for medium and lowest tensile and flexural strength are due to the weak bonding of the interlayer and no uniform struts coming out from the nozzle head. This might be due to the unsuitable printing temperature and printing speed set of parameters.

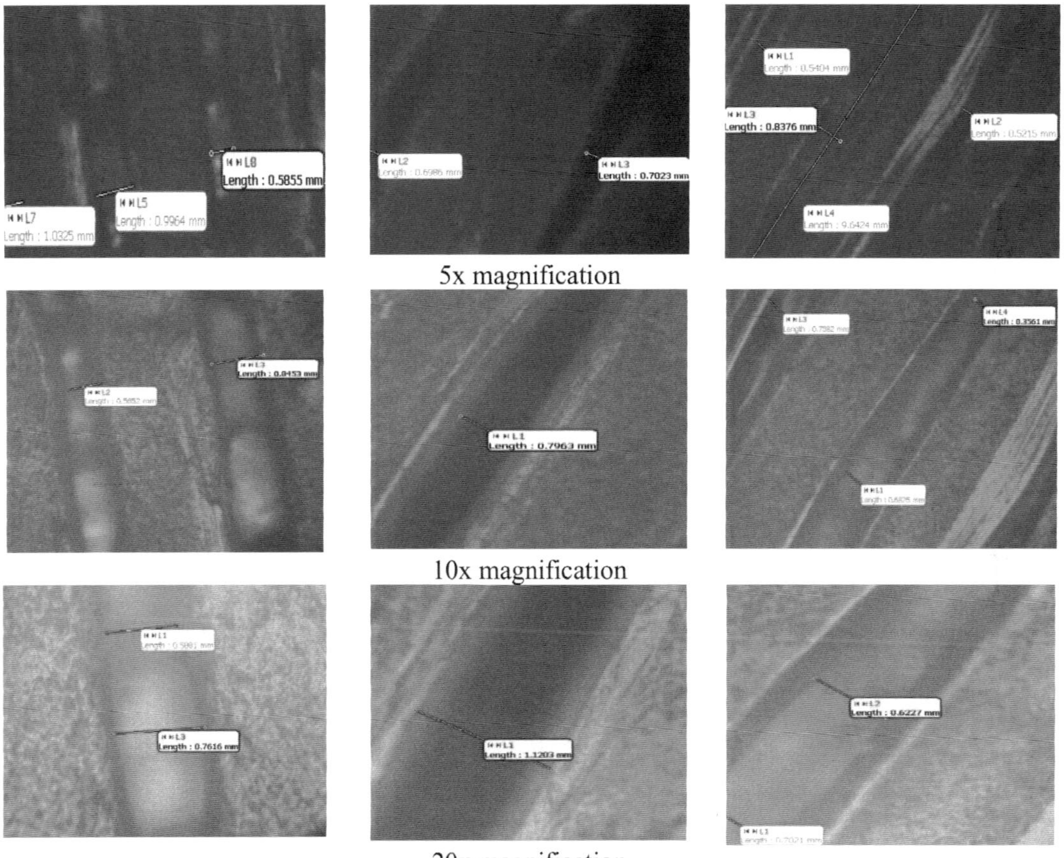

A. Highest **B.Medium** **C. Lowest**

Figure 1: Morphological characteristic of tensile test specimens at varied magnification, A. Highest tensile strength, B. Medium tensile strength, and C. Lowest tensile strength

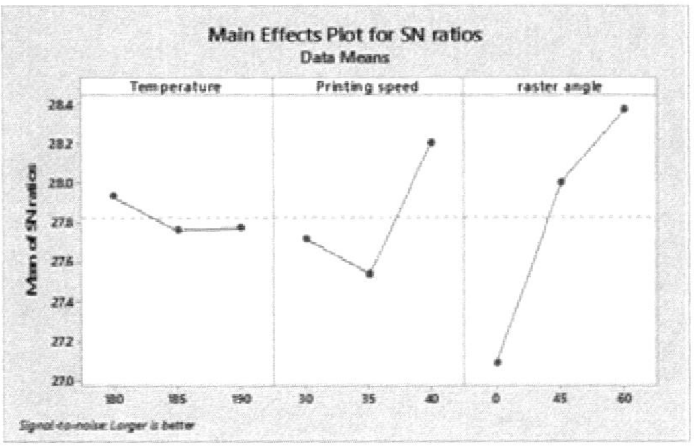

Figure 2: Graph of S/N ratio versus level of parameter for tensile test.

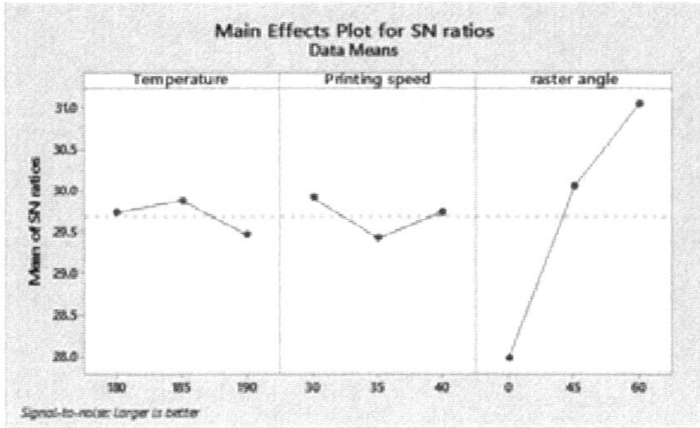

Figure 3: Graph of S/N ratio versus level of parameter for flexural test..

From the data obtain, the yield tensile and flexural strength were recorded and the S/N ratio for each level are calculated. Type of S/N ratio is the bigger the better. The optimum parameter for the tensile test is 180-degree printing temperature, 40mm/s printing speed and 30/60-degree raster angle as shown in Figure 1 and Table 5, while optimum parameter for flexural test is 185-degree printing temperature, 30mm/s printing speed and 30/60-degree raster angle as shown in Figure 2 and at Table 6.

Table 5. Response table (Tensile)

Level	Printing Temperature	Printing Speed	Raster Angle
1	180*	30	-45/45
2	185	35	0/90
3	190	40*	30/60*

Table 6. Response table (Flexural)

Level	Printing Temperature	Printing Speed	Raster Angle
1	180	30*	-45/45
2	185*	35	0/90
3	190	40	30/60*

3.2 ANOVA Analysis

The most significant parameter is raster angle. The F value for both experiments is higher than f table of 90 confident interval. The experiment not influence by Printing temperature for tensile and fatigue test. Table 7 and Table 8 show the ANOVA table for tensile and fatigue test respectively.

Table 7. ANOVA (Tensile)

Source	DF	Adj SS	Adj MS	F-value	P-value	Contribution %
Printing Temperature	2	0.05665	0.02822	0.14	0.880	1.49
Printing Speed	2	0.71690	0.35845	1.73	0.366	18.88
Raster Angle	2	2.60887	1.30444	6.30	0.137	68.72
Error	2	0.41393	0.20696			
Total	8	3.79636				

Table 8. ANOVA (Flexural)

Source	DF	Adj SS	Adj MS	F-value	P-value	Contribution %
Printing Temperature	2	0.2628	0.1314	0.82	0.549	1.67
Printing Speed	2	0.3604	0.1802	1.13	0.471	2.28
Raster Angle	2	14.8392	7.4196	46.30	0.021	94.02
Error	2	0.41393	0.2070			
Total	8	3.79636				

IOP Conf. Series: Materials Science and Engineering **670** (2019) 012066 doi:10.1088/1757-899X/670/1/012066

4. Conclusion

From the result of this research the influence of three parameter and the mechanical test are determined. The result is:

1) The highest tensile yield strength is 23.16MPa from level 3 with -45/45 raster angle, 40mm/s printing speed and 190-degree printing temperature. The highest flexural yield strength is level 5 with 0/90 raster angle, 35mm/s printing diameter and 185-degree printing temperature with 63.48MPa strength.
2) The optimum parameter for the tensile test is 180-degree printing temperature, 40mm/s printing speed and 30/60-degree raster angle while optimum parameter for flexural test is 185-degree printing temperature, 30mm/s printing speed and 30/60-degree raster angle.
3) The most significant printing parameter is raster angle for both tensile and three-point bending test.
4) The printing temperature did not give great influence in the printing parameter and mechanical properties.
5) Other factor for instance air gap, humidity, printing orientation, printing head size, different type of support structure, number of perimeter and another parameter must be conducted. So that the information and the result from the testing can be simplified in order to find the standard parameter of 3D printing and desirable quality, lower cost, high mechanical properties in the 3D printing.

Acknowledgement
The authors would like to express profound gratitude and deep appreciation to the School of Mechatronics Engineering, School of Manufacturing Engineering and Universiti Malaysia Perlis in providing researching platform well as providing the facilities and equipment.

References
[1] Favi M M C and Germani M 2016 Design for manufacturing and assembly vs. design to cost: toward a multi-objective approach for decision-making strategies during conceptual design of complex products *Procedia CIRP* **50** 275–280.
[2] Boothroyd G 1994 Product design for manufacture and assembly *Computer-Aided Design* **26** (7) 505–520.
[3] Marini D and Corney J 2017 A methodology for assessing the feasibility of producing components by flow forming *Prod. Manuf. Res.* **5** (1) 210–234.
[4] Steven Kalpakjian K s V S and Steven R S 2003 *Manufacturing processes for engineering materials*, Seven edit. Upper Saddle River: N.J: Prentice Hall.
[5] Gomez-gras C, Jerez-mesa R, Travieso-rodriguez J A and Lluma-fuentes J 2018 Fatigue performance of fused fi lament fabrication PLA specimens *Mater. Des.* **140** 278–285, 2018.
[6] Boenzi N *Chapter 9 : Mechanical Failure* **30**.
[7] Fadell T *The 3D Printing Handbook*
[8] Matweb 2017 *Overview of materials for Polylactic Acid (PLA) Biopolymer* 1.
[9] Polymer C, Polymer A B S, Butadiene A, Abs S and Grade P 2011 Overview of materials for Acrylonitrile Butadiene Styrene (ABS) Plating Grade 10–12.
[10] Manner K J 2016 *Engineering Design project* college of Engineering, University of wisconsin madison.
[11] Harpool T D *Observing the effects of infill shape on tensile characteristic of 3D printing part* Wichita State University.
[12] Agarwala M K, Jamalabad V R, Langrana N A, Safari A and Whalen P J 1996 Structural quality of parts processed by fused deposition *Rapid Prototyp. J.* 4–19.
[13] Es-Said B A, Foyos O S, Noorani J, Mendelson R M, Marloth R and Pregger *Effect of layer orientation on mechanical properties of rapid prototyped samples*. 2002.

[14] Hernandez R, Slaughter D, Whaley D, Tate J and Asiabanpour B 2016 Analyzing the tensile, compressive, and flexural properties of 3D printed ABS p430 plastic based on printing orientation using fused deposition modeling 939–950.
[15] Ahn S -H, Montero M, Odell D, Round S and Wright P K 2002 Anisotropic material properties of fused deposition modeling ABS *Rapid Prototyp. J.* **8** (4) 248-257.
[16] Escola J A R, Santana A, Afonso P, Zanin A and Wernke R 2017 Fatigue lifespan study of PLA Conference parts obtained by additive manufacturing Costing models optimization in Industry Puig operational efficiency between used capacity Fatigue lifespan study of PLA parts obtained by additiv *Procedia Manuf.* **13** 872–879.
[17] Aaron M Materials testing standards for additive manufacturing of polymer materials.
[18] Salleh A F et al. 2007 Preliminary study on the influence of slice thickness and support parameters on part defects of FDM prototype 25–26.

IOP Conf. Series: Materials Science and Engineering **670** (2019) 012067 doi:10.1088/1757-899X/670/1/012067

Effect of surface treatment on drag coefficient of free-falling solid sphere in water

***M Sofwan Mohamad[1,2], C M Mackenzie Dover[1] and K Sefiane[1]**

[1]Institute for Multiscale Thermofluids, School of Engineering, University of Edinburgh, Faraday Building, King's Buildings, Edinburgh United Kingdom. EH9 3DW
[2]School of Mechatronic Engineering, Universiti Malaysia Perlis (UniMAP), Kampus Pauh Putra, 02600 Arau, Perlis Malaysia.

*corresponding author: Muhammad.Mohamad@ed.ac.uk

Abstract. The study aims to examine the influence of different surface treatments on the drag coefficient of free-falling spheres in water. The spheres used in the experiment are classified into four categories and labelled as: ref – no surface treatment, FDTS – coated with hydrophobic perfluorodecyltrichlorosilane (FDTS) coating, E8 – 8 minutes etched in ferric chloride ($FeCl_3$) and E32 – 32 minutes etched in $FeCl_3$. No significant difference is observed in drag coefficient of the etched spheres compare with the reference sphere. However, the FDTS sphere's drag coefficient was increased by 13%.

1. Introduction

There has been recent, significant progress in the design as well as fabrication of surface treatments that can potentially reduce skin friction drag [1], [2]. For example, the modification of surface chemistry by the addition of a hydrophobic coating. By definition, a hydrophobic surface is low-energy and characterized by a water contact angle (WCA) higher than 90° [3]. A surface can be easily coated with a hydrophobic material such as perfluorodecyltrichlorosilane (FDTS) to reduce the substrate's energy in order to engineer a water-repellent interface [4], [5]. Surface topography (and roughness) can also have a significant effect on the hydrophobicity of the surface [6]. Moreover, a superhydrophobic surface (WCA higher than 150°) can be achieved by combining these two modifications (surface chemistry and surface roughness) appropriately [7]. Furthermore, the ability of a hydrophobic surface to sustain an air layer on the surface when immersed in water has attracted substantial research interest because it can promote a slip boundary condition at the interface which can lead to drag reduction [8]–[11]. These could have benefit to various applications such as enhanced liquid mobility at interfaces presented by ships [12] , submarines [13] and piping systems [14].

The focus of the present study is to use free-falling spheres to investigate the effect of several surface treatment techniques on the motility through a fluidic environment. This is achieved by measurement of the velocity of the falling sphere in water. The measured velocity is then used to calculate the drag coefficient of the sphere.

IOP Conf. Series: Materials Science and Engineering **670** (2019) 012067 doi:10.1088/1757-899X/670/1/012067

2. Methodology

An experimental setup was designed to capture the behaviour of a free-falling sphere in water (Figure 1). The setup comprises a 1000 mm high square cross section, straight-walled column with inner side width of 70.45 mm made of clear Perspex. These dimensions were selected such that a falling sphere with a diameter of less than 7 mm could reach terminal velocity with negligible wall effects and end effects [15], [16]. The motion of the falling sphere was recorded by a high speed camera which was mounted on vertical and horizontal sliders to give the camera 2 degrees of freedom. A custom made high power LED light was used to illuminate the test section. An electromagnetic release mechanism was integrated with the top lid. The mechanism was designed such that the sphere's motion is induced by the effect of gravity only, at the centre of the column with zero initial velocity directly above the water's surface. The bottom part of the column was attached to a retrieval mechanism which allowed the sphere to be removed from the column easily. The temperature of the water was controlled by an electric heater and a proportional-integral-derivative (PID) controller connected to two thermocouple probes. However, all experiments were conducted at room temperature for the present paper. The spheres used in the experiments are commercially available stainless steel spheres with diameters of 7 mm. The density of these spheres is 7750 kg/m^3. The captured video was processed using an in-house ImageJ macro to obtain the relative position and velocity of the sphere.

The experimental samples were prepared by coating the sphere with hydrophobic FDTS coating (labelled as FDTS), and roughening the sphere's surface via an etching process. The coating was applied over the sphere by means of molecular vapour deposition (MVD). The implementation of the MVD coating technique resulted in a mono-layer coating on the surface. Thus, the molecularly thin coating (in the order of nanometers) layer has no significant effect on the diameter of the sphere and it can still be considered to be 7 mm. The etching process was done by simply immersing the spheres in $FeCl_3$ for two different interval times; 8 minutes and 32 minutes (labelled as E8 and E32, respectively).

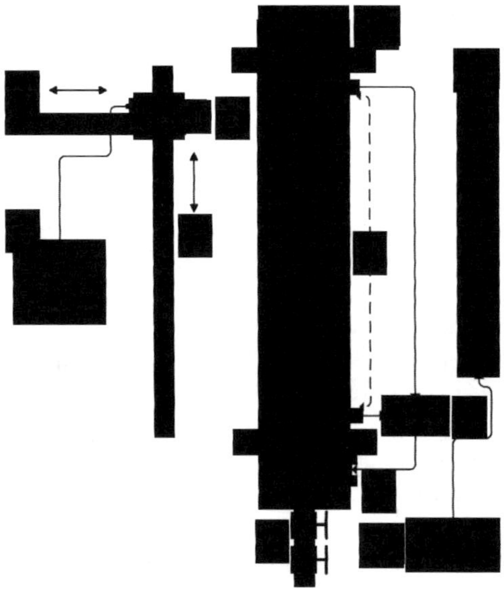

Figure 1. The schematic diagram of experimental set-up.

IOP Conf. Series: Materials Science and Engineering **670** (2019) 012067 doi:10.1088/1757-899X/670/1/012067

3. Results

3.1. Velocity

The spheres described in Section 2 were tested in free-fall in water. The evolution of the velocity of the spheres are inferred from the time derivative of the falling trajectory and this is depicted in Figure 2(a). Each curve illustrates the ensemble average of the instantaneous velocity over 3 trials. Error bars are not displayed in the figure for the sake of clarity. Surprisingly, instead of the skin-friction reducing effect of hydrophobic surface, it can be seen that the FDTS sphere falls the most slowly. On the other hand, no significant difference in velocity is observed between the etched spheres and the reference sphere. The velocity of the spheres reaches a plateau value at a time of about 0.25s, indicating a steady state condition. The terminal velocity for each sphere is calculated by averaging the instantaneous velocity in the steady state region. In the case of the FDTS sphere, due to the limited experimental data points, an empirical exponential model proposed by Mordant & Pinton [17] was used to estimate its terminal velocity (dashed black line in Figure 2(a)):

$$v(t) = v_T(1 - e^{-t/\tau}) \tag{1}$$

where v_T (terminal velocity) and τ (characteristic time) in the equation are independent fitting parameters. Indeed, the FDTS sphere reached steady state condition within the field-of-view of the camera. The average terminal velocity of those spheres in the case of 7 mm diameter for 3 repeated experiments are shown in Figure 2(b). The error bars in the figure represent one standard deviation of the measured mean values.

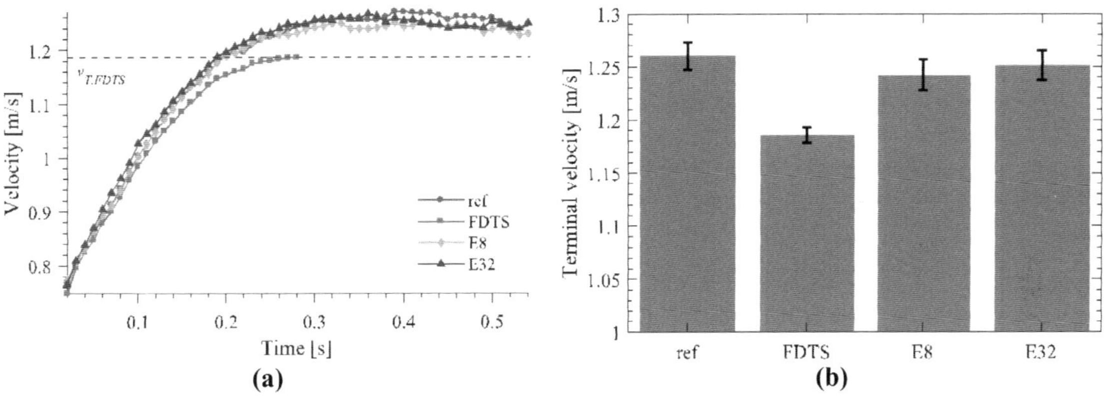

Figure 2. The velocity of the free-falling spheres for all different surface treatment: (a) at transient state and (b) at steady state.

3.2. Drag coefficient

Knowing the terminal velocity, the steady-state drag coefficient for a sphere can be determined by the following equation:

$$C_D = \frac{4}{3}\frac{(\rho_s - \rho_f)gd}{\rho_s v_T{}^2} \tag{2}$$

where ρ_s is the sphere density, ρ_f is the surrounding fluid (water) density, g is gravitational acceleration and v_T is the terminal velocity of the free-falling sphere. Setting the unmodified sphere as the reference, the effect of the surface treatment on drag coefficient can be estimated from $\Delta C_D = 1 - \left(C_D / C_D^{ref}\right)$ (where superscript ref is referring to unmodified sphere) and depicted in Figure 3 as a function of Reynolds number. It can be seen that the drag coefficient of the etched spheres are identical with the reference sphere with no significant differences (less than 3%). In contrast, the FDTS sphere's drag coefficient is higher than the reference sphere by 13%.

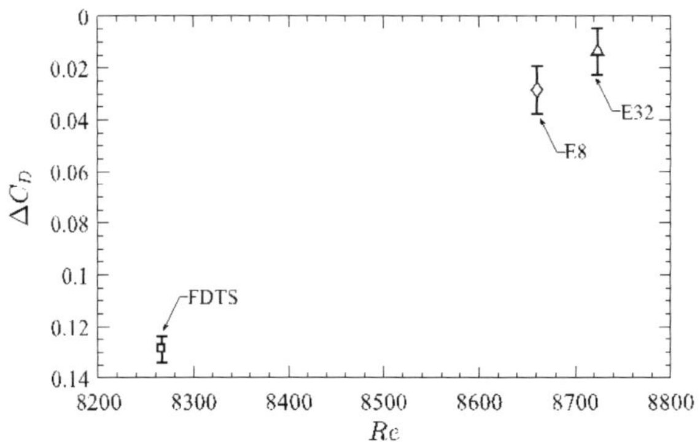

Figure 3. The ratio of steady-state drag coefficient of spheres with various surface treatments to the reference sphere.

4. Conclusion

An experimental study is conducted on the effect of surface treatment on the drag coefficient of free-falling spheres in water. The surface of the spheres were modified by two methods; coating the spheres with FDTS and roughening the surface of the spheres by a wet-etching process. The drag coefficient of the etched spheres were almost identical to the unmodified spheres. Surprisingly, the drag coefficient of the FDTS sphere was 13% higher than that of the reference sphere. There are extant experimental investigations previously reported that are discordant in their conclusions about whether an increase in surface hydrophobicity reduces [18] or enhances [19] drag coefficient. Interestingly, both researches relate the reduction and enhancement to the retaining air on the surface of the sphere. In their analytical study, McHale *et al.* suggest that the air layer must be at an optimum thickness so that the drag reduction effect due to the slip condition outweigh the increase in buoyancy due to presence of air layer on the surface [11]. Therefore, higher-resolution images need to be captured in order to evaluate the thickness of the air layer retained on the sphere's surface and further detailed analysis need to be done to understand the underlying physical mechanism of the drag enhancement in the present investigation.

Acknowledgement

The author acknowledge the financial support from Kementerian Pendidikan Malaysia and Universiti Malaysia Perlis for the sponsorship which enable me to pursue this study. MEMSstar Ltd. is gratefully acknowledged for the coating. The author would like to thank Mr. Alan Ross at the Scottish Microelectronics Centre (SMC) for the help with sample preparation.

References

[1] Luo Y, Wang J, Sun G, Li X and Liu Y 2016 Experimental investigations on manufacturing different-shaped bio-inspired drag-reducing morphologies and hydrodynamic testing *Exp. Tech.* **40** (3) 1129–1136.

[2] Abdulbari H A, Mahammed H D and Hassan Z B Y 2015 Bio-Inspired Passive Drag Reduction Techniques: A Review *ChemBioEng Rev.* **3** 185–203.

[3] Law K Y 2014 Definitions for hydrophilicity, hydrophobicity, and superhydrophobicity: Getting the basics right *J. Phys. Chem. Lett.* **5** (4) 686–688.
[4] Kobrin B et al. 2006 An improved chemical resistance and mechanical durability of hydrophobic FDTS coatings *J. Phys. Conf. Ser.* **34** (1) 454–457.
[5] Gnanappa A K et al. 2008 Effect of annealing on improved hydrophobicity of vapor phase deposited self-assembled monolayers *J. Phys. Chem. C* **112** (38) 14934–14942.
[6] Öner D and McCarthy T J 2000 Ultrahydrophobic surfaces. Effects of topography length scales on wettability *Langmuir* **16** (20) 7777–7782.
[7] Barati Darband G, Aliofkhazraei M, Khorsand S, Sokhanvar S and Kaboli A 2018 Science and engineering of superhydrophobic surfaces: review of corrosion resistance, chemical and mechanical stability *Arab. J. Chem.*
[8] Ou J, Perot B and Rothstein J P 2004 Laminar drag reduction in microchannels using ultrahydrophobic surfaces *Phys. Fluids* **16** (12) 4635–4643.
[9] Rothstein J P 2010 Slip on Superhydrophobic surfaces *Annu. Rev. Fluid Mech.* **42** (1) 89–109.
[10] Panchanathan D, Rajappan A, Varanasi K K and McKinley G H 2018 Plastron regeneration on submerged superhydrophobic surfaces using in situ gas generation by chemical reaction *ACS Appl. Mater. Interfaces* **10** (39) 33684–33692.
[11] McHale G, Flynn M R and Newton M I 2011 Plastron induced drag reduction and increased slip on a superhydrophobic sphere *Soft Matter* **7** (21) 10100–10107.
[12] Dong H, Cheng M, Zhang Y, Wei H and Shi F 2013 Extraordinary drag-reducing effect of a superhydrophobic coating on a macroscopic model ship at high speed *J. Mater. Chem. A* **1** (19) 5886–5891.
[13] Zhang S et al. 2015 Underwater drag-reducing effect of superhydrophobic submarine model *Langmuir* **31** (1) 587–593.
[14] Karami H R, Keyhani M and Mowla D 2016 Experimental analysis of drag reduction in the pipelines with response surface methodology *J. Pet. Sci. Eng.* **138** 104–112.
[15] Uhlherr P and Chhabra R 1995 Wall effect for the fall of spheres in cylindrical tubes at high Reynolds number *Can. J. Chem.* **73**.
[16] Brown P P and Lawler D F 2003 Sphere Drag and Settling Velocity Revisited *J. Environ. Eng.*, **129** (3) 222–231.
[17] Mordant N and Pinton J F 2000 Velocity measurement of a settling sphere *Eur. Phys. J. B* **18** (2) 343–352.
[18] McHale G, Shirtcliffe N J, Evans C R and Newton M I 2009 Terminal velocity and drag reduction measurements on superhydrophobic spheres *Appl. Phys. Lett.* **94** (6) 1–6.
[19] Su B, Li M and Lu Q 2010 Toward understanding whether superhydrophobic surfaces can really decrease fluidic friction drag *Langmuir* **26** (8) 6048–6052.

Mechanical properties of commercial PLA filament as 3D printed parts utilizing fused filament fabrication

H Zakaria[1], S F Khan[1,2], S N Sabtu[1] and M Ibrahim[3]

[1] School of Mechatronic Engineering, Universiti Malaysia Perlis, Malaysia.
[2]Additive Manufacturing and Design Lab, School of Mechatronic Engineering, Universiti Malaysia Perlis, Malaysia.
[3]School of Manufacturing Engineering, Universiti Malaysia Perlis, Malaysia
coresponding author, hafizawati@unimap.edu.my

Abstract. In the field of Additive Manufacturing, Fused Filament Fabrication (FFF) is one of the methods used for prototyping and production applications. However, the performance of the printed parts produced can be affected by the various parameters used. The purpose of this research is to evaluate the bonding between layers of 3D printed parts from 3D printer by using Polylactic Acid (PLA) with printing parameters, namely printing temperature, printing speed and raster angle. The specimens were printed by using Monoprice Maker Select 3D printer. Compression test and fatigue test were carried out to determine the behavior of true stress-strain and the lifespan of sample subjected to cycling loading. The influence of printing temperature, printing speed and raster angle is determined by using Analysis of Variance (ANOVA). The results show that printing temperature affect the bonding between layers of samples more than the printing speed and raster angle.

1. Introduction

Manufacturing has dependably been tied in with handling crude material into segments or parts that can be collected into utilitarian products. Different inquires about have been done to additionally enhance the numerous parts of ordinary assembling, for example, materials, feasibility, process optimization, and sustainability [1]-[4]. Fused Filament Fabrication (FFF) is an additive manufacturing which combined individual layer of extruded thermoplastic filament into a 3D geometry form [5]-[8]. It is a complex process with a large number of processing parameters. The combination of these parameters which influence product quality and material properties is difficult to understand [9, 10]

Polylactic Acid (PLA) will be utilized for the technique that specifically manage a thermoplastic polymer through spout since Polylactic acid is other printable material besides has mechanical properties altogether greater than utmost different plastics [11], [12] and it is appropriate eco-accommodating choices of effectively existing polymers and in addition growing amazing failure rate polymers which have aggressive execution things from feed stocks that are accessible from inexhaustible stock properties [13]. Generally, the bonding molten polymer is controlled by a diffusion progression which is reliant on the temperature also contact pressure in the bonding zone, as well as time allowed for diffusion to take place. Given that bonding in the FFF method happens on ambient pressure, it acts which printing is major influence evolving bond strength. On the other hand, temperature difference too shows a foremost character. Therefore, this study on printing parameter in example printing temperature, printing speed and raster angle is conducted to investigate their effect on the mechanical properties of the printed parts."

IOP Conf. Series: Materials Science and Engineering **670** (2019) 012068 doi:10.1088/1757-899X/670/1/012068

2. Methodology

The material that used is polylactic acid (PLAPolylactic acid is a thermoplastic, which means can be transform into molten in its melting point of 150°C to 160°C. Monoprice Maker Select 3D printer is used to print out the samples which are 18 samples of Model 1 and 54 samples of Model 2. Figure 1 shows the sample of Model 1 and Figure 2 shows the sample of Model 2. Before printing, CATIA v5 software is used to draw the models of the sample and converted into STL files.

Model 1 is for compression test with diameter of 1 mm and height of 2 mm. The purpose of compression test is to determine the true stress-strain. Model 2 is for fatigue test with total length of 115 mm and 0.02 mm of thickness. The purpose of fatigue test is to determine the total number of cycles that sample can be subjected to under a single loading scheme.

Figure 1: Specimen for compression test

Figure 2: Specimen for fatigue test .

3. Result and discussion

3.1. Compression Strength

A force of 10 kN and speed of 1.5 mm/s are applied and data obtained from the test is the data of force (kN) and stroke (mm). Figure 3 shows the deformation sample of Model 1. From the data obtained, stress and strain are calculated. Figure 4 shows the graph of stress versus strain.

Figure 3: Specimen deformation after compression test

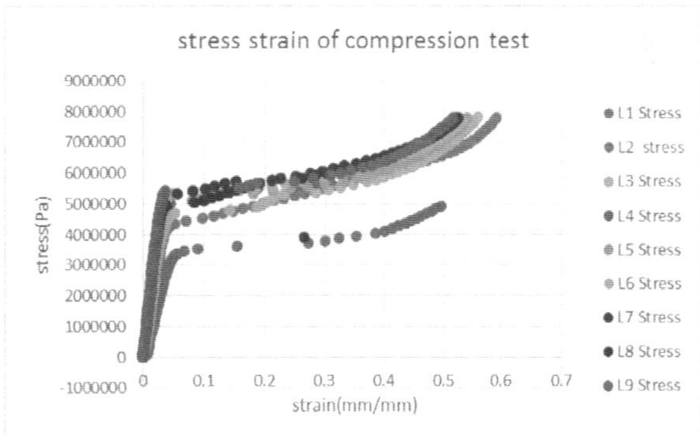

Figure 4: Graph of stress vs strain for compression test .

S/N ratio is calculated to maximize the response, so the larger is better. Figure 5 shows the graph mean of S/N ratio obtained. Table 1 shows the response table of printing parameters.

Table 1. Response Table of Printing Parameters

IOP Conf. Series: Materials Science and Engineering **670** (2019) 012068 doi:10.1088/1757-899X/670/1/012068

Level	Temperature A (dB)	Printing Speed B (dB)	Printing Angle C (dB)
1	130.6	132.2	132.2
2	132.6	131.4	131.6
3	133.1	132.7	132.5

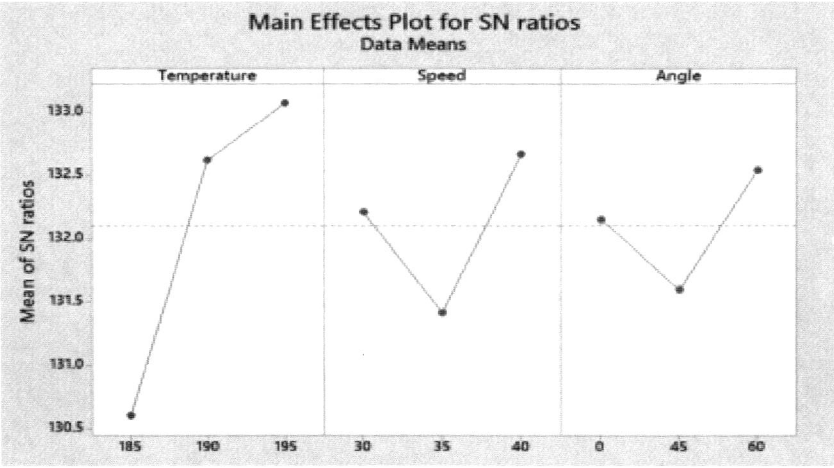

Figure 5: Graph mean of S/N ratio vs level of each printing parameters

3.2. Fatigue strength

A 50% from value of UTS is applied to conduct the fatigue test. Figure 6 shows the graph mean of S/N ratio and Table 2 shows the response table of printing parameters.

Table 2. Response table of printing parameters

Level	Temperature A (dB)	Printing Speed B (dB)	Printing Angle C (dB)
1	88.10	99.20	95.06
2	88.00	103.25	97.45
3	109.69	83.34	92.38

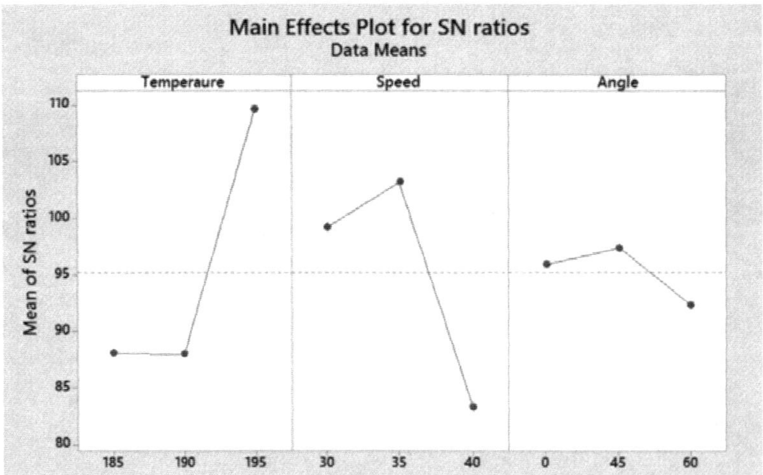

Figure 6: Graph mean of S/N ratio vs level of each printing parameters

3.3. Analysis of Variances (ANOVA)

Analysis of variance (ANOVA) is a collection of statistical models and associated estimation procedures used to analyze the differences among group means in a sample. The values are obtained based on the S/N ratio. Table 3 and 4 show the ANOVA table analysis form compression test and fatigue test.

Table 3. ANOVA for compression test

Printing Parameter	Factor	DOF f_n	Sum of Square S_n	Variance V_n	F-value	P-value	Contribution %
Printing Temperature	A	2	10.4070	5.2035	11.63	0.079	69.02
Printing Speed	B	2	2.4069	1.2035	2.69	0.271	15.96
Raster Angle	C	2	1.3686	0.6843	1.53	0.395	9.08
Error	e	2	0.8948	0.4474			5.93
Total		8	15.0774				100

Table 4. ANOVA for fatigue test

Printing Parameter	Factor	DOF f_n	Sum of Square S_n	Variance V_n	F-value	P-value	Contribution %
Printing Temperature	A	2	936.72	468.36	2.20	0.313	45.30
Printing Speed	B	2	664.51	332.25	1.56	0.391	32.13
Raster Angle	C	2	40.71	20.36	0.10	0.913	1.97
Error	e	2	426.09	213.05			20.60
Total		8	2068.03				100

4. Conclusion

In this research, the effect of printing temperature, printing speed and raster angle on mechanical properties of compression test and fatigue test are examined for PLA. These are conclusions from this study:

i. All the experiment results showed that the printing temperature, printing speed and raster angle have a marked effect on compression test and fatigue test.

ii. Compression and fatigue test revealed that the printing temperature is the most effect on the samples compared to printing speed and raster angle.

iii. The significant parameters for compression test 195˚C of printing parameter, 40 mm/s of printing speed and 60° of raster angle and for fatigue test is 195 ˚C of printing temperature, 35 mm/s of printing speed and 45° of raster angle.

Acknowledgments

The authors would like to express profound gratitude and deep appreciation to the School of Mechatronics Engineering, School of Manufacturing Engineering and Universiti Malaysia Perlis in providing researching platform well as providing the facilities and equipment.

References

[1] Marini D and Corney J 2017 A methodology for assessing the feasibility of producing components by flow forming *Production and Manufacturing Research* **5** (1) 210–234.

[2] Ye X, Liu H, Ding Y, Li H and Lu B 2009 Research on the cast molding process for high quality PDMS molds *Microelectronic Engineering* **86** (3) 310–313.

[3] Favi C, Germani M and Mandolini M 2016 Design for manufacturing and assembly vs. design to cost: toward a multi-objective approach for decision-making strategies during conceptual design of complex products *Procedia CIRP* **50** 275–280.

IOP Conf. Series: Materials Science and Engineering **670** (2019) 012068 doi:10.1088/1757-899X/670/1/012068

[4] Boothroyd G 1994 Product design for manufacture and assembly *Computer-Aided Design* **26** (7) 505–520.
[5] Godec D, Spoerk M, Kukla C and Holzer C 2018 3D printing conditions determination for feedstock used in fused filament fabrication (FFF) of 17-4Ph **57** 117–120.
[6] Sood A K, Ohdar R K and Mahapatra S S 2010 Parametric appraisal of mechanical property of fused deposition modelling processed parts *Mater. Des.* **31** 287–295.
[7] Ahn S H, Lee C S, Kim S G and Kim H J 2007 Measurement of anisotropic compressive strength of rapid prototyping parts *J. Mater. Process. Technol.* **187–188** 627–630.
[8] Gajdos I and Slota J 2013 Influence of printing conditions on structure in FDM prototypes *Tehnicki Vjesnik.* **20** 231–236.
[9] Casavola C, Cazzato A, Moramarco V and Pappalettere C 2016 Orthopedic mechanical properties of fused deposition modelling parts described by classical laminate theory *Mater. Des* **90** 453-458.
[10] Mohamed O A, Masood S H and Bhowmik J L 2015 Optimization of fused deposition modelling process parameters: a review of current research and future prospects *Advances in Manufacturing* **3** 42-53.
[11] Tymrak B M, Kreiger M and Pearce J M 2014 Mechanical properties of components fabricated with open-source 3-D printers under realistic environmental conditions *Mater. Des* **58** 242–246.
[12] Lanzotti A, Grasso M, Staiano G and Martorelli M 2015 The impact of process parameters on mechanical properties of parts fabricated in PLA with an open-source 3-D printer *Rapid Prototyp. J.* **21** 604–617.
[13] Hillmyer M A and Tolman W B 2014 Aliphatic Polyester Block Polymers: Renewable, Degradable, and Sustainable *Acc. Chem. Res* **47** 2390–2396.

IOP Conf. Series: Materials Science and Engineering **670** (2019) 012069 doi:10.1088/1757-899X/670/1/012069

Pet food autofeeder by using Arduino

***M Ibrahim[1] , H Zakaria[2] and EE Wei Xian[1]**

[1]School of Manufacturing Engineering, Universiti Malaysia Perlis, Malaysia.
[2]School of Mechatronic Engineering, Universiti Malaysia Perlis, Malaysia.
[3]Design for Sustainable Research Group, School of Manufacturing Engineering, Universiti Malaysia Perlis, Malaysia.

***coresponding author: maliki@unimap.edu.my**

Abstract. Automated pet feeder is a product that would replace manual feeding method which can be set at a required feeding amount and feeding time. Innovation of pet feeder is used to overcome forgetful of the pet owner to feed their pet and to avoid extra spending expenses by leaving them at a pet hotel. The objective of this project is to design and develop a mechanism for pet feeder. In this project, Arduino is used as the microcontroller to control the mechanism and to operate the system with a given of set time. Finite element analysis is used to analyze the maximum stress that the designed mechanism can withstand. Furthermore prototype testing on the mechanism with experimental and theoretical was conducted, the result of the analysis is analyzed. The analysis is providing the path to make sure the final concept of the pet feeder is reached.

1. Introduction

Pet is good accompany partner to human. It helps to reduce the stress from living life style and as a precaution to keep away burglars [1]. Nowadays, pet treats as one part of the family member. Therefore, healthy level of the pet need to consider and focus on balance diet. Different size of the pet will have different amount on consuming and different nutrient needed. From the survey, the obesity problem among LI the pet is happened due to pet owner is busy with work [2]. In additional to that, pet doesn't have ability to get the food itself, and it has to depend on the pet owner to feed them. Therefore, Automated Pet Feeder is created and used to solve the problems feeding the pet on time with required portion.

Automated pet feeder is started popular among the choices of pet owner used to replace the traditional feeding method. The product brings convenience to pet owner by setting the feeding time and portion of the food, which it helps to make sure the pet receives the required diet through a day [3]. The automated pet feeder trains the pet how to consume food properly and consume at certain time. The result of using pet feeder is the pet will receive their diet and pet owner can be no worry about feeding their pet [4].

From the survey of Association for Pet Obesity Prevention, there are 56% of dogs and 60% of cats are classified as the obesity by their veterinary healthcare professional [2]. The obesity problems happened due to the pet owners is busy with their working and lack of time to bring their pet go for exercise. Not only lack of time bring LI pet to exercise, pet owner worry forgetting to feed their pet, they will pour the big amount of the pet food on the bowl which causes overfeeding their pet [5]. To overcome the obesity and starve problem, the design on automated pet feeder is produced and bring the convenience to the pet owners.

The study on mechanism that used to pour out the pet food out from container has divided into motion, there are linear motion and rotational to linear motion [6]. The rotational motion of mechanism

placed linear under container, it moves pet food out by using the rotational motion [7]. From the researched, the advantages of using rotational to liner motion can be avoiding the pet food be blocked in the hopper [8].
Objective of doing this research is to design and develop a mechanism for automated pet feeder which able to give appropriate portion of pet food on a given set time.

2 Methodology
Market survey is the initial method used to collect the data for design specification. Three method on collecting data are online survey, interview session and questionnaire. After collecting the customer needs, table of screening and scoring will use to finalize the specification that needed on final design. Market survey on existing product is being analyse based on the product specification.

Final concept design is to designed and detailed drawing of each component is generated by using Solid Works. Finite element analysis is used to analyze the mechanism of the pet feeder. The mechanism is applied 40N force and the analysis that carried out are total deformation analysis, equivalent elastic strain, minimum principal stress, maximum principal stress and maximum shear stress.

Prototype is fabricated and undergoes the experiment testing on the mechanism to ensure that the feeding process run smoothly without any obstruction involving mechanism of pet food and designated Arduino program for the prototype.

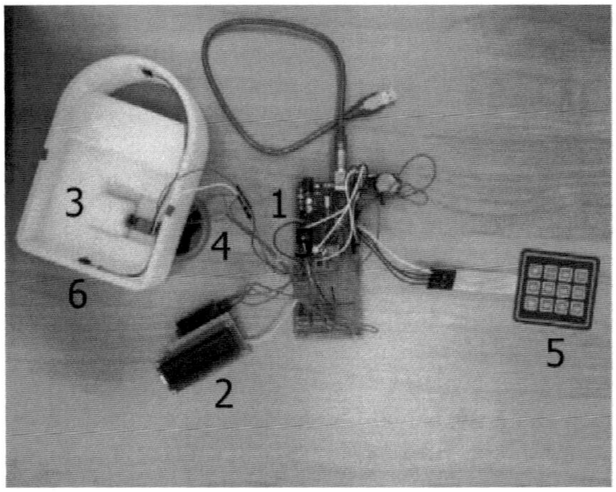

Figure 1. Arduino system of Automated Pet Feeder.

Figure 1 shown the pet feeder Arduino system. The component is Real Time Clock(1), Liquid Crystal Display(2), Dc Motor(3), Buzzer(4), 4x3 Keypad(5) and a limit switch(6). Arduino is being used as the microcontroller in which it can adjust the feeding time and number of portions.

IOP Conf. Series: Materials Science and Engineering **670** (2019) 012069 doi:10.1088/1757-899X/670/1/012069

3 Result and Discussion

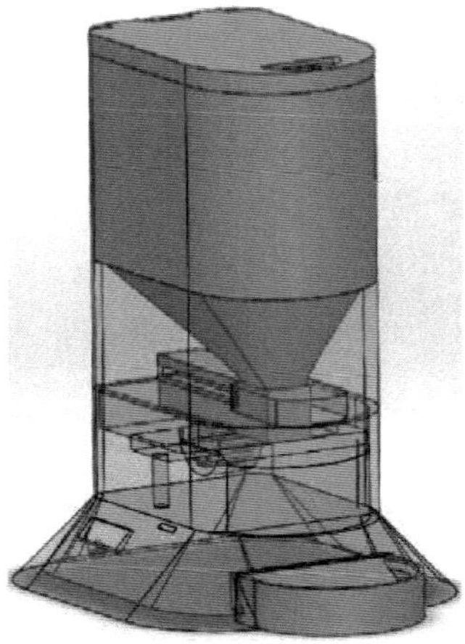

Figure 2. Final design of Pet Feeder.

The design of the pet feeder as shown in Figure 2. The mechanism that used in the pet feeder is rotational to linear motion. The mechanism is installed in between the base part of the component.

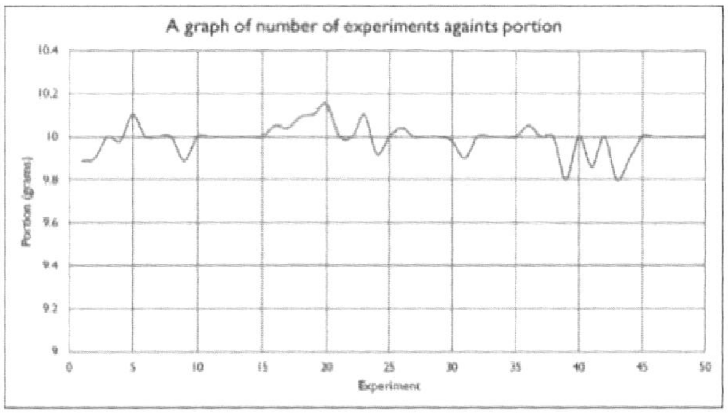

Figure 3. Result for number of experiment against portion.

From the prototype testing, the feed size for one period is 10 grams, due to the prototype is scale down to 1:5. In real scale of prototype, it almost will be 20 grams of the amount. Figure 3 shows the consistency of the experimental feed size for each periods. It is fall between the range 9.8 and 10.1 g, it considered as constant feed size.

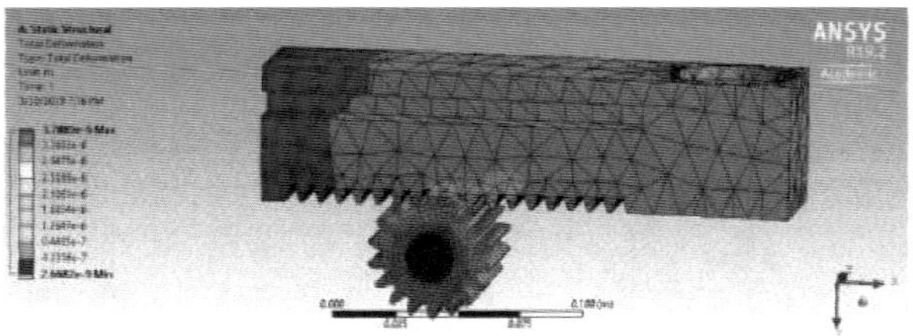

Figure 4. Result of mechanism analysis.

From Figure 4, ANSYS is used to analyze the maximum stress can be withstanding at the critical point. The analysis that applied is static structural, whereas rectangular rack is the moving object while the spur gear is fixed at the base part of pet feeder.nozzle diameter and 0.1mm layer thickness. While the lowest (level8) have 60-degree orientation angle, 0.3mm nozzle diameter and 0.1mm layer thickness.

Table 1. Summary result of mechanism analysis.

	Total deformation (m)	Von-Mises Stress (Pa)	Maximum Shear Stress (Pa)	Maximum Principal Stress (Pa)	Minimum Principal Stress (Pa)
PET	1.6768 e-006	34024	19272	18111	- 20433
HDPE	3.7242 e-006	34034	19293	18130	- 20457

From Table 1, it shows that HDPE has higher probability that can withstand higher force applied on the mechanism. Total deformation of HDPE can be reach 3.7242 e-006 which higher than PET. The total deformation is included the elastic deformation and plastic deformation. Among the stress analysis that going through, HDPE shown the higher value that can withstand more stress and more force to the mechanism of pet feeder. The Von-Mises stress of HDPE is higher than PET, it known that the theoretical measure of stress used to estimate yield failure criteria in the material. While the principal stress is directly measurable stress by adding the load exceeds the principal stress which the failure occurs in one dimensional tensile test.

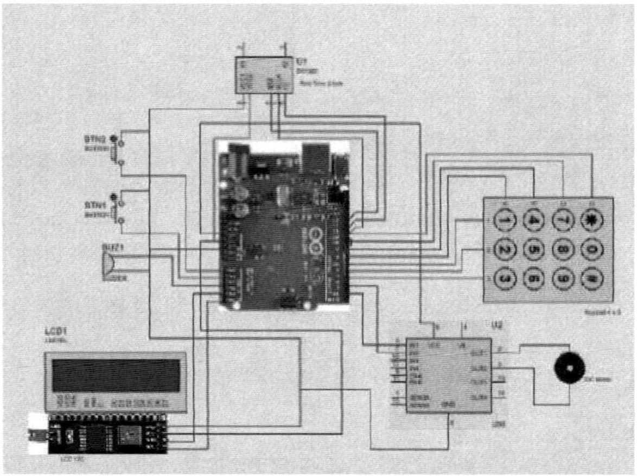

Figure 5. Schematic Diagram of Automated Pet feeder

The system that design is the automated pet feeder able to adjust three main criteria. By pressing '*', the user is able to adjust the real time based on the time when they start to use. Secondly, by pressing '1' and '2', the user is able to set the first and second feeding time respectively. Lastly, by pressing '3', the user able to adjust the feed size up to 9 periods. One period means that the rectangular rack with measuring cup moved forward and backward. Forward action means that move the pet food drop to pet bowl, the backward action of rectangular rack also known as the original position of the rectangular rack which is the filled the measuring cup with the pet food.

IOP Conf. Series: Materials Science and Engineering **670** (2019) 012069 doi:10.1088/1757-899X/670/1/012069

4. Conclusion

Overall, the objective of automated pet feeder by using the Arduino is achieved. The automated pet feeder can be adjusted the feeding time and the motor can rotate according to the required time. The material that used to fabricate the part of automated pet feeder is HDPE. The Manufacturing process that will be used is extrusion molding process. Besides that, the mechanism analysis is analyzing by using ANSYS. It has mention the critical part of the mechanism and it shows the maximum stress that can be applied on the mechanism. Recommendation for improvement on the design of the pet food container is adding the rotary bladder to rotate the pet food when the mechanism is moved. It is to reduce the probability caused the pet food stuck at the end of the container due to the compact of the pet food.

5. Acknowledgement

The authors would like to express profound gratitude and deep appreciation to the School of Manufacturing Engineering, School of Mechatronics Engineering and Universiti Malaysia Perlis in providing research platform as well as providing the facilities and equipment.

References

[1] McConnell, Allen R, Brown, Christina M, Shoda, Tonya M, Stayton, Laura E, Martin and Colleen E 2011 Friends With Benefits: On the Positive Consequences of Pet Ownership *Journal of Personality and Social Psychology* **101** (6) 1239-1252.
[2] Shiham K 2017 *Pet obesity survey result.*
[3] Pedigree 2016 *Dog Feeding Guidelines* 70 201 206
[4] James F 2017 *Automated Pet Feeder* **63** 21-24
[5] Jenks S 2018 *Our Fat Pets* The New York Times 9-10
[6] Layosa C 2018 *How to Convert from Rotary to Linear Motion*
[7] Robert L Pitchford Jr., 1418 W, & 134th St., Compton C 90222 (1985). Timed automatic pet feed and water dispenser.
[8] Hernandez R, Slaughter D, Whaley D, Tate J and Asiabanpour B 2016 Analyzing the Tensile, Compressive, and Flexural Properties of 3D Printed ABS P430 Plastic Based on Printing Orientation Using Fused Deposition Modeling 939–950.

IOP Conf. Series: Materials Science and Engineering **670** (2019) 012070 doi:10.1088/1757-899X/670/1/012070

Safety and risk evaluation implementation at sheet metal stamping company using HIRARC model

N L A Othman[1], N S Hussain, D Ibrahim[1] and Y Yaacob[1]

[1]Faculty of Mechanical Engineering, Universiti Teknologi MARA, Cawangan Pulau Pinang, Jalan Permatang Pauh, 13500 Pulau Pinang, Malaysia.

*corresponding author: norliawati086@uitm.edu.my

Abstract. Sheet metal stamping has been identified as one of the most hazardous and high-risk process in the world of manufacturing industries. This project was carried out at a sheet metal stamping company located in Perak which is a precision metal stamping company and is focused on 150 tons and 200 tons stamping only. A systematic analysis based on Occupational Safety and Health (OSHA) standard was used to create a model of Hazard Identification, Risk Assessment and Control (HIRARC) at the stamping department to determine hazardous process/areas for department operation and maintenance of observation, safety analysis, workplace inspection, safety checklist and high-risk investigation. Initial risk assessment was conducted by classifying the hazards in three types of level, such as High, Medium and Low. The severity of hazards was calculated to estimate the likelihood it will happen and to decide which process/area needs to be applied control measures implemented in the departments. The risk assessments were repeated, and the results were analyzed. From the analysis of risk assessment shows that a reduction of 17% from the high-risk level of 150 tons and 27% at 200 tons. Safety and health are very important to prevent the occurrence of accidents.

1. Introduction

Manufacturing industry has been identified as one of the most hazardous industries in many parts of the world manufacturing [1]. Industrial workers have to face the risks of fatality or major accident with a low level of safety in the industry [2]. This project will be required in advanced in a big company of the sheet metal industry to assist in order to bring success in this research. The study and experiences contribute to the identification of the safety problem in production safety layout of the department. This project will be held at a sheet metal stamping company located in Perak. The company performs prototyping, precision metal stamping as well as mechanical and product assembly. The factory is 12,891 square meters consist of 500 workers and 8 departments: Stamping, Numerical Control Turret (NCT), Assembly, Tooling, Supply Chain, Warehouse, Finance and Sales. For this project, the scope of work will cover stamping department, which have 150 tons and 200 tons (same department) and 600 tons stamping machine. The case study will cover only for layout 150 and 200 tons of stamping machine which the production line area is less size than 600 tons. The safety of the workers must give an attention to provide the best facilities and productivity of the industries [3]. In addition, the cases are general in the manufacturing line [4]. This research focuses on the sheet metal industries which have its safety level and risk requirement for the material handling. This project is proposed to improve the risk by observing before implement safety requirement and others safety factors at the production line and comparing the performances at stamping department to achieve the project goals for improvement of safety facility performance for the long term of the production.

2. Methodology

2.1. *HIRARC Assessment*
There are several stages involved to complete this project. First stage is the factory visit to sheet metal stamping company. The main manufacturing process is stamping process. Suitable department had been chosen to continue the project and go through the process and anything that related to the objective of the project. Survey and observation of safety level have been done to record the data and identify the hazard analysis.

2.2. *Classification of Work Activities*
Next stages for this project are going through the flowchart of HIRARC process. First step in this process was observing and classifying the work activities. The department was observed to collect the data of their work activities. Thus, identifying the specific area which has a high level of risk. Work activities are classified by its stage of production, moving vehicle, production space area or workers' space area, machine handling and worker safety protection. The task of work is defined such as loading, packing or others that related to the working department.

2.3. *Identify Hazards*
Hazards can identify their safety, environmental, health and analyze of estimating risk. To gain the data hazards, the working machine identifies its process, transportation of dies to the machine, packaging and manual handling of machines. The observation through 150 and 200 tons can be used to calculate the level of risk. Hazardous condition is highlighted to the activities performed. Table 1 shows work activities by its hazard's identification at the department.

Table 1: Work Activities of Hazard Identification

Hazard Identification	Work Activities
Mechanical	• Vehicle movement, such as a forklift • Moving head crane
Manual handling	• Passing sheet metal to another stamping machine • Defect check and data output recorded
Energy	• Expose to the electricity
Work Environment	• No safety lines • Moving overhead crane
Equipment	• Unprotected machine • Unseen movement machine

2.4. *Risk Assessment*
A risk assessment will be represented in varying ways to show the results of analysis for making a decision on risk control. The analysis of the risk uses likelihood and severity rating of the situation will be represented in a percent graph effective communication of the data collected by the department. Likelihood value is based on surveys or questions at the end the categories of likelihood as shown in Table 2. In Table 3, Severity is divided into five and the analysis uses likelihood and severity, the method and the results are represented in percentage. Equation 1 shows the formula to calculate risk percentage of risk level at the department.

Table 2: Rating Situation of Severity

Likelihood	Situation	Rating
Most likely	The most likely result of hazard	5
Possible	As good a chance of occurring and is not unusual	4
Conceivable	Might be occur at sometimes in the future	3
Remote	Has not been known to occur after many years	2
Inconceivable	It's practically impossible and has never occurred	1

Table 3: Rating Situation of Likelihood

Severity	Situation	Rating
Catastrophic	Numerous fatalities, irrevocable property damage and productivity	5
Fatal	Approximately one single fatality major property damage if the hazard is realized	4
Serious	Nonfatal injury, permanent disability	3
Minor	Disabling but not permanent injury	2
Negligible	Minor abrasions, bruises, cuts, first aid type injury	1

$$\text{Risk percentage} = (\text{Total Risk Calculated}/\text{Total Maximum Risk}) \times 100\% \quad (1)$$

Next observations, calculation and surveys at the department were identified to calculate percent of risk level. The process were observed by week, and the risk are calculated again based on the performances of workers. All the data were recorded by HIRARC form to see how many risk minimization can be increase. The observation data were finalized to identify the risk and hazards for the next step of the methodology. All the activities are recorded before implement the safety precaution at the department. Based on Figure 1, the HIRARC form is used to recognize the name of the company, location and the date review of project process. This documentation is referring to calculate the risk of the workstation.

HIRARC FORM			
COMPANY		CONDUCTED BY (Name, designation) Date: (From and until)	
PROCESS/LOCATION	Stamping Department 200 Tons		
APPROVED BY (Name, designation)	(Assistant Production Manager)		
DATE	5 February 2017	REVIEW DATE: 29 Jan 2017	NEXT REVIEW DATE: 3 Feb 2017

HAZARD IDENTIFICATION				RISK ANALYSIS				RISK CONTROL	
No.	**Work Activity**	**Hazard**	**Which Can Cause/Effect**	**Existing Risk Control (If Any)**	**Likelihood**	**Severity**	**Risk**	**Recommended Control Measures**	**Pic (Due Date/Status)**
1.	Lifting die using a heavy overhead crane	Heavy load	Object fall down or hit the workers	The person in charge wears gloves and a safety helmet	4	5	20 (High)	Wear safety helmet and remind other workers	2 Feb 2017 **(Progress)**
2.	Auto machine moves the parts to be stamped on another machine	Fast movement of auto machine	Can hit the worker's head	Workers alert with the moving machine	4	4	16 (High)	Lower speed limit of the auto machine	2 Feb 2017 **(Progress)**
3.	Barely seen movement of auto machine in the department	Unprotected movement	Can hit the worker's head while move around the area	Workers alert with the moving machine	3	4	12 (Medium)	Color the part of auto machine with bright color	2 Feb 2017 **(Complete)**
4.	Worker's walkway around the department	Unprotected movement	The workers are not alert and realize movement machine	Only color the line of the walkway	3	3	8 (Medium)	Install gate around the auto machine	2 Feb 2017 **(Complete)**
5.	Sound above 90 dB	Too loud	Hearing loss	Workers wear ear plugs	2	2	4 (Low)	Wear earplug	2 Feb 2017 **(Complete)**

High **Medium** **Low**

Figure 1: HIRARC Form for Risk Control Record and Progress

3. Result and Discussion

After the process implementation and data collection, the risk was calculated and the level of risk at the workplaces was identified. Based on the HIRARC form, it shows the progress of the observation until it is completed. The data were interpreted by using graph of 150 tons, 200 tons and also graph of total percentage of improvement at the department to shows the percent improvement. From Figure 2, the graph is describing the improvement of hazard identification at 150 tons. This workplace is not in serious of risk level, it is calculated because the workplace is in the same department with 200 tons which is the place was small and crowded. The department sharing walkway line with heavy moving vehicles and others regular movement. Based on the graph, sound above 90 dB gives the high percent of improvement which is 65%. This is due to the workers applied and follows instruction to wear an earplug for hearing protection.

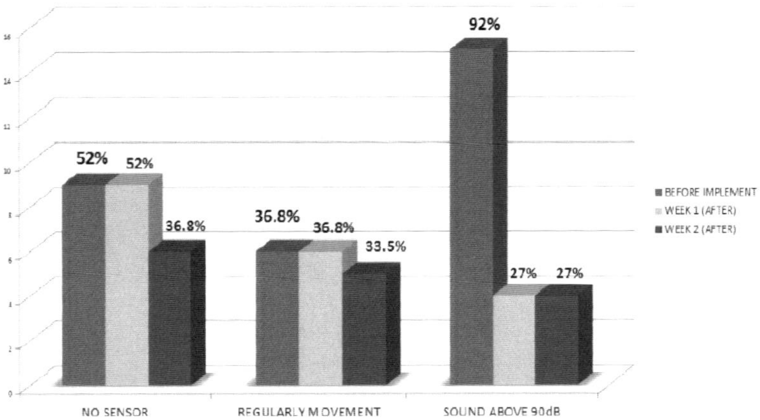

Figure 2: Graph of Improvement of Hazard Identification (150 tons)

Based on Figure 3, the graph of improvement at 200 tons shows percent improvement of heavy load 35.9%, unprotected machine 39.6% and sound above 90 dB 42.7% gives good results of improvement. This is due to the all the moving machines are protected by the workers. The auto machine was protected by the fences while it is running and the workers wearing earplug with standard personal protective equipment.

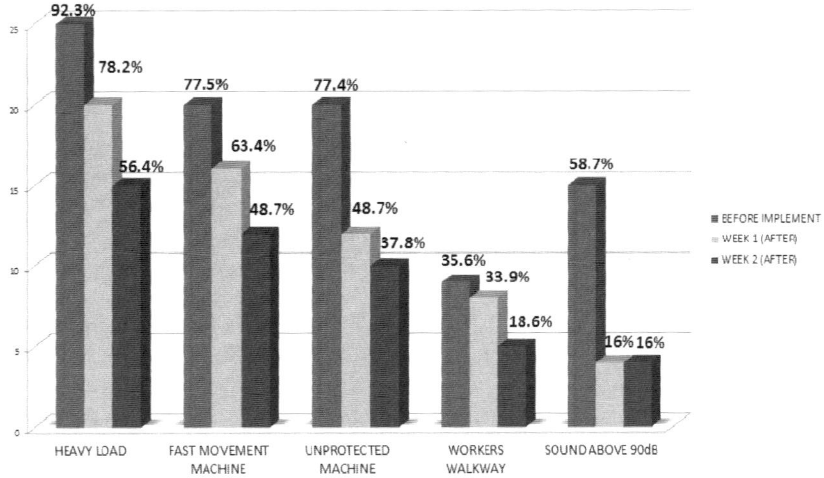

Figure 3: Graph of Improvement of Hazard Identifications (200 tons)

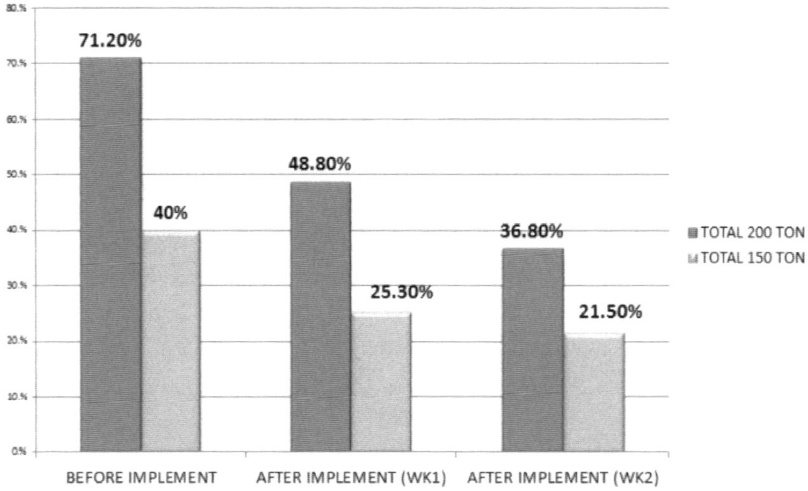

Figure 4: Graph Percentage Of Total Percent Improvement In Stamping Department

The data from all aspect at the workstation and department was interpreted in graph to make it clear and easy to read the improvement. Based on Figure 4, the graph of percent of total percent improvement in stamping department was calculated. The data of overall is made to make it precise because 150 tons and 200 tons are sharing in the same department. Before implement, the risk shows at high percent which is 40% at 150 tons while 71.2% at 200 tons. After several investigation and observation at the department, it is drops to 25.3% and 48.8% at 150 and 200 tons respectively for week 1. From the calculation until week 2, it drops again and gives the best result which is below 50% and level of risk. Last result shows 21.5% at 150 tons and 36.8% at 200 tons.

4. Conclusion

As a conclusion, by doing this research, it gives a lot of information about the HIRARC process of safety. The objectives of this project were achieved after a couple of months, several studies and research about OSHA, safety and risk in industries. After doing this project, workers now alert and knows the importance of safety inspection at their workplace and it helps to minimize the risk and improve the level of safety.

Acknowledgements
The authors would like to thank Universiti Teknologi MARA Cawangan Pulau Pinang for research support and facilities. The authors also are grateful for the sheet metal stamping company for providing us with data required to complete our case study.

References

[1] Chong H Y and Low T S 2014 Accidents in Malaysian Construction Industry : Statistical Data and Court Cases *Int. J. Occup. Saf. Ergon.* **20** (3) 503–513.
[2] Ajslev J et al. 2017 Safety climate and accidents at work: Cross-sectional study among 15,000 workers of the general working population *Saf. Sci.* **91** 320–325.
[3] Wagner P, Prinz C, Wannöffel M and Kreimeier D 2015 Learning factory for management, organization and workers' participation *Procedia CIRP* **32** 115–119.
[4] Gabryelewicz I, Sadłowska-Wrzesińska J and Kowal E 2015 Evaluation of safety climate level in a production facility *Procedia Manuf.* **3** 5822–5829.

IOP Conf. Series: Materials Science and Engineering **670** (2019) 012071 doi:10.1088/1757-899X/670/1/012071

Dynamic oblique crushing of single and bi-tubular aluminium conical tubes with imperfection

Fauziah Mat[1], Mohammad Syarul Nazreen Hani[1] and Md Noor Arib Md Rejab[1]

[1] School of Mechatronic Engineering, Universiti Malaysia Perlis, Arau, Perlis, Malaysia.

Abstract. This paper aims to study the crashworthiness response of single and bi-tubular aluminium conical tubes subjected to oblique loading. The influence of imperfection on crashworthiness criteria is studied by introducing circular cutouts to the single and bi-tubular conical tubes. The explicit finite element (FE) code LS DYNA is employed to develop the FE model of the tubes. Satisfactory agreement between the developed FE model and previous experimental work has been obtained. The validated model is then employed to investigate the influence of imperfection on the tubes based on internal energy at maximum deformation. The finding shows that with the introduction of cutouts, the energy absorption capability and Specific Energy Absorption (SEA) of bi-tubular tubes are improved as high as 3.43 % and 3.44 %, respectively compared to the single tubes. Compared to the tubes without cutouts, there is a reduction of the internal energy and the SEA of the tubes with cutouts. For bi-tubular tubes, the reduction of the energy absorbed is only 6.41 % thus implies that bi-tubular tubes are better than single tubes when subjected to oblique loading. The similar pattern is also found in the SEA. Furthermore, the increase in loading angle, α reduces the internal energy and the SEA.

1. Introduction

Recently, the amount of road vehicles has increased rapidly year by year when the development of the automotive and transport industries was improved. In the matter of fact, vehicle crash accident has become a major health problem worldwide. An expansion interest about vehicle safety or crashworthiness has been analytical, numerical and experimentally conducted in evaluating the response of energy absorbers during crash [1], [2].

Various types of thin wall tubes are frequently used especially in automotive industries as energy absorbers. A single tube consists one piece of the tube, while the bi-tubular tube consists of two single tubes which are combined. Tran et al. used a multicell with triangular section to increase the energy absorption capability of the tubes [3]. Djamaluddin et al. [4] performed numerical analysis on bi-tubular circular tubes, with or without aluminium foam as filler. It is found that foam filled double tube exhibit higher energy absorption capability compared to foam filled single tube and empty double tubes.

To further enhance the capability of the tubes, the imperfection has been introduced. Various types of imperfection have been examined such as cutouts and grooves [5], [6]. During impact, energy absorber collapses in complex ways. The tube collapses by the global bending when the oblique load is applied and hence reduces the capacity of energy absorption [7]. Conical tubes have shown lower tendency for global bending collapse to occur during oblique loading [8]. As such, the overall design of crushable structures should also consider the oblique loading. Therefore in the present work, the crashworthiness performance of single and bi-tubular tubes is compared with the introduction of imperfection under oblique loading condition.

2. Finite element development

The model for the straight tube was developed based on previous study [9]. The radius of the model for tube is 69.82 mm with 0.60 mm thickness. The tube's length was 50.94 mm. Once the model validated, the model is employed in later analysis for single and bi-tubular conical tubes. The radius and the length of the single tube are 34.715 mm and 50.53 mm, respectively. For the bi-tubular tubes, the inner and outer radiuses are 17.3575 mm and 34.715 mm, respectively. The length for single and bi-tubular tubes is kept constant at 50.53 mm. Figure 1 shows the finite element (FE) model for the tubes.

Auto-node-to-surface was incorporated into the model to account for the contact between the rigid bodies. Auto-single-surface was introduced to initiate fold formation during shell buckling. A piecewise linear plasticity (MAT024) is used to represent the aluminium 6061-T6. For aluminium, the material properties employed are Modulus Young, E = 68.9GPa, Yield strength, Y = 276.0 MPa, density, ρ= 2.70 g/cm^3, Poisson ratio, v= 0.33 g/cm^3 and the strain rate parameter for C = 6.5 and P = 4.

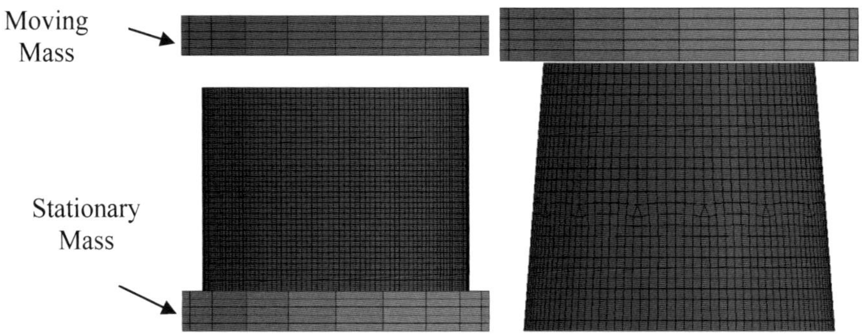

(a) (b)

Figure 1: The finite element model for the (a) straight and (b) conical tubes

3. Validation

To ensure that the developed FE model was satisfactory and sufficient, it was validated by experimental in the literature [9]. Based on the Table 1, the model shows good correlation with the existing experimental results.

Table 1: Internal energy of empty straight tube.

Outer Diamater (mm)	Length (mm)	Energy Absorption (J) at 13 mm deformation (J)		Error (%)
		Mat et al. [9]	Simulation	
69.82	50.94	77.89	86.04	10.5

4. Results and discussion

The energy absorption response of obliquely loaded tubes with and without imperfection for various loading angles, α of 10° and 30° are compared. Figures 2 and 3 show the force-deformation curves of single and bi-tubular tubes. Based on the results, it can be seen that the introduction of cutouts to the conical tubes reduces the deformation length of the single tubes. Similar observation can be observed on the bi-tubular tubes. On top of that, the increase of load angle from 10° to 30° demonstrates further reduction in impact force and hence reduces the energy absorbed. To compare between the tubes with imperfection and without the imperfection, it is profound that the tubes with cutouts exhibit higher impact force but smaller deformation length. In contrast, the tubes without cutouts show lower impact force but larger deformation length.

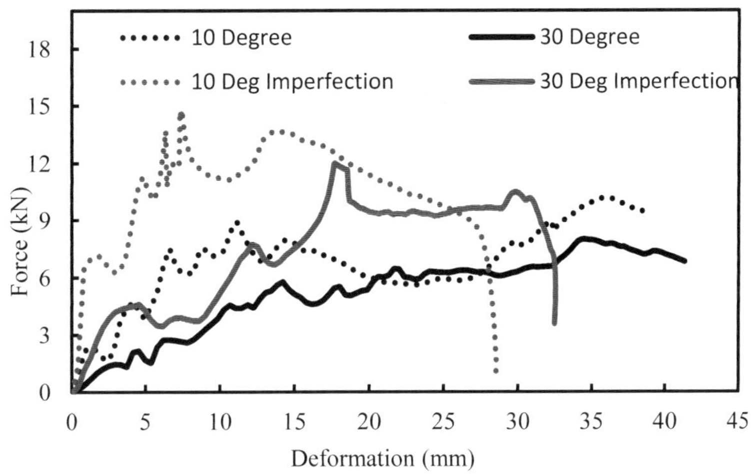

Figure 2: Effect of imperfection for single tube

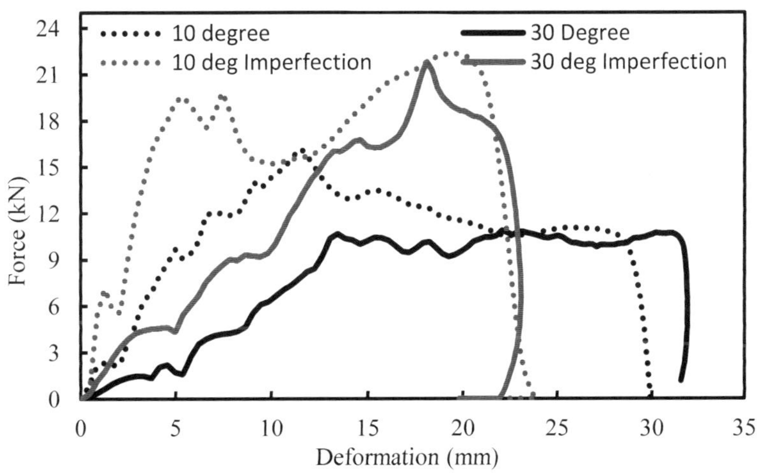

Figure 3: Effect of imperfection for bi-tubular tube

As shown in Table 2, bi-tubular tubes demonstrate higher internal energy compared to single tubes for tubes with and without cutouts. In fact, the bi-tubular tubes with cutouts demonstrate as high as 3.43% increase of internal energy. However, there is a reduction of internal energy for both single and bi-tubular tubes with cutouts compared to the tubes without the cutouts. The reduction of the energy absorbed is higher for single tube with cutouts as high as 9.07% when comparison is made between tubes with (169.52 J) and without imperfection (186.42 J).

For bi-tubular tubes, the reduction of the energy absorbed is only 6.41 % thus implies that bi-tubular tubes is better than single tube when subjected to oblique loading. The reduction of the energy

IOP Conf. Series: Materials Science and Engineering **670** (2019) 012071 doi:10.1088/1757-899X/670/1/012071

absorbed is also observed in previous research [10]. The introduction of rectangular and slotted window reduces the energy absorbed of the rectangular tubes studied.

To consider the lightweight criteria, Specific Energy Absorption (SEA) is tabulated in Table 3. As expected, the SEA of the tubes demonstrates similar patterns as observed for internal energy since the SEA is calculated based on internal energy gained by the tubes. The bi-tubular tubes with cutouts exhibit 3.44% higher SEA compared to the single tubes. On the other hand, the SEA reduces when comparison is made between tubes with and without cutouts for single and bi-tubular tubes. This scenario implies that the use of multiple tubes able to enhance the energy absorption capability of the tubes under oblique loading.

In top of that, the influence of load angle is more significant at higher load angle. As the load angle increases, the internal energy and SEA reduces. However, Cheng et al. has shown that the introduction of circular cut-outs increases the energy absorbed as well as the SEA [11]. The opposite patterns might be contributed by the geometry of the tubes as square tubes are used in the study. In fact, tube's geometry has shown significant effect on the dynamic response of the tubes under axial and oblique loading conditions.

Table 2: Internal energy at maximum deformation

Load Angle , α	Without cut-out (J)		Percentage Difference (%)	With cut-outs (J)		Percentage Difference (%)
	Single	Bi-tubular		Single	Bi-tubular	
10 °	186.42	187.02	0.32	169.52	175.35	3.43
30 °	172.73	175.75	1.72	161.66	167.01	3.31

Table 3: Specific Energy Absorption (SEA) of single and bi-tubular tubes

Load Angle , α	Without imperfection (J/kg)		Percentage Difference (%)	With imperfection (J/Kg)		Percentage Difference (%)
	Single	Bi-tubular		Single	Bi-tubular	
10 °	11387.90	11424.56	0.32	10355.53	10711.67	3.44
30 °	10551.62	10736.10	1.75	9875.38	10202.19	3.21

5. Conclusions

This study examined the dynamic response under oblique impact loading of single and bi-tubular tubes with the introduction of cutouts as imperfection. Good agreements have been observed between the developed FE model and experimental results from previous literature. The validated FE model is used to gain insight into the oblique crush response and to quantify the internal energy and specific energy absorption (SEA) for load angle variations. With the introduction of imperfection, bi-tubular tubes exhibits higher increase in energy absorption compared to single tube. While smaller reduction of the internal energy is observed for the bi-tubular tubes with imperfection compared to the ones without imperfection. Similar observations can be seen for SEA. These findings show that bi-tubular conical tubes are effective energy absorbing devices as they can withstand an oblique impact load. This study has shown that in impact applications where oblique impact load was expected, conical tubes with imperfection appear to be advantageous and this research information can be used in applications involving oblique loading such as vehicle crashworthiness.

References

[1] Zhang X, Zhang H and Wen Z 2015 Axial crushing of tapered circular tubes with graded thickness *Int. J. Mech. Sci.* **92** 12–23.
[2] Altin M, Güler M A and Mert S K 2017 The effect of percent foam fill ratio on the energy absorption capacity of axially compressed thin-walled multi-cell square and circular tubes *Int. J. Mech. Sci.* **131–132** 368–379.
[3] Tran T N and Baroutaji A 2018 Crashworthiness optimal design of multi-cell triangular tubes under axial and oblique impact loading *Eng. Fail. Anal.* **93** 241–256.
[4] Djamaluddin F, Abdullah S, Ariffin A K and Nopiah Z M 2015 Non-linear finite element analysis of bitubal circular tubes for progressive and bending collapses *Int. J. Mech. Sci.* **99** 228–236.
[5] Taştan A, Acar E, Güler M A and Kılınçkaya U 2016 Optimum crashworthiness design of tapered thin-walled tubes with lateral circular cutouts *Thin-Walled Struct.* **107** 543–553.
[6] Rawat S, Narayanan A, Upadhyay A K and Shukla K K 2017 Multiobjective optimization of functionally corrugated tubes for improved crashworthiness under axial impact *Procedia Eng.* **173** 1382–1389.
[7] Xiang X M, Lu G, Li Z X and Lv Y 2017 Large deformation of tubes under oblique lateral crushing *Int. J. Impact Eng.* **110** 138–148.
[8] Ahmad Z and Thambiratnam D P 2009 Dynamic computer simulation and energy absorption of foam-filled conical tubes under axial impact loading *Comput. Struct.* **87** (3–4) 186–197.
[9] Mat F, Ismail K A and Ahmad M 2019 Geometrical parameters and foam filling effect of aluminium thin-walled tubes under axial and oblique loading *Int. J. Mech. Eng. Robot. Res.* **8** (3) 347–353.
[10] Zhou C et al. 2018 The energy absorption of rectangular and slotted windowed tubes under axial crushing *Int. J. Mech. Sci.* **141** 89–100.
[11] Cheng Q, Altenhof W and Li L 2006 Experimental investigations on the crush behaviour of AA6061-T6 aluminum square tubes with different types of through-hole discontinuities *Thin-Walled Struct.* **44** (4) 441–454.

IOP Conf. Series: Materials Science and Engineering **670** (2019) 012072 doi:10.1088/1757-899X/670/1/012072

Displacement response of femur with various deformity angles under vertical load: FEA and experiment

S Zainul, K S Basaruddin, S F K Mohamad Khan, M H Mat Som, M J M Ridzuan and R Daud

School of Mechatronic Engineering, Universiti Malaysia Perlis, Pauh Putra Campus, 02600 Arau, Perlis.

Abstract. Osteogenesis Imperfecta (OI) is a disease that affect the bone in terms of deformity and brittleness. It is known that after certain deformation angles, OI bone cannot sustain the excessive stress that resulting to bone fracture. Although several studies have been undertaken to analyse the fracture prediction on OI bone due to deformity angles using finite element analysis, however none of them has validated experimentally. Hence, the main objective of this study was to investigate the validity of finite element analysis in predicting the fracture risk of OI-affected femur. For experimental works, the bone specimens were prepared using polylactic acid material through Fuse Deposition Material method at 15°, 20° and 25°deformation angles. Findings of this study show a promising result as the deformation response for the FEA was agreed to experimental result. Hence, it proves that prediction of finite element analysis on bone fracture risk of OI is acceptable. This study also found that femur bone with 25° deformation angle tend to fracture under minimum load. Therefore, patients that have femur bone at 25°deformation angles should be advised to undergo surgical intervention.

1. Introduction

Osteogenesis imperfecta (OI) is a disease that can affect human bone characteristics. Bone with OI is brittle due to low amount of type I collagen which is normally consist of 95% from entire human bones [1]. Approximately, 6 to 7 per 100,000 peoples suffer from OI and it is occurring in approximately 1 in 20,000 births without gender or race dependence [2]. Cross-link between collagen 1 alpha 1 and collagen 1 alpha 2 will produce type I collagen. Abnormal condition of any one of these two genes will lead to distorting the structure of type I collagen in the bone. This condition causes the bone mass decrease and contribute to brittleness and fragility [3]. OI patient suffers several bone break throughout their lifetime due to the brittle condition. Currently, there are no special treatments that can cure OI disease. Alternatively based on severity of the disease, surgeon will be advised to patients for treatment such as fracture care, physical therapy and bracing parallel with normal medication. Due to lack of studies on OI bone fracture, surgeon do not have any guidance to predict the fracture quantitatively. Surgeon will depend on their experience as there are no quantitative indicators to do the surgical intervention [4]. The past decade has seen the rapid development of finite element analysis (FEA) [5]–[7] in many engineering area as it produces fairly accurate result. Hence, this study was carry out to validate the result from FEA with experiment by utilizing the rapid prototyping technique. Although the number of cases on OI in Malaysia is low, however there is need to address the OI bone fracture problem as its secondary effect to the patients is severe. The focus of this research will be on the femur bone as the femur bone fracture incident is high due to this bone endures the highest mechanical stress during performing daily activities.

2. FEA modelling of OI bone and experimental work

2.1 Reconstruction of OI femur model

A standard femur model of healthy person was obtained from Biomedtown website in .igs format. ANSYS was used in this study to analyse the OI deform femur bone. First, the downloaded standard femur model was imported to ANSYS software. Then, the femur shaft length was measured to determine the displacement required at the middle of the shaft to create OI femur model. The shaft length was found as 347 mm. An adequate displacement was then applied to the middle of the shaft as shown in Figure 1. Three angles of deformation which is 15°, 20° and 25° was created to mimic femur bone of OI patients by using tangential rule.

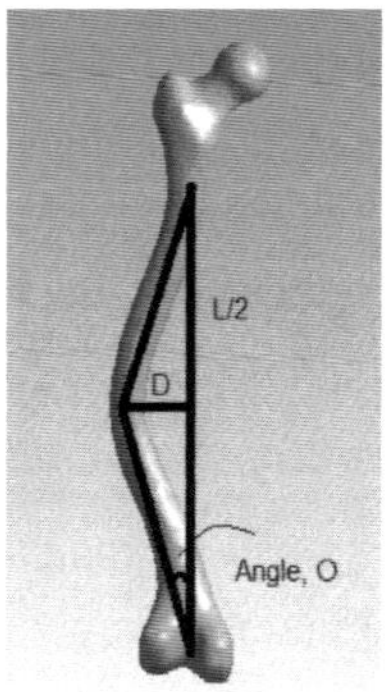

Figure 1. Deformed model of the femur

Once the deformation shape as showed in Figure 2 were obtained, the shape was scaled to 1/4 in order to accommodate FDM machine before being export to .stl file format for rapid prototyping process.

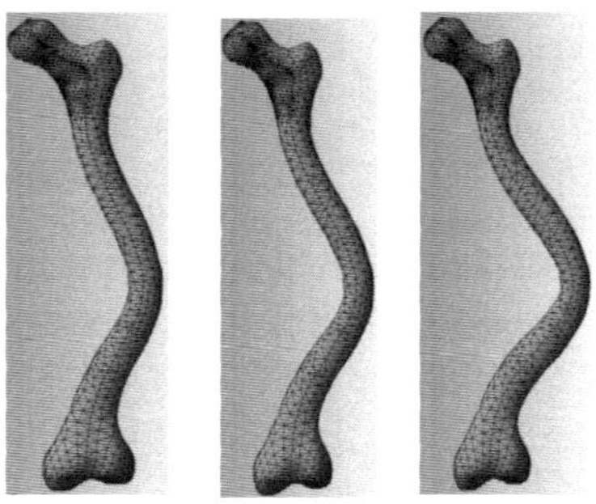

Figure 2. OI femur models with three deformity angles of (a) 15 degree, (b) 20 degree, and (c) 25 degree.

IOP Conf. Series: Materials Science and Engineering **670** (2019) 012072 doi:10.1088/1757-899X/670/1/012072

2.2 FEA on OI bone

All OI femur models were then reset its stress value to create a situation where the OI femurs are under zero loading. Instead of using bone properties, the material properties of the FEA model was assigned with PLA in order to ensure consistency between experiment and FEA. Then, boundary conditions were set as a force at the femoral head was applied in the vertical direction at the load same with the experiment. The load was increased at increment of 5 N for every step until 50 N. The load was limited to 50 N as to maintain the analysis within the elastic region of PLA material. In this FEA, displacement of the femur head at the highest point was recorded to be compared with the experimental results.

2.3 Experimental work on OI

All three types of OI femur model in the form of .stl file were transported to CURA software. The model was sliced using a certain algorithm to create a numerical control coding that will be used by the FDM machine. The coding was required to move its nozzle in three axes to create the physical OI bone specimen, as shown in Figure 3. A set of PLA print cap was fabricated to partially encase both ends of the printed OI femur bone in order to make sure vertical force that will be exerted on the specimen spread along a wider area. This is to prevent concentration of force on a small area resulting local material failure and the specimen from slipping during experimental work.

Figure 3. Preparation of OI bone specimen using rapid prototyping.

For the experimental work, the test specimen was mounted on the UTM vertically with both ends was capped with prefabricated PLA material as shown in Figure 4. A small preload was applied to the specimen at 10N to prevent any mechanical damage such as hysteresis. Then the specimen was compressed for 10 mm at speed of 3 mm per minute. The specimen material begins to yield when the machine recorded decrease of load value. The displacement of the highest points of the femur head was recorded to build a displacement versus load graph.

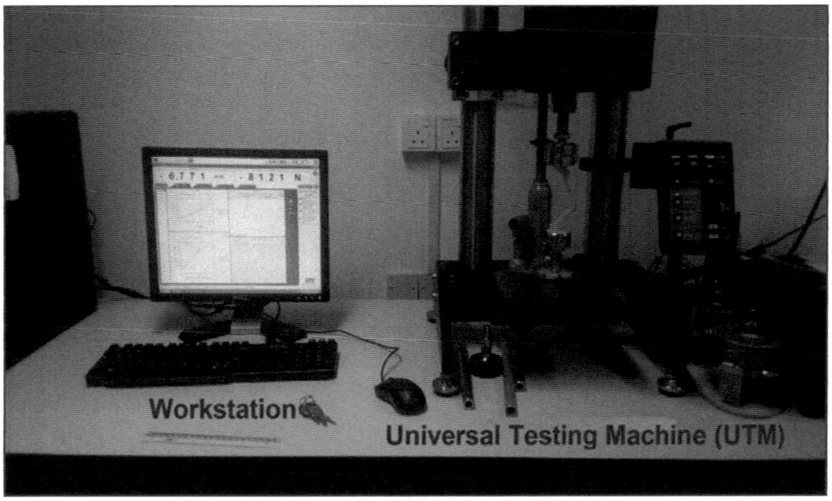

Figure 4. Experimental setup of OI bone specimen under compression test.

3. Results & Discussion

Figure 5 shows the deformation contour distribution under compression load. It was found that the maximum displacement values are increase with respect to the increasing of OI deformity angles. The maximum displacement was located at the femoral head for all cases.

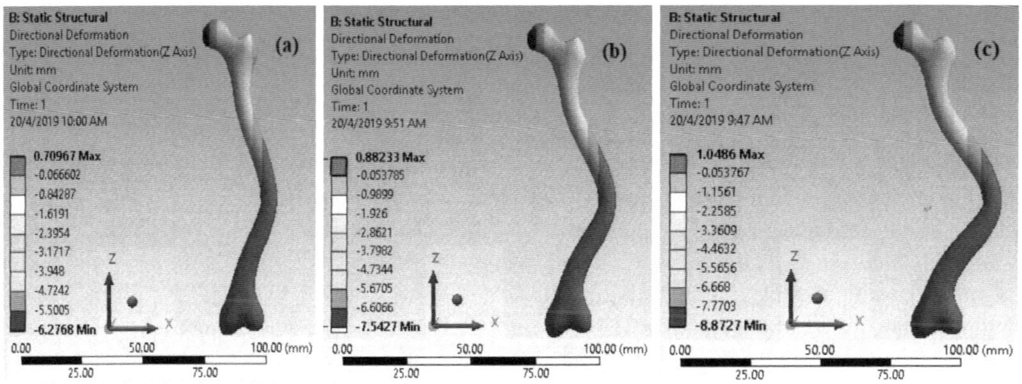

Figure 5. Displacement contour distribution on OI femur models for (a) 15 degree, (b) 20 degree, and (c) 25 degree.

Figure 6 shows the displacement response of OI bone for the selected deformity angles under experiment and FEA. According to the both graphs, it indicates that OI bone model with deformation angle of 25° is the weakest among all models as its femur head displacement was found highest at all load values. As the bone deformation angle decrease, the displacement of the femur head decrease as expected. Furthermore, the displacement value for the experiment is nonlinear as the slope changing over load values, whereas the FEA results show almost linear. The results also show that the risk for bone fracture was increased as the deformation angle increased. It seems that both results are agreed well.

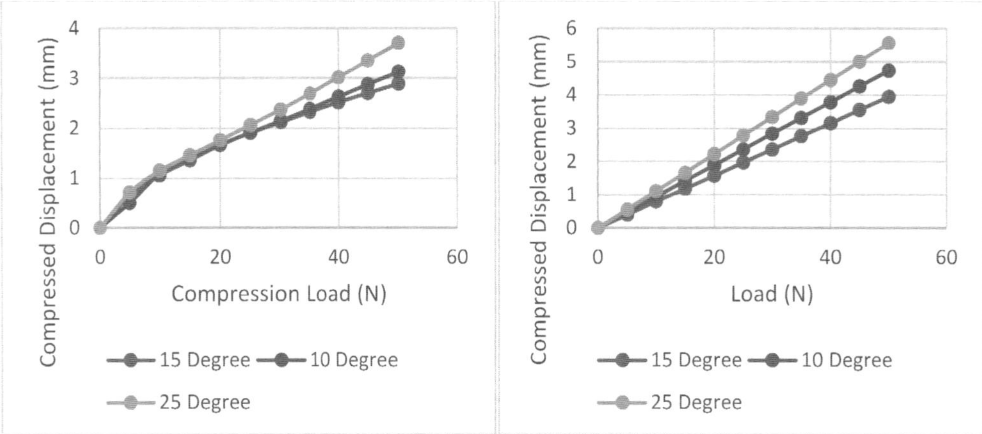

(a) (b)

Figure 6. Displacement response under compression load for all deformity models based on (a) experiment and (b) FEA.

4. Conclusion

In general, the results of displacement response of OI bone under compression load was in agreement between experiment and FEA. The maximum displacement was found increased with respect to the increase of deformity angles. Therefore, the finding of this study shows OI bone analysis using FEA has predicted very well since it was agreed with the experimental results.

Acknowledgement

This work was financially supported by the Ministry of Education Malaysia under Fundamental Research Grant Scheme, FRGS/1/2016/TK03/UNIMAP/02/6.

References

[1] Fratzl-Zelman N, Misof B M, Klaushofer K and Roschger P 2015 Bone mass and mineralization in osteogenesis imperfecta. *Wien. Med. Wochenschr.* **165** (13–14) 271–277.

[2] Oakley I and Reece L P 2010 Anesthetic implications for the patient with osteogenesis imperfecta *AANA J.* **78** (1) 47–53.

[3] Primorac D, Anticević D, Barisić I, Hudetz D and Ivković A 2014 Osteogenesis imperfecta--multi-systemic and life-long disease that affects whole family *Coll. Antropol.* **38** (2) 767–72.

[4] Fritz J M, Guan Y, Wang M, Smith P a and Harris G F 2009 A fracture risk assessment model of the femur in children with osteogenesis imperfecta (OI) during gait *Med. Eng. Phys.* **31** (9) 1043–1048.

[5] Wanna S B C et al. 2018 Fracture risk prediction on children with Osteogenesis Imperfecta subjected to loads under activity of daily living *IOP Conference Series: Materials Science and Engineering* **429** 012004.

[6] Wanna S B C et al. 2017 Prediction on fracture risk of femur with Osteogenesis Imperfecta using finite element models : Preliminary study Prediction on fracture risk of femur with Osteogenesis Imperfecta using finite element models : Preliminary study *Journal of Physics: Conf. Series* **908** 012022.

[7] Fan Z, Smith P, Reiners K, Hassani S and Harris G 2004 Biomechanics of femoral deformity in osteogenesis imperfecta (OI): a quantitative approach to rehabilitation *Conf. Proc. IEEE Eng. Med. Biol. Soc.* **7** 4884–4887 G..

IOP Conf. Series: Materials Science and Engineering **670** (2019) 012073 doi:10.1088/1757-899X/670/1/012073

Fracture prediction on patient-specific tibia model with Osteogenesis Imperfecta under various loading direction

C J Mok[1], K S Basaruddin[1], M H Mat Som[1], M S Abd Majid[1], A. R. Sulaiman[2], and A. Shukrimi[3]
[1]School of Mechatronic Engineering, Universiti Malaysia Perlis (UniMAP), Pauh Putra Campus, 02600 Arau, Perlis, Malaysia.
[2]Department of Orthopaedics, School of Medical Science, Universiti Sains Malaysia, 16150 Kubang Kerian, Kelantan, Malaysia.
[3]Department of Orthopaedics, Kulliyyah of Medicine, International Islamic University Malaysia, 25710 Kuantan, Pahang, Malaysia, Malaysia.

Abstract. This study aims to predict the fracture of bone with osteogenesis imperfecta (OI) by considering the homogenization properties of real patient. A Type-III of osteotomy in OI femur was used as bone specimen. Nine representative volume element (RVE) models were developed based on μCT-images of bone specimen. Homogenized properties particularly the Young's moduli of the RVEs was obtained based on homogenization theory in Voxelcon software. The obtained homogenized properties were then assigned to the OI patient-specific model that was developed from CT-images of real patient. The fracture of OI bone was predicted based on linear static analysis and finite element method under loadings of activity daily living (ADL). The results found that the fracture might be happen to the patient under jumping load case, whereas the subject is expected to be safe under standing still and walking load case.

1. Introduction

Bone generally is known as porous, having complicated hierarchical design and heterogeneous material. The bone properties is dependent on the morphology of bone, bone tissue properties and bone volume fraction in microscopic scale [1]–[4]. However, most of the studies of OI bone were carried out in macroscopic level [5], [6]. The effective elastic properties of a bone is representing the stiffness at microscopic scale that can be calculated using homogenization method [7]–[9]. Homogenization method was originally used for predicting the effective elastic properties for periodic microstructure such as unidirectional fibre-matrix composite or cellular materials, but it has been validated that the method could also applicable for random microstructure such as bone and porous ceramics [10], [11]. Hence, the method could possibly use for predicting the effective elastic properties of bone with Osteogenesis Imperfecta (OI).

OI (also known as brittle bone disease) is a congenital and genetic disorder disease. Several studies on OI, particularly on fracture prediction using finite element analysis have been undertaken quite extensive recently[5], [12], [13]. Fan et al. [12] applied mathematical function to reflect different deformities of OI femur in model reconstruction. They concluded that deformity could deteriorate the stress and strain distribution. On the other hand, Caouette at al. [13] developed a reconstruction

IOP Conf. Series: Materials Science and Engineering **670** (2019) 012073 doi:10.1088/1757-899X/670/1/012073

method of the tibia for OI patients that used as input of a comprehensive finite element model to examine fracture risks. The FEM results showed the importance of patient-specific geometry modelling for fracture risk assessment. Recently, Albert et al. [5] characterized the macroscopic anisotropic cortical bone material properties and investigate the relationships with bone density measures in children with severe OI. They found that cortical bone in children with severe OI yields at the same strain as normal bone, and that their decreased bone material strength is associated with reduced volumetric bone mineral density.

Although several studies have investigated the fracture assessment on OI, none has applied the bone properties that considering the porosity of the microstructure. Hence, this study was undertaken to predict the fracture risk of femur with OI that considering the influence of bone porosity, by applying the homogenization method to obtain the bone elastic properties.

2. Finite Element Analysis

2.1 Reconstruction of Representative Volumetric Element (RVE) Model

In order to obtain the homogenized properties of OI bone model, RVE models were developed based on μCT-images of a patient (male, 19 years old). The series of images were binarized using Otsu method in MATLAB.

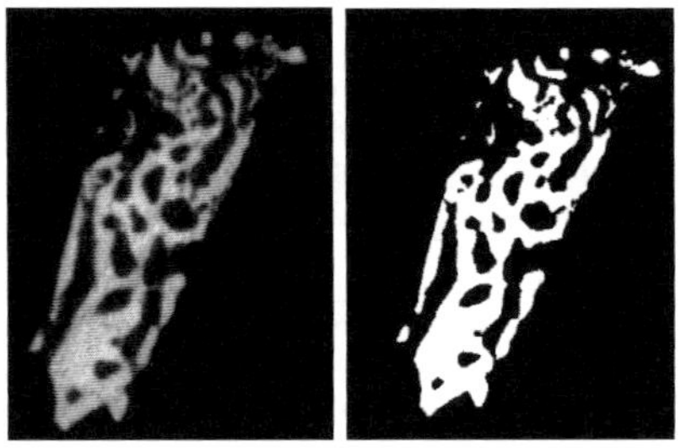

Figure 1. A sample μCT image of OI bone specimen (left) before segmentation and (right) after segmentation.

Nine RVEs were developed from three pieces of bone specimen using Voxelcon (Quint Corp., Tokyo) as shown in Figure 2. All RVEs were extracted from different locations of bone specimens with the same dimension of approximately $4 \times 4 \times 4$ mm^3. Bone volume fraction (BV/TV) for each RVE model was then calculated based the number of element over the total RVE volume.

Figure 2. Nine RVE models that were extracted from bone speciemens.

2.2 Homogenized Properties

In order to determine the effective elastic (homogenized) properties of each RVE, the homogenization method was employed. Young's modulus of 22 GPa and Poisson's ratio of 0.3 was assigned as bone tissue properties. The influence of BV/TV on the effective elastic properties for the developed RVE was then investigated.

2.3 Development of OI Patient-specific Femur Model
OI patient-specific femur model was developed based on CT-images of real patient as shown in Figure 3. The element size was selected same as the image resolution in order to avoid argument on mesh size.

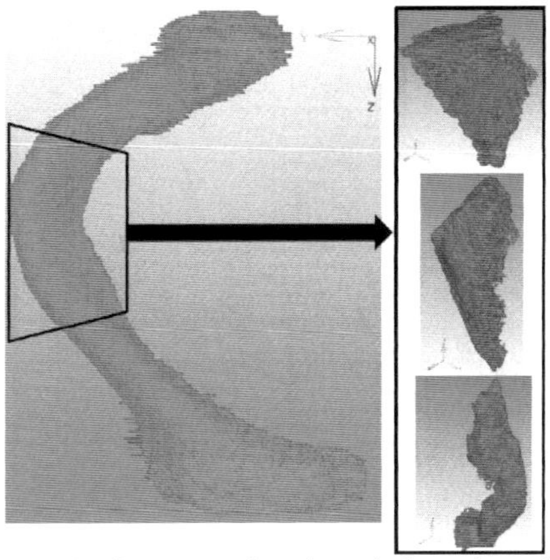

Figure 3. A patient-specific model of OI was developed (based on the CT-image before surgery). Three voxel models (right) were the osteotomy that been used for reconstruction of RVE models.

2.4 Fracture risk analysis
Fracture risk analysis was conducted based on static stress analysis. The obtained effective elastic property was then assigned to the OI femur. Analysis under various loading directions (as shown in Figure 4) and load cases that represent standing, walking and jumping was performed in Ansys [14], [15]. Body weight (BW) was taken as 62 kg. Fracture strength of 115 MPa was assumed [16] and the fracture risk for each load case was calculated as the fracture strength over the obtained von-misses stress.

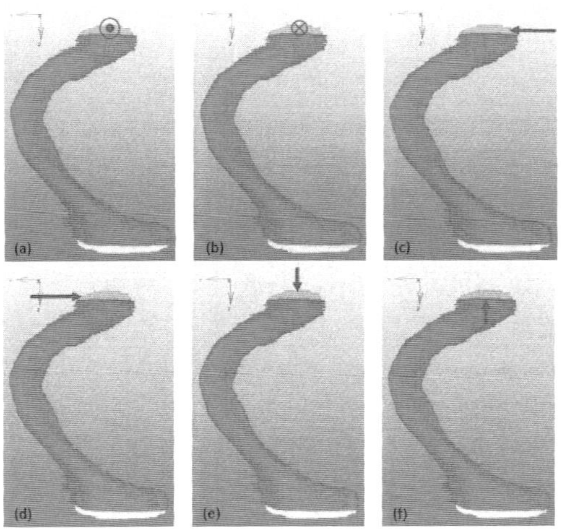

Figure 4. The fracture load is in the direction of (a) lateral (b) medial (c) anterior (d) posterior (e) compression (f) tension.

3. Results and Discussion

Figure 5 shows the effect of BV/TV on the Young's moduli in orthogonal axes. Obviously, the results show the increase of BV/TV has increased the average Young's moduli. The effective elastic properties of the OI bone that consists of Young's moduli, Poisson's ratio and shear moduli are listed in Table 1. It seems that the effective elastic properties were strongly dependant on BV/TV.

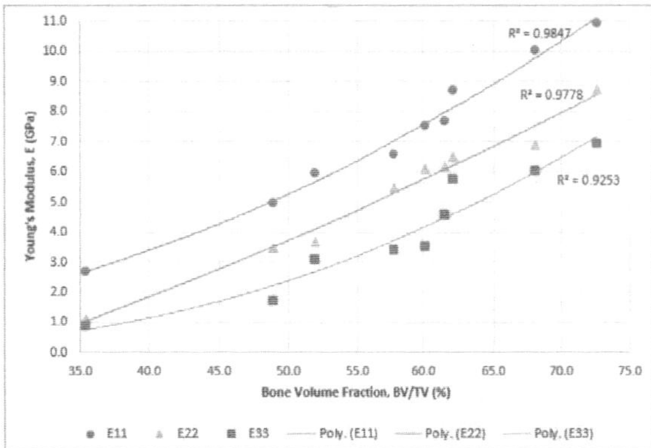

Figure 5. The effect of BV/TV on the Young's moduli.

Table 1. Effective elastic properties of OI bone.

Young's Modulus, E (GPa)	
E_{11}	7.22233 ± 2.5419
E_{22}	5.34594 ± 2.2548
E_{33}	3.98186 ± 2.0127
Poisson's Ratio, v	
V_{12}	0.27105 ± 0.037
V_{23}	0.20070 ± 0.655
V_{31}	0.18092 ± 0.0483
Shear Modulus, G (GPa)	
G_{12}	2.07578 ± 1.0703
G_{23}	2.11612 ± 0.7325
G_{31}	1.84222 ± 0.9941

In the analysis of fracture risk, the OI femur was found safe (> 1.0) under standing and walking load cases, meanwhile it was expected to be fractured (< 1.0) under jumping condition. This indicates that the OI femur can bear high load under stationary condition like sitting and standing because it does not have excessive force. The fracture load in lateral/medial, anterior/posterior and compression/tension directions were obtained as 904.7 N, 1278.6 N and 2028.1 N respectively. Figure 6 shows the von-Mises stress distribution of the OI-affected femur under three orthogonal directions. The results found that the mid of femoral shaft is subjected to stress concentration and it is expected to initiate fracture under excessive load.

Table 2: Fracture risk under activity of four loading cases.

Load case	Standing	Walking (3 × BW)	Walking (7 × BW)	Jumping (15 × BW)	Jumping (20 × BW)
Fracture risk	33.08	3.68	1.58	0.74	0.55

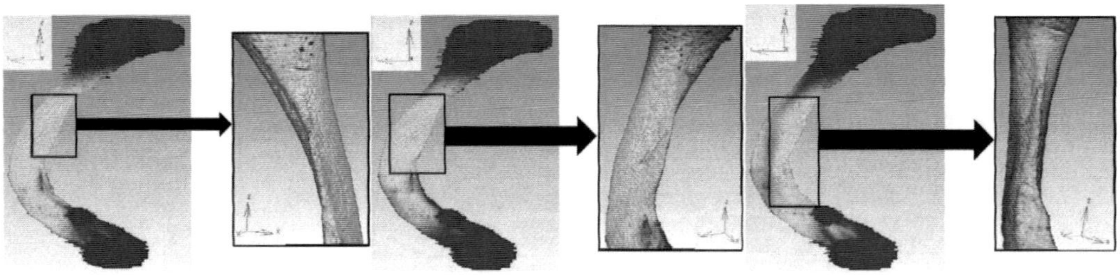

(a) stress under load in axis-1 (b) stress under load in axis-2 (c) stress under load in axis-3

Figure 6. Von Mises stress distribution of femur under various loading directions.

4. Conclusion

In this study, the RVE is well developed and the effective elastic properties of OI bone was obtained to be assigned as bone tissue properties of OI femur. The results found that the OI bone stiffness was strongly dependent on the BV/TV of RVE. Meanwhile in macroscopic scale analysis, the femoral shaft was spotted as the weakest area of the OI femur under all load cases and tend to experience fracture initiation. Loads under standing and walking activities were predicted to be safe for OI patient, whereas OI bone was expected to be failed under jumping activity.

5. Acknowledgement

This work was financially supported by the Ministry of Education Malaysia under Fundamental Research Grant Scheme, FRGS/1/2016/TK03/UNIMAP/02/6.

References

[1] Wu D, Isaksson P, Ferguson S J and Persson C 2018 Young's modulus of trabecular bone at the tissue level: A review *Acta Biomater.* **78** 1–12.
[2] Yoshiwara Y, Clanche M, Basaruddin K S, Takano N and Nakano T 2011 Numerical study on the morphology and mechanical role of healthy and osteoporotic vertebral trabecular bone *J. Biomech. Sci. Eng.* **6** (4) 270–285.
[3] Basaruddin K, Takano N, Yoshiwara Y and Nakano T 2012 Morphology analysis of vertebral trabecular bone under dynamic loading based on multi-scale theory *Med. Biol. Eng. Comput.* **50** (10) 1091–1103.
[4] Basaruddin K S, Takano N, Akiyama H and Nakano T 2013 Uncertainty modeling in the prediction of effective mechanical properties using stochastic homogenization method with application to porous trabecular bone *Mater. Trans.* **54** (8) 1250–1256.
[5] Albert C, Jameson J, Tarima S, Smith P and Harris G 2017 Macroscopic anisotropic bone material properties in children with severe osteogenesis imperfecta *J. Biomech.* **64** 103–111.
[6] Imbert L, Aurégan J -C, Pernelle K and Hoc T 2015 Microstructure and compressive mechanical properties of cortical bone in children with osteogenesis imperfecta treated with bisphosphonates compared with healthy children *J. Mech. Behav. Biomed. Mater.* **46** 261–270.
[7] Abdul Halim S A et al. 2015 The Relationship between Strain Energy Release Rate with Crack-to-Width Ratio of Human Phalanx Bone *Appl. Mech. Mater.* **786** 141–146.
[8] Hamed E, Lee Y and Jasiuk I 2010 Multiscale modeling of elastic properties of cortical bone *Acta Mech.* **213** (1–2) 131–154.
[9] Basaruddin K S, Kamarrudin N S and Ibrahim I 2014 Stochastic multi-scale analysis of homogenised properties considering uncertainties in cellular solid micros- tructures using a first-order perturbation *Lat. Am. J. Solids Struct.* **11** 755–769.
[10] Takano N, Zako M, Kubo F and Kimura K 2003 Microstructure-based stress analysis and evaluation for porous ceramics by homogenization method with digital image-based modeling *Int. J. Solids Struct.* **40** (5) 1225–1242.
[11] Tawara D et al. 2008 High-resolution micro-mechanical analysis of cancelluos bone in vertebra considering bone quality *Japanese J. Clin. Biomech.* **29** 7–14.
[12] Fan Z, Smith P, Reiners K, Hassani S and Harris G 2004 Biomechanics of femoral deformity in osteogenesis imperfecta (OI): a quantitative approach to rehabilitation *Conf. Proc. IEEE Eng. Med. Biol. Soc.* **7** 4884–4887.
[13] Caouette C, Ikin N, Villemure I, Arnoux P J, Rauch F and Aubin C É 2016 Geometry reconstruction method for patient-specific finite element models for the assessment of tibia fracture risk in osteogenesis imperfecta *Med. Biol. Eng. Comput.* 1–12.
[14] Wanna S B C et al. 2017 Prediction on fracture risk of femur with Osteogenesis Imperfecta using finite element models : Preliminary study Prediction on fracture risk of femur with Osteogenesis Imperfecta using finite element models : Preliminary study *Journal of Physics: Conf. Series* **908** 012022.
[15] Wanna S B C et al. 2018 Fracture risk prediction on children with Osteogenesis Imperfecta subjected to loads under activity of daily living *IOP Conference Series: Materials Science and Engineering* **429** 012004.
[16] Keyak J H, Rossi S a, Jones K a and Skinner H B 1998 Prediction of femoral fracture load using automated finite element modeling *J. Biomech.* **31** (2) 125–33.

IOP Conf. Series: Materials Science and Engineering **670** (2019) 012074 doi:10.1088/1757-899X/670/1/012074

Evaluating water resource sustainability of paddy for one planting schedule in Perlis

K L Fang[1], N S Kamarrudin[1], M S M Hashim[1], A A A Manap[2], M R Ramli[2], M S M Azmi[1], MA Rojan[1], A B Shahriman[1] and Z M Razlan[1]

[1] School of Mechatronic Engineering, University Malaysia Perlis, 02600 Arau, Perlis, Malaysia
[2] Jabatan Pengairan dan Saliran (JPS) Negeri Perlis, Jalan Pengkalan Asam, 01000 Kangar, Perlis

Abstract. The increase industrial and agricultural development, caused by the population growth on the one hand and the limited water resources available on the other hand has caused the groundwaters to be considered as a valuable water supply more than ever. Optimal operating policies are very much essential for sustainable management of reservoir as poor operating policies may lead to over or underuse of water. The purpose of this study was to investigate the water availability of Timah Tasoh Reservoir, water demand for minor scheme paddy, and hence balance the water inflow and outflow from Timah Tasoh Reservoir. The study area for agriculture is the minor scheme for whole Perlis state, and inflow areas are Lubok Sireh, Padang Besar at Titi Keretapi, Wang Kelian and Kaki Bukit. The result shows that the water inflow to Timah Tasoh Reservoirs is sufficient to supply for the needs of minor scheme agriculture. The highest water balance among the first planting season is in May. Total inflows calculated are much more significant than the total outflows; therefore, it concludes that the water available in Timah Tasoh Reservoir is enough to supply for minor scheme paddy in Perlis for a whole year.

1. Introduction

Water demand keeps increasing as the population become more significant from time to time. Large quantities of water are required for agricultural to produce the food needed. Timah Tasoh Reservoir is one of water supply zone existed in Malaysia which utilised as a water catchment for many usages. It is a project created by the Drainage and Irrigation Department (DID) in the state of Perlis. The store zone is partitioned into two classes, which are life and dead reservoirs and the region covered 13.3 million m3 and 6.7 million m3 respectively. Life store zone is the storage of water that can be used while the dead store zone is the storage that required keeping in the reservoir [1]. The agriculture segment is the fundamental driver of the state's financial development, with 48,127ha (58% of the land in Perlis) committed to cultivating in 2014 . Based on the of the gross domestic product (GDP) by division 2016 from Department of Statistics Malaysia, the agriculture area in Perlis took 21.6% of the flow of economic activity which is higher than industrial segment.

Water demand of agriculture will be completely fulfilled when the reservoir storage quantity is higher than water provide operation rule curve, while water demand of agriculture has to be rationed when the reservoir storage quantity is lesser than water supply operation rule curve [2]. Moreover, the problem of imbalance of water distribution inside every area may also take place in the conventional water allocation pattern, which distributes water assets primarily based on the current situation and style of water use.

Conceptually, the water balance approach was clear and was an accounting of the inputs and outputs of water. I. Hussain utilized the technique of continuity equation where the difference between ∑inflow and ∑outflow is the rate of change in storage. This equation depends on the assumption of the straightforward and highly restrictive system, which was impenetrable slanted plane surface, limited on each of the four sides with an outlet at one corner. The water balance has separated into two fundamental areas, which are water balance over the surface and water balance underneath the surface. The water requirement of agriculture has separated into net water requirement (the actual water demand is the product of net crop water requirement and the basin's crop area) and gross water demand, the water that required from gross surface water supply and gross groundwater supply which is considering the proficiency of the water system [3]. T. Yong et al. also proposed the similar water balance which water accessible is the distinction between the inflow (Qt) and the outflow (E+Rt). Evaporation (E) is determined by increasing the water surface zone (At) and the dissipation rate (et). There are two limitations in the research, where the reservoir storage must be higher than its dead storage and should not surpass its standard storage. Second, the water discharged should not surpass the reservoir's greatest release ability [4].

In this study, the water resource sustainability of minor scheme for 1 planting season in Timah Tasoh Reservoir was evaluated by mathematical modelling.

2. Mathematical modelling

2.1. Water availability of Timah Tasoh Reservoir

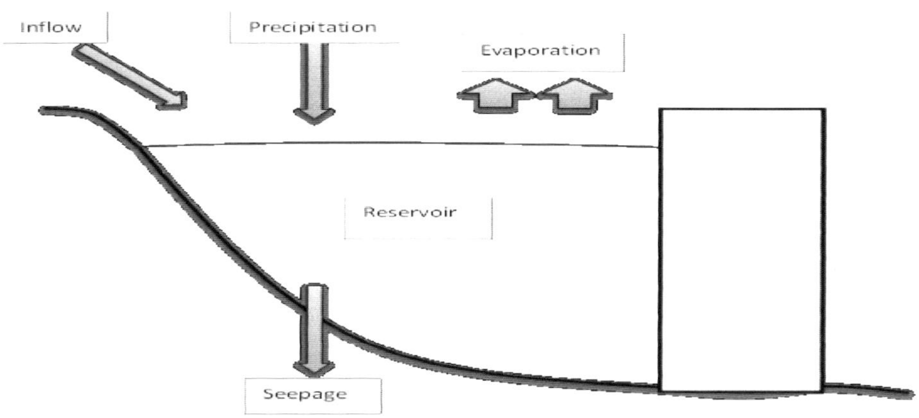

Figure 1. Resrvoir system in Timah Tasoh [1]

Figure 1 shows the water component of Timah Tasoh reservoir. In this research, only inflow, precipitation and evaporation is considered, while seepage from the reservoir has been ignored.

Water availability equation is given in the following expression:

$$Q_A = Q_{Begin} + Q_R + Q_I - (Q_L + Q_D + Q_{End}) \qquad (1)$$

where

Q_A = water availability in Timah Tasoh Reservoir
Q_{Begin} = water storage at the beginning of the month
Q_{End} = water storage at the end of the month
Q_R = rainfall catch by the Timah Tasoh Reservoir,

Q_I = inflow from river
Q_L = total losses of Timah Tasoh Reservoirs (evaporation, seepage, leakage)
Q_D = water required from domestic

2.2. Irrigation water requirement

The main crop requires water supply from the Timah Tasoh reservoir is paddy. There are two sources of inflow which are irrigation and precipitation while outflow are crop evapotranspiration, surface runoff, seepage and seepage-percolation. Assuming the standing water depth is constant all the time, the following equation shows the relationship between inflow and outflow:

$$IWR+PREP=EVPT+SURF+SEEP+SEPR \quad (2)$$

where

IWR = irrigation water requirement
$PREP$ = precipitation capture by agriculture area
$EVPT$ = evapotranspiration of crop
$SURF$ = surface runoff
$SEEP$ = seepage
$SEPR$ = seepage percolation

2.3. Water balance equation

In order to balance the water level in the reservoir and water supply for agriculture, a general equation is needed to calculate it. Based on the continuity equation, total inflow must be equal to total outflow:

$$\sum Q_{in} = \sum Q_{out} \quad (3)$$

where Q_{in} is for total inflow and Q_{out} is for total outlflowof TImah Tasoh reservoir.

The total inflow of reservoir is river inflow and precipitation:

$$\sum Q_{in} = Q_I + Q_P \quad (4)$$

The total outflow is the evaporation, gross water requirement from agriculture:

$$\sum Q_{out} = Q_E + IWR_{gross} \quad (5)$$

where IWR_{gross} is for irrigation gross water requirement.

Substituting Equation (4) and Equation (5) into Equation (3), the water balance equation is expressed as below;

$$Q_I + Q_p = Q_E + IWR_{gross} \quad (6)$$

Considering the situation when inflow is less than outflow or in reverse conditions, an additional variable was added in Equation (5):

$$Q_I + Q_p = Q_E + IWR_{gross} \mp x \quad (7)$$

where x is representing the other resource or river, depending on the situation.

The value of x in the equation might be positive or negative depending on the situation. When the inflow of the reservoir is less than outflow of the reservoir, x will become a negative value representing the volume of water required from underground water to balance the water in the reservoir. On the other hand, if the inflow of the reservoir is more than outflow, x will become a positive value indicates that the excess water will be flowing out from the reservoir through a river.

3. Results

Based on Table 1 and Figure 2(a), the amount of water at the end of the month (final water available) are definite from March to July indicates that the water in Timah Tasoh Reservoir is sufficient to supply to outflow. Besides that, the amount of water shows an increasing trend throughout the month. This may be due to the large increase in rainfall from the river into the Timah Tasoh Reservoir due to heavy rainfall in the basin.

Table 1. Results of the amount of water available in the Timah Tasoh reservoir for the first planting season

Type of data	February	March	April	May	June	July
Current water available (million m^3)		18.73	25.44	38.88	265.21	342.04
Water losses (million m^3)		3.50	3.50	3.50	3.50	3.50
Domestic (million m^3)		1.74	1.74	1.74	1.74	1.74
Agriculture irrigation (million m^3)		6.59	9.57	5.36	4.17	2.00
Rainfall (million m^3)		0.107	0.080	2.419	1.970	4.606
Inflow from the river (million m^3)		18.43	28.17	234.51	84.27	184.43
Final water available (million m^3)	18.73	25.44	38.88	265.21	342.04	523.83

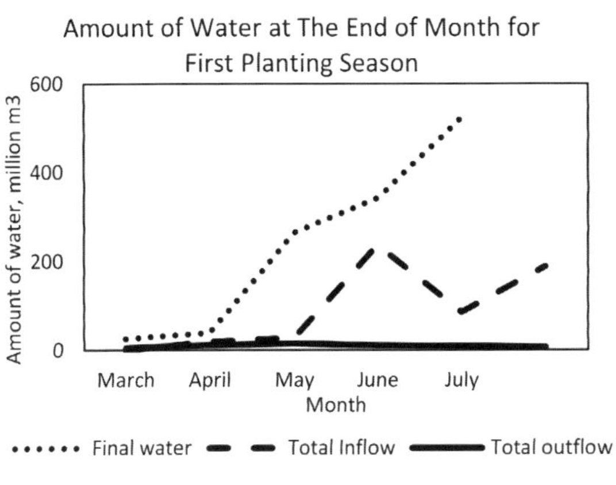

Figure 2(a). Amount of water at the end of the month for first planting season

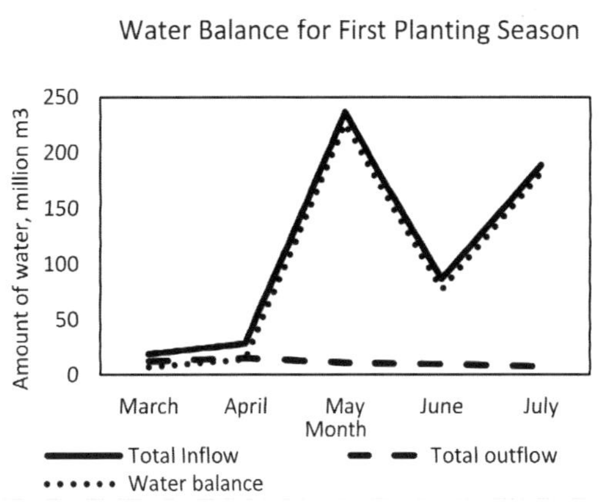

Figure 2(b). Water balance for first planting season

According to Figure 2(b), the positive water balance from March to July shows that the water flow is sufficient to supply the outflow without any additional water resources.

It can be seen that the water balance reaches its maximum in May which is 226.33 million m^3. This is because the total inflow is much more significant than the total outflow. The total inflow is generally mainly determined by the river inflow, and the river inflow reaches the maximum value of 234.51 million m^3 in May. However, the maximum total outflow is in April, which is 14.81 million m^3. Therefore, the water balance shows a sudden increase in the trend in May.

In short, the water resources of the Timah Tasoh Reservoir are not only sufficient to meet domestic demand, and cover for the water loss caused by evaporation, but also enough to meet the agricultural irrigation needs of small paddy fields.

4. Conclusion

Based on presented results, it can be concluded that the water availability of Timah Tasoh Reservoir is sufficient to supply for the needs of domestic and agriculture irrigation (Minor Scheme). However, the amount of water calculated at the end of the month exceeded expectations. According to information provided by DID, the Timah Tasoh Reservoir has an emergency level of 29.6 meters and can hold up to 50 million m3 of water at a time. The calculation result shows much more abundant than this value, due to an incomplete part of the equation that solves the water balance. Excess water will be pumped or flowed out to rivers or oceans to avoid excessive water stored. To sum up, the amount of water that exceeds the emergency limit will eventually be discharged to the river or ocean.

Besides that, in the real situation, the purpose served by Timah Tasoh Reservoir is not limited to paddy irrigation and domestic. It covered the industrial supply, other crop irrigation, tourism use, etc. Therefore, the actual total outflow of Timah Tasoh Reservoir is supposed to be higher than the calculated value. However, since the total inflow calculated is much higher than the total outflow, so it has strong evidence to said that the water available in Timah Tasoh Reservoir is sufficiently enough to supply for all the purpose of water supply.

References

[1] Hasan M Z, Abdul Hamid H S, bin Idris H and Bahar S 2015 Development of a monitoring system for timah tasoh dam *Appl. Mech. Mater.* **793** 590–594.

[2] Chu J, Zhang C, Fu G, Li Y and Zhou H 2015 Improving multi-objective reservoir operation optimization with sensitivity-informed problem decomposition *Hydrology and Earth System Sciences Discussions* **12** (4) 3719-3752.

[3] Hussain I, Hussain Z, Sial M H, Akram W and Farhan M F 2011 Water balance, supply and demand and irrigation efficiency of Indus Basin *Pak. Econ. Soc. Rev.* **49** (1) 13-8.

[4] Tian Y et al. 2018 Joint operation of surface water and groundwater reservoirs to address water conflicts in arid regions: An integrated modeling study *Water* **10** (8) 1–18.

IOP Conf. Series: Materials Science and Engineering **670** (2019) 012075 doi:10.1088/1757-899X/670/1/012075

Determination of blind spot zone for motorcycles

***M S M Hashim[1], A A Al Hamati[1], Mohd Hafzi M I[2], Shahriman A B[1], A H Ismail[1], M S Muhamad Azmi[1], Zuradzman M R[1], A Harun[1], N S Kamarrudin[1], I Ibrahim[1], M K Faizi[1], M A M Saad[1] and M A Rojan[1]**

[1]Motorsport Technology Research Unit (MoTECH), School of Mechatronic Engineering, Universiti Malaysia Perlis (UniMAP), 02600 Arau, Perlis, Malaysia.
[2]Malaysian Institute of Road Safety Research (MIROS), Lot 125-135, Jalan TKS 1, Taman Kajang Sentral, 43000 Kajang, Selangor Darul Ehsan, Malaysia

*corresponding author: sanihashim@unimap.edu.my

Abstract. The problem of the blind spot zone (BSZ) for motorcycles is common, as it causes many accidents that occur between motorcycles and cars, or motorcycles with other vehicles. The problem of BSZ is occurring for many reasons, such as if the motorcyclist wants to change the lane or manoeuvre or turn without realizing the presence of other vehicle which may cause a terrible collision and leads to casualties, either because of darkness, the full dependence on side mirrors that give a limited scope of vision, or due to a malfunction in the front lights of the car that prevented the motorcyclists from recognizing it. However there were limited research on identifation of BSZ for motorcycle, even though most vehicle accidents in Malaysia involved motorcycles. This paper discusses the initial works on the identification of BSZ for motorcyles. Three types of motorcycles were used to determine the BSZ using grid-based technique. From the data collected, the BSZ was identified for the motorcycles.

1. Introduction

In the late 1800s, the first motorcycle has been taken to the street by inventor Sylvester Roper in America [1], and it was only a few years before it became famous all around the world. The motorcycles were competing with cars in terms of popularity because of its fuel economy and relatively cheap price compared to the car, and ease of driving in narrow and rugged roads. However, motorcycle accidents are much more risk than car accidents as the motorcyclist is not surrounded by any structure. Unlike the car driver, the motorcyclist is expose to physical injuries at a very high rate as the motorcyclist only depends on the balance during riding. Globally, 23 % of accidents deaths are motorcycle riders [2], and more than 24 % of motorcycle accidents were caused by maneuvers or lane changes [3]. This percentage shows the motorcyclist was unable to spot any car around him. Thus, there is a need for identifying the blind spot zone for motorcycles to study the risk and further to develop the crash prevention system for the motorcycles.

Most researchers had used ultrasonic sensor [4-6] and camera [7] for detecting other vehicles around the motorcycles. Nazrul Islam et al. [4] had invented a blind spot detection system for cars based on ultrasonic sensors and microcontroller to calculate the distance between the motorcycle and a vehicle in a blind spot. The sensor data is then sent to the application by Bluetooth technology. After receiving and analyzing the data, the driver is alerted by the application if necessary. Later, Yaashwanth et al. [5] had developed a detection system uses Arduino and ultrasonic to prevent blind

IOP Conf. Series: Materials Science and Engineering **670** (2019) 012075 doi:10.1088/1757-899X/670/1/012075

spot accidents for motorcycles. When a sensor detects a vehicle in blind zone, it will send signals to the microcontroller, and then the microcontroller will receive and analyze the signals, and display it on the LCD to alert the driver if necessary.They called the system as GUARD and successfully tested the system at the stationary bike. Then Pineda [8] had focused on solving the problem of detecting the blind spot independently of the lean angle of the motorcycle. This system includes several parts that have contributed to solving this problem such as an accelerometer, a gyroscope, and a detection device for detecting the presence of a vehicle. For this system, the accelerometer detects a gravity force vector, then the gyroscope detects the position of the motorcycle relative to the gravity force vector to calculate the lean angle of the motorcycle. After that the detection device is configured to keep the position of the motorcycle relative to the gravity force vector exactly same. Thus, the system can detect the vehicles in blind spot zone accurately regardless the lean of the motorcycle.

In 2012, a group of researchers were developed a system to detect the vehicles in blind spot zone for motorcycles by using a single camera in the visible spectrum that has been designed to detect a vehicles in both daytime and nighttime conditions, during daytime vehicle detection is carried out using optical flow features and Support Vector Machine-based (SVM) classification, and during night time vehicle detection is based on head lights detection. This system alerts the driver if the there is any vehicle in his blind spot area, this system also provides information about position and type of vehicle to the driver [7].

Therefore, the blind spot problem is a main reason of accidents between motorcycles and other vehicles. However, sometimes the accident happens because of the car driver does not see the motorcyclist around him, or realises the motorcyclist only at the last moment [8]. Thus, the important point is to identify the blind spot zone for the motorcycles and then this information can be used to develop the system for detecting and preventing other vehicles in blind spot zone. This paper proposes a method for identifying the blind spot zone for motorcycles.

2. Experiment Setup

2.1. Grid-based technique

Blind spot zone for motorcycles was determined by using grid-based technique [9,10] as shown in Figure 1. The shaded area indicates the area of blind spot and θ is angle of blind spot. Each grid was set to 500 mm × 500 mm. The area covered is 2000 mm × 4500 mm for each side of the motorcycle. The motorcycle was placed in the middle of the grid and data collection by manually observation from the motorcyclist.

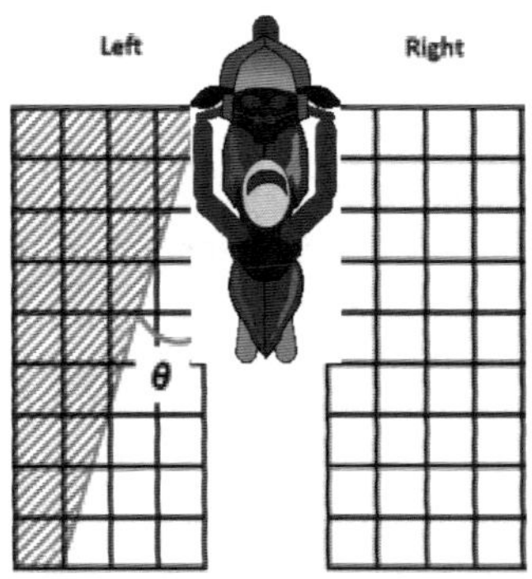

Figure 1: Grid-based technique for BSZ

Each data was recorded for each grid based on area and angle for left and right side of the motorcycle. Various heights of the motorcyclist and types of motorcycles were used for data collection.

2.2. Experimental works

The experimental works were conducted on three motorcycles as shown in Figure 2. The motorcycle were selected based on engine capacity and types of motorcycle; which are underbone (110cc and 115cc) and roadster (200cc). There were eight motorcyclists volunteered for data collection and the height of the motorcyclists were ranging from 172 cm to 189 cm.

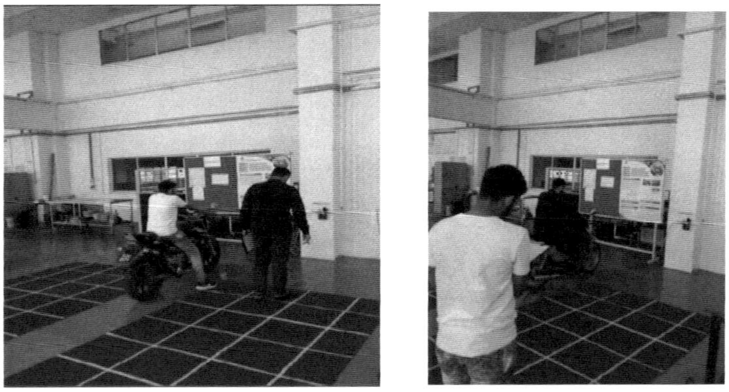

Figure 2: Data collection for BSZ

The experiment began by adjusting the motorcycle to the middle of the grid and adjusting the side mirror. Then the data was collected by observing the present of a person on grid-by-grid using the side mirror. Each grid was marked if motorcyclist can see the person or not.

3. Result and Discussion

The data collected were the plotted as in Figure 3. From the results, it was observed that the angle of BSZ is linearly decrease with the increment of motorcyclist height. In contrast to angle, the area of BSZ is linearly increase with the increment of motorcyclist height. Furthermore, it was observed that the left and right side had the identical results for angle and area. This is because of the location of the motorcyclist, which is in the middle of the motorcycle. The results for BSZ of motorcycles are slightly different from the BSZ for the passenger car, which the car had different results for left and right side [9].

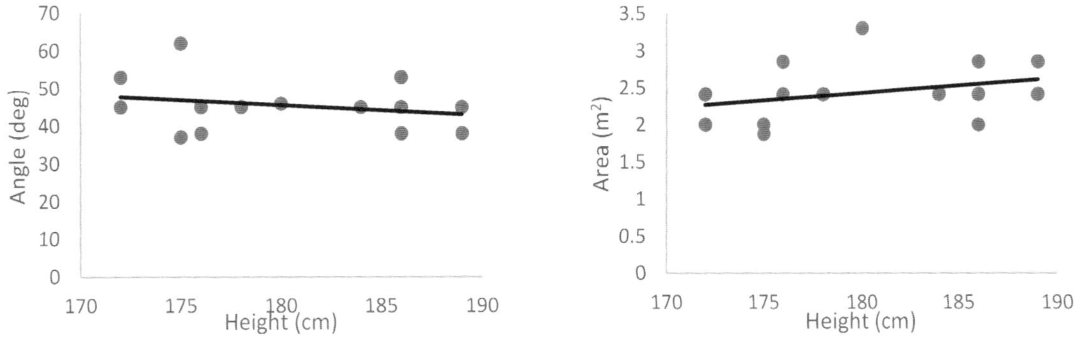

(a) (b)

Figure 3: Results for various height of motorcyclists for (a) angle of the BSZ and (b) area of BSZ.

The average blind spot angle for the left and right side of the motorcycle were 45.3° and the average blind spot area for the left and right side of the motorcycle were 2.4 m^2. The identified BSZ for motorcycle is depicted in Figure 4.

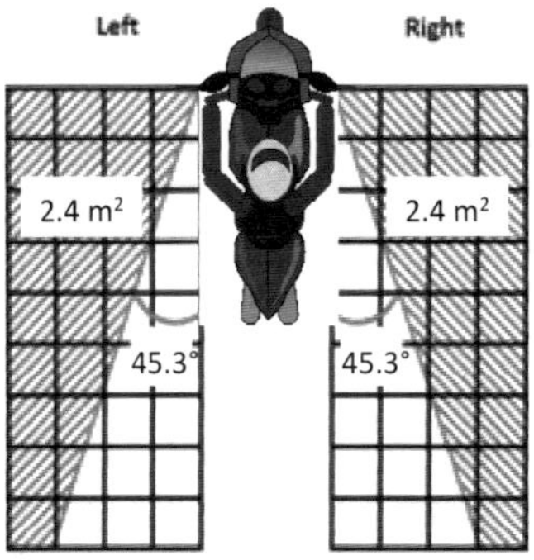

Figure 4: Blind spot zones for motorcycle

4. Conclusion

The problem of the blind spot was one of the reasons for many accidents involving motorcycles that resulted a large number of deaths. Many researchers tried to solve this problem by developing systems that can prevent the impact of the blind spot area or reduce the probability of collision and accidents that happen because of it. In this paper, the blind spot zone for motorcycles was established as an initial work before developing the detection and prevention system for blind spot. From the results, the blind spot area for motorcycle was identified for further reference. It shows that the blind spot zone for both sides are equal for angle and area, respectively. Furthermore, the relationship between the blind spot angle and area with the height of motorcyclists can be shown in linear relationship.

Acknowledgments

The authors would like to acknowledge the UniMAP Motorsport Technology Research Unit (MoTECH) for funding this study and all the team members who supported this study.

REFERENCES

[1] Ensanian A 2016 *Discovering the Motorcycle: The History. The Culture. The Machines.* Equus Potentia Publishing.

[2] Sharples S and Shorrock S 2014 Contemporary Ergonomics and Human Factors 2014 *Proceedings of the international conference on Ergonomics & Human Factors 2014*, Southampton.

[3] Grzyb K M, Hoffmann P and Osman J A 2017 Advanced Driver Assistance Systems for Motorcycles : Concept of a Lane Change Assist. *25th International Technical Conference on the Enhanced Safety of Vehicles (ESV)*, Detroit Michigan, United States.

[4] Islam N, Shamsher U, Nayyeri D and Kulesza W 2013 Motorbike Crash Avoidance System with Ultrasonic Sensor and Android Application *Canadian Journal on Electrical and Electronics Engineering* **4** (2).

[5] Yaashwanth L, Shanmuganathan R, Vishal K, Roja R and Anto Rovin K S 2015 GUARD Preventing Blind Spot Accident using Arduino *International Journal of Innovative Research in Science, Engineering and Technology* **4** (10).

[6] Deom D P 2017 *Motorcycles blind spot detection system and rear collision alert using mechanically aligned radar* Patent application publication.

[7] Fernandez C, Llorca D F, Sotelo M A, Daza I G, Hellin A M and Álvarez S 2013 Real time vision based blind spot warning system: Experiments with motorcycles in day time /night times conditions *International Journal of Automotive Technology* **14** (1) 113–122.

[8] Turner T R 2010 *Color-Coded Motorcycle Safety: Spotting the Accident Before It Happens*. America Star Books.

[9] Hashim M S M, Ismail A H, Shahriman A B, Muhamad Azmi M S, Zuradzman M R, Harun A, Kamarrudin N S, Ibrahim I, Faizi M K, Saad M A M, Rani M F H, Mahayadin A R, Rojan M A, Azizizol M S and Mohd Hafzi M I 2018 Identifying Blind Spot Zone for Passenger Cars using Grid-Based Technique *Journal of the Society of Automotive Engineers Malaysia* **2** (3) 245-251.

[10] Muhamad Azmi M S, Ismail A H, Hashim M S M, Azizizol M S, Shahriman A B, Kamarrudin N S, Ibrahim I I, Zuradzman M R, Harun A, Ibrahim I, Faizi M K, Saad M A M, Rani M F H, Rojan M A and Mohd Hafzi M I 2018 Parametric Study of a Blind Spot Zone: Case Studies Of Perodua Viva, Perodua MyVi and Proton Suprima *2nd Conference on ASEAN Road Safety (CARS2018)* Selangor, Malaysia.

IOP Conf. Series: Materials Science and Engineering **670** (2019) 012076 doi:10.1088/1757-899X/670/1/012076

Simulation of a diesel generator - battery energy system for domestic applications at Pulau Tuba, Langkawi, Malaysia

Azizul Mohamad[1*], Nasrul Amri Mohd Amin[1], Zuradzman Mohamad Razlan[1]

[1]School of Mechatronic Engineering, Universiti Malaysia Perlis, Perlis, Malaysia

*Corresponding author: azizul@unimap.edu.my

Abstract. Design of micropower system that able to fulfil electrical energy demand is challenging due several factors, thus comprehensive and reliable simulation of such system is essential. This paper focuses on simulation of a diesel generator – battery energy system for domestic applications by using HOMER software. The simulation is based on electric load demand for 10 houses in Pulau Tuba, Langkawi, Malaysia. The simulation shows that the total cost of the energy (COE) of the system is $179,363 with more than half of it is due to the fuel cost. This system is enough to fulfil the load demand, and the battery could sustain the load independently for about 23.5 hours. Pollutant produces of the system is estimated to be as high as 32,874 kg of carbon dioxide emitted annually. These values would serve as baseline reference to compare the renewable energy systems subjected to the same electric load demands.

Keywords: diesel – battery energy system; simulation; domestic application; HOMER

1. Introduction

Micropower system is a system that generate electricity to serve a nearby load, by utilizing various types of energy generation and storage technologies [1]. Most of the micropower system has been designed and developed to satisfy the energy demands of communities beyond the reach of national electricity grid due to geographic and financial constraints. The design of micropower system is challenging due to high level of uncertainty in key parameters, for instance load size and fuel price, thus a comprehensive and reliable simulation of such system is essential to predict its feasibility and practicality. This paper is focusing on simulation of a conventional diesel generator - battery energy system for domestic applications at Pulau Tuba, Langkawi, Malaysia.

2. Simulation of Energy System Using HOMER Software

Simulation process determines how a certain system configuration, (i.e. mixture of system components of different sizes, as well as an operational strategy that describes how those components working together) would perform in a given setting over a long period of time. HOMER Micropower Optimization Model was developed by the United States' National Renewable Energy Laboratory (NREL) as a tool to design micropower systems as well as to assist in comparing various power generation technologies and applications [2]. HOMER notably enables engineers to compare and consider various design options based on their technical and economic performances.

Simulation process in HOMER serves two main purposes. The first objective of the simulation is to determine whether a specified system is feasible. In exchange for the system to be feasible, it should be able to adequately fulfil the electric loads, including to be able to satisfy all the constraints set by the modeller. Another purpose of the simulation is to determine the life-cycle cost of the system, which somewhat the total cost of overall system installation and operation over its lifespan.

Upon completion of the calculation for one year (i.e. 8760 hours), HOMER software will check if the system satisfies the constraints as set by the modeller, in terms of the renewable energy fraction, capacity shortage and level of pollutants emission. The software also calculates the life-cycle cost of the system based on annual fuel consumption, annual generator operating duration and expected battery lifespan.

3. Components of Diesel Generator-Battery Energy System

Diesel generator-battery system is one of the most common conventional energy systems for electrical power generation. The cycle charging dispatch strategy is typically be used for such system whereby the diesel generator runs at full power and subsequently charged the battery. The currency used in this research is United States Dollar with currency exchange rate of 4 Malaysian Ringgit to 1 US Dollar.

3.1. System Overview

In this research, diesel generator is used to generate alternating current electricity to fulfil the electric load. Above and beyond the electric load demand, the electricity generated is transformed into direct current electricity and stored in the battery. Similarly, when the diesel generator is not running, the battery, together with the inverter, are used instead to satisfy the electric load until it depleted to a specified discharge level. **Figure 1** shows the system layout of diesel generator-battery system.

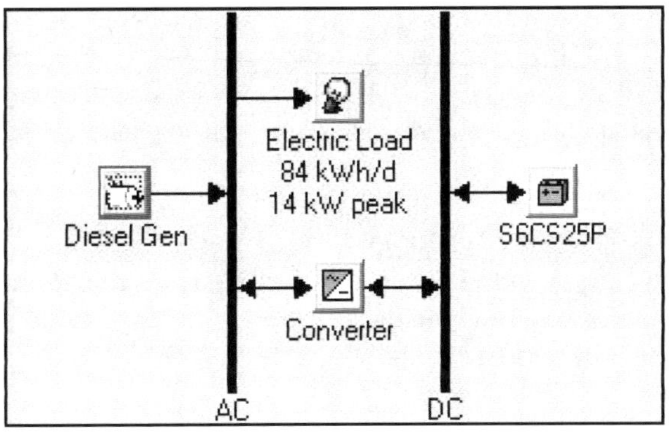

Figure 1. System layout of diesel-battery system

3.2. Electric load

Proper determination of energy load is crucial in any energy system since it will dictate the type of equipment to be used, the complexity of the system, and the overall cost of the system. Load variation throughout the day needs to be taken into consideration and the system requires to be designed based on maximum load for any month of the year to ensure sufficiency and no power shortages.

The electric load used throughout this research is based on average hourly electric demands for 10 houses in Pulau Tuba, Langkawi. In this case, the scale annual average of the load is given as 84 kWh/day and the scale peak load are 14.0 kW. Daily as well as hourly noises are given as 15% and 20%, respectively [2]. Hourly load profile is provided in **Figure 2**. Based on the figure, the load demand is shown to be the highest at late evening and early night, with spikes around sunrise and afternoon [3-4].

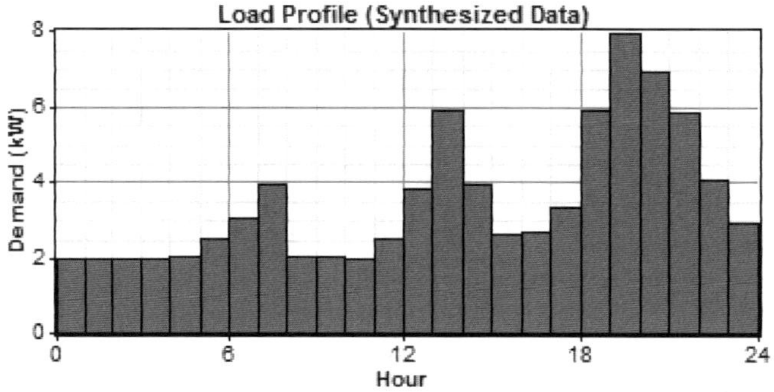

Figure 2. Hourly electric load profile for 10 houses

3.3. Diesel Generator

Diesel generators with unit rated capacity of 1 kW has been chosen as conventional alternating current (AC) prime mover. Its expected lifetime approximately around 15,000 hours and its capital and replacement costs are both \$500 per unit. The cost for operation and maintenance is estimated at \$0.05/hour [5]. The efficiency of diesel generators is depending on their output, with optimized efficiency of 30% obtained at full load. Emission factors of the diesel generator for every litre of diesel are given as 6.5 g of carbon monoxide, 0.72 g of unburned hydrocarbons, 0.49 g of particulate matter, and 58 g of nitrogen oxides.

3.4. Surrette 6CS25P Battery

Surrette 6CS25P battery with rated voltage of 6V, nominal capacity 1,156 ampere-hour, and lifetime throughput of 9,645 kWh was picked as energy storage. Battery banks of two batteries per string was used to provide 12V bus. Round-trip efficiency of the battery is 80% with minimum state of charge of 40% [5]. The capital cost of the battery is \$1,150 each, and its replacement cost is expected to be around \$1,000 each. The battery will also require \$10 per year for maintenance purposes.

3.5. Converter

In this analysis, a converter that could act as inverter as well as rectifier was used to convert direct current (DC) to alternating current and vice versa. The capital and replacement costs of the converter are \$800 each, and it virtually has zero maintenance cost. Its expected lifetime is about 15 years. Efficiency of the inverter and the rectifier are both set at 90% [6].

3.6. Operation Parameters

Due to noise emissions from the diesel generator, the operating hours are somehow limited from 6.00 am to 10.00 pm every day. Beyond this specified time, the electrical load is served solely by the batteries. Dispatch strategy used in this setting is cycle charging type where the diesel generator will run at full power to fulfil the load demand as well as in charging the battery bank up to its specified state of charge, which has been set at 80% for this study.

4. Results and Analysis of Diesel Generator – Battery Energy System Simulation

This section describes some of significant results of the simulation, with emphasis on economic, technical and environment analysis of the system simulated performances.

4.1. Economic Analysis

Diesel generator-battery energy system consisted of 10 kW diesel generator, twenty Surrette S6CS25P batteries and 5 kW converter was simulated by using HOMER software. HOMER simulation calculated that the levelized cost of energy (COE) for this system was \$0.511/kWh which is significantly higher than the COE tariff for domestic utilization from Tenaga Nasional Berhad's grid connection at \$0.07 to \$0.20/kWh, based on the amount of monthly consumption. Despite its higher price, off-grid system as such may become a feasible option to electrify remote and isolated areas when connection to national grid is technically no longer practical and/or excessively expensive.

Per the economic values obtained from HOMER simulation, as shown in **Table 1**, the total net present cost of this system is $179,363 with 56% of this cost is contributed by the diesel cost, set at $0.65 per litre, given as $100,231. With the current trend of increasing fuel price, it is expected that the overall cost of the system will increase over time. Annual operating and maintenance (O & M) costs are expected at $25,607 per year.

On the other hand, the capital cost for the system is $32,000 as compared to its replacement cost of $25,339; with major portion of the replacement cost is contributed by diesel generator since its expected life is less than four years, and it needs to be replaced several times throughout the 20-year duration of the project.

Table 1. Net present costs of diesel-battery system

Component	Quantity	Capital ($)	Replacement ($)	O&M ($)	Fuel ($)	Salvage ($)	Total ($)
Diesel Generator	10	5,000	13,730	23,313	100,231	-904	141,370
Surrette 6CS25P	20	23,000	9,939	2,294	0	-2,079	33,155
Converter	5 kW	4,000	1,669	0	0	-831	4,838
System		32,000	25,339	25,607	100,231	-3,814	179,363

4.2. Technical Analysis

Since there is only one type of generator, all the electric production is produced by diesel generator. Unmet electric load is calculated at 66.9 kWh/year while the capacity shortage is reported as 119 kWh/year. These values are within the specified 2% capacity shortage tolerance, with renewable energy fraction is 0% as expected. The diesel generator operates for 4,065 hours annually with daily start-up at 6.00 am and shut down by 10.00 pm. Its operational life is 3.69 years; thus, needing replacement approximately five times throughout the entire project duration.

The capacity factor recorded is around 42.2 %, indicating that the diesel generator only operating around two-fifth of its nominal capacity. Its electrical output is 36,928 kWh/year with mean output of 9.08 kW. Each year the diesel generator consumes 12,484 litres of diesel with specific fuel consumption calculated at 0.338 litre for every kWh of electrical energy produced, estimated at 30.1% mean electrical efficiency.

A total of 20 batteries consist of 10 parallel strings of 2 batteries each, are used in the system. The bus voltage is set at 12 V. Usable nominal capacity of the batteries is 83.8 kWh and its lifetime throughput are 192,904 kWh. The batteries can sustain the electric load demands without the diesel generator for approximately 23.5 hours, thus any maintenance or replacement work on the diesel generator should be completed within this duration to avoid any potential of energy disruption.

The state of charge of the batteries is maintained above 40% for most of the time. There are several times where the state of charge of the batteries reduces to below 50%, particularly around 12.00 at midnight to 6.00 in the morning since the batteries are discharged to sustain the load demand and the diesel generator at that moment is shut down to prevent its noise disturbing the people that live nearby.

A 5-kW converter is used in the analysis that acts as both; the inverter as well as the rectifier. The mean output of inverter and rectifier is 1.35 kW and 1.86 kW respectively, and their corresponding capacity factor is 26.9% and 37.2%. The converter operates for 4,751 hours as inverter and 3,998 hours as rectifier in annual term.

4.3. Environmental Analysis

Table 2 lists the mass of pollutants emits by the energy system every year. Considering the project lifetime of 20 years, the quantity of pollutant emissions will be evidently substantial and may impose damaging impacts to the surroundings. Aiming to eliminate, or at least significantly reduce these harmful pollutions is one of major steps in helping to save the environment.

Table 2. Annual rate of pollutant emissions for diesel-battery system

Pollutant	Emissions (kg/year)
Carbon dioxide	32,874
Carbon monoxide	81.1
Unburned hydrocarbons	8.99
Particulate matter	6.12
Sulphur dioxide	66
Nitrogen oxides	724

5. Conclusion

Based on the simulation results of the diesel generator – battery energy system from technical, economic and environmental considerations, the system is found to be technically feasible, expensive and polluting. While the system is considered good from technical perspective, its economic and environmental aspects could be improved further. Reduction in the overall cost of the energy system could be achieved by using renewable energies such as solar, wind and hydrogen energies. Similarly, the pollutant emissions could be substantially decrease by replacing fossil fuels such as diesel with their renewable energy alternatives.

The values obtained from this simulation would serve as baseline reference in designing the renewable energy systems for the same electric load demands to determine the technical, economic and environmental improvements over the conventional system.

References

[1] Türkay B E and Telli A Y 2011 Economic analysis of standalone and grid connected hybrid energy systems *Renewable Energy* **36** (7) 1931-1943.
[2] Lambert T, Gilman P and Lilienthal P 2006 Micropower system modelling with HOMER. In Farret F A and Simoes M G *Integration of alternative sources of energy* (379-418). New Jersey: John Wiley & Sons, Inc.
[3] Ho W S, Hashim H and Lim J S 2014 Integrated biomass and solar town concept for a smart eco-village in Iskandar Malaysia (IM) *Renewable Energy* **69** 190-201.
[4] Haidar A M A, John P N and Shawal M 2011 Optimal configuration assessment of renewable energy in Malaysia *Renewable Energy* **36** (2) 881-888.
[5] Lau K Y, Tan C W and Yatim A H M 2015 Photovoltaic systems for Malaysian islands: Effects of interest rates, diesel prices and load sizes *Energy* **83** 204-216.
[6] Ashourian M H, Cherati S M, Mohd Zin A A, Niknam N, Mokhtar A S and Anwari M 2013 Optimal green energy management for island resorts in Malaysia *Renewable Energy* **51** 36-45.

IOP Conf. Series: Materials Science and Engineering **670** (2019) 012077 doi:10.1088/1757-899X/670/1/012077

Pre-forming evaluation of dissimilar aluminium alloys blank fabricated using friction stir welding technique.

A F Pauzi, A B Abdullah* and M F Jamaludin

Metal Forming Research Laboratory, School of Mechanical Engineering, Universiti Sains Malaysia, 14300 Nibong Tebal, Penang, MALAYSIA.

***Corresponding author: mebaha@usm.my**

Abstract. Friction stir welding (FSW) is a solid-state welding that is using to weld the relatively difficult to weld materials such as aluminium alloy. FSW also can weld different material type and thickness to produce tailor welded blanks (TWBs). Although FSW capable to join dissimilar material or similar material hence from different grades, the properties of welded blanks may affect as compared to base material. In this study, performance of three aluminium alloy grades i.e. AA6061, AA5052, and AA1100 blanks that were combined among them and evaluated. Furthermore, the effect rolling direction was also considered. The welding was performed at constant speed, welding speed and tool angle using flat pin tool on a conventional milling machine. The performance is based on few mechanical properties such as hardness and tensile strength. In addition, crack pattern on the welded blank was also observed using three-point bending test. Based on the result, the hardness reduced at the stir zone (SZ) as compared to the base material, even though from different rolling direction. As expected, for tensile behaviour, the maximum load can be found on combination of similar material i.e. AA6061-T6 compared than other combinations. While based on the crack pattern, there are unsuitable combinations because of inappropriate welding parameters, and it is complying with the previous observation on tensile properties.

1. Introduction

Recently, automotive industries have high demand of light weight sheet metal material such as aluminium. Many European automotive companies had implemented tailor welded blank (TWBs) technology to achieve lighter vehicle structure. Aluminium alloy has got great attention due to its advantages. Each of the alloys have been developed to meet certain criteria and characteristic. For example, AA5052 is the strongest non-heat treatable aluminium that usually used as an alternative to steel in aerospace, marine and automotive industries because of their light weight, good formability. Good strength and high corrosion resistance [1-2]. There are numbers of studies and findings made on friction stir welding of various dissimilar aluminium alloys. Lin et al., [3] found that dissimilar aluminium alloy is difficult to be joined using friction stir welding because of the difference in material properties as it is difficult to choose the suitable welding parameters. Although difficult, there are number of success, for example, Balaji et al., [4] found that the most suitable rotational speed to weld AA2024 and AA7075 is between 600-1200 RPM. They also discovered that the tensile test shows the strength reduced by only 10% as compared to the based materials. This is in agreement to Liu et al., [5], as they claimed that the maximum tensile strength to joining different materials is approximately 85% of the base Al alloy. Hasan et al.,[6] indicated that weld hardness drop at the HAZ on the softer material. This is based their investigated-on friction stir welding of AA7075-T651 and AA2024-T351. Tool profile also influence the performance of the weld, this is based on study by Ilangovan et al., [7] on AA6061 and AA5086 and Krisna et al., [8] on AA2024 combined with AA7075. Furthermore, welding parameters such as welding speed and tool rotation speed effect on tensile strength and micro hardness also carried out on AA2024 and AA7075 and they found that higher tensile strength obtain when 31.5mm/min and under 900rpm rotation speed. In other work, Piccini and Svoboda,[9] studied the effect of tool depth to blank made by combination of AA5052 and AA6063 on the lap joint. They discovered that when the AA6063 positioning at upper material the fracture load will increase. In this paper, tensile and hardness test of few combinations from commonly used aluminium alloys will be carried out. The present work would form the basis for subsequent studied in the fabrication of aluminium-based TWBs using friction stir welding technology

2. Methodology

In this study, the specimen in 3mm thickness from three aluminium alloy grades i.e. AA6061, AA5052 and AA1100 plates were prepared in 200mm x 80mm size. Pair of specimens arrange in a butt-weld configuration were clamped on a fixed mounted on a conventional milling machine as shown in figure 1. The feed rates (welding speed), tool angle and spindle speed (rotational speed) were set constant. The clamp is as shown in Figure 2(a), was fabricated based on the design recommendation by Kamble [10].The fixture was assembly, was designed with left and right fixture plate that can ensured firm hold of the specimens, irrespective of their thicknesses. In addition, prevent specimen from lifting or bending during welding. The cylindrical FSW tool is made from hardened tool steel. Figure 2(b) show the dimensions and it consists of 2mm pin and 10mm shoulder profile, both are unthreaded.

Figure 1. Experimental set-up on the conventional milling machine

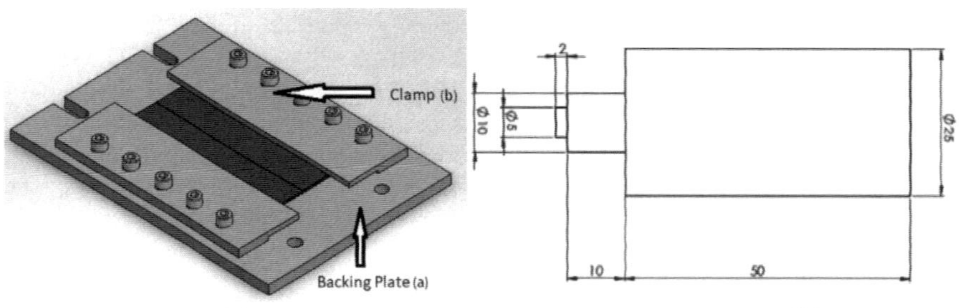

(a) (b)

Figure 2. (a) Design of the Fixture and (b) schematic diagram of the tool (unit in mm).

The two plates of AA6061 and AA6061, measuring 200 mm x 80 mm x 3 mm, were arranged in a butt-weld configuration onto the clamp, as shown in figure 3. Then repeat with AA6061 and AA5052, and AA6061 and AA1100. The feed rates (welding speeds) and the spindle speeds (rotational speed) are set on the control panel of the milling machine.

2.1. Hardness Test
The hardness values were measured along and perpendicular to the welding direction, as shown in figure 4. A Rockwell harness tester was used to determine the hardness using a steel ball with a diameter 1.588mm and 100kg of loading.

2.2. Tensile Test
Tensile tests were conducted in accordance to ASTM E8. The specimens were cut perpendicular to the weld and carried on three tensile specimens for each combination.

3. Result and Discussion
3.1. Hardness pattern
The discussion will be focus on the effect of the studied parameters to the springback pattern. Before that, material properties such as UTS, yield stress and hardness distribution need to be determined first. Figure 3 show the hardness distribution for all combinations. In general hardness will reduce at the weld area and start to increase from the distance of 5mm from the centre. The hardness of the weld zone was found to be lower than the hardness of the base materials. From the three-combinations tested, hardness of the combination AA6061-T6 – AA1100-H14 with 45° rolling direction has the highest value on the centre of weld line which is 65 RHB. It similar with figure 6 (b) the combination of AA6061-T6 – AA5052-H32 in this research that the hardness starts to decrease at the AA6061-T6 to the weld zone then increase at AA5052-H32.

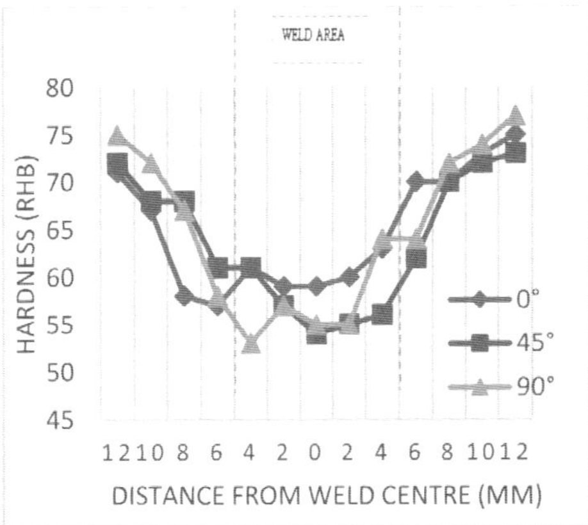

(a) AA6061-T6 -AA6061-T6

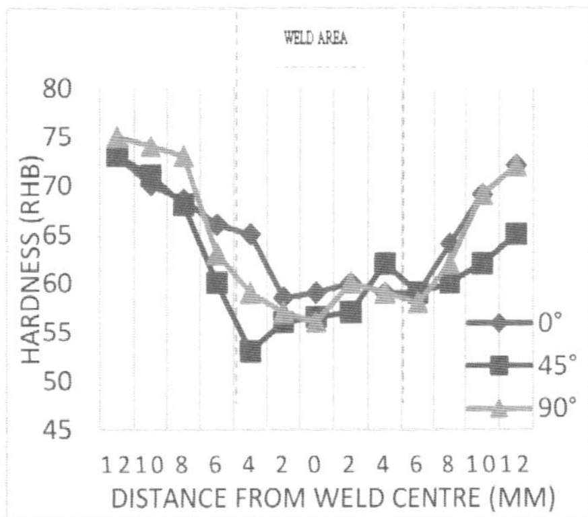

(b)AA6061-T6 -AA5052-H32

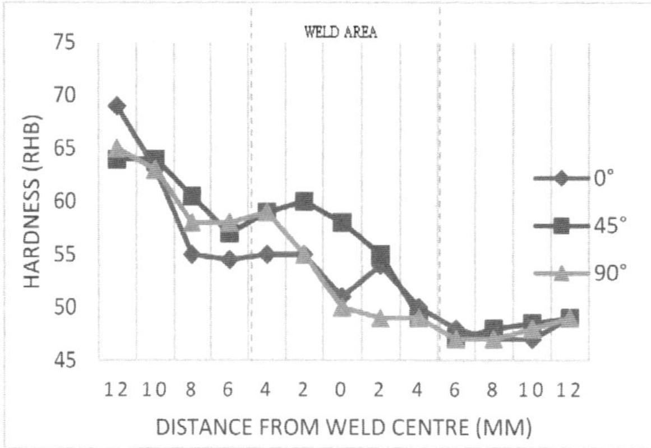

(c) AA6061-T6 -AA1100-H14

Figure 3. Hardness pattern from centre of weld line

3.2. *Tensile Result*

Figure 4 shows the stress-strain graph for three combination specimens with different rolling direction. From the graph shows that the combination of AA6061-T6 – AA6061-T6 with rolling direction 0° has the highest tensile strength of 172 MPa. It can be seen that the 0°of rolling direction give highest value of tensile strength for the combination of AA6061-AA6061 and AA6061-AA1100. But for the combination of AA6061-AA5052, the 90° shows the highest. Najib et al. [11] said in their research that 0° orientations will give high strength and higher Young's Modulus if it is properly heat treated but the ultimate tensile stress increase as rolling direction increase which mean 90° has the highest UTS. From the three combination, the combination with both AA6061-T6 has the highest tensile strength followed by AA6061-T6 – AA5052-H32 and the combination AA6061-T6 – AA1100-H14 has the lowest strength. This is because according to Hasan et al. [12] the join efficiency in dissimilar FSW is based on the strength of softer material. In this research AA1100-H14 has the lowest strength from the three grades that affecting the combination of AA6061-T6 – AA1100-H14 had the lowest tensile strength. This is because necking and fracture will happened on the heat affected zone (HAZ) on the softer grades [5].

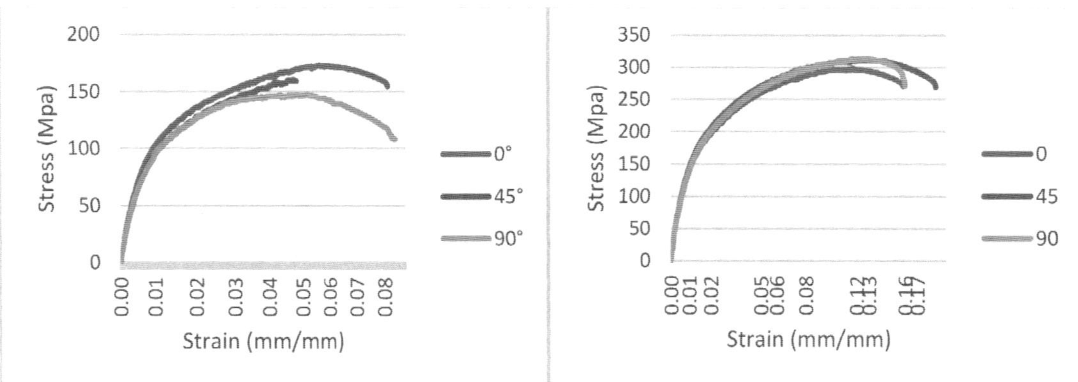

(a) (b)

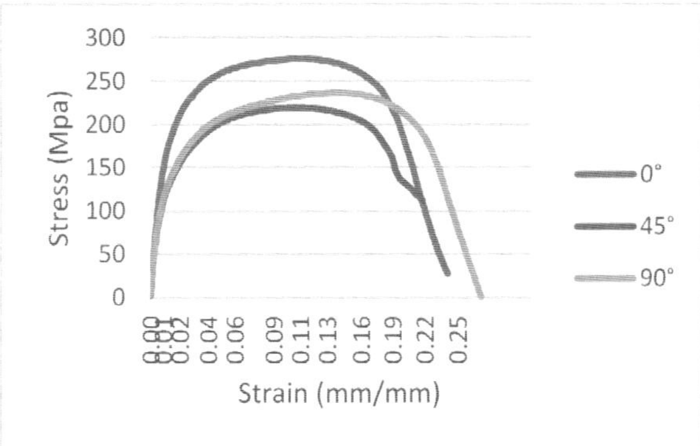

(c)

Figure 4. Stress-Strain curve of the specimen at different combinations, (a) AA6061-T6 – AA6061-T6, (b) AA6061-T6 – AA5052-H14 and (c) AA6061-T6 - AA1100-H14.

3.3. Crack Pattern

Figure 5 shows the crack pattern that occur on the combination AA6061-AA1100 when bending test were done at different stroke. The crack happened on 90° of rolling direction (RD) and parallel bending direction (BD). The crack happened due the undesirable defect that exists in the welding, namely as void or wormhole defect. The wormhole defect is a cavity below the weld surface that cannot be detect by human eye. Abnormality in term of material flow cause by the unsuitable welding speed i.e. feed rate and rotational speed i.e. spindle speed are the primary reason for wormhole appearance [13] as shown in Figure 8. The wormhole will reduce the mechanical properties of the joint so that is one of the reasons why cracking happened as expected. It also could be a detect that is refer as kissing bond defects or zigzag defect [14], as can be seen in figure 5(b) and (c) from top view. This defect is believed cause by high welding speed but low rotational speed, that lead to insufficient stirring of metal that can lead to cracking or material break. Furthermore, incomplete root penetration[15]is another defect that typically found on this friction stirred blank. This type of defect is located below the stir zone at the interface of the faying surface of the joint (Figure 5). This is because of the impropriate plunge depth or the pin was too short. So that the bottom of the joint of the specimen was not properly weld.

Based in this experiment as well, the effect of selected joining materials and also BD could result cracking on the specimen. In this experiment, the cracking only happened on the combination AA6061-AA1100 with the 90° of RD and when parallel BD was performed on the specimen. But for the perpendicular BD or the others RD, there was no cracking happened. This is could be because of the defect stated above that happened during the welding process. The combination of AA6061-T6 - AA1100-H14 might be unsuitable for joining with the following welding parameters.

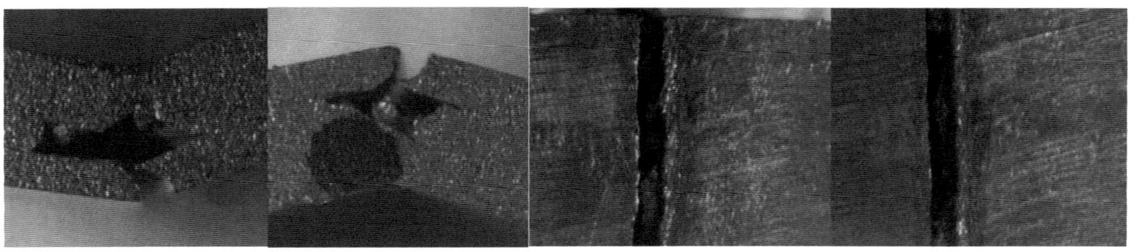

Side View *Top View*

(a) AA6061-AA1100; RD: 90°; BD: PAR; SD: 3mm

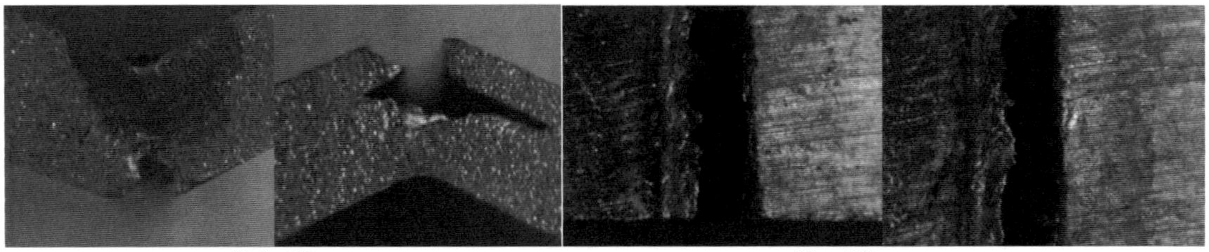

Side View *Top View*

(b) AA6061-AA1100; RD: 90°; BD: PAR; SD: 6mm

Side View *Top View*

(c) AA6061-AA1100; RD: 90°; BD: PAR; SD: 9mm

Figure 5. Crack/breakage pattern based on the macro graphic image using microscope USB on combination AA6061-T6 - AA1100-H14

4. Conclusions and Future Work

The friction stir welding process was successfully join the dissimilar grades of aluminium alloys. The effect of rolling direction and combined materials were evaluated on the hardness and tensile test. From the study, the hardness will be reduced on the stir zone (SZ) and start to increase from heat affected zone (HAZ) towards the based materials. The result shows that with 45° of rolling direction on the AA6061-T6 – AA1100-H14 give the highest hardness value on the welding zone. From the tensile result, it was notices that the combination on both AA6061-T6 with 0° rolling direction has the highest maximum load which is 3100N. The defect in FSW such as wormhole and kissing bond also need to be considered in this experiment to prevent cracking. The most important thing to be controlled to avoid defect is the welding parameters such as welding speed and rotational speed. In the future, the formability test or post forming test can be carried out on the dissimilar grades of aluminium to further explore on the FSW research.

Acknowledgements

This work is partly supported by the Universiti Sains Malaysia RUI Grant (Grant No: 1001/PMEKANIK/8014031).

References

[1] Shanavas S and Raja Dhas J E 2017 *IOP Conf. Ser. Mater. Sci. Eng.* **247** (1) 012016.
[2] Shanavas S, Edwin Raja Dhas J and Murugan N 2018 *Int. J. Adv. Manuf. Technol.* **95** (9–12) 4535–4546.
[3] Lin H, Wu Y, Liu S and Zhou X 2018 *Mater. Charact.* **141** 2116–2120.
[4] Balaji N, Kannan S and Arun S 2017 *Int. J. Eng. Res. Adv. Technol.* **3** (4) 10–16.
[5] Liu X, Lan S and Ni J *Mater. Des.* **59** 50–62 2014.
[6] Mehta K P and Badheka V J 2016 *Procedia Technol.* **23** 513–518.
[7] Ilangovan M, Rajendra Boopathy S and Balasubramanian 2015 V *Def. Technol.* **11** (2) 174–184.
[8] Krishna K G, Devaraju A and Manichandra B 2017 *Int. J. Nanotechnol. Appl.* **11** (3) 285–291.
[9] Piccini J M and Svoboda H G 2015 *Procedia Mater. Sci.* **8** 868–877.
[10] Kamble L V, Soman S N and Brahmankar P K 2017 *Mater. Today Proc.* **4** (2) 1277–1284.
[11] Najib L M, Alisibramulisi A and Amin N M 2015 *InCIEC 2014*.

[12] Hasan M M, Ishak M and Rejab M R M 2018 *Int. J. Adv. Manuf. Technol.* **98** (9–12) 2747–2758.
[13] Boldsaikhan E, Corwin E, Mcgough J and Arbegast W 2009 *Midwest instruction and computing symposium.*
[14] Kah P, Rajan R, Martikainen J and Suoranta R 2015 *Int. J. Mech. Mater. Eng.* **10** (26).
[15] Khan N Z, Khan Z A, Siddiqee A N, Al-Ahmari A M and M. H. Abidi 2017 *Trans. Nonferrous Met. Soc. China* **27** (7) 1507–1516.

IOP Conf. Series: Materials Science and Engineering **670** (2019) 012078 doi:10.1088/1757-899X/670/1/012078

The effect of support structure on RANS actuator disc for shallow water application

A Rahman, N B Yahaya, M T A Rahman

School of Mechatronic Engineering, Universiti Malaysia Perlis, Pauh Putra Main Campus, 02600 Arau, Perlis, Malaysia.

Abstract. The effect of support structure on the wake formation of the tidal turbine needs to be considered when investing in a tidal energy farm. The wake resulted from the support structure may vary with the array of turbine. There is various type of support structure that are under development and each type of the support structure exhibit a different behaviour of the wakes.. Four types of geometry of the support structure such as cylinder, ellipse, square and diamond were analyzed using Computational Fluid Dynamic (CFD). The k-epsilon turbulence model was used to analyze the effects of the support structure. From this study, the flow of turbulence on each support structure exhibits different behaviour. For a single turbine, it was found that ellipse geometry of the support structure indicates the shortest turbulence form. However, in terms of velocity deficit the diamond geometry illustrates the highest value followed by the square, cylinder and ellipse geometry. For the array configuration, it proves that the 1.5D for distances between two turbines is acceptable since turbulence occurrence is separated from each of the turbine. In addition, for three turbines arrangement, the separation between the first and second rows is agreeable on 3D because the wakes are well merged and almost return to ambient condition at 1D downstream. Based on the results of this research it can be concluded that ellipse support structure shows promising results due to the wake formation recovers back to the inlet velocity the fastest among the other support structure when passing through the tidal turbine.

1. Introduction

The growth of the population is proportional to the demanding of electricity usage. Electricity can generate by a various source of energy. Nowadays, natural energy such as water, solar, wind, wave, and tide are widely used to gain electricity. The usage of natural energy is increasing from time to time. Before this, fossil fuel such as coal, oil and natural gas are used to generate electricity [1]. The changes of the technologies happen because the negatives impact on the environment. This study will be focusing on the support structure. The support structure is important in order to make the turbine stand stiffly. In order to choose the suitable support structure, there are several characteristics need to be considered. The support structure is choosing based on the size of the turbine, the depth of the water and the suitability of the soil condition. Thus, the type of the support structure for shallow water needs to be considered based on these characteristics.

2. Software used

In this project, there are two types of software have been used. In order to model the 3D modelling, this project used Computer-Aided Design (CAD) software which is SOLIDWORKS. Then, to make a meshing and run the simulation for the model, this project used ANSYS software.

2.1. SOLIDWORKS
SOLIDWORKS is one of the CAD software that is widely used in industry. It is easy to use for parametric design and to edit the design. Since this software has been teaching in class session, it makes easier to use this software.

2.2. ANSYS
ANSYS software is recognized as the most powerful engineering design and analysis software. It also used to simulate interactions of all disciplines of physics, structural, vibration, fluid dynamics, heat transfer and electromagnetic for engineers. This project will use the FLUENT fluid flow model to simulate the effect of the support structure.

3. Results and Discussion
This chapter discusses the results that obtained throughout the whole process. The results were compared with the published data. The results are contained of single turbine, two turbines and three turbines. Then, the results are compared with the M.E. Harrison et al technical paper[2] and Muchala et al technical paper[3]. Based on the results, the discussion was carried out.

3.1. Single turbine
The results of this study were compared with the previous study in terms of velocity magnitude. The comparison was made with M.E. Harrison et al technical paper. The comparisons are between the actuator disc without support structure from previous study and actuator disc with support structure from this study. Based on the M.E. Harrison et al technical paper, the study is about the far wake prediction between the experimental and the CFD simulation on actuator disc.

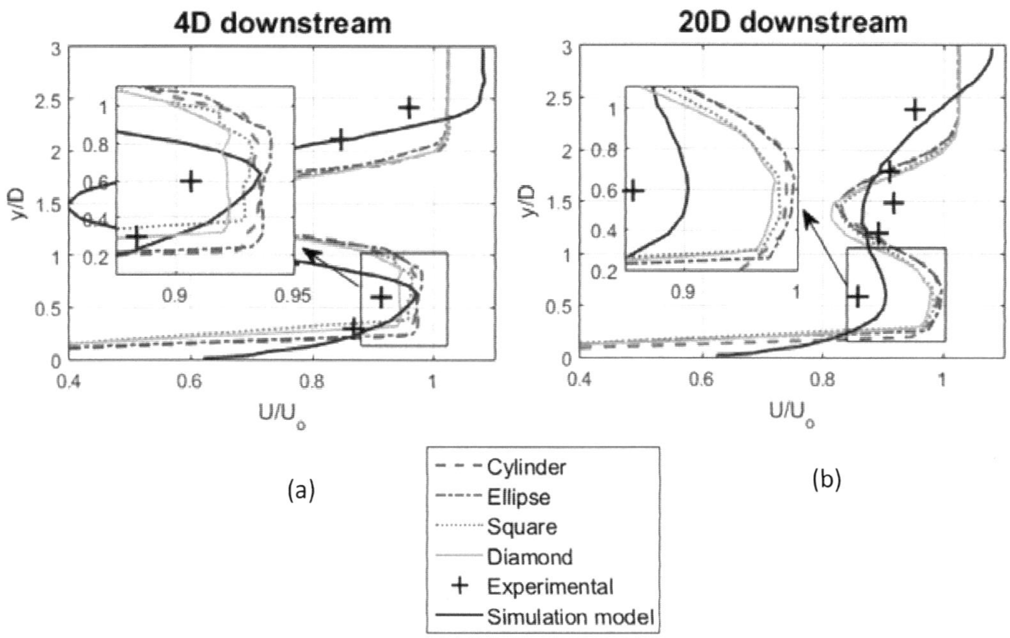

Figure 1: The results of velocity magnitude against the published data

Figure 1 displays the velocity magnitude of two different downstream which are 4D downstream and 20D downstream. At 4D downstream, it shows the actuator disc with support structure have higher wake than the actuator disc without support structure. Among all the support structure geometry, ellipse support structure shows the highest velocity followed by cylinder, square and diamond geometry. However, the wakes decreasing gradually in the absence of the support structure while increasing in the presence of the support structure as shown in **Figure 1(b)**. This is due to the higher blockage in the domain with the presence of the support structure. The geometry of the support structure, the different model use and the inlet velocity used also contributes to the different results of the wake formation.

3.2. Two turbines

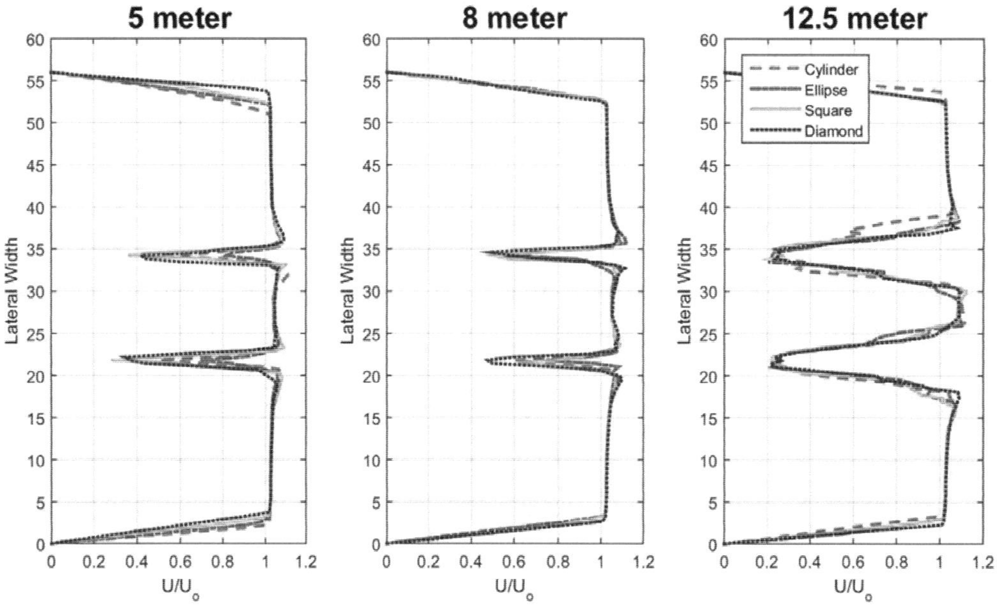

Figure 2: The velocity magnitude at different height from seabed taken at **0.125D downstream.**

Figure 2 shows the velocity magnitude at 0.125D downstream with three differences distance on support structure which is 5 meters, 8 meters and 12.5 meters from the seabed. The purpose of taken three difference place on support structure is because to investigate the behaviour of the wake formation based on the sea depth. As can be seen, at 5 meters and 8 meters, the wake recovery on support structure is faster than at 12.5 meters. The center line of wake velocity is increasing with the distance of the of the support structure from the seabed. This is because the presence of the free surface on the top of the disc. The presence of free surfaces allows the depth to change and require gravity considerations. At 5 meters and 8 meters, the wake of each support structure merges at between 20 to 35 meters lateral width. However, at the 12.5 meters it merges at 25 to 30 meters of the lateral width. Along the downstream, the wake formation start to recovery and almost returned to ambient condition at 2D downstream as shown in **Figure 3**. However, at 12.5 meters the wake formed is still in recovery state. This is because at this location the surface of the support structure and the disk is in contact. Thus, the surface roughness between two objects is affecting the turbulence occurrence.

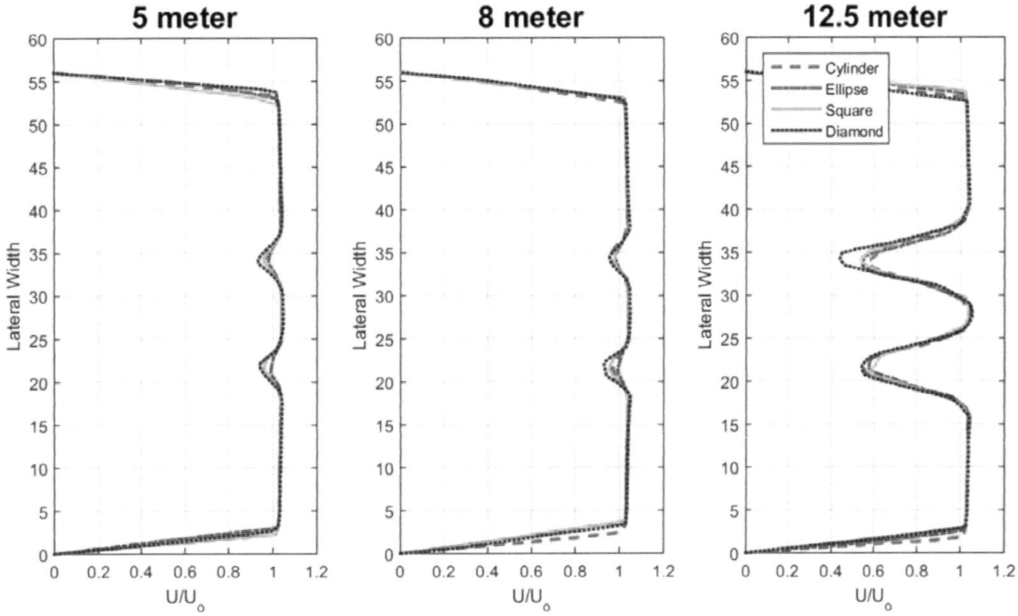

Figure 3: The velocity magnitude at different height from seabed taken at **2D downstream.**

3.3. Three turbines

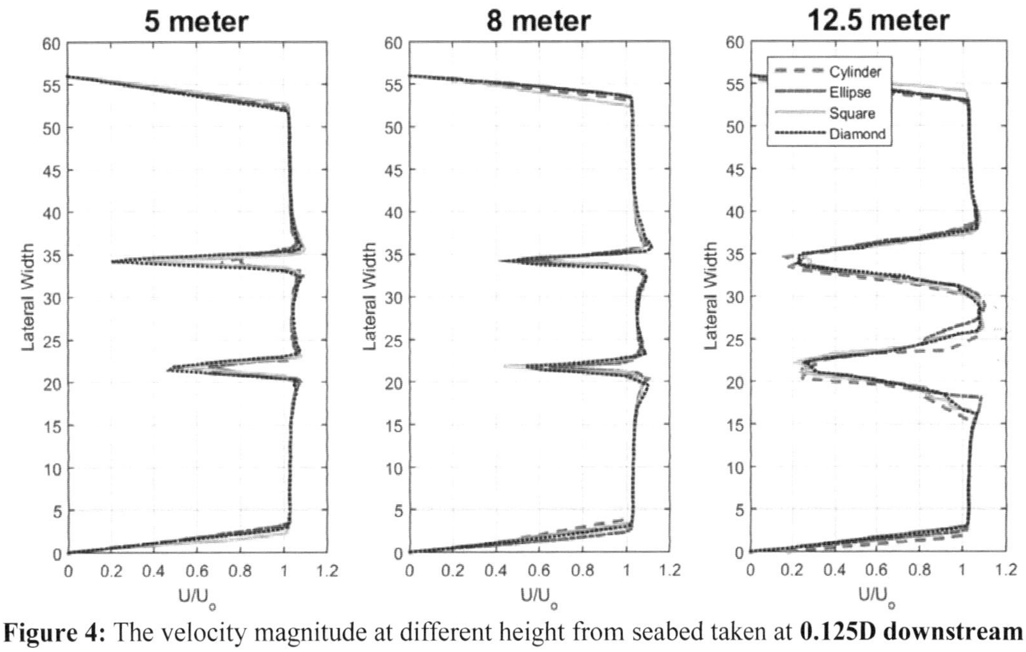

Figure 4: The velocity magnitude at different height from seabed taken at **0.125D downstream**.

Figure 4 shows the behaviour of the wake formation from first rows. The first row of the turbine shows the similar behaviour as the analysis on the two turbines. The wakes start to merge after passing through the support structure. The wake formation start to recover at 1D downstream and almost returned to ambient condition. However, at 3D downstream, the wake formation starts to meet the second row of the disk as shown in **Figure 5**. Thus, the wake formed start to spilt up by the support structure in between. The blockage effect from the second rows of the support structure affects the acceleration of the flow of the turbulence.

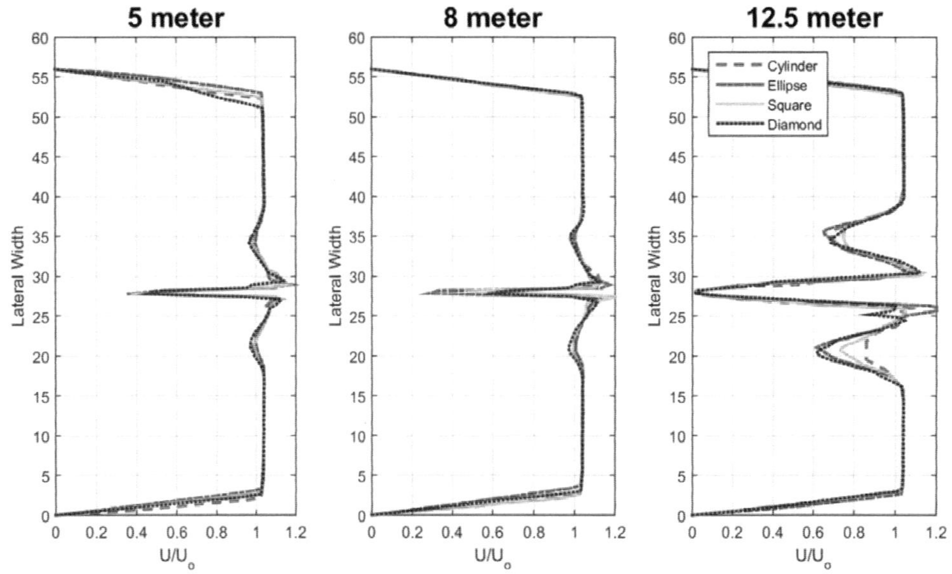

Figure 5: The velocity magnitude at different height from seabed taken at **3D downstream**

4. Conclusion

This paper found that the best design of the support structure for shallow water is mono-pile support structure. This is because the suitable water depth to plant the mono-pile support structure is 30 meter and above which is suitable with the Malaysia's open water depth. Next, this paper also analysed the influence of support structures on the actuator disc wake formation and examined the impact of the support structure on flow interactions between devices in the array. It was found that the flow of the turbulence on each support structure showed different behaviour. For a single turbine, it is clearly shown that ellipse support structure has the shortest turbulence form and it is the fastest among the support structure that recover back into inlet velocity after passing through the tidal turbine.

For the array of support structure, the 1.5D for the distance between two turbines was proven to be acceptable due to the turbulence occurrence being separated from each of the turbine. Next, for three turbines the separation between the first and second rows is agreeable to be 3D on a basis that after passing through the support structure the wakes are well merged and almost return to ambient condition at the 1D downstream. Among all the support structure, the first geometry to recover to the inlet velocity is ellipse geometry in array condition. To summarize, in comparison with previous research it is can be concluded that the results of this research approximately follow similar pattern of published data for a deeper site depth.

Acknowledgement

The first author would like to acknowledge the School of Mechatronic Engineering, UniMAP for the financial support. This project would never be possible without the help of my student, Nuur Baizura, who is tireless in the pursuit of knowledge.

References

[1] Khan N, Kalair A, Abas N and Haider A 2017 Review of ocean tidal, wave and thermal energy technologies *Renew. Sustain. Energy Rev.* **72** 590–604.

[2] Harrison M E, Batten W M J, Myers L E and Bahaj A S 2010 Comparison between CFD simulations and experiments for predicting the far wake of horizontal axis tidal turbines *IET Renew. Power Gener.* **4** (6) 613–627.

[3] Muchala S and Willden R H J 2017 Influence of support structures on tidal turbine power output *Renew. Energy* **114** 588–599.

Evaluation of square deep drawn AA6061-T6 blank based on thinning pattern

***N M Yusop and A B Abdullah**

Metal Forming Research Laboratory, School of Mechanical Engineering, Universiti Sains Malaysia, 14300 Nibong Tebal, Penang, MALAYSIA.

*Corresponding author: mebaha@usm.my

Abstract. Deep drawing can be generally described as a process of converting a metal sheet into a desired shape and geometry. Material flow and formability of the blank depends on the die and the punch geometries and it is important to predict the material behavior during square deep drawing to reduce possibility of defect and part rejection. In this research, a 3D Finite Element (FE) model is developed to simulate the square deep drawing process of an AA6061-T6 material using ANSYS Explicit Dynamics. The developed model can be used to predict the thinning pattern of the blank especially at the corner area. This research focus on the effect of design parameters namely punch corner radius (R_1), punch nose radius (R_2), die corner radius (R_3) and die shoulder radius (R_4) towards thinning pattern.

Keywords: Punch radius, die radius, square deep drawing, thinning, FEA

1. Introduction

Deep drawing is one of the most important processes in the production of sheet metal parts. Automotive industry is one of the sectors that utilized this process to manufacture various parts such as structural and outer panels. Deep drawing is complex deformation affected by the geometrical and process parameters including blank holder force, blank thickness, friction, die radius and punch radius [1].Typically thinning, tearing and wrinkling (flange and wall) are the most frequent defect can be found in deep drawn part [2]. The effects of all these parameters are not independent. According to Prasad and Rao, tearing effect will occur when lower the die corner radii and wrinkling will take part when die corner radii is higher [3]. Gowtham et al., also supported that, a decreased die radius created stretching marks and earring type quality problems [4]. Study on prediction of the forming results such as determination of the thickness distribution and the thinning of the sheet metal blank may increase the productivity through less defect and part rejection. It was first studied by Colgan and Monaghan using statistical analysis and design of experiments, they manage to determine the most important factors influencing a drawing process. It seems that the punch/die radii have the greatest effect on the thickness of the deformed mild steel cups compared to blank-holder force or friction [5]. Zein et al. also presented a Finite Element (FE) model for the 3-D numerical simulation of deep drawing process to predict the thickness distribution and thinning of the blank with the die design parameters (geometrical and physical parameters) [6]. The simulation was successful as the forming loads can be predicted to complete the drawing process. Hassan et al. have proposed a newly process to increase the formability of square cups using a flat headed square punch through a conical die with a square aperture. The effects of die fillet radius, die corner radius, die throat length, punch profile radius; punch corner radius, punch shape factor, relative die clearance and blank thickness on the draw ability of square cup were mainly investigated using FEA [7]. Based on the predicted thickness distribution of the deep drawn circular cup and analysis of variance test, it is evident that die radius has the greatest influence on the deep drawing of stainless steel blank sheet followed by the blank holder force and the friction coefficient [8]. However, Venkateswarlu et al. using the combination of finite element method and Taguchi Analysis observed that blank temperature has greatest influence on the formability of aluminium material followed by punch velocity and die arc radius [9]. Material flow during the deep drawing process is differ between circular and square cups. Younis and Jaber using Low carbon steel AISI (1008) shows that excessive earing appears in the square cup. This is due to excessive material flow in the corner and minimum material flow at the flat area [10].

2. Methodology

The investigation will utilize the finite element analysis software namely Ansys to study the effect of few design parameters such as die profile radius and punch nose radius to thinning pattern. The finite element (FE) model is developed based on a quarter of 3-D square deep drawing of AA 6061-T6 using ANSYS18.2/Explicit dynamics. The modeling is as the following steps:

2.1 Basic configuration

For this study, the geometrical details of the punch, blank and die are schematically shown in Figure 1. The AA6061-T6 blank with thickness of 0.8 mm is prepared in 60 mm diameter size.

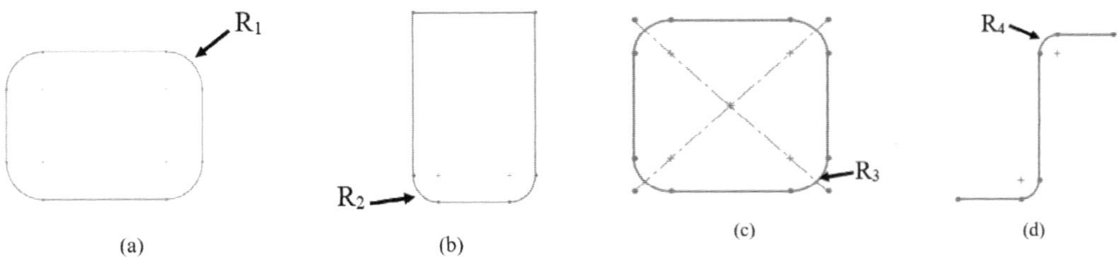

Figure 1. Varied geometrical features definition. (a) top view of a punch, (b) side view of a punch, (c) top view of a die and (d) side view of a die. (Note: R_1 = Punch corner radius, R_2 = Punch nose radius, R_3 = Die corner radius and R_4 = Die shoulder radius)

2.2 Numerical Model

Material Properties

The material is modeled as an elastic-plastic material with isotropic elasticity, using the Hill anisotropic yield criterion for the plasticity to describe the anisotropic characteristics of the blank. Table 1 shows the material properties of the blank.

Table 1. Physical and Mechanical Properties (Ramulu et al., 2013)

Young modulus (E)	66 G Pa
Poisson ratio (υ)	0.33
Density (ρ)	2.703×10^{-6} kg mm^{-3}
Yield strength (σ)	269 MPa
Friction coefficient (μ)	0.2

IOP Conf. Series: Materials Science and Engineering **670** (2019) 012079 doi:10.1088/1757-899X/670/1/012079

Boundary Condition

The quarter model of the deep drawing process is shown in Figure 2. For simplifying the simulation of the deep drawing processes, the following setup was made. The punch moved down at drawing force, F = 21510 N. No change in the temperature during the forming process, no heat transfer between tool and blank. Fixed support is provided to die and blank holder force. Blank is fully constrained along X and Z direction. Punch is fully constrained along X and Z direction.

Element Type and Meshing

The punch, blank holder and die were modeled as rigid bodies and blank was modeled as flexible body. Know that mesh density of the blank and other tools affects the accuracy of the result. Therefore, the meshes in the blank are finer than the mesh in punch and die to get the accurate results for deformation occurs in the blank. Mesh was generated by providing the following mesh parameters for blank is 1.0mm and for punch and die mesh size were set as default. Element type set was linear and default. Following boundary condition were applied in the Analysis Process using solver target AUTODYN.

Contact

Automatic contact procedures in ANSYS 18.2 were utilized to model complex interaction between the tooling and sheet and were set as frictional contact. The target and contact surfaces constitute contact pair which represents contact and sliding between the surfaces of blank with punch and die.

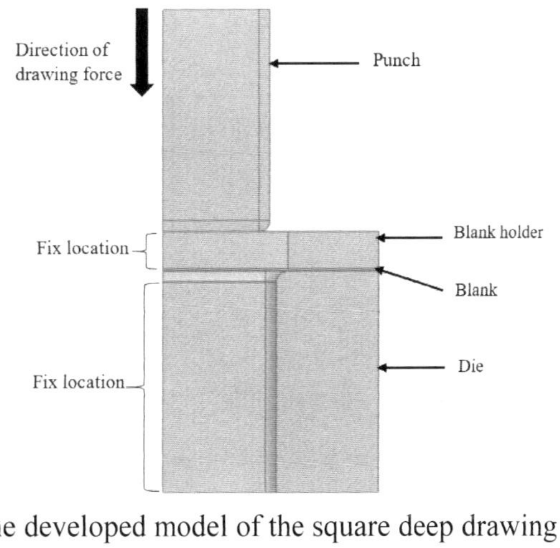

Figure 2. The developed model of the square deep drawing process components

2.3 Thickness Measurement

Figure 5(a) shows sectioning made on the drawn blank, while Figure 3(b) illustrate the locations with a distance 5mm of thickness measured on the drawn square cup from the center of the cup. For consistency of the location and to reduce error, the resulted profile will be exported to Space Claim Design Modeler available in Ansys Explicit Dynamics for thickness measurement and location plane was cut at the corner of the cup drawn.

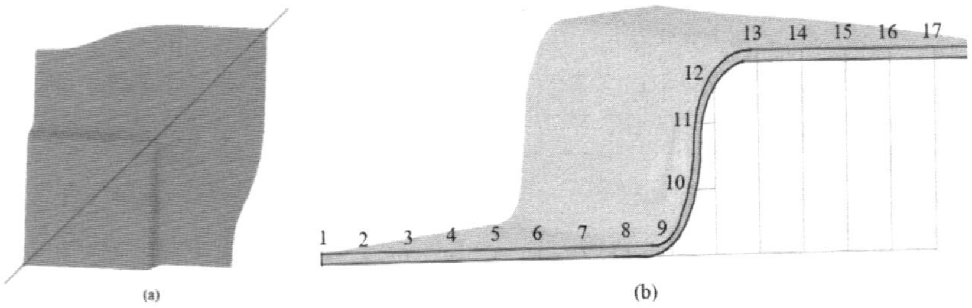

Figure 3. (a) Sectioning of the drawn blank and (b) locations for the thickness measurement

3. RESULTS AND DISCUSSION

3.1 Effect of Punch Radius to Thinning

The geometry of punch corner and punch shoulder radius influences the thickness distribution and thinning in blank in the deep drawing processes. Figure 4(a) and 4(b) show the result of thickness distribution for different values of punch corner and nose radius. The thickness was measured at the corner side of square cup. From the results, the thinning start occurs at position 8 and 9 from the center of the cup. The more thinning occurs because of stretching in the punch radius region. Both graphs show the maximum thinning at position 9 and the blank start thickening gradually until at position 11 and 12. Figure 4(a) shows that the maximum thinning and thickening of the cup occurred when the value of punch corner radius is 3 mm. The thickness of the cup in base region is equal to initial thickness of the blank 0.8mm and then start to be thinning and reach maximum value of thinning at position 9. The thinning at the wall cup is maximum due to stresses produced in this region. When punch corner radius is smallest the area at the corner side is the smallest, so thinning is maximum due to excessive bending and unbending at this area.

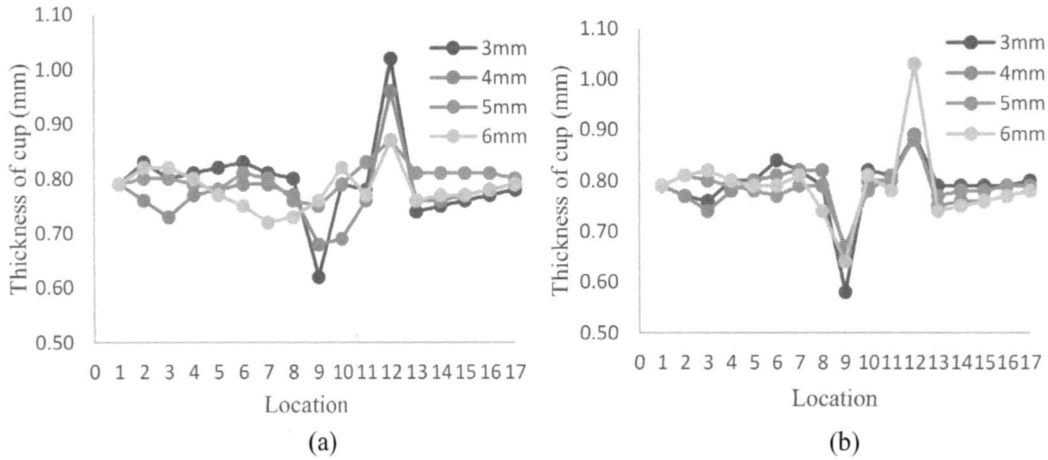

Figure 4. (a). Effect of punch corner radius (R_1) on blank thickness distribution and (b) Effect of punch nose radius (R_2) on blank thickness distribution.

3.2 Effect of Die Radius to Thinning

The geometry of die corner and die shoulder radius influences the thickness distribution and thinning in blank in the deep drawing processes. Figure 4(a) and 4(b) show the graph of thickness distribution for different values of die corner and die shoulder radius. Figure 4(a) shows that thinning occurred at position 9 and 11. Then the blank start thickening at position 12. The maximum thinning occurred when the value of die corner radius is 6 mm. The more thinning occurs because of stress and compressive force in the die corner region. Figure 8(b) shows thinning of the cup occurred at position 9 and 11 whereas the thickening occurred at position 12. The maximum thinning of cup occurred when the value of die shoulder radius is 6 mm and the maximum thickening occurred when the value of die shoulder radius is 3mm. When die corner radius is highest the area at the corner side is smallest so thinning is maximum due to excessive bending and unbending at this area. When the die shoulder radius is smallest the area at the corner die is wider, so less material of blank will flow at that area and thus thickening happen.

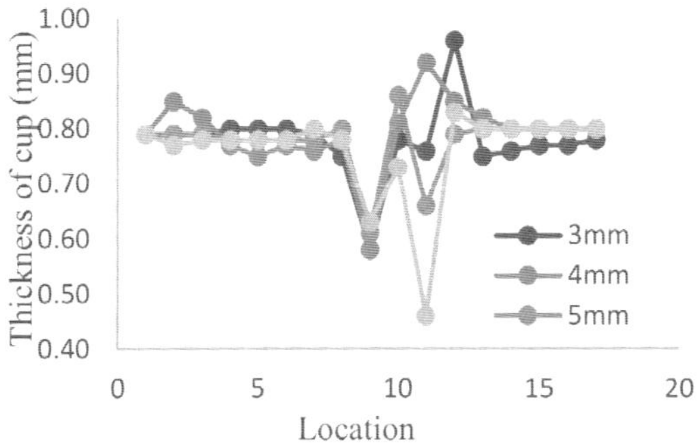

(a)

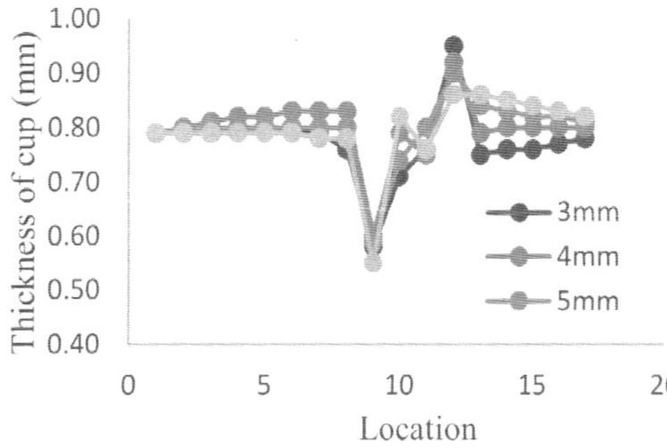

(b)

Figure 5. (a) Effect of die corner radius (R_3) on blank thickness distribution and (b) effect of die shoulder radius (R_4) on blank thickness distribution.

3.3 Depth of the cup

Figure 6 shows the effect of punch and die radius to the depth of the drawn square cup. When all the radius of punch and die are 3 mm the height of the cup drawn are same 15.36 mm. The height of the cup drawn differ when the radius of punch and die changing. The maximum height of cup drawn is when the radius of punch corner, R_2 = 6 mm and minimum height of cup drawn occurred when punch nose radius, R_1 = 4 mm. When the punch corner radius increases the area in that region is bigger thus more material flow will be occurred as a result the height of the cup is maximum. When the punch nose radius smaller the area in that region is smaller thus less material flow will be occurred due to high compressive stress as a result the height of the cup is minimum.

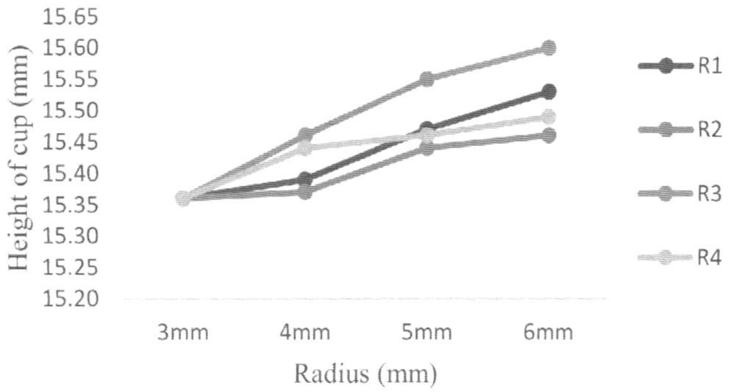

Figure 6. Effect of punch and die radius on height of the square cup.

4. Validation

Figure 7 shows the validation graph of thickness distribution from the cup center of AA6061 between the experimental method and FE simulation. The result from the FE simulation was compared with result from experiment conducted by Raju et al. (2010) for validation. From the result, it is observed that thinning is maximum at position 9 and 10 with the value 0.74mm. Thickness increases gradually, and both graphs show nearly depict similar pattern. The result shown that maximum percentage of thinning for experimental method is 7.34 while 7.50% for simulation. The pattern of both graphs is nearly same compared to each other but the percentage of thinning with simulation is less compared to experiment. Result showed a reasonable agreement with experimental work of some researchers with an average standard deviation of **1.86%** for thickness location #1 to #8, **0.12%** for location #9 to #10 and **1.11%** for location #11 to #17.

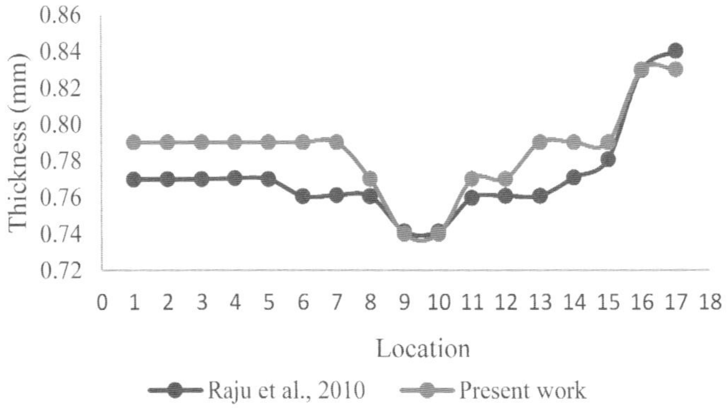

Figure 7. Validation (measured at flat side with R_2 = 6mm)

5. Conclusions

A finite element analysis-based simulation has been done using ANSYS Explicit Dynamic for the square cup deep drawing process. The effects of die corner, die shoulder, punch corner and punch shoulder radii on the drawability and quality characteristics has been done. It has been found that when the die radius reduced the amount of force needed to draw the blank is increased by 0.57%. An optimum die and punch radius were an important parameter to obtain minimum thinning is 0.01mm.

Acknowledgements

This work is partly supported by the Universiti Sains Malaysia RUI Grant (Grant No: 1001/PMEKANIK/8014031).

References

[1] Dwivedi R and Agnihotri G 2017 *Mater Today: Proceeds* **4** (2) 820–826.
[2] Zoesch A, Wiener T and Kuhl M 2015 *Procedia CIRP* **33** 179–184.
[3] Prasad K S and Rao C S 2014 *STM J* **4** (2) 22–32.
[4] Gowtham K, Srikanth K V N S and Murty K L N 2012 *Int J AppRes Mech Eng* (1) 17–23.
[5] Colgan M and Monaghan J 2003 *J Mater Process Technol* **32** 35–41
[6] Zein H M, Abd-Rabou M and El Shazly M 2013 *American J Mech Eng* **1** (2) 20–29.
[7] Hassan M, Hezam L, El-Sebaie M and Purbolaksono J 2014 *Procedia Eng* **81** 873–880.
[8] Padmanabhan R, Oliveira M C, Alves J L and Menezes L F 2007 *Fin Ele Ana Des* **43** (14) 1062–1067.
[9] Venkateswarlu G, Davidson M J and Tagore G R N 2010 *Int J Eng, Sci Technol* **2** (11) 41–49.
[10] Muhsin Younis D and Shbeeb Jaber A 2011 *Eng. Tech J* **29** (12) 2456-2467.

AUTHOR INDEX